MHCC WITHDRAWN

TK 4058 .K73 2001
Krishnan, R.
Electric motor drives

ELECTRIC MOTOR DRIVES

Modeling, Analysis, and Control

ELECTRIC MOTOR DRIVES

Modeling, Analysis, and Control

R. Krishnan
Virginia Tech, Blacksburg, VA

Upper Saddle River, New Jersey 07458

Library of Congress Catalog-in-Publication Data

Krishnan, R. (Ramu)
   Electric motor drives: modeling, analysis, and control / R. Krishnan.
     p. cm.
   Includes bibliographical references and index.
   ISBN 0-13-091014-7
   1. Electric driving. I. Title.

TK4058.K73 2001
621.46'—dc21
                                                                     00-050216

Vice President and Editorial Director, ECS: *Marcia J. Horton*
Acquisitions Editor: *Eric Frank*
Editorial Assistant: *Jessica Romeo*
Vice President of Production and Manufacturing, ESM: *David W. Riccardi*
Executive Managing Editor: *Vince O'Brien*
Managing Editor: *David A. George*
Production Editor: *Lakshmi Balasubramanian*
Creative Director: *Jayne Conte*
Art Editor: *Adam Velthaus*
Manufacturing Manager: *Trudy Pisciotti*
Manufacturing Buyer: *Pat Brown*
Marketing Manager: *Holly Stark*
Marketing Assistant: *Karen Moon*

 © 2001 Prentice Hall
Prentice Hall, Inc.
Upper Saddle River, New Jersey 07458

The author and publisher of this book have used their best efforts in preparing this book. These efforts include the development, research, and testing of the theories and programs to determine their effectiveness. The author and publisher make no warranty of any kind, expressed or implied, with regard to these programs or the documentation contained in this book. The author and publisher shall not be liable in any event of incidental or consequential damages in connection with, or arising out of, the furnishing, performance, or use of these programs.

All rights reserved. No part of this book may be reproduced, in any form or by any means, without permission in writing from the publisher.

Printed in the United States of America

10 9 8 7 6 5 4 3 2 1

ISBN 0-13-0910147

Prentice-Hall International (UK) Limited, *London*
Prentice-Hall of Australia Pty. Limited, *Sydney*
Prentice-Hall of Canada Inc., *Toronto*
Prentice-Hall Hispanoamericana, S.A., *Mexico*
Prentice-Hall of India Private Limited, *New Delhi*
Prentice-Hall of Japan, Inc., *Tokyo*
Prentice-Hall Asia Pte. Ltd., *Singapore*
Editora Prentice-Hall do Brasil, Ltda., *Rio de Janeiro*

In fond memory of:
My grandfather, **Mr. A. Duraiswami Mudaliar,** *an Indian Freedom Fighter,*
My grandmother, **Mrs. D. Rajammal,**
and
My father-in-law, **Prof. S. K. Ekambaram,** *B.Sc.(Hons.)., M.A.(Cantab).*

Preface

Electronic control of machines is a course taught *as an elective at senior level and as an introductory course for the graduate-level course on motor drives.* Many wonderful books address individually the senior-level or graduate-level requirements, mostly from the practitioner's point of view, but some are not ideally suited for classroom teaching in American universities. This book is the result of teaching materials developed over a period of twelve years at Virginia Tech. Parts of the material have been used extensively in seminars both in the United States and abroad.

The area of electric motor drives is a dependent discipline. It is an applied and multidisciplinary subject comprising electronics, machines, control, processors/computers, software, electromagnetics, sensors, power systems, and engineering applications. It is not possible to cover all aspects relevant to motor drives in one text. Therefore, this book addresses mainly the *system-level modeling analysis, design and integration of motor drives.* In this regard, knowledge of electrical machines, power converters, and linear control systems is assumed at the junior level. The modeling and analysis of electrical machines and drive systems is systematically derived from first principles. The control algorithms are developed, and their implementations with simulation results are given wherever appropriate.

The book consists of nine chapters. Their contents are briefly described here.

Chapter 1 contains the introduction and discusses the motor-drive applications, the status of power devices, classes of electrical machines, power converters, controllers, and mechanical systems.

Chapter 2 is on dc machines, their principle of operation, the steady-state and dynamic modeling, block-diagram development, and measurement of motor parameters.

Chapter 3 describes phase-controlled dc motor drives for variable-speed operation. The principle of dc machine speed control is developed, and four-quadrant operation is introduced. The electrical requirements for four-quadrant operation are derived. Realization of these voltage/current requirements for four-quadrant operation with phase-controlled converters is studied. In that process, the operation and control of phase-controlled converters are developed. The closed-loop speed-controlled dc motor drive system is considered for analysis and controller synthesis. The synthesis of current and speed controllers is developed. The dynamic simulation procedure is derived for the motor drive system from the subsystem differential equations and functional relationships. The same procedure is adopted throughout the book. Impact of harmonics both on the utility and on the machine is analyzed. Supply-side harmonics can cause resonance in the case of interaction with the power systems, as is illustrated with an example, and machine-side harmonics cause increased resistive losses, resulting in derating of the machine. Some application considerations for the drive are given. An application of the drive is described. A more or less similar approach is taken for all other drives in this book.

Chopper-controlled dc motors are described in Chapter 4. The principle of operation of a four-quadrant chopper, realization of dc input supply, regeneration, modeling of a chopper, the closed-loop speed-controlled drive and its current- and speed-controller synthesis, harmonics, and their impact on electromagnetic torque, losses, and derating are developed and presented.

The principle of operation of induction machines and their steady-state and dynamic modeling are presented in Chapter 5. The concept of space-phasor modeling is also introduced, to enable readers to follow literature mainly from Germany. A number of illustrative examples are included.

The principle of speed control of induction motors is introduced in Chapter 6. The rest of the chapter is devoted to the stator-phase control and slip-energy-recovery control of induction motors. Only steady-state aspects are covered. Their dynamic analysis is left to the interested reader. Emphasis is placed on efficiency, energy savings, speed-control range, harmonics, and application for these drives.

Variable-frequency control of induction machines with both variable voltage and variable current is introduced in Chapter 7. The realization of variable voltage and variable frequency with two-stage controllable converters and single-stage pulse-width-modulated (PWM) inverters is introduced. Reducing harmonics with multiple inverters or with one PWM inverter is discussed and is illustrated with examples. Steady-state analysis using fundamental and harmonic equivalent circuits, direct steady-state evaluation, and the use of a dynamic model with boundary-matching conditions is systematically developed in this chapter. Various control strategies for variable-voltage, variable-frequency drives are explained. They are V/Hz, constant-slip-speed, and constant-air gap-flux controls. The limitations and merits of each control scheme and relevant modeling to evaluate their dynamic performance are developed. Effects of harmonics on the machine losses, with the resultant derating and torque pulsation due to six-step voltage input to the machine, are quantified. Torque pulsations are calculated by using harmonic equivalent circuits of the induction motor fed from voltage-source inverters. The concept of current source is introduced by using a PWM inverter with voltage-source and current-feedback control. A two-stage current-source inverter drive with thyristor converter front end and autosequentially commutated inverter is considered for both steady-state and dynamic performance evaluation. The design aspects of the current-source drive control are considered in adequate detail.

The high-performance induction motor drive is considered from the control point of view in Chapter 8. The principle of vector control and design and its various implementations, their strengths and weaknesses, impact of parameter sensitivity, parameter-compensation methods, flux-weakening operation, and the design of a speed controller are explained with detailed algorithms and illustrated with dynamic simulation results and example problems. Tuning of the vector controller and position-sensorless operation are not dealt in detail. Interested readers will be helped by the cited references.

Chapter 9 deals with permanent magnet (PM) synchronous and brushless dc motor drives. The salient differences between these two types of motors are derived. Vector control and various control strategies within the scope of vector control, such as constant-torque-angle control, unity-power-factor control, constant-mutual-flux-linkages control, and maximum-torque-per-unit-current control are derived from the dynamic and steady-state equations of the PM synchronous motor. Two types of flux-weakening operation and their implementations, speed-controller design, position-sensorless control, and parameter sensitivity and its compensation are developed with control algorithms. Ample dynamic simulation results are included to enhance the understanding of the PM synchronous motor drive operation. Similar coverage is carried out for the PM brushless dc motor drive. In addition, an analytical method to evaluate torque pulsation and various unipolar/half-wave inverter topologies is included for low-cost but high-reliability motor drive systems.

A list of symbols has been given to enable readers who skip some sections to follow the text they are interested in. The material in this book is recommended for two semesters. In the author's experience, the subject matter that can be covered for each semester is given only to serve as a guideline. Depending on the strength of the program in each school, course instructors can flexibly choose the material from the book for their lectures. The introductory course that can be taught at senior elective or at graduate level includes the following:

> Chapters 1, 2, 3, and 4, Chapter 5 (covering only the steady-state operation and modeling of induction motors), Chapter 6, and Chapter 7 (excluding the dynamic performance evaluation or parts that use dynamic model of the induction motor).

The advanced graduate course can consist of the following:

> Dynamic modeling of induction machines from Chapter 5, voltage- and current-source drive dynamic performance from Chapter 7, and Chapters 8 and 9.

The author's many graduate and undergraduate students have enriched this book in the course of its development. Some of them deserve my special thanks and gratitude. They are Mr. Praveen Vijayraghavan, Dr. Shiyoung Lee, Dr. A. S. Bharadwaj, Dr. Byeong-Seok Lee and Dr. Ramin Monajemy. Without their crucial help, this endeavor would still be in the manuscript stage. The Chapter 5 development draws heavily from the work of one of my doctoral supervisors, Dr. J. F. Lindsay's material taught in 1980; the portion on the space-phasor model is from Prof. Dr. Ing–J. Holtz's research work. I have been fortunate to have Drs. J. F. Lindsay and V. R. Stefanovic as my doctoral supervisors, and a long-time professional and personal association with them helped me in my understanding of the subject matter. I am in eternal debt to them. Prof. B. K. Bose's evaluation and encouragement of the early manuscript in 1988 was inspiration to carry out this task. Prof. S. Bolognani of University of Padova provided opportunities to stay and lecture in University of Padova, which advanced the refinement of Chapters 5 and 8, and his help is gratefully acknowledged. Prof. M. Kazmierkowski of Warsaw University of Technology and Prof. Frede Blaabjerg of Aalborg University in Denmark arranged a Dan Foss Professorship to complete Chapter 9, and their assistance is gratefully acknowledged. Prof. J. G. Sabonnadiere, INPG, LEG, Grenoble hosted my sabbatical and provided me a lively environment for writing, and I am deeply indebted to him.

My editor at Prentice-Hall, Eric Frank smoothed all the glitches and provided advice on many aspects of the book. Lakshmi Balasubramanian, production editor, interfaced pleasantly during copy editing, proofreading and production. Brian Baker/WriteWith Inc. copy edited in a very short time. Robert Lentz took care of the proofreading. Laserwords provided the graphic work. All of them made the process of bringing the book to reality with such ease, pleasantness and clockwork schedule. I am grateful to this incredible team for working with me.

But for my wife, Vijaya's, encouragement and help, this book would not have been possible. To her, I owe the most.

R. KRISHNAN

Contents

Symbols   xxi

1 Introduction   1

    1.1   Introduction   1
    1.2   Power Devices and Switching   2
         1.2.1   Power Devices   2
         1.2.2   Switching of Power Devices   6
    1.3   Motor Drive   7
         1.3.1   Electric Machines   8
         1.3.2   Power Converters   12
         1.3.3   Controllers   13
         1.3.4   Load   14
    1.4   Scope of the Book   16
    1.5   References   17

2 Modeling of DC Machines   18

    2.1   Theory of Operation   18
    2.2   Induced Emf   19
    2.3   Equivalent Circuit and Electromagnetic Torque   21
    2.4   Electromechanical Modeling   22
    2.5   State-Space Modeling   22
    2.6   Block Diagram and Transfer Functions   23
    2.7   Field Excitation   24
         2.7.1   Separately-Excited DC Machine   24
         2.7.2   Shunt-Excited DC Machine   27
         2.7.3   Series-Excited DC Machine   27

xii   Contents

      2.7.4   DC Compound Machine   30
      2.7.5   Permanent-Magnet DC Machine   30

2.8   Measurement of Motor Constants   31
      2.8.1   Armature Resistance   31
      2.8.2   Armature Inductance   32
      2.8.3   Emf Constant   32

2.9   Flow Chart for Computation   33

2.10   Suggested Readings   35

2.11   Discussion Questions   35

2.12   Exercise Problems   35

3 Phase-Controlled DC Motor Drives   36

3.1   Introduction   36

3.2   Principles of DC Motor Speed Control   37
      3.2.1   Fundamental Relationship   37
      3.2.2   Field Control   37
      3.2.3   Armature Control   38
      3.2.4   Armature and Field Controls   38
      3.2.5   Four-Quadrant Operation   43

3.3   Phase-Controlled Converters   47
      3.3.1   Single-Phase-Controlled Converter   47
      3.3.2   Three-Phase-Controlled Converter   51
      3.3.3   Control Circuit   54
      3.3.4   Control Modeling of the Three-Phase Converter   55
      3.3.5   Current Source   56
      3.3.6   Half-Controlled Converter   57
      3.3.7   Converters with Freewheeling   58
      3.3.8   Converter Configuration for a Four-Quadrant DC Motor Drive   59

3.4   Steady-State Analysis of the Three-Phase Converter-Controlled DC Motor Drive   60
      3.4.1   Average Analysis   60
      3.4.2   Steady-State Solution, Including Harmonics   64
      3.4.3   Critical Triggering Angle   67
      3.4.4   Discontinuous Current Conduction   67

3.5   Two-Quadrant, Three-Phase Converter-Controlled DC Motor Drive   71

3.6   Transfer Functions of the Subsystems   73
      3.6.1   DC Motor and Load   73
      3.6.2   Converter   75
      3.6.3   Current and Speed Controllers   75
      3.6.4   Current Feedback   75
      3.6.5   Speed Feedback   75

3.7   Design of Controllers   76
      3.7.1   Current Controller   76
      3.7.2   First-Order Approximation of Inner Current Loop   78
      3.7.3   Speed Controller   79

- 3.8 Two-Quadrant DC Motor Drive with Field Weakening 88
- 3.9 Four-Quadrant DC Motor Drive 89
- 3.10 Converter Selection and Characteristics 91
- 3.11 Simulation of the One-Quadrant DC Motor Drive 92
  - *3.11.1 The Motor Equations 92*
  - *3.11.2 Filter in the Speed-Feedback Loop 93*
  - *3.11.3 Speed Controller 93*
  - *3.11.4 Current Reference Generator 94*
  - *3.11.5 Current Controller 94*
  - *3.11.6 Flowchart for Simulation 95*
  - *3.11.7 Simulation Results 97*
- 3.12 Harmonics and Associated Problems 98
  - *3.12.1 Harmonic Resonance 98*
  - *3.12.2 Twelve-Pulse Converter for DC Motor Drives 102*
  - *3.12.3 Selective Harmonic Elimination and Power-Factor Improvement by Switching 104*
- 3.13 Sixth-Harmonic Torque 107
  - *3.13.1 Continuous Current-Conduction Mode 107*
  - *3.13.2 Discontinuous Current-Conduction Mode 110*
- 3.14 Application Considerations 114
- 3.15 Applications 115
- 3.16 Parameter Sensitivity 118
- 3.17 Research Status 119
- 3.18 Suggested Readings 119
- 3.19 Discussion Questions 120
- 3.20 Exercise Problems 121

4 Chopper-Controlled DC Motor Drive 124

- 4.1 Introduction 124
- 4.2 Principle of Operation of the Chopper 124
- 4.3 Four-Quadrant Chopper Circuit 126
  - *4.3.1 First-Quadrant Operation 126*
  - *4.3.2 Second-Quadrant Operation 129*
  - *4.3.3 Third-Quadrant Operation 130*
  - *4.3.4 Fourth-Quadrant Operation 131*
- 4.4 Chopper for Inversion 132
- 4.5 Chopper With Other Power Devices 133
- 4.6 Model of the Chopper 133
- 4.7 Input to the Chopper 133
- 4.8 Other Chopper Circuits 135
- 4.9 Steady-State Analysis of Chopper-Controlled DC Motor Drive 136
  - *4.9.1 Analysis by Averaging 136*
  - *4.9.2 Instantaneous Steady-State Computation 137*

    4.9.3 Continuous Current Conduction 137
    4.9.4 Discontinuous Current Conduction 140
 4.10 Rating of the Devices 143
 4.11 Pulsating Torques 144
 4.12 Closed-Loop Operation 151
    4.12.1 Speed-Controlled Drive System 151
    4.12.2 Current Control Loop 151
    4.12.3 Pulse-Width-Modulated Current Controller 152
    4.12.4 Hysteresis Current Controller 155
    4.12.5 Modeling of Current Controllers 156
    4.12.6 Design of Current Controller 157
    4.12.7 Design of Speed Controller by Symmetric Optimum Method 158
 4.13 Dynamic Simulation of the Speed-Controlled DC Motor Drive 160
    4.13.1 Motor Equations 161
    4.13.2 Speed Feedback 161
    4.13.3 Speed Controller 162
    4.13.4 Command Current Generator 162
    4.13.5 Current Controller 163
    4.13.6 System Simulation 163
 4.14 Application 167
 4.15 Suggested Readings 170
 4.16 Discussion Questions 171
 4.17 Exercise Problems 172

5 Polyphase Induction Machines   174

 5.1 Introduction 174
 5.2 Construction and Principle of Operation 175
    5.2.1 Machine Construction 175
    5.2.2 Principle of Operation 180
 5.3 Induction Motor Equivalent Circuit 181
 5.4 Steady-State Performance Equations of the Induction Motor 184
 5.5 Steady-State Performance 188
 5.6 Measurement of Motor Parameters 193
    5.6.1 Stator Resistance 193
    5.6.2 No-Load Test 193
    5.6.3 Locked-Rotor Test 194
 5.7 Dynamic Modeling of Induction Machines 196
    5.7.1 Real-Time Model of a Two-Phase Induction Machine 197
    5.7.2 Transformation to Obtain Constant Matrices 200
    5.7.3 Three-Phase to Two-Phase Transformation 203
    5.7.4 Power Equivalence 209
    5.7.5 Generalized Model in Arbitrary Reference Frames 209
    5.7.6 Electromagnetic Torque 212
    5.7.7 Derivation of Commonly Used Induction Motor Models 213

　　　　　5.7.7.1　Stator Reference Frames Model　*213*
　　　　　5.7.7.2　Rotor Reference Frames Model　*214*
　　　　　5.7.7.3　Synchronously Rotating Reference Frames Model　*215*
　　　5.7.8　Equations in Flux Linkages　218
　　　5.7.9　Per-Unit Model　220
　5.8　Dynamic Simulation　223
　5.9　Small-Signal Equations of the Induction Machine　226
　　　5.9.1　Derivation　226
　　　5.9.2　Normalized Small-Signal Equations　234
　5.10　Evaluation of Control Characteristics of the Induction Machine　236
　　　5.10.1　Transfer Functions and Frequency Responses　236
　　　5.10.2　Computation of Time Responses　238
　5.11　Space-Phasor Model　242
　　　5.11.1　Principle　242
　　　5.11.2　DQ Flux-Linkages Model Derivation　243
　　　5.11.3　Root Loci of the DQ Axes-Based Induction Machine Model　244
　　　5.11.4　Space-Phasor Model Derivation　246
　　　5.11.5　Root Loci of the Space-Phasor Induction Machine Model　248
　　　5.11.6　Expression for Electromagnetic Torque　249
　　　5.11.7　Analytical Solution of Machine Dynamics　252
　　　5.11.8　Signal-Flow Graph of the Space-Phasor-Modeled Induction Motor　253
　5.12　Control Principle of the Induction Motor　254
　5.13　References　257
　5.14　Discussion Questions　258
　5.15　Exercise Problems　259

6　Phase-Controlled Induction Motor Drives　　　　　　　　　　262

　6.1　Introduction　262
　6.2　Stator-Voltage Control　263
　　　6.2.1　Power Circuit and Gating　263
　　　6.2.2　Reversible Controller　263
　　　6.2.3　Steady-State Analysis　265
　　　6.2.4　Approximate Analysis　267
　　　　　6.2.4.1　Motor Model and Conduction Angle　*267*
　　　　　6.2.4.2　Fourier Resolution of Voltage　*269*
　　　　　6.2.4.3　Normalized Currents　*271*
　　　　　6.2.4.4　Steady-State Performance Computation　*272*
　　　　　6.2.4.5　Limitations　*273*
　　　6.2.5　Torque–Speed Characteristics with Phase Control　273
　　　6.2.6　Interaction of the Load　273
　　　　　6.2.6.1　Steady-State Computation of the Load Interaction　*275*
　　　6.2.7　Closed-Loop Operation　279
　　　6.2.8　Efficiency　279
　　　6.2.9　Applications　282
　6.3　Slip-Energy Recovery Scheme　283
　　　6.3.1　Principle of Operation　283

    6.3.2 *Slip-Energy Recovery Scheme* 283
    6.3.3 *Steady-State Analysis* 285
      6.3.3.1 Range of Slip 287
      6.3.3.2 Equivalent Circuit 287
      6.3.3.3 Performance Characteristics 289
    6.3.4 *Starting* 296
    6.3.5 *Rating of the Converters* 296
      6.3.5.1 Bridge-Rectifier Ratings 297
      6.3.5.2 Phase-Controlled Converter 297
      6.3.5.3 Filter Choke 297
    6.3.6 *Closed-Loop Control* 298
    6.3.7 *Sixth-Harmonic Pulsating Torques* 299
    6.3.8 *Harmonic Torques* 303
    6.3.9 *Static Scherbius Drive* 304
    6.3.10 *Applications* 305
 6.4 References 308
 6.5 Discussion Questions 309
 6.6 Exercise Problems 311

7 Frequency-Controlled Induction Motor Drives 313

 7.1 Introduction 313
 7.2 Static Frequency Changers 313
 7.3 Voltage-Source Inverter 317
    7.3.1 *Modified McMurray Inverter* 317
    7.3.2 *Full-Bridge Inverter Operation* 319
 7.4 Voltage-Source Inverter-Driven Induction Motor 320
    7.4.1 *Voltage Waveforms* 320
    7.4.2 *Real Power* 323
    7.4.3 *Reactive Power* 323
    7.4.4 *Speed Control* 324
    7.4.5 *Constant Volts/Hz Control* 325
      7.4.5.1 Relationship Between Voltage and Frequency 325
      7.4.5.2 Implementation of Volts/Hz Strategy 328
      7.4.5.3 Steady-State Performance 330
      7.4.5.4 Dynamic Simulation 330
      7.4.5.5 Small-Signal Responses 338
      7.4.5.6 Direct Steady-State Evaluation 340
    7.4.6 *Constant Slip-Speed Control* 346
      7.4.6.1 Drive Strategy 346
      7.4.6.2 Steady-State Analysis 347
    7.4.7 *Constant-Air Gap-Flux Control* 350
      7.4.7.1 Principle of Operation 350
      7.4.7.2 Drive Strategy 350
    7.4.8 *Torque Pulsations* 354
      7.4.8.1 General 354
      7.4.8.2 Calculation of Torque Pulsations 354
      7.4.8.3 Effects of Time Harmonics 360

7.4.9   *Control of Harmonics*   *362*
   7.4.9.1   General   362
   7.4.9.2   Phase-Shifting Control   362
   7.4.9.3   Pulse-Width Modulation (PWM)   365
7.4.10   *Steady-State Evaluation with PWM Voltages*   *369*
   7.4.10.1   PWM Voltage Generation   369
   7.4.10.2   Machine Model   370
   7.4.10.3   Direct Evaluation of Steady-State Current Vector by Boundary-Matching Technique   372
   7.4.10.4   Computation of Steady-State Performance   373
7.4.11   *Flux-Weakening Operation*   *377*
   7.4.11.1   Flux Weakening   377
   7.4.11.2   Calculation of Slip   379
   7.4.11.3   Maximum Stator Frequency   380

7.5   Current-Source Induction Motor Drives   381
   7.5.1   *General*   *381*
   7.5.2   *ASCI*   *382*
      7.5.2.1   Commutation   382
      7.5.2.2   Phase-Sequence Reversal   383
      7.5.2.3   Regeneration   384
      7.5.2.4   Comparison of Converters for AC and DC Motor Drives   385
   7.5.3   *Steady-State Performance*   *385*
   7.5.4   *Direct Steady-State Evaluation of Six-Step Current-Source Inverter-Fed Induction Motor (CSIM) Drive System*   *389*
   7.5.5   *Closed-Loop CSIM Drive System*   *396*
   7.5.6   *Dynamic Simulation of the Closed-Loop CSIM Drive System*   *398*

7.6   Applications   405
7.7   References   406
7.8   Discussion Questions   407
7.9   Exercise Problems   409

8 Vector-Controlled Induction Motor Drives   411

8.1   Introduction   411
8.2   Principle of Vector Control   412
8.3   Direct Vector Control   415
   8.3.1   *Description*   *415*
   8.3.2   *Flux and Torque Processor*   *416*
      8.3.2.1   Case (i): Terminal voltages   417
      8.3.2.2   Case (ii): Induced emf from flux sensing coils or Hall sensors   420
   8.3.3   *Implementation with Six-Step Current Source*   *422*
   8.3.4   *Implementation with Voltage Source*   *425*
   8.3.5   *Direct Vector (Self) Control in Stator Reference Frames with Space-Vector Modulation*   *426*
8.4   Derivation of Indirect Vector-Control Scheme   446
8.5   Indirect Vector-Control Scheme   448
8.6   An Implementation of an Indirect Vector-Control Scheme   450

8.7 Tuning of the Vector Controller 454
8.8 Flowchart for Dynamic Computation 457
8.9 Dynamic Simulation Results 458
8.10 Parameter Sensitivity of the Indirect Vector-Controlled Induction Motor Drive 461
  8.10.1 Parameter Sensitivity Effects When the Outer Speed Loop Is Open 461
    8.10.1.1 Expression for Electromagnetic Torque 461
    8.10.1.2 Expression for the Rotor Flux Linkages 463
    8.10.1.3 Steady-State Results 463
    8.10.1.4 Transient Characteristics 466
  8.10.2 Parameter Sensitivity Effects on a Speed-Controlled Induction Motor Drive 468
    8.10.2.1 Steady-State Characteristics 469
    8.10.2.2 Discussion on Transient Characteristics 473
    8.10.2.3 Parameter Sensitivity of Other Motor Drives 474
8.11 Parameter Sensitivity Compensation 475
  8.11.1 Modified Reactive-Power Compensation Scheme 476
  8.11.2 Parameter Compensation with Air Gap-Power Feedback Control 477
    8.11.2.1 Steady-State Performance 478
    8.11.2.2 Dynamic Performance 482
8.12 Flux-Weakening Operation 484
  8.12.1 Flux-Weakening in Stator-Flux-Linkages-Controlled Schemes 485
  8.12.2 Flux-Weakening in Rotor-Flux-Linkages-Controlled Scheme 486
  8.12.3 Algorithm to Generate the Rotor Flux-Linkages Reference 487
  8.12.4 Constant-Power Operation 490
8.13 Speed-Controller Design for an Indirect Vector-Controlled Induction Motor Drive 492
  8.13.1 Block-Diagram Derivation 492
    8.13.1.1 Vector-Controlled Induction Machine 492
    8.13.1.2 Inverter 495
    8.13.1.3 Speed Controller 495
    8.13.1.4 Feedback Transfer Functions 495
  8.13.2 Block-Diagram Reduction 496
  8.13.3 Simplified Current-Loop Transfer Function 496
  8.13.4 Speed-Controller Design 499
8.14 Performance and Applications 502
  8.14.1 Application: Centrifuge Drive 503
8.15 Research Status 504
8.16 References 505
8.17 Discussion Questions 509
8.18 Exercise Problems 511

9 Permanent-Magnet Synchronous and Brushless DC Motor Drives 513

9.1 Introduction 513
9.2 Permanent Magnets and Characteristics 514

- 9.2.1 Permanent Magnets 514
- 9.2.2 Air Gap Line 514
- 9.2.3 Energy Density 517
- 9.2.4 Magnet Volume 518

9.3 Synchronous Machines with PMs 518
- 9.3.1 Machine Configurations 519
- 9.3.2 Flux-Density Distribution 521
- 9.3.3 Line-Start PM Synchronous Machines 522
- 9.3.4 Types of PM Synchronous Machines 523

9.4 Vector Control of PM Synchronous Motor (PMSM) 525
- 9.4.1 Model of the PMSM 525
- 9.4.2 Vector Control 527
- 9.4.3 Drive-System Schematic 529

9.5 Control Strategies 531
- 9.5.1 Constant ($\delta = 90°$) Torque-Angle Control 531
- 9.5.2 Unity-Power-Factor Control 534
- 9.5.3 Constant-Mutual-Flux-Linkages Control 536
- 9.5.4 Optimum-Torque-Per-Ampere Control 537

9.6 Flux-Weakening Operation 539
- 9.6.1 Maximum Speed 539
- 9.6.2 Direct Flux-Weakening Algorithm 540
  - 9.6.2.1 Control Scheme 542
  - 9.6.2.2 Constant-Torque-Mode Controller 543
  - 9.6.2.3 Flux-Weakening Controller 544
  - 9.6.2.4 System Performance 545
- 9.6.3 Indirect Flux-Weakening 548
  - 9.6.3.1 Maximum Permissible Torque 548
  - 9.6.3.2 Speed-Control Scheme 549
  - 9.6.3.3 Implementation Strategy 550
  - 9.6.3.4 System Performance 552
  - 9.6.3.5 Parameter Sensitivity 552

9.7 Speed-Controller Design 555
- 9.7.1 Block-Diagram Derivation 555
- 9.7.2 Current Loop 556
- 9.7.3 Speed-Controller 558

9.8 Sensorless Control 562

9.9 Parameter Sensitivity 567
- 9.9.1 Ratio of Torque to Its Reference 568
- 9.9.2 Ratio of Mutual Flux-Linkages to Its Reference 569
- 9.9.3 Parameter Compensation Through Air Gap-Power Feedback Control 569
  - 9.9.3.1 Algorithm 571
  - 9.9.3.2 Performance 572

9.10 PM Brushless DC Motor (PMBDCM) 577
- 9.10.1 Modeling of PM Brushless DC Motor 578
- 9.10.2 The PMBDCM Drive Scheme 580
- 9.10.3 Dynamic Simulation 581
- 9.10.4 Commutation-Torque Ripple 582

       9.10.5 Phase-Advancing 586
       9.10.6 Normalized System Equations 587
       9.10.7 Half-Wave PMBDCM Drives 588
              9.10.7.1 Split-Supply Converter Topology 589
              9.10.7.2 C-Dump Topology 601
              9.10.7.3 Variable-DC-Link Converter Topology 607
       9.10.8 Sensorless Control of PMBDCM Drive 611
       9.10.9 Torque-Smoothing 614
       9.10.10 Design of Current and Speed Controllers 614
       9.10.11 Parameter Sensitivity of the PMBDCM Drive 614

9.11 References 615

9.12 Discussion Questions 616

9.13 Exercise Problems 620

Index 621

Symbols

A	State-transition matrix
a	Turns ratio
a_n	Product of L_{sn} and σ
B_1, B_2	Bearing friction coefficients, N·m/(rad/sec)
B_l	Load constant
B_n	Normalized friction coefficient, p.u.
B_t	Total friction coefficient, N·m/(rad/sec)
C	Capacitance of the filter, F
c_p	Winding pitch factor
c_d	Winding distribution factor
C_f	Filter capacitance, F
C_o	Value of the C-dump capacitor, F
d	Duty cycle
d_c	Critical duty cycle
e	Instantaneous induced emf, V
E	Steady-state induced emf and also used as C-dump capacitor voltage, V
E_1	RMS air gap induced emf in AC machines, V
E_2	RMS rotor-induced emf in AC machines, V
e_{as}	Induced emf in phase *a* (instantaneous), V
E_{as}	Steady-state rms phase *a* induced emf, V
e_{bs}	Induced emf in *b* phase (instantaneous), V
e_{cs}	Induced emf in *c* phase (instantaneous), V
e_n	Instantaneous normalized induced emf, p.u.
E_n	Normalized steady-state induced emf, p.u.
E_p	Peak stator phase voltage in PM brushless DC machine, V
F, F^*	Modified reactive power and its reference, VAR
$f_{as}(\theta_r)$	Position-dependent function in the induced emf in *a* phase
$f_{bs}(\theta_r)$	Position-dependent function in the induced emf in *b* phase
f_c	Control frequency, and PWM carrier frequency, Hz
$f_{cs}(\theta_r)$	Position-dependent function in the induced emf in *c* phase
f_n	Natural frequency of the power system, Hz
f_o	System frequency, Hz
f_s	Utility supply frequency or stator supply frequency, Hz
f_s^*	Stator frequency command, Hz
f_{sn}	Normalized stator frequency, p.u.
$G_{wl}(s)$	Speed-to-load–torque transfer function
$G_{wv}(s)$	Speed-to-applied-voltage transfer function
h	Average duty cycle of phase switches in half-wave converter topologies in Ch. 9
h	Harmonic number in Ch. 7
H	Inertial constant in normalized unit, p.u.
H_c	Gain of the current transducer, V/A
H_f	Field-current-transducer gain, V/A
H_ω	Gain of the speed filter, V/(rad/sec)
i_0	Zero-sequence current, A
I_6	Sixth-harmonic armature current, A
I_{6np}	Normalized peak sixth-harmonic armature current, p.u.
i_a	DC machine armature current, A
i_a^*	Armature current reference, A
I_{a0}	Steady-state minimum armature current, A
i_{a1}	First-harmonic armature current (instantaneous), A
I_1	First-harmonic armature current (steady state), A

xxii Symbols

Symbol	Description
I_{a1}	Steady-state maximum armature current, A
i_{abc}	*abc* phase current vector
i_{ai}	Initial armature current, A
I_{a11}	Fundamental system input phase current, A
i_{an}	Normalized armature current in dc machine, p.u.
i_{an}^*	Normalized armature current reference, p.u.
I_{an}	Normalized phase *a* stator current, p.u.
i_{ar}	Rated armature current in dc machine, A
i_{as}	Instantaneous stator phase *a* current, A
i_{asn}	Normalized phase *a* stator current (instantaneous), p.u.
I_{avn}	Normalized average armature current, A
I_b	Base current in 2-phase system, A
I_b	Base current, A
I_{b3}	Base current in 3-phase system, A
i_{bs}, i_{cs}	Instantaneous *b* and *c* stator phase currents, A
$i_{as}^*, i_{bs}^*, i_{cs}^*$	*a,b,c* phase-current commands, A
I_c	Core-loss current per phase, A
I_{cn}	n^{th}-harmonic capacitor current, A
I_d	Average diode current, A
i_{dc}	Instantaneous dc link current or inverter input current, A
I_{dc}	Steady-state dc link current, A
I_{dcn}	Normalized dc link current, p.u.
$\bar{i}_{dr}^e, \bar{i}_{qr}^e$	Conjugates of i_{dr}^e and i_{qr}^e.
i_{drr}, i_{qrr}	Fictitious rotor currents in *d* and *q* axis, A
i_{ds}, i_{qs}	*d* and *q* axis stator currents, A
i_f	Instantaneous field current in dc machines or flux-producing component of the stator–current phasor in *ac* machines, A
I_f	Steady-state field current or flux-producing component of the stator-current phasor in ac machines, A
i_f^*	Field-current reference, A
I_{fr}, I_{frn}	Rated values of I_f in SI and normalized units
I_m	Magnetizing current per phase, A
I_{m1}	Fundamental magnetizing current, A
I_n	n^{th}-harmonic current, A
I_o	No-load RMS phase current, A
i_{os}, i_{or}	Stator and rotor zero-sequence currents, A
I_p	Peak stator-phase current in PM brushless dc machine, A
I_{ph}	Fundamental phase current, A
I_{ps}	Peak stator-phase current in PM synchronous machine, A
i_{qdo}	*qdo* current vector
i_r	Current in the dc link, A
I_r	RMS rotor current referred to the stator, A
I_{r1}	Stator-referred fundamental rotor-phase current, A
I_{rms}	RMS armature current, A
I_{rn}	Normalized stator-referred rotor-phase current, p.u.
I_{rr}	Actual RMS rotor-phase current, A
I_{rr1}	Fundamental rotor-phase current, A
I_s	RMS stator-phase current, A
I_s	Steady-state source current to phase-controlled converter, A
I_{sc}	RMS stator-phase current when the rotor is locked, A
I_{sc}	Short-circuit current, A
i_{sn}^e, i_{rn}^*	Normalized stator- and rotor-current phasors in the arbitrary reference frame, p.u.
i_T	Torque-producing component of the stator-current phasor, A
I_T^*	Reference-torque-producing component of stator-current phasor, A

Symbols xxiii

Symbol	Description
i_T, I_T^*	Steady-state values of i_T, and i_T^* in Chapters 8 and 9
I_T	RMS current of the chopper switch, A, in Ch. 4
I_{Ti}, I_{Tm}	Rated values of I_T in SI and normalized units
I_{ts}	Power-switch current (RMS), A
i_{xyz}^r	Current with superscript s for stator, r for rotor, e for synchronous and c for arbitrary reference frames, and subscript xy for q and d axes and n for normalized unit. Without subscript n, the variable is in SI units.
i_α, i_β	Two-phase instantaneous currents, A
$i_{\alpha m}, i_{\beta m}$	Currents in the new rotor reference frames, A
J	Total moment of inertia, Kg-m^2
J_l	Moment of inertia of load, Kg-m^2
J_m	Moment of inertia of motor, Kg-m^2
K^*	Ratio between control-voltage and rotor-speed references, V/rad/sec
K_b	Induced-emf constant, V/(rad/sec)
K_f	Control-voltage to stator-frequency converter, Hz/V
K_c	Gain of the current controller
K_i	Equivalent gain of the current to its reference
K_{ii}	Integral gain of the current controller
K_{is}	Integral gain of the speed controller
k_m	Ratio between mutual and selfinductances per phase in PMBDC machine
K_{pi}	Proportional gain of the current controller
K_{ps}	Proportional gain of the speed controller
K_r	Converter gain, V/V
k_r	Rotor coupling factor
K_s	Gain of the speed controller
k_s	Stator coupling factor
K_t	Torque constant, N·m/A
K_{te}	Torque constant of induction motor, N·m/A
K_{tg}	Gain of the tachogenerator, V/rad/s
K_{vf}	Ratio between stator-phase voltage and stator frequency, V/Hz
k_w	Winding factor
k_{w1}	Stator winding factor
k_{w2}	Rotor winding factor
L, M	Stator self and mutual inductances in PM brushless machines, H
L_a	DC machine armature inductance, H
L_b	Base inductance, H
L_{eq}	Total leakage inductance, H
L_f	DC machine field inductance, H
L_f	Filter inductance, H
L_i	Interphase inductance, H
L_{lr}	Stator-referred rotor-leakage inductance per phase, H
L_{lrr}	Actual rotor leakage inductance per phase, H
L_{ls}	Stator-leakage inductance per phase, H
L_m	Magnetizing Inductance per phase, H
L_o	Value of inductor in energy-recovery chopper in C-dump topology, H
L_q, L_d	Quadrature- and direct-axis selfinductances, H
L_{qn}, L_{dn}	Normalized quadrature- and direct-axis selfinductances, p.u.
L_{qq}, L_{dd}	Selfinductance of the stator q and d axis windings, H
L_r	Stator-referred rotor selfinductance per phase, H
L_r^*	Vector-controller instrumented L_r
L_s	Stator selfinductance per phase, H
L_s^*	Vector-controller instrumented L_s
L_{sc}	Short circuit inductance, A

Symbol	Description
L_{se}	DC series machine field inductance, H
L_{sh}	DC shunt machine field inductance, H
L_{xy}	Mutual inductance between windings given by the subscripts x and y, H
$L_{\alpha\alpha}$	Selfinductance of the rotor α axis windings, H
$L_{\beta\beta}$	Selfinductance of the rotor β axis windings, H
M	Mutual inductance between field and armature windings in dc machine, H
m	Modulation ratio
N_1	Number of turns/phase with half-wave converter control
N_f	Number of turns in the field winding in dc machines
n_r	Rotor speed, rpm
n_s	Stator field speed or synchronous speed, rpm
n_t	Turns ratio of the transformer
o	subscript ending with o in a variable indicates its steady-state operating point value
p	Differential operator
P	Number of Poles
P_1	Air gap power with half wave converter, rad/sec
$_p1n$	Dominant-harmonic resistive losses, W
P_a	Air gap power, W
P_a^*	Air gap power reference, W
P_{an}	Normalized air gap power, p.u.
P_{av}	Average input power, W
P_b	Base power, W
P_c	Armature resistive losses, W
P_{c1}	Machine copper loss with half wave converter operation, W
P_{co}	Core losses, W
P_{c6n}	Normalized sixth harmonic armature copper losses, p.u.
P_{cn}	Normalized armature resistive losses, p.u.
P_{ex}	Energy savings using phase control, W
P_{fw}	Friction and windage losses, W
P_i	Input power, W
p_i	Instantaneous input power, W
P_m	Mechanical power output, W
P_o	Output power, W
P_{on}	Normalized power output, p.u.
P_{rc}	Per-phase rotor resistive losses in ac machines, W
P_s	Shaft power output, W
P_{sc}	Per-phase copper losses in ac machines, W
P_{sl}	Slip power, W
P_{st}	Stray losses in ac machines, W
P_{VA}	Apparent power, VA
Q_i	Reactive power, VAR
R_1	Stator resistance/phase with half-wave converter operation, W
R_a	DC machine armature resistance, Ω
R_{ac}	Changed armature resistance, Ω
R_{an}	Normalized armature resistance, p.u.
R_c	Core-loss resistance, Ω
R_d, R_q	Stator d and q axis winding resistances, Ω
R_{ex}	Braking resistance, Ω
R_f	DC machine field resistance, Ω Ch. 2 and 3
R_f	Dc link filter resistance, Ω
R_{im}	Induction motor equivalent resistance per phase, Ω
R_r	Stator-referred rotor-phase resistance, Ω
R_{rn}	Normalized stator-referred rotor resistance per phase, p.u.
R_s	Stator resistance per phase, Ω

Symbol	Description
R_{se}	DC series machine field resistance, Ω
R_{sh}	DC shunt machine field resistance, Ω
R_{sn}	Normalized stator resistance per phase, p.u.
R_α, R_β	Rotor α and β axis winding resistances, Ω
s	Laplace operator and slip in induction machines
S_a, S_b, S_c	a,b,c phase-switching states of the inverter
s_n	n^{th}-harmonic slip
s_r	Rated slip
T	Carrier period time, s
t	Time, s
T_1	Number of turns in phase A
T_1, T_2	Electrical time constants of the motor, s
T_{1e}	Stator effective turns/phase
T_{2e}	Rotor effective turns/phase
T_a	DC machine armature time constant, s
T_{abc}	Transformation from abc to qdo axes
T_{abc}^s	Transformation from abc to qdo variables in the stationary reference frames
T_{av}	Average torque, N·m
T_b	Base torque, N·m
T_c	Time constant of the current controller
Tc	Transformation from stator arbitrary qd variables to stationary qd variables
T_e	Air gap or electromagnetic torque, N·m.
T_e^*	Torque reference, N·m
T_{e1}	First-harmonic air gap torque, N·m
T_{e6}	Sixth-harmonic torque, N·m
T_{e6n}	Normalized sixth-harmonic torque, p.u.
T_{ec}	Torque reference generated by speed error, N·m
T_{ef}	Maximum air-gap torque generated with the voltage and current constraints, N·m
T_{ehn}	Normalized n^{th} harmonic torque, p.u.
T_{en}	Normalized air gap torque, p.u.
T_{en}^*	Normalized torque reference, p.u.
T_{er}	Rated air gap torque, N·m
T_l	Load torque, N·m.
T_{ln}	Normalized load torque, p.u.
T_m	Mechanical time constant, s
T_{max}	Maximum torque limit, N·m
T_o	Output torque, N·m.
t_{off}	Chopper off-time, s
t_{on}	Chopper on-time, s
T_r	Converter time delay, s
T_r	Rotor time constant in Ch.8, s
T_r^*	Vector-controller instrumented rotor time constant, s
T_s	Time constant of the speed controller
t_x	Armature current-conduction time, s
$T_{\alpha\beta}$	Transformation from $\alpha\beta$ axes to dq axes
T_ω	Time constant of the speed filter, s
u	Input vector
v	Applied voltage to armature of the DC machine, V
v_a^*	Peak value of the carrier signal, V
V_{6p}	Peak sixth-harmonic voltage, V
V_{ab}, V_{bc}, V_{ca}	RMS line-to-line voltages between a and b, b and c and c and a phases, V
v_{ab}, v_{bc}, v_{ca}	Instantaneous line-to-line voltages between a and b, b and c and c and a phases, V
v_{abc}	abc voltage vector

xxvi Symbols

Symbol	Description
V_{as}	Stator rms voltage input per phase
V_{asn}	Normalized phase a stator voltage, p.u.
V_{ao}, V_{bo}, V_{co}	Inverter midpole voltages, V
V_{ar}	RMS phase a rotor voltage, V
V_{as}	RMS stator phase a voltage, V
V_{av}	Average voltage, V
V_b	Base voltage, V
v_b	Blade velocity, m/s
V_{b3}	Base voltage in 3-phase system, V
v_c	Control voltage, V
V_c	Voltage across the capacitor, V
v_{cf}	Control voltage in the DC field current loop, V
V_{cm}	Maximum control voltage, V
V_{cn}	n^{th} harmonic voltage across the capacitor, V
V_d	Diode peak voltage, V
V_{dc}	DC-link Voltage, V
V_{dc6}	Sixth-harmonic DC-link voltage input, V
V_{dcn}	Normalized DC-link voltage, p.u.
v_{drr}, v_{qrr}	Fictitious rotor voltages in d and q axis, V
v_{ds}, v_{qs}	d and q axis stator voltages, V
V_i	Inverter input voltage, V
V_{i6}	Sixth-harmonic input voltage to the filter, V
V_{i6}	Sixth-harmonic inverter input voltage, V
V_{i6m}	Maximum sixth-harmonic inverter input voltage, V
v_L	Voltage drop across the inductor, V
V_{ll}	Line-to-line AC source voltage, V
V_m	Peak supply voltage, V
v_m	Velocity of the metal flow, m/s
V_n	Normalized applied DC armature voltage, p.u.
V_o	Load voltage, V
V_o	Offset voltage to counter the stator resistive voltage drop, V
V_{o6}	Sixth-harmonic output voltage from the filter, V
V_{on}	Normalized offset voltage, p.u.
v_{os}, v_{or}	Stator and rotor zero-sequence voltages, V
V_{ph}	Fundamental rms phase voltage for the six-stepped voltage, V
v_r	Controlled-rectifier output voltage, V
V_r	Rated armature voltage in DC machines, V
v_R	Voltage drop across the resistor, V
V_{rt}	RMS rotor line voltage, V
V_s	RMS stator-voltage phasor, V
V_{sc}	RMS stator-phase voltage applied when rotor is locked, V
v_{sl}	Slip-speed signal, V
V_{sm}	Maximum slip-speed signal, V
V_{sn}	Normalized stator-voltage phasor, V
v_{tg}	Tachogenerator output voltage, V
V_{ts}	Peak power-switch voltage, V
V_{vs}	Sensor output voltage, V
V_α, V_β	Rotor voltages in α and β axis windings, V
v_{rn}^c	Normalized rotor-voltage phasor in the arbitrary reference frame, p.u.
v_{sn}^c	Normalized stator-voltage phasor in the arbitrary reference frame, p.u.
v_{xyn}^r	Voltage with superscript s for stator, r for rotor, e for synchronous and c for arbitrary reference frames, and subscript xy for q and d axes and n for normalized unit. Without subscript n, the variable is in SI units.

Symbol	Description
X	State-variable vector
X(0)	Initial steady-state vector
X_{an}	Normalized armature reactance, p.u.
X_c	Capacitive reactance, Ω
X_{eq}	Total leakage reactance per phase
X_{im}	Induction-machine equivalent reactance per phase, Ω
X_l	Inductive reactance, Ω
X_{ln}	n^{th} harmonic reactance, Ω
X_{lr}	Stator-referred rotor-leakage reactance per phase, Ω
X_{lrn}	Normalized stator-referred rotor-leakage reactance per phase, p.u.
X_{ls}	Stator-leakage reactance per phase, Ω
X_{lsn}	Normalized stator-leakage reactance per phase, p.u.
X_m	Magnetizing reactance per phase, Ω
X_{mn}	Normalized magnetizing reactance per phase, p.u.
X_{rn}	Normalized stator-referred rotor selfreactance per phase, p.u.
X_{sn}	Normalized stator selfreactance per phase, p.u.
Z	Armature conductors
Z_a	Armature impedance of DC machine, Ω
Z_6	Sixth-harmonic impedance, Ω
Z_{an}	Normalized armature impedance, p.u.
Z_{eq}	Equivalent impedance, Ω
Z_{eqn}	Normalized equivalent impedance
Zim	Induction machine equivalent impedance per phase, Ω
Z_n	n^{th}-harmonic armature impedance, Ω
Z_{sc}	Stator-phase short-circuit impedance, Ω
Δi	Hysteresis-current window, A
ΔI_{dc}	Ripple current in DC link, A
α	Triggering-angle delay, rad
α	Ratio of actual to controller-instrumented rotor time constant in Chapters 8 and 9
α_c	Critical triggering-angle delay, rad
α_f	Triggering angle in the DC field converter, rad
β	Impedance angle of DC machine, rad in Chapter 3; current-conduction angle, rad in Chapter 6
β	Ratio of actual to controller-instrumented mutu inductance in Chapters 8 and 9
δ	Preceding a variable indicates small-signal variation
δ	Torque angle in synchronous machine, rad in Ch. 9
$\delta_\alpha, \delta_\beta$	Error in stator α and β currents in PMSM, A
δR_a	Change in armature resistance, Ω
$\delta\theta$	Error in rotor position, rad
$\delta(\theta)$	Impulse function
δT_e	Air gap torque hysteresis window, N·m
$\delta\lambda_s$	Stator flux-linkages hysteresis window, V-s
$\delta\omega_m$	Change in rotor speed, rad/sec
η	Efficiency
ϕ	Power-factor angle, rad
ϕ_1	Fundamental power-factor angle or displacement angle, rad
ϕ_f	Field flux, Wb
ϕ_{fn}	Normalized field flux, p.u.
ϕ_{fr}	Rated field flux in DC machine, Wb
ϕ_m	Peak Mutual flux, Wb
ϕ_{mr}	Angle between the mutual flux and rotor current, rad
ϕ_o	No-load power-factor angle, rad
γ	Current-conduction angle, rad
λ_1, λ_2	Poles of the DC machine

Symbol	Description
λ_{af}	Mutual flux linkages due to rotor magnets, V-s
λ_b	Base flux linkages, V-s
λ_{er}	Stator flux error, V-s
λ_m	Mutual air gap flux linkages, V-s
λ_{mn}	Normalized mutual flux linkages, p.u.
$\lambda_{os}, \lambda_{or}$	Stator and rotor zero sequence flux linkages, V-s
λ_p	Peak mutual flux linkages from rotor magnets, V-s
$\lambda_{qr}, \lambda_{dr}$	Rotor flux linkages in q and d axes, V-s
$\lambda_{qs}, \lambda_{ds}$	Stator flux linkages in q and d axes, V-s
λ_r	Rotor flux-linkages phasor, V-s
γ_{rn}^c	Normalized rotor flux linkages in the arbitrary reference frame, p.u.
γ_{sn}^c	Normalized stator flux linkages in the arbitrary reference frame, p.u.
μ	Overlap angle, rad
θ_c	Arbitrary lag angle between q axis and a phase windings, rad
θ_f	Rotor flux linkages or field angle, rad
θ_{fs}	Stator flux-linkages phasor angle, rad
θ_r	Rotor position with respect to d-axis, rads
θ_s	Stator-current phasor angle, rad
θ_{sl}	Slip angle, rad
θ_T	Torque angle, rad
ρ	Saliency ratio, i.e, between q and d axes selfinductances
τ_{hw}	Stator time constant with half-wave converter operation, s
τ	Normalized time, s
τ_c	Time lag between stator current and its command, s
τ_r	Rotor time constant, s
τ_r'	Transient rotor time constant, s
τ_s	Stator time constant, s
τ_s'	Transient stator time constant, s
ω_1	Speed with half-wave converters in Chap. 9, rad/sec
ω_b	Base angular frequency, rad/sec
ω_c	Carrier angular frequency, rad/sec; also, the speed of arbitrary reference frames in induction machines
ω_{cn}	Normalized arbitrary reference frames speed, p.u.
ω_m	Rotor speed, rad/sec
ω_{rn}^*	Normalized speed reference, p.u.
ω_{max}	Normalized maximum rotor speed, p.u.
ω_{mc}	Steady-state speed with increased armature resistance, rad/sec
ω_{mn}	Normalized rotor speed, p.u.
ω_{mo}	Steady-state operating speed, rad/sec
ω_{mr}	Speed signal from the output of speed filter, V
ω_r	Electrical rotor speed, rad/sec
ω_r^*	Speed reference, V
ω_{rm}	Speed of model rotor frames in Chap. 9, rad/sec
ω_{rn}	Normalized rotor speed, p.u.
ω_s	Supply angular velocity, rad/sec
ω_s^*	Stator angular frequency reference, rad/sec
ω_{sl}	Slip speed or slip angular frequency, rad/sec
ω_{sl}^*	Slip-speed reference, rad/sec
ω_r	Rated stator angular frequency, rad/sec
ψ_{os}, ψ_{or}	Modified zero-sequence stator and rotor flux linkages, V-s
ψ_{qr}, ψ_{dr}	Modified rotor flux linkages in q and d axes, V-s
ψ_{qs}, ψ_{ds}	Modified stator flux linkages in q and d axes, V-s
σ	Leakage coefficient of the induction machine
$\mathfrak{R}_m$	Mutual reluctance

ELECTRIC MOTOR DRIVES

Modeling, Analysis, and Control

CHAPTER 1

Introduction

1.1 INTRODUCTION

The utility power supply is of constant frequency, and it is 50 or 60 Hz. Since the speed of ac machines is proportional to the frequency of input voltages and currents, they have a fixed speed when supplied from power utilities. A number of modern manufacturing processes, such as machine tools, require variable speed. This is true for a large number of applications, some of which are the following:

   (i) Electric propulsion
  (ii) Pumps, fans, and compressors
 (iii) Plant automation
  (iv) Flexible manufacturing systems
   (v) Spindles and servos
  (vi) Aerospace actuators
 (vii) Robotic actuators
(viii) Cement kilns
  (ix) Steel mills
   (x) Paper and pulp mills
  (xi) Textile mills
 (xii) Automotive applications
(xiii) Underwater excavators, mining equipment, etc.
 (xiv) Conveyors, elevators, escalators, and lifts
  (xv) Appliances and power tools
 (xvi) Antennas

The introduction of variable-speed drives increases the automation and productivity and, in the process, efficiency. Nearly 65% of the total electric energy produced in the USA is consumed by electric motors. Decreasing the energy input or increasing the efficiency of the mechanical transmission and processes can reduce the energy consumption. The system efficiency can be increased from 15 to 27% by the introduction of variable-speed drive operation in place of constant-speed operation. It would result in a sizable reduction in the annual energy bill of approximately $90 billion in USA. That many companies' profits, in recent times, stem mainly from saving in their energy bills is to be noted. The energy-saving aspect of variable-speed drive operation has the benefits of conservation of valuable natural resources, reduction of atmospheric pollution through lower energy production and consumption, and competitiveness due to economy. These benefits are obtained with initial capital investment in variable-speed drives that can be paid off in a short time. The payback period depends on the interest rate at which money is borrowed, annual energy savings, cost of the energy, and depreciation and amortization of the equipment. For a large pump variable-speed drive, it is estimated that the payback period is nearly 3 to 5 years at the present, whereas the total operating life is 20 years. That amounts to 15 to 17 years of profitable operation and energy savings with variable-speed drives.

This section briefly introduces power devices, device switching and losses, motor drive and its subsystems, efficiency and machine rating computation, and power converters, controllers, and load. The scope of the book is included.

1.2 POWER DEVICES AND SWITCHING

1.2.1 Power Devices

Since the advent of semiconductor power switches, the control of voltage, current, power, and frequency has become cost-effective. The precision of control has been enhanced by the use of integrated circuits, microprocessors, and VLSI circuits in control circuits. Some of the popular power-switching devices, their symbols, and their capabilities are described below. The device physics and their functioning in detail are outside the scope of the text, and the interested reader is referred to other sources.

(i) Power Diode: It is a PN device. When its anode potential is higher than the cathode potential by its on-state drop, the device turns on and conducts current. The device on-state voltage drop is typically 0.7 V. When the device is reverse biased, i.e., the anode is less positive than the cathode, the device turns off and goes into blocking mode. The current through the diode goes to zero and then reverses and then resurfaces to zero during the turn-off mode, as shown in Figure 1.1. The reversal of current occurs because the reverse bias leads to the reverse recovery of charge in the device. The minimum time taken for the device to recover its reverse voltage blocking capability is t_{rr}, and the reverse recovery charge contained in the diode is Q_{rr}, shown as the area during the reverse current flow. The diode does not have forward voltage blocking capability beyond its on-state drop. The power diode is available in ratings of

Section 1.2   Power Devices and Switching   3

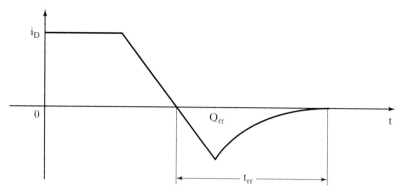

Figure 1.1   Diode current during turn-off

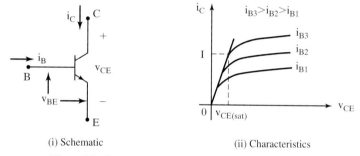

(i) Schematic   (ii) Characteristics

Figure 1.2   Power-transistor schematic and characteristics

kA and kV, and its switching frequency is usually limited to line frequency. Power diodes are used in line rectifier applications.

For fast switching applications, fast-recovery diodes with reverse recovery times in tens of nanoseconds with ratings of several 100 A, at several 100 V, but with a higher on-state drop of 2 to 3 V are available. They are usually used in fast-switching rectifiers with voltages higher than 60 to 100 V and in inverter applications. In case of low-voltage switching applications of less than 60 to 100 V, Schottky diodes are used. They have on-state drop of 0.3 V, thus enabling higher efficiency in power conversion compared to the fast-recovery diodes and power diodes.

(ii) Power Transistor: It is a three-element device, with NPN being the more prevalent. The device can be turned on and off with base current. The device schematic and its characteristics are shown in Figure 1.2. The preferred mode of operation in power circuits is that the transistor be in quasi-saturation, i.e., at knee operating point, during its conduction state. Then it can be pulled back into nonconduction state in shorter time. This device does not have reverse voltage blocking capability. The maximum available rating at present for the device is 1000 A, 1400 V with on-state drop of 2 V. Its switching frequency is very high for the bipolar power transistor, with a lower current gain of 4 to 10, and much in the region of 2 to 6 kHz for Darling power transistors

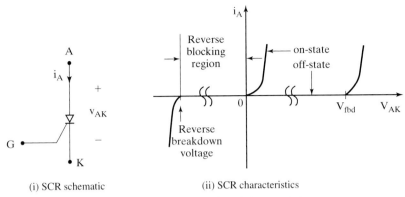

Figure 1.3 SCR schematic and characteristics

with current gain of 100 to 200. These devices are not used very much in newer products.

(iii) Silicon Controlled Rectifier (SCR) or Thyristor: It is a four-element (PNPN) device with three junctions. The power electronics revolution started with the invention of this device in 1956. Its symbol and characteristics are shown in Figure 1.3. The device is turned on with a current signal to the gate, and when the device is forward biased, i.e., the anode is at a higher potential than the cathode by the on-state drop of 1 to 3 V. The device can be turned off only by reverse biasing the device, i.e., reversing the voltage across its anode and cathode. During the reverse biasing, known as commutation, the device behaves like a diode. A negative voltage has to be maintained across the device for a period greater than the reverse recovery time for it to recover its forward blocking voltage capability. The device can also block negative voltages and, beyond a certain value, the device will break down and conduct in the reverse direction. This device is very similar to a power diode but has the capability to hold off its conduction in the forward-biased mode until the gate signal is injected. Its maximum ratings are 6 to 8 kA, 12 kV, with on-state voltage drop ranging from 1 to 3 V. The device is used only in HVDC rectifiers and inverters and in large motor drives with ratings higher than 30 MW. The switching frequency of the device is very limited, to 300 to 400 Hz, and the auxiliary circuit for its turn-off has caused other power devices to displace this device in all applications other than those mentioned. Variations of the device are plentiful; some that belong to this family are the following:

(a) Inverter-grade SCR
(b) Light-activated thyristors
(c) Asymmetrical SCR
(d) Reverse-conducting thyristor
(e) MOS-controlled thyristor
(f) Gate turn-off thyristor (GTO)

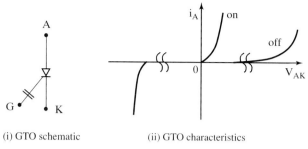

(i) GTO schematic   (ii) GTO characteristics

Figure 1.4   The schematic and characteristics of the GTO

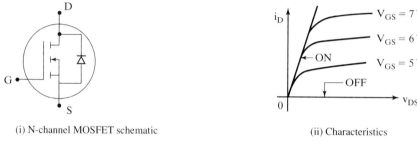

(i) N-channel MOSFET schematic   (ii) Characteristics

Figure 1.5   Schematic and characteristics of the N-channel MOSFET

(iv) Gate Turn-Off Thyristor (GTO): It is a thyristor device with gate turn-on and gate turn-off capability. Its symbol and characteristics are shown in Figure 1.4. The device comes with maximum ratings of 6 kA, 6 kV, with an on-state voltage drop of 2 to 3 V. The maximum switching frequency is 1 kHz, and the device is used mainly in high-power inverters.

(v) MOSFET: This device is a class of field-effect transistor requiring lower gate voltages for turn-on and turn-off and capable of higher switching frequency in the range of 30 kHz to 1 MHz. The device is available at 100 A at 100 to 200 V and at 10 A at 1000 V. The device behaves like a resistor when in conduction and therefore can be used as a current sensor, thus eliminating one sensor device in a drive system. The device always comes with an anti-parallel body diode, sometimes referred to as a *parasitic* diode, that is not ultra-fast and has a higher voltage drop. The device symbol for an N-channel MOSFET and its characteristics are shown in Figure 1.5. The device has no reverse voltage blocking capability.

(vi) Insulated Gate Bipolar Transistor: It is a three-element device with the desirable characteristics of a MOSFET from the viewpoint of gating, transistor in conduction, and SCR/GTO in reverse voltage blocking capability. Its symbol is given in Figure 1.6. The currently available ratings are 1.2 kA at 3.3 kV and 0.6 kA at 6.6 kV, with on-state voltage drop of 5 V. Higher currents at reduced voltages with much lower on-state voltage drops are available. It is expected that further augmentation in the maximum current and voltage ratings will

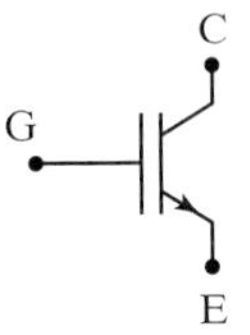

6 Chapter 1   Introduction

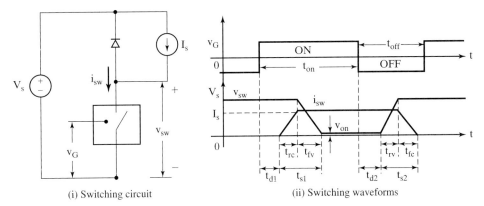

Figure 1.6   IGBT symbol

(i) Switching circuit   (ii) Switching waveforms

Figure 1.7   Switching circuit and waveforms

occur in the future. The switching frequency is usually around 20 kHz for many of the devices, and its utilization at high power is at low frequency, because of switching loss and electromagnetic-interference concerns.

1.2.2 Switching of Power Devices

The understanding of device switching in transient is of importance in the design of the converter as it relates to its losses, efficiency of the converter and the motor drive system, and thermal management of the power converter package. The transient switching of the devices during turn-on and turn-off is illustrated in this section by considering a generic device. Ideal current and voltage sources and power devices are assumed for this illustration. The circuit for illustration is shown Figure 1.7. The switching device is gated on and the current in the device increases from zero to I_s after a turn-on delay time t_{d1}. The current transfers from the diode linearly in time t_{rc}, which is the rise time of the current. During this rise time, the diode is conducting, and therefore the voltage across the switching device is the source voltage, V_s. When the current is completely transferred from the diode to the switching device, the voltage drop across the diode rises from zero to the source voltage and the voltage falls at the same time across the switching device in t_{fv}. The sum of the current rise and voltage fall times is the turn-on switching transient time; note that, during this time, the device losses are very high. During the conduction, the voltage across the device is its on-state voltage drop and the power loss is smaller.

When the gating signal goes to the turn-off condition, the switching device responds with a turn-off delay time of t_{d2}. Then, the device voltage rises to V_s in t_{rv}, which forward biases the diode and initiates the current transfer from the switching device to the diode. The current transfer is completed in t_{fc}. The sum of the voltage rise time across the switching device and current fall time is the turn-off transient time, during which the device loss is very high.

From this illustration, the losses in the switching device are approximately derived as follows:

(a) Conduction energy loss, $E_{sc} = I_s V_{on} [t_{on} + t_{d2} - t_{s1} - t_{d1}]$ (1.1)

(b) Sum of turn-on and turn-off energy loss, $E_{st} \cong 0.5\, V_s I_s [t_{s1} + t_{s2}]$ (1.2)

(c) Total power loss, $P_{sw} = \dfrac{E_{st} + E_{sc}}{t_{on} + t_{off}} = f_c(E_{sc} + E_{st})$ (1.3)

(d) Switching frequency, $f_c = \dfrac{1}{t_{on} + t_{off}}$ (1.4)

Note that the power loss is averaged over a period of the switching period, and total power loss is the sum of the conduction and switching losses in the switching device. Similarly, the power losses in the diode can be derived. The switching losses are proportional to the switching frequency and to the product of the source voltage and load current. Note that, in general, the switching times are much smaller than the conduction time, and therefore the switching losses are less than (or at most equal to) the conduction loss in the switching device. Low switching frequency is preferred because of lower switching losses but contributes to poor power quality. Higher switching frequency enables voltage and current waveform shaping and reduces the distortion but invariably is followed by higher switching losses.

The switching illustrated is known as hard switching: current and voltage transitions occur in the device at full source voltage and current, respectively, during turn-on and turn-off periods. Resonant and soft-switching circuits enable switching device transitions at zero voltage and current, reducing or almost eliminating the switching losses, but these circuits are not economical at present and so are not considered in motor-drive applications or in this text any further.

The power switches are used in the circuits to control energy flow from source to load and vice versa. This is known as static power conversion. The details of power conversion are not the objectives of this text, and many good texts are available on this topic. Some references are listed at the end of this chapter. As and when necessary, a brief description of a power converter is included in this text.

1.3 MOTOR DRIVE

A modern variable-speed system has four components:

(i) Electric machines—ac or dc
(ii) Power converter—Rectifiers, choppers, inverters, and cycloconverters

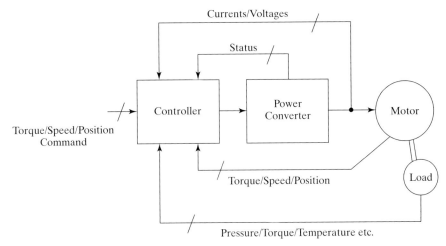

Figure 1.8 Motor-drive schematic

(iii) Controllers—Matching the motor and power converter to meet the load requirements
(iv) Load

This is represented in the block diagram shown in Figure 1.8. A brief description of the system components is given in the following sections.

1.3.1 Electric Machines

The electric machines currently used for speed control applications are the following:

(i) dc machines—Shunt, series, compound, separately-excited dc motors, and switched reluctance machines;
(ii) ac machines—Induction, wound-rotor synchronous, permanent-magnet synchronous, synchronous reluctance, and switched reluctance machines.
(iii) Special machines—Switched reluctance machines.

All the machines are commercially available from fractional-kW to MW ranges except permanent-magnet synchronous, synchronous reluctance, and switched reluctance machines, which are available up to 150 kW level. The latter machines are available at higher power levels but would be expensive from a commercial point of view, because they require custom designs. A number of factors go into the selection of a machine for a particular application:

(i) Cost
(ii) Thermal capacity
(iii) Efficiency
(iv) Torque–speed profile

(v) Acceleration
(vi) Power density, volume of the motor
(vii) Ripple, cogging torques
(viii) Availability of spare and second sources
(ix) Robustness
(x) Suitability for hazardous environment
(xi) Peak torque capability

They are not uniformly relevant to any one application. Some could take precedence over others. For example, in a position-servo application, the peak torque and thermal capabilities together with ripple and cogging torques are preponderant characteristics for application consideration. High peak torques decrease the acceleration/deceleration times, small cogging and ripple torques help to attain high positioning repeatability, and high thermal capability leads to a longer motor life and a higher loading. In this text, only dc, induction, and permanent-magnet synchronous and brushless dc machines are considered, and their steady-state and dynamic models are derived. Wound-rotor synchronous machines and drives technology is well established, and very little change has occurred in them over the last 40 years. The synchronous reluctance and switched reluctance machines and their drive systems are fairly recent but yet to establish themselves in the market. Very few engineers are involved with these topics, whereas a large number of engineers are employed in the dc, induction, permanent-magnet synchronous, and brushless dc machines industrial sector, and, therefore, only these topics are covered in this text.

Efficiency computation. The interest in energy savings is one of the major motivational factors in the introduction of variable-speed drives in some industries. Therefore, it is prevalent to encounter the efficiency computation for electric motors whenever variable-speed operation is considered. A brief introduction of efficiency computation for electric motors is given in the following.

The power, voltage and speed given in the nameplate details of the motor are its rated values, i.e., for continuous steady-state operation. Notice how the power corresponds to shaft output power. The output or shaft torque, T_o, of the machine is calculated as

$$\left. \begin{array}{l} T_o = \dfrac{\text{Power, W}}{\text{Speed, rad/sec}} = \dfrac{P_o \times 745.6}{\omega_m}, \text{N·m} \\ \omega_m = \dfrac{2\pi N_r}{60}, \text{rad/sec} \end{array} \right\} \quad (1.5)$$

where P_o is the output power in hp, N_r is speed in rpm, and ω_m is the speed in rad/sec.

The internal torque, T_e, of the motor is the sum of the output torque and the shaft torque losses, viz., friction and windage torques. Friction is contributed by the bearings (usually proportional to speed) and windage by the effects of the circulating air on the rotating parts (proportional to square of the speed). Let the shaft power losses be denoted by P_{sh} and loss torque by T_{lo}. The internal torque then is given by

$$T_e = T_o + T_{lo}, \text{N·m} \tag{1.6}$$

This internal torque is known as air gap or electromagnetic torque, because it is the torque developed by the motor in the air gap through electromagnetic coupling. The air gap torque is developed from the power crossing over to the air gap from the armature (usually the rotor in dc machines, the stator in ac machines) and is known as air gap power, P_a. In general, it is obtained from the armature input, P_i, and its loss, P_l, due to armature resistance (known as copper losses), core losses, and unidentifiable losses (known as stray losses, ranging from 0.5 to 1% of output power).

Note that air gap torque from air gap power is calculated as

$$T_e = \frac{P_a}{\omega_m}, \text{N·m} \tag{1.7}$$

The efficiency, then, is computed as

$$\text{Efficiency}, \eta = \frac{\text{Output power}}{\text{Input power}} = \frac{P_0}{P_i} \tag{1.8}$$

For particular motors, the details of the computation are shown in respective chapters. Note that the output torque is equal to the load torque, T_l. If shaft loss torque is neglected, note that the electromagnetic torque is equal to load torque.

Motor rating. A typical electric train's torque and speed profiles for one cycle are shown in Figure 1.9. It is seen that neither the load torque nor the speed is

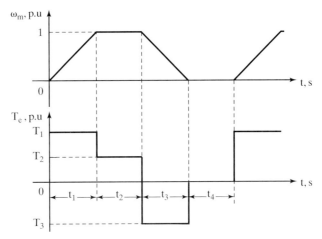

Figure 1.9 An electric train's torque and speed profiles for one cycle

constant; both vary. Such loads are encountered in practice. The selection of motor torque and power ratings, then, is not straightforward, as in the case of constant-speed drives. In the varying-load case, the motor torque and power are selected on the basis of effective torque and power. The effective torque is calculated from the load profile, such as the one shown in Figure 1.9. For example, the effective electromagnetic torque for the load cycle shown in Figure 1.9, discounting the rest period, is obtained as

$$T_e = \sqrt{\frac{T_1^2 t_1 + T_2^2 t_2 + T_3^2 t_3}{t_1 + t_2 + t_3}} \qquad (1.9)$$

Similarly, the effective power is calculated from its time history. These calculations yield effective torque and power. Ensuring the effective torque and output power to be well below or equal to the rated values of the machine does not guarantee that the machine is thermally within its designed limits. The thermal rating influences the motor operation. The thermal rating is dependent on the motor losses, and they, in turn, are preponderantly influenced by the effective current in the case of a dc machine but also by additional factors, such as stator frequency and applied voltages, in ac machines. In order to determine whether the machine is operating within the thermal limits of rated power losses, it is necessary to calculate the effective losses in the machine for a load cycle. If the effective losses are lower than the rated losses, then the machine is thermally sound for operation.

Example 1.1

Two electrical machines are considered to drive a load. The torque and speed ratings of the machines are same. The thermal limits for the machines correspond to their rated armature resistive losses. The machines 1 and 2 have the following torque-to-armature current relationships:

(1) $T_e \propto i^2$

(2) $T_e \propto i$

The load is twice the rated torque for half the time and zero for the other half of the time in a load cycle. Find which machine is suitable for the application.

Solution By ignoring the shaft-loss torque, the load torque is equal to the electromagnetic torque. Let T_b be the base (rated) torque, I_b the base current, and $I_b^2 R$ the base losses. The effective electromagnetic torque is

$$T_e = 2T_b \sqrt{d} = 2T_b \sqrt{\frac{1}{2}} = \sqrt{2} T_b$$

where d is the duty cycle.

Machine 1: From the relationship that the electromagnetic torque is proportional to square of the current, the current for the load is found as follows:

$$T_b \propto I_b^2$$
$$2T_b \propto I_1^2$$

The armature current for the load is $I_1 = I_b \sqrt{2}$

RMS armature current is $I_{1r} = I_b \sqrt{2} \sqrt{d} = I_b \sqrt{2} \sqrt{\frac{1}{2}} = I_b$

Losses $= I_{1r}^2 R = I_b^2 R =$ Rated losses

Machine 2: As electromagnetic torque is proportional to current, the current for the load is found as follows for this machine:

$$T_b \propto I_b$$
$$2T_b \propto I_2$$

The armature current for the given load is $I_2 = 2I_b$

RMS value of the armature current is $I_{2r} = 2I_b \sqrt{d} = 2I_b \sqrt{\frac{1}{2}} = I_b \sqrt{2}$

Loss $= I_{2r}^2 R = 2I_b^2 R = 2 *$ Rated losses

Therefore, machine 1 alone satisfies the thermal limit of rated losses and hence is suitable for the given application. Note that the effective torque is more than the rated torque, but, still, thermally safe operation is possible for the considered case.

1.3.2 Power Converters

The power converters driving the electrical machines are

   (i) Controlled rectifiers: They are fed from single- and three-phase ac mains supply and provide a dc output for control of the dc machines or sometimes input dc supply to the inverters in the case of ac machines.
   (ii) Inverters: They provide variable alternating voltages and currents at desired frequency and phase for the control of ac machines. The dc supply input to the inverters is derived either from a battery in the case of the electric vehicle or from a rectified ac source with controlled or uncontrolled (diode) rectifiers. Because of the dc intermediary, known as dc link, between the supply ac source and the output of the inverter, there is no limitation to the attainable output frequency other than that of the power device switching constraints in the inverters.
   (iii) Cycloconverters: They provide a direct conversion of fixed-frequency alternating voltage/current to a variable voltage/current variable frequency for the control of ac machines. The output frequency is usually limited from 33 to 50% of the input supply frequency, to avoid distortion of the waveform, and therefore they are used only in low-speed but high-power ac motor drives.

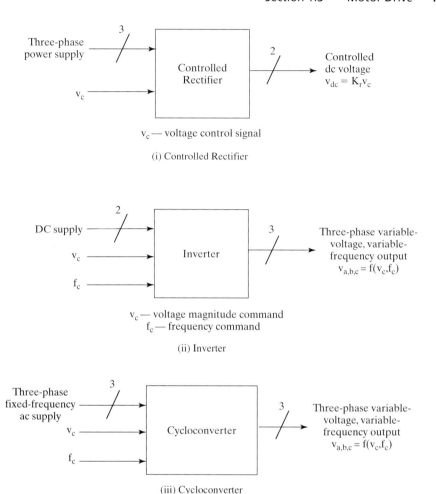

Figure 1.10 Schematics of power converters

These power converters can be treated as black boxes with certain transfer functions. In that case, the referred converters are symbolically represented as shown in Figure 1.10. The transfer functions are derived for various converters in this text.

1.3.3 Controllers

The controllers embody the control laws governing the load and motor characteristics and their interaction. To match the load and motor through the power converter, the controller controls the input to the power converter. Very many control strategies have been formulated for various motor drives and the controllers implement their algorithms. For instance, the control of flux and torque requires a coordination of the field and armature currents in a separately excited dc motor. In the case of an ac induction motor, the same is implemented by coordinating the three stator currents, whereas the

14 Chapter 1 Introduction

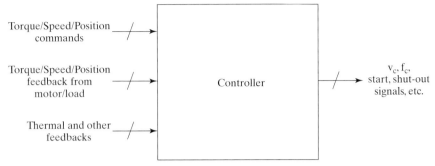

Figure 1.11 Controller schematic

synchronous motor control requires the control of the three stator currents and its field current too. The laws governing their control are complex and form the core of this text.

The schematic of the controller is shown in Figure 1.11. Its input consists of the following:

(i) Torque, flux, speed, and/or position commands
(ii) Their rate of variations, to facilitate soft start and preserve the mechanical integrity of the load
(iii) The measured torque, flux, speed, and/or position for feedback control
(iv) Limiting values of currents, torque, acceleration, and so on
(v) Temperature feedback and instantaneous currents and/or voltages in the motor and/or converter
(vi) The constants in the speed and position controllers, such as proportional, integral, and differential gains.

The controller output determines the control signal for voltage magnitude, v_c in the case of inverters, and the control signal for determining the frequency, f_c. These functions can be merged, and only the final gating signals might be directly issued to the bases/gates of the power converter. It may also perform the protection and other monitoring functions and deal with emergencies such as sudden field loss or power failure.

The controllers are realized with analog and integrated circuits. The present trend is to use microprocessors, single-chip microcontrollers, digital signal processors [DSPs], VLSI, and special custom chips also known as application specific ICs [ASICs] to embody a set of functions in the controller. The real-time computational capability of these controllers allows complex control algorithms to be implemented. Also, they lend themselves to software and remote control, hence paving the way to flexible manufacturing systems and a high degree of automation.

1.3.4 Load

The motor drives a load that has a certain characteristic torque-vs.-speed requirement. In general, the load torque is a function of speed and can be written as $T_1 \propto \omega_m^x$, where x can be an integer or a fraction. For example, the load torque is

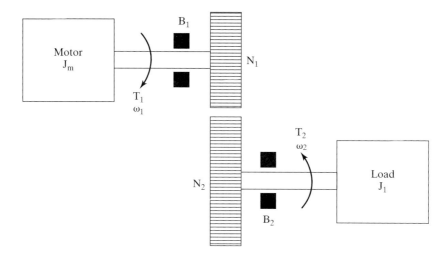

N_1, N_2 — Teeth numbers in the gear
B_1, B_2 — Bearings and their friction coefficients
J_1, J_m — Moment of inertia of the motor and load

Figure 1.12 Schematic of the motor–load connection through a gear

proportional to speed in frictional systems such as a feed drive. In fans and pumps, the load torque is proportional to the square of the speed.

In some instances, the motor is connected to the load through a set of gears. The gears have a teeth ratio and can be treated as torque transformers. The motor–gear–load connection is schematically shown in Figure 1.12. The gears are primarily used to amplify the torque on the load side that is at a lower speed compared to the motor speed. The motor is designed to run at high speeds because it has been found that the higher the speed, the lower is the volume and size of the motor. But most of the useful motion is at low speeds, hence the need for a gear in the motor-load connection. The gears can be modeled from the following facts [5]:

(i) The power handled by the gear is the same on both the sides.
(ii) Speed on each side is inversely proportional to its tooth number.

Hence,

$$T_1 \omega_1 = T_2 \omega_2 \tag{1.10}$$

$$T_2 = T_1 \left(\frac{\omega_1}{\omega_2} \right) \tag{1.11}$$

and

$$\frac{\omega_1}{\omega_2} = \frac{N_2}{N_1} \tag{1.12}$$

Substituting equation (1.12) into (1.11), we get

$$T_2 = \left(\frac{N_2}{N_1}\right) T_1 \tag{1.13}$$

Note that this is obtained with the assumption of zero losses in the gear, and for greater accuracy the losses have to be modeled. In that case, the difference between T_1 and the loss torque will have to be multiplied by the turns ratio, N_2/N_1, to obtain the output gear torque T_2. Similarly to the case of a transformer, the constants of the load as reflected to the motor are written as

$$J_1 \text{ (reflected)} = \left(\frac{N_1}{N_2}\right)^2 J_1 \tag{1.14}$$

$$B_2 \text{ (reflected)} = \left(\frac{N_1}{N_2}\right)^2 B_2 \tag{1.15}$$

The ratio between N_1 and N_2 is very small, so the reflected inertia and friction constant are negligible in many systems.

Hence, the resultant mechanical constants are

$$J = J_m + \left(\frac{N_1}{N_2}\right)^2 J_1 \tag{1.16}$$

$$B = B_1 + \left(\frac{N_1}{N_2}\right)^2 B_2 \tag{1.17}$$

The torque equation of the motor–load combination is described by

$$J\frac{d\omega_1}{dt} + B\omega_1 = T_1 - T_2 \text{(reflected)} = T_1 - \left(\frac{N_1}{N_2}\right) T_2 \tag{1.18}$$

To model the motor drive, it is essential to have a physical model of its load, with the characteristics of friction, inertia, torque-speed profile, and gears and backlash. Modeling of the physical systems can be found in references [8, 11].

1.4 SCOPE OF THE BOOK

This book addresses the study of steady-state and dynamic control of ac and dc machines supplied from power converters and their integration to the load. In due course, the design of controllers and their implementation are considered. The system analysis and design of the motor drive are kept in perspective with regard to current practice. Control algorithms and analysis have been developed to facilitate dynamic simulation with personal computers. Applications of motor drives are illustrated with selections from industrial environment.

1.5 REFERENCES

1. B. D. Bedford and R. G. Hoft, *Principles of Inverter Circuits,* John Wiley and Sons, New York, 1964.
2. Bimal K. Bose, *Power Electronics and AC Drives,* Prentice Hall, New York, 1985.
3. Samir K. Datta, *Power Electronics and Controls,* Reston Publishing Company, Inc., Reston, VA, 1985.
4. V. Del Toro, *Electric Machines and Power Systems,* Prentice Hall, 1985.
5. S. B. Dewan, G. R. Slemon, and A. Straughen, *Power Semiconductor Drives,* John Wiley and Sons, New York, 1984.
6. S. B. Dewan and A. Straughen, *Power Semiconductor Circuits,* John Wiley and Sons, New York, 1975.
7. A. E. Fitzgerald, C. Kingsley, and S. D. Umans, *Electric Machinery,* McGraw-Hill Book Co., 1983.
8. Hans Gross, *Electrical Feed Drives for Machine Tools,* John Wiley and Sons Limited, Berlin, 1983.
9. John D. Harnden, Jr., and Forest B. Golden, *Power Semiconductor Application,* Vols. I and II, IEEE Press, New York, 1972.
10. R. G. Hoft, *Semiconductor Power Electronics,* Van Nostrand Reinhold Company, New York, 1986.
11. Benjamin C. Kuo, *Automatic Control Systems,* Prentice–Hall Inc., Englewood Cliffs, New Jersey, 1982.
12. W. Leonhard, *Electric Drives,* Springer-Verlag, New York, 1984.
13. J. F. Lindsay and M. H. Rashid, *Electromechanics and Electrical Machinery,* Prentice Hall, 1986.
14. Matsch and Morgan, *Electromagnetic and Electromechanical Machines,* MIT Press, Cambridge, MA, 1972.
15. W. McMurray, *The Theory and Design of Cycloconverters,* MIT Press, Cambridge, MA, 1972.
16. Gottfried Moltgen, *Line Commutated Thyristor Converters,* Pitman Publishing, London, 1972.
17. S. A. Nasar and Unnewher, *Electromechanics and Electric Machines,* John Wiley and Sons, New York, 1979.
18. B. R. Pelly, *Thyristor Phase-Controlled Converters and Cycloconverters,* John Wiley and Sons, New York, 1971.
19. G. R. Slemon and A. Straughen, *Electric Machines,* Addison–Wesley Publishing Company, 1980.
20. P. C. Sen, *Principles of Electric Machines and Power Electronics,* John Wiley and Sons, 1989.
21. N. Mohan, T. M. Undeland, and W. P. Robbins, *Power Electronics: Converters, Applications and Design,* John Wiley and Sons, 1989.

CHAPTER 2

Modeling of DC Machines

Direct current (dc) machines have been in service for more than a century. Their fortune has changed a great deal since the introduction of the induction motor, sometimes called the ac shunt motor. DC motors have staged a comeback with the advent of the silicon-controlled rectifier used for power conversion, facilitating a wide-range speed control of these motors.

This chapter contains a brief description of the theory of operation of separately-excited and permanent-magnet dc brush motors and of their modeling and transfer functions, an evaluation of steady state and transient responses, and a flow chart for their computation.

2.1 THEORY OF OPERATION

It is known that maximum torque is produced when two fluxes are in quadrature. The fluxes are created with two current-carrying conductors. The flux path is of low reluctance with steel. Figure 2.1 shows such a schematic representation. Coil 1 (called the field winding) wound on the pole produces a flux ϕ_f with an input current i_f. Coil 2 (called the armature winding) on a rotating surface produces a flux ϕ_a with a current i_a. For the given position, the two fluxes are mutually perpendicular and hence will exert a maximum torque on the rotor, moving it in clockwise direction. If it is assumed that the rotor has moved by 180°, coils B_1, B_2, and B_3 will be under the south pole and carrying negative current. The torque will be such as to move the rotor in the counterclockwise direction, keeping the rotor in oscillation. In order to have a uniform torque (i.e. unidirectional torque) and a clockwise (or counterclockwise) direction of rotation, the armature winding needs to carry a current of the same polarity underneath a field pole. That is arranged by segmenting the armature coils and connecting them to separate copper bars, called a commutator, mounted on the same shaft as the armature and fed from brushes. The brushes are stationary and hence supplied with currents of fixed polarity. Thus, the commutator segments under a brush will continue to receive a current of fixed polarity.

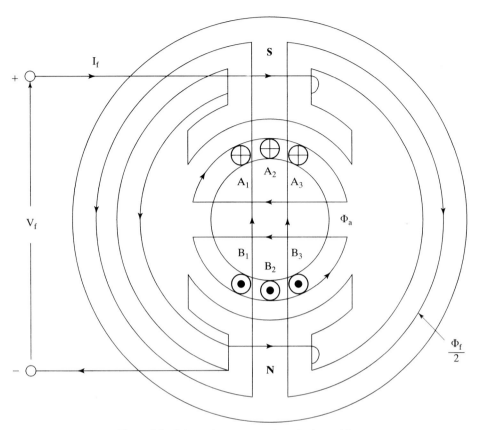

Figure 2.1 Schematic representation of a dc machine

2.2 INDUCED EMF

The expression for induced emf and torque is derived for a machine with P poles, Z armature conductors in a field with a flux per pole of ϕ_f and rotating at n_r rpm. From Faraday's law, the induced emf (neglecting the sign) is

$$e = Z\frac{d\phi_f}{dt} = Z\frac{\phi_f}{t} \qquad (2.1)$$

where t is the time taken by the conductors to cut ϕ_f flux lines. Therefore

$$t = \frac{1}{2 \times \text{frequency}} = \frac{1}{2\left(\dfrac{P}{2}\right)\left(\dfrac{n_r}{60}\right)} \qquad (2.2)$$

The flux change occurs for each pole pair.
By substituting equation (2.2) in equation (2.1),

$$e = \frac{Z\phi_f P n_r}{60} \qquad (2.3)$$

20 Chapter 2 Modeling of DC Machines

If the armature conductors are divided into 'a' parallel paths, then

$$e = \frac{Z\phi_f P n_r}{60a} \quad (2.4)$$

There are two possible arrangements of conductors in the armature, wave windings and lap windings. The values of a for these two types of windings are

$$a = \begin{cases} 2 \text{ for wave winding} \\ P \text{ for lap winding} \end{cases} \quad (2.5)$$

It is usual to write the expression (2.4) in a compact form as

$$e = K\phi_f \omega_m \quad (2.6)$$

where $\omega_m = 2\pi n_r/60$ rad/sec and $K = (P/a)Z(1/2\pi)$.

If the field flux is constant, then the induced emf is proportional to the rotor speed and the constant of proportionality is known as the *induced emf* or *back emf* constant. Then the induced emf is represented as

$$e = K_b \omega_m \quad (2.7)$$

where K_b is the induced emf constant, given by

$$K_b = K\phi_f \text{ volt/(rad/sec)} \quad (2.8)$$

and K is a proportionality constant which will be obvious from the following. The field flux is written as the ratio between the field mmf and mutual reluctance,

$$\phi_f = \frac{N_f i_f}{\mathcal{R}_m} \quad (2.9)$$

where N_f is the number of turns in the field winding, i_f is the field current, and $\mathcal{R}_m$ is the reluctance of the mutual flux path. The mutual flux is the resultant of the armature and field fluxes.

By substituting (2.9) into (2.7), the emf constant is obtained as

$$K_b = \frac{K N_f i_f}{\mathcal{R}_m} = M i_f \quad (2.10)$$

where M is the fictitious mutual inductance between armature and field windings given by

$$M = \frac{K N_f}{\mathcal{R}_m} = \frac{P}{\pi} \frac{Z}{2a} \frac{N_f}{\mathcal{R}_m} \quad (2.11)$$

Note that $Z/(2a)$ is the number of turns in the armature per parallel path and together in product with the field winding turns N_f gives the familiar mutual inductance definition. The factor P/π makes it a fictitious inductance. By substituting equation (2.10) in equation (2.7), the induced emf is obtained as

$$e = M i_f \omega_m \quad (2.12)$$

The mutual inductance is a function of the field current and must be taken note of to account for saturation of the magnetic material (stator laminations). For operation within the linear range, mutual inductance is assumed to be a constant in the machine.

2.3 EQUIVALENT CIRCUIT AND ELECTROMAGNETIC TORQUE

The equivalent circuit of a dc motor armature is based on the fact that the armature winding has a resistance R_a, a self-inductance L_a, and an induced emf. This is shown in Figure 2.2. In the case of a motor, the input is electrical energy and the output is the mechanical energy, with an air gap torque of T_e at a rotational speed of ω_m. The terminal relationship is written as

$$v = e + R_a i_a + L_a \frac{di_a}{dt} \tag{2.13}$$

In steady state, the armature current is constant and hence the rate of change of the armature current is zero. Hence the armature voltage equation reduces to

$$v = e + R_a i_a \tag{2.14}$$

The power balance is obtained by multiplying equation (2.14) by i_a:

$$v i_a = e i_a + R_a i_a^2 \tag{2.15}$$

The term $R_a i_a^2$ denotes the armature copper losses and $v i_a$ is the total input power. Hence $e i_a$ denotes the effective power that has been transformed from electrical to

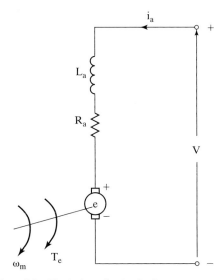

Figure 2.2 Equivalent circuit of a dc motor armature

mechanical form, hereafter called the air gap power, P_a. The air gap power is expressed in terms of the electromagnetic torque and speed as

$$P_a = \omega_m T_e = ei_a \tag{2.16}$$

Hence, the electromagnetic torque or air gap torque is represented as

$$T_e = \frac{ei_a}{\omega_m} \tag{2.17}$$

By substituting for the induced emf from (2.7) into (2.17), it is further simply represented as

$$T_e = K_b i_a \tag{2.18}$$

Note that the torque constant is equal to the emf constant if it is expressed in volt-sec/rad for a constant-flux machine. The dc motor drives a load, the modeling of load is considered next, to complete the analysis for the motor–load system.

2.4 ELECTROMECHANICAL MODELING

For simplicity, the load is modeled as a moment of inertia, J, in kg-m^2/sec^2, with a viscous friction coefficient B_1 in N·m/(rad/sec). Then the acceleration torque, T_a, in N·m drives the load and is given by

$$J\frac{d\omega_m}{dt} + B_1 \omega_m = T_e - T_l = T_a \tag{2.19}$$

where T_l is the load torque. Equations (2.13) and (2.19) constitute the dynamic model of the dc motor with load.

2.5 STATE-SPACE MODELING

The dynamic equations are cast in state-space form and are given by

$$\begin{bmatrix} pi_a \\ p\omega_m \end{bmatrix} = \begin{bmatrix} -\dfrac{R_a}{L_a} & -\dfrac{K_b}{L_a} \\ \dfrac{K_b}{J} & -\dfrac{B_1}{J} \end{bmatrix} \begin{bmatrix} i_a \\ \omega_m \end{bmatrix} + \begin{bmatrix} \dfrac{1}{L_a} & 0 \\ 0 & -\dfrac{1}{J} \end{bmatrix} \begin{bmatrix} V \\ T_l \end{bmatrix} \tag{2.20}$$

where p is the differential operator with respect to time. Equation (2.19) is expressed compactly in the form given by

$$\dot{X} = AX + BU \tag{2.21}$$

where $X = [i_a \; \omega_m]^t$, $U = [V \; T_l]^t$, X is the state variable vector, and U is the input vector.

Even though the load torque is a disturbance, for sake of a compact representation it is included in the input vector in this text.

Section 2.6 Block Diagrams and Transfer Functions

$$A = \begin{bmatrix} -\dfrac{R_a}{L_a} & -\dfrac{K_b}{L_a} \\ \dfrac{K_b}{J} & -\dfrac{B_1}{J} \end{bmatrix}, B = \begin{bmatrix} \dfrac{1}{L_a} & 0 \\ 0 & -\dfrac{1}{J} \end{bmatrix}$$

The roots of the system are evaluated from the A matrix; they are

$$\lambda_1, \lambda_2 = \frac{-\left(\dfrac{R_a}{L_a} + \dfrac{B_1}{J}\right) \pm \sqrt{\left(\dfrac{R_a}{L_a} + \dfrac{B_1}{J}\right)^2 - 4\left(\dfrac{R_a B_1}{JL_a} + \dfrac{K_b^2}{JL_a}\right)}}{2} \quad (2.22)$$

It is interesting to observe that these eigenvalues will always have a negative real part, indicating that the motor is stable on open-loop operation.

2.6 BLOCK DIAGRAMS AND TRANSFER FUNCTIONS

Taking Laplace transforms of equations (2.13), (2.19) and neglecting initial conditions, we get

$$I_a(s) = \frac{V(s) - K_b \omega_m(s)}{R_a + sL_a} \quad (2.23)$$

$$\omega_m(s) = \frac{K_b I_a(s) - T_l(s)}{(B_1 + sJ)} \quad (2.24)$$

In block-diagram form, the relationships are represented in Figure 2.3. The transfer functions $\dfrac{\omega_m(s)}{V(s)}$ and $\dfrac{\omega_m(s)}{T_l(s)}$ are derived from the block diagram. They are

$$G_{\omega v}(s) = \frac{\omega_m(s)}{V(s)} = \frac{K_b}{s^2(JL_a) + s(B_1 L_a + JR_a) + (B_1 R_a + K_b^2)} \quad (2.25)$$

$$G_{\omega l}(s) = \frac{\omega_m(s)}{T_l(s)} = \frac{-(R_a + sL_a)}{s^2(JL_a) + s(B_1 L_a + JR_a) + (B_1 R_a + K_b^2)} \quad (2.26)$$

The separately-excited dc motor is a linear system, and hence the speed response due to the simultaneous voltage input and load torque disturbance can be written as a sum of their individual responses:

$$\omega_m(s) = G_{\omega v}(s) V(s) + G_{\omega l}(s) T_l(s) \quad (2.27)$$

Laplace inverse of $\omega_m(s)$ in equation (2.27) gives the time response of the speed for a simultaneous change in the input voltage and load torque. The treatment so far is

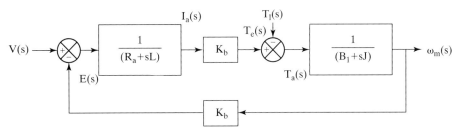

Figure 2.3 Block diagram of the dc motor

based on a dc motor obtaining its excitation separately. There are various forms of field excitation, discussed in the following section.

Example 2.1

A dc motor whose parameters are given in example 2.3 is started directly from a 220-V dc supply with no load. Find its starting speed response and the time taken to reach 100 rad/sec.

Solution

$$\frac{\omega(s)}{V(s)} = G_{\omega v}(s) = \frac{K_b}{s^2(JL_a) + s(B_1 L_a + JR_a) + (B_1 R_a + K_b^2)} = \frac{15{,}968}{s^2 + 167s + 12874}$$

$$V(s) = \frac{220}{s}$$

$$\omega(s) = \frac{3.512 \times 10^6}{s(s^2 + 167s + 12874)}$$

$$\omega(t) = 272.8(1 - 1.47 e^{-83.5t} \sin(76.02t + 0.744))$$

The time to reach 100 rad/sec is evaluated by equating the left-hand side of the above equation to 100 and solving for t, giving an approximate value of 10 rad/ms.

2.7 FIELD EXCITATION

The excitation to the field is dependent on the connections of the field winding relative to the armature winding. A number of choices open up, and they are treated briefly in the following sections.

2.7.1 Separately Excited dc Machine

If the field winding is physically and electrically separate from the armature winding, then the machine is known as a separately-excited dc machine, whose equivalent circuit is shown in Figure 2.4. The independent control of field current and armature current endows simple but high performance control on this machine, because the torque and flux can be independently and precisely controlled. The

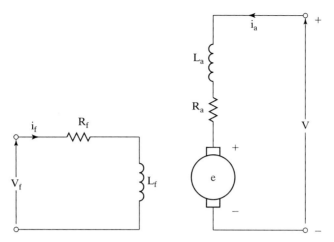

Figure 2.4 Separately-excited dc machine

field flux is controlled only by the control of the field current. Assume that the field is constant. Then the torque is proportional only to the armature current, and, hence, by controlling only this variable, the dynamics of the motor drive system can be controlled. With the independence of the torque and flux-production channels in this machine, it must be noted that it is easy to generate varying torques for a given speed and hence make torque generation independent of the operating speed. This is an important operating feature in a machine: the speed regulation can be zero. The fact that such a feature comes only with feedback control and not simply on open-loop operation is to be recognized.

Example 2.2

A separately-excited dc motor is delivering rated torque at rated speed. Find the efficiency of the motor at this operating point. The details of the machine are as follows: 1500 kW, 600 V, rated current = 2650 A, 600 rpm, Brush voltage drop = 2 V, Field power input = 50 kW, Ra = 0.003645 Ω, L_a = 0.1 mH, Machine frictional torque coefficient = 15 N·m/(rad/sec). Field current is constant and the armature voltage is variable.

Solution To find the input power, the applied voltage to the armature to support a rated torque and rated speed has to be determined. In steady state, the armature voltage is given by

$$V_a = R_a I_{ar} + K_b \omega_{mr} + V_{br}$$

where I_{ar} is the rated armature current, given as 2650A, ω_{mr} is the rated speed in rad/sec, and V_{br} is the voltage drop across the brushes in the armature circuit and is equal to 2V (given in the problem). To solve this equation, the emf constant has to be solved for from the available data. Recalling that the torque and emf constants are equal, the torque constant can be computed from the rated electromagnetic torque and the rated current as

$$K_t = \frac{T_{er}}{I_{ar}} = \frac{T_s + T_f}{I_{ar}}$$

where the rated electromagnetic torque generated in the machine, T_{er}, is the sum of the rated shaft torque T_s and friction torque T_f. The rated shaft or output torque is obtained from the output power and rated speed as follows:

$$\text{Rated speed, } \omega_{mr} = \frac{2\pi * 600}{60} = 62.83 \text{ rad/sec}$$

$$\text{Rated shaft torque, } T_s = \frac{P_m}{\omega_{mr}} = \frac{1500 * 10^3}{62.83} = 23{,}873 \text{ N·m}$$

Friction torque, $T_f = B_1 \omega_{mr} = 15 * 62.83 = 942.45$ N·m
The electromagnetic torque, $T_{er} = T_s + T_f = 23{,}873 + 942.45 = 24{,}815.45$

Therefore, the torque constant is

$$K_t = \frac{T_{er}}{I_{ar}} = \frac{24{,}815.45}{2650} = 9.364 \text{ N·m/A}$$

$K_b = 9.364$ V/(rad/sec)

Hence the input armature voltage is computed as

$$V_a = 0.003645 * 2650 + 9.364 * 62.83 + 2 = 600 \text{ V}$$

Armature and field power inputs $= V_a I_{ar}$ + Field power input $= 600 * 2650 + 50{,}000 = 1640$ kW

Output power, $P_m = 1500$ kW

$$\text{Efficiency, } \eta = \frac{P_m}{P_i} = \frac{1500 * 10^3}{1640 * 10^3} * 100 = 91.46\%$$

Example 2.3

A separately-excited dc motor with the following parameters: $R_a = 0.5 \, \Omega$, $L_a = 0.003$H, and $K_b = 0.8$ V/rad/sec, is driving a load of $J = 0.0167$ kg-m², $B_1 = 0.01$ N·m/rad/sec with a load torque of 100 N·m. Its armature is connected to a dc supply voltage of 220 V and is given the rated field current. Find the speed of the motor.

Solution The electromagnetic torque balance is given by

$$T_e = T_l + B_1 \omega_m + J \frac{d\omega_m}{dt}$$

In steady state, $\dfrac{d\omega_m}{dt} = 0$

$$T_e = T_l + B_1 \omega_m = 100 + 0.01 \omega_m$$

$$T_e = K_b i_a = 100 + 0.01 \omega_m$$

$$i_a = \frac{(100 + 0.01 \omega_m)}{K_b} = (125 + 0.0125 \omega_m)$$

$$e = V - R_a i_a = 220 - 0.5 \times (125 + 0.0125 \omega_m) = 157.5 - 0.00625 \omega_m = K_b \omega_m$$

Rearranging in terms of ω_m,

$$\omega_m (0.8 + 0.00625) = 157.5$$

Hence $\omega_m = \dfrac{157.5}{0.80625} = 195.35$ rad/sec

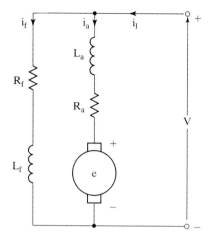

Figure 2.5 dc shunt machine

2.7.2 Shunt-Excited dc Machine

If the field winding is connected in parallel to the armature winding, then the machine goes by the name of *shunt-excited dc machine* or simply by *dc shunt machine*. The equivalent circuit of the machine is shown in Figure 2.5. Note that, in this machine, the field winding does not need a separate power supply, as it does in the case of the separately-excited dc machine. For a constant input voltage, the field current and hence the field flux are constants in this machine. While it is good for a constant-input-voltage operation, it has troubling consequences for variable-voltage operation. In variable-input voltage operation, an independent control of armature and field currents is lost, leading to a coupling of the flux and torque production channels in the machine. This is in contrast to the control simplicity of the separately-excited dc machine. For a fixed dc input voltage, the electromagnetic torque vs. speed characteristic of the dc shunt machine is shown in Figure 2.6. As torque is increased, the armature current increases, and hence the armature voltage drop also increases, while, at the same time, the induced emf is decreased. The reduction in the induced emf is reflected in a lower speed, since the field current is constant in the machine. The drop in speed from its no-load speed is relatively small, and, because of this, the dc shunt machine is considered a constant-speed machine. Such a feature makes it unsuitable for variable-speed application.

2.7.3 Series-Excited dc Machine

If the field winding is connected in series with the armature winding, then it is known as the *series-excited dc machine* or *dc series machine,* and its equivalent circuit is shown in Figure 2.7. It has the same disadvantage as the shunt machine, in that there is no independence between the control of the field and the armature

28 Chapter 2 Modeling of DC Machines

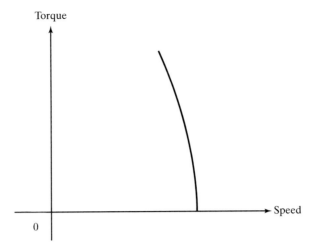

Figure 2.6 Torque–speed characteristics of a dc shunt motor

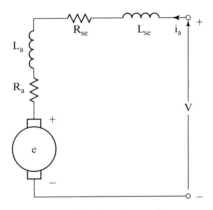

Figure 2.7 dc series machine

currents. The electromagnetic torque of the machine is proportional to the square of the armature current, because the field current is equal to the armature current. At low speeds, a high armature current is feasible, with a large difference between a fixed applied voltage and a small induced emf. This results in high torque at starting and low speeds, making it an ideal choice for applications requiring high starting torques, such as in propulsion. The torque-vs.-speed characteristic for the dc series machine for a fixed dc input voltage is shown in Figure 2.8. With the dependence of the torque on the square of the armature current and the fact that the armature current availability goes down with increasing speed, torque-vs.-speed characteristic resembles a hyperbola. Note that, at zero speed and low speeds, the torque is large but somewhat curtailed from the square current law because of the saturation of the flux path with high currents.

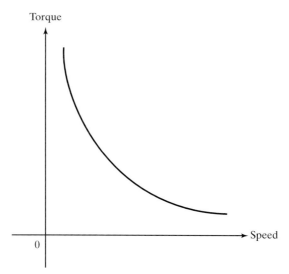

Figure 2.8 Torque–speed characteristic of a dc series motor

Example 2.4

A series-excited dc machine designed for a variable-speed application has the following name-plate details and parameters:

$$3 \text{ hp}, 230 \text{ V}, 2000 \text{ rpm}$$

$R_a = 1.5 \, \Omega, R_{se} = 0.7 \, \Omega, L_a = 0.12 \text{ H}, L_{se} = 0.03 \text{ H}, M = 0.0675 \text{ H}, B_1 = 0.0025 \text{ N·m/(rad/sec)}$

Calculate (i) the input voltage required in steady state to deliver rated torque at rated speed and (ii) the efficiency at this operating point. Assume that a variable voltage source is available for this machine.

Solution (i) The name-plate details give the rated speed and rated power output of the machine, from which the rated torque is evaluated as follows:

$$\text{Rated speed}, \omega_{mr} = \frac{2\pi N_r}{60} = \frac{2\pi * 2000}{60} = 209.52 \text{ rad/sec}$$

$$\text{Rated output torque}, T_s = \frac{P_m}{\omega_{mr}} = \frac{3 * 745.6}{209.52} = 10.675 \text{ N·m}$$

$$\text{Friction torque of the machine}, T_f = B_1 \omega_{mr} = 0.0025 * 209.52 = 0.52 \text{ N·m}$$

$$\text{Air gap torque}, T_e = T_s + T_f = 10.675 + 0.52 = 11.195 \text{ N·m}$$

The voltage equation of the dc series machine from the equivalent circuit is derived as

$$v = R_a i_a + R_{se} i_f + M i_f \omega_m + L_a \frac{di_a}{dt} + L_{se} \frac{di_f}{dt}$$

where the armature and field current are equal to one another in the series dc machine ($I_f = I_a$) and, in steady state, the derivatives of the currents are zero, resulting in the following expression:

$$V = (R_a + R_{se} + M\omega_m)I_a$$

and air gap torque is given by

$$T_e = Mi_f i_a = Mi_a^2 = MI_a^2 \quad (\text{N·m})$$

The air gap torque is computed as 11.195 N·m, and the steady-state armature current is found from the expression above as

$$I_a = \sqrt{\frac{T_e}{M}} = \sqrt{\frac{11.195}{0.0675}} = 12.88 \text{ A}$$

which, upon substitution in the steady-state input voltage equation at the rated speed, gives

$$V = (1.5 + 0.7 + 0.0675 * 209.52)\, 12.88 = 210.46 \text{V}$$

(ii) The input power is $P_i = VI_a = 210.46 * 12.88 = 2{,}710.45$ W
The output power is $P_m = 3 * 745.6 = 2236.8$ W
Efficiency is $\eta = \dfrac{P_m}{P_i} = \dfrac{2236.8}{2710.45} * 100 = 82.5\%$

2.7.4 DC Compound Machine

Combining the best features of the series and shunt dc machines by having both a series and shunt field in a machine leads to the *dc compound machine* configuration shown in Figure 2.9. The manner in which the shunt-field winding is connected in relation to the armature and series field provides two kinds of compound dc machine. If the shunt field encompasses the series field and armature windings, then that configuration is known as a *long-shunt compound dc machine*. The *short-shunt compound dc machine* has the shunt field in parallel to the armature winding. In the latter configuration, the shunt-field excitation is a slave to the induced emf and hence the rotor speed, provided that the voltage drop across the armature resistance is negligible compared to the induced emf. Whether the field fluxes of the shunt and series field are opposing or strengthening each other gives two other configurations, known as *differentially* and *cumulatively* compounded dc machines, respectively, for each of the long and short shunt connections.

2.7.5 Permanent-Magnet dc Machine

Instead of an electromagnet with an external dc supply, the excitation can be provided by permanent magnets such as ceramic, alnico, and rare earth varieties: samarium–cobalt and boron–iron–neodymium magnets. The advantage of such an excitation consists in the compactness of the field structure and elimination of resistive losses in the field winding. These features contribute to a compact and cool machine, desirable features for a high-performance motor. The cross section of a permanent-magnet dc machine is shown in Figure 2.10. Note that the armature is similar to other dc motors in construction and performance.

To analyze a dc motor, the constants R_a, L_a, and K_b and the resistance and inductance of field windings are required. Some of them are given in the manufacturer's data sheet. In case of nonavailability of the data, it is helpful to have knowledge of procedures to measure these constants. The next section deals with the measurement of motor parameters.

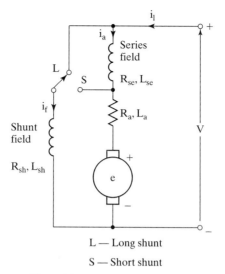

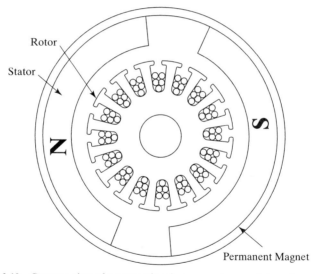

Figure 2.9 DC compound machine

Figure 2.10 Cutaway view of a conventional permanent-magnet dc motor assembly

2.8 MEASUREMENT OF MOTOR CONSTANTS

The following test methods apply to a separately excited dc motor; this type is the most widely used motor for variable-speed applications.

2.8.1 Armature Resistance

The dc value of the armature resistance is measured between the armature terminals by applying a dc voltage to circulate rated armature current. Care should be

32   Chapter 2   Modeling of DC Machines

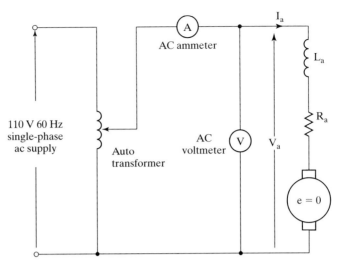

Figure 2.11   Measurement of armature inductance

taken to subtract the brush and contact resistance from the measurement and to correct for the temperature at which the motor is expected to operate at steady state.

2.8.2 Armature Inductance

By applying a low ac voltage through a variac to the armature terminals, the current is measured. The motor has to be at a standstill, keeping the induced emf at zero. Preferably, the residual voltage in the machine is wiped out by repetitive application of positive and negative dc voltage to the armature terminals. The test schematic is shown in Figure 2.11. The inductance is

$$L_a = \frac{\sqrt{\frac{V_a^2}{I_a} - R_a^2}}{2\pi f_s} \qquad (2.28)$$

where f_s is the frequency in Hz and the armature resistance has to be the ac resistance of the armature winding. Note that this is different from the dc resistance, because skin effect produced by the alternating current.

2.8.3 EMF Constant

Specified field voltage (called rated voltage) is applied and kept constant, and the shaft is rotated by a prime mover (another dc motor) up to the speed given in the name plate (called rated speed or base speed). The armature is open-circuited, with

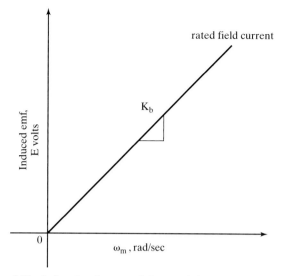

Figure 2.12 Induced emf-vs.-speed characteristic at rated field current

a voltmeter connected across the terminals. The voltmeter reads the induced emf, and its readings are noted for various speeds and are plotted as shown in Figure 2.12. The slope of this curve at a specific speed gives the emf constant in volt-sec/rad as seen from equation (2.4). The relationship shown in Figure 2.12 is known as the open-circuit characteristic of the dc machine.

Similar procedures for the field-circuit parameters, given in sections 2.8.1 and 2.8.2, are used to evaluate the resistance and inductance of the field circuit. Note that, for a permanent-magnet machine, this procedure is not applicable.

2.9 FLOW CHART FOR COMPUTATION

The basic steps involved in the computation of dc motor response and other quantities of interest are as follows:

(i) Reading of machine parameters
(ii) Specification of computational needs, such as time response, frequency response, eigenvalues for stability
(iii) Use of transfer functions to evaluate the responses
(iv) Printing/displaying of the output results

An illustrative flow chart containing these steps for a separately-excited dc motor is shown in Figure 2.13.

34 Chapter 2 Modeling of DC Machines

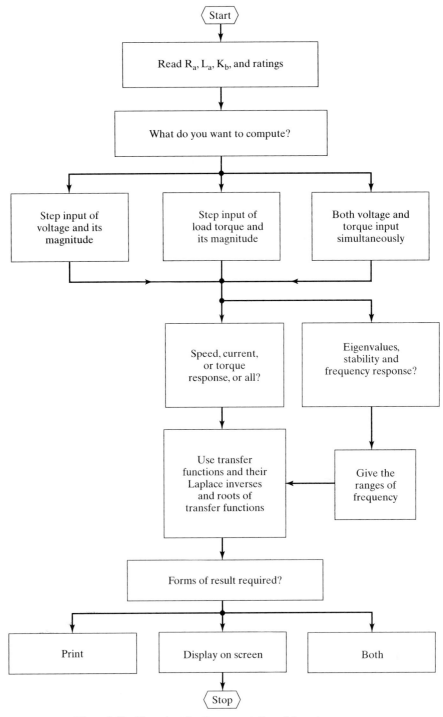

Figure 2.13 Flow chart for the computation of dc motor response

2.10 SUGGESTED READINGS

1. Fitzgerald, Kingsley, and Kusko, *Electric Machinery*, McGraw-Hill, 1971.
2. B.C. Kuo, *Automatic Control Systems*, Prentice-Hall, New York, 1982.

2.11 DISCUSSION QUESTIONS

1. A separately-excited dc motor has a nonlinear characteristic between the field current and field flux due to the saturation of the iron core in the stator and rotor. How will this saturation affect the derived model of the dc motor?
2. How is one to account for saturation in dc-machine modeling?
3. The armature resistance of the dc motor is sensitive to temperature variations. Will this adversely affect the stability of the dc motor?

2.12 EXERCISE PROBLEMS

1. Determine the torque vs. speed and torque vs. current characteristics for a separately-excited dc motor with the following parameters:

    2.3 hp, 220 V, 6000 rpm, $R_a = 1.39 \Omega$, $L_a = 0.00182$ H, $K_b = 0.331$ volt/rad/sec.

    The machine has rated field excitation and its armature is fed a constant voltage of 220 V dc.

2. Derive the dynamic equations of a dc series motor and find its characteristic equation. (Hint: To find the characteristic equation, perturb the dynamic equations to get a set of small-signal dynamic equations, and then find the characteristic equation, or directly calculate the eigenvalues of the A matrix.)

3. The machine given in 2.12.1 has $J = 0.002$ kg-m^2, $B = 0.005$ N·m/rad/sec. Determine the time taken to accelerate the motor from standstill to 6000 rpm when started directly from a 220-V dc supply. The field is maintained at its rated value.

4. The field current of a separately-excited motor is variable from zero to rated value. Derive the dynamic equations of the dc motor, its block diagram, and the small-signal armature-current response for a simultaneous voltage input to the field and torque disturbance, keeping the voltage applied to the armature constant. The motor is running at 1000 rpm, delivering 10 N·m torque, with half the rated flux. The machine details are given below:

    400 V dc, 22.75 hp, 3600 rpm, $R_a = 0.34 \Omega$, $J = 0.035$ kg-m^2, $L_a = 1.13$ mH, $B_1 = 0$ N·m/rad/sec, $K_b = 1.061$ volt/rad/sec, $M = 0.2122$ H, $R_f = 20 \Omega$, $L_f = 17.7$ H, $\Delta V_f = 5$ V, $\Delta T_1 = 1$ N·m.

5. Using a computer program, find the transfer function between rotor speed and load torque and plot its frequency response. The separately-excited dc motor's details are as follows:

    $R_a = 0.027 \Omega$, $L_a = 0.9$ mH, $K_b = 0.877$ volt/rad/sec, $J = 0.29$ kg-m^2, $B = 0.1$ N·m/rad/sec.

6. Determine the stability of a dc series motor with the following parameters:

    $R_a = 1.5 \Omega$, $R_{se} = 0.7 \Omega$, $L_a = 0.12$ H, $L_{se} = 0.03$ H, $M = 0.0675$ H, $J = 0.02365$ kg-m^2, $B_1 = 0.0025$ N·m/rad/sec, and the operating point is given by $V_a = 200$ V, $\omega_m = 209.52$ rad/sec, and $T_1 = 10$ N·m.

CHAPTER 3

Phase-Controlled DC Motor Drives

3.1 INTRODUCTION

The principle of speed control for dc motors is developed from the basic emf equation of the motor. Torque, flux, current, induced emf, and speed are normalized to present the motor characteristics. Two types of control are available: armature control and field control. These methods are combined to yield a wide range of speed control. The torque–speed characteristics of the motor are discussed for both directions of rotation and delivering both motoring and regenerating torques in any direction of rotation. Such an operation, known as four-quadrant operation, has a unique set of requirements on the input voltage and current to the dc motor armature and field. These requirements are identified for specifying the power stage.

Modern power converters constitute the power stage for variable-speed dc drives. These power converters are chosen for a particular application depending on a number of factors such as cost, input power source, harmonics, power factor, noise, and speed of response. Controlled-bridge rectifiers fed from single-phase and three-phase ac supply are considered in this chapter. Chapter 4 deals with another converter, fed from a dc source, for dc motor control.

The theory, operation and control of the three-phase controlled-bridge rectifier is considered in detail, because of its widespread use. A model for the power converter is derived for use in simulation and controller design. Two- and four-quadrant dc motor drives and their control are developed. The design of the current and speed controllers is studied with an illustrative example. The interaction of the converter and motor is also discussed, and an illustrative example of their interaction with power system is presented. An industrial application of the motor drive is described.

3.2 PRINCIPLES OF DC MOTOR SPEED CONTROL

3.2.1 Fundamental Relationship

The dependence of induced voltage on the field flux and speed has been derived in Chapter 2 and is given as

$$e = K\phi_f \omega_m \quad (3.1)$$

The various symbols in equation (3.1) have been explained earlier. The field flux is proportional to the field current if the iron is not saturated and is represented as

$$\phi_f \propto i_f \quad (3.2)$$

By substituting (3.2) into (3.1) the speed is expressed as

$$\omega_m \propto \frac{e}{\phi_f} \propto \frac{e}{i_f} \propto \frac{(v - i_a R_a)}{i_f} \quad (3.3)$$

where v and i_a are the applied voltage and armature current, respectively.

From equation (3.3), it is seen that the rotor speed is dependent on the applied voltage and field current. Since the resistive armature voltage drop is very small compared to the rated applied voltage, the armature current has only a secondary effect. To make its effect dominant, an external resistor in series with armature can be connected. In that case, the speed can be controlled by varying stepwise the value of the external resistor as a function of operational speed. Power dissipation in the external resistor leads to lower efficiency; therefore, it is not considered in this text. Only two other forms of control, using armature voltage and field current, are considered in this chapter.

3.2.2 Field Control

In *field control*, the applied armature voltage v is maintained constant. Then the speed is represented by equation (3.3) as

$$\omega_m \propto \frac{1}{i_f} \quad (3.4)$$

The rotor speed is inversely proportional to the field current; by varying the field current, the rotor speed is changed. Reversing the field current changes the rotational direction. By weakening the field flux, the speed can be increased. The upper speed is limited by the commutator and brushes and the time required to turn off the armature current from a commutator segment. It is not possible to strengthen the field flux beyond its rated (nominal) value, on account of saturation of the steel laminations. Hence, field control for speed variation is not suitable below the rated (nominal) speed. At rated speed, the field current by design is at rated value, and, hence, the flux density is chosen to be near the knee of the magnetization curve of the steel laminations.

3.2.3 Armature Control

In this mode, the field current is maintained constant. Then the speed is derived from equation (3.3) as

$$\omega_m \propto (v - i_a R_a) \quad (3.5)$$

Hence, varying the applied voltage changes speed. Reversing the applied voltage changes the direction of rotation of the motor.

Armature control has the advantage of controlling the armature current swiftly, by adjusting the applied voltage. The response is determined by the armature time constant, which has a very low value. In contrast, the field time constant is at least 10 to 100 times greater than the armature time constant. The large time constant of the field causes the response of a field-controlled dc motor drive to be slow and sluggish.

Armature control is limited in speed by the limited magnitude of the available dc supply voltage and armature winding insulation. If the supply dc voltage is varied from zero to its nominal value, then the speed can be controlled from zero to nominal or rated value. Therefore, armature control is ideal for speeds lower than rated speed; field control is suitable above for speeds greater than the rated speed.

3.2.4 Armature and Field Control

By combining armature and field control for speeds below and above the rated speed, respectively, a wide range of speed control is possible. For speeds lower than that of the rated speed, applied armature voltage is varied while the field current is kept at its rated value; to obtain speeds above the rated speed, field current is decreased while keeping the applied armature voltage constant. The induced emf, power, electromagnetic torque, and field–current-vs.-speed characteristics are shown in Figure 3.1. The armature current is assumed to be equal to the rated value for the present. The power and torque curves need some elaboration. It can be deduced that

$$T_e = K\phi_f i_a \quad (3.6)$$

Equation (3.6) can be normalized if it is divided by rated torque, which is expressed as

$$T_{er} = K\phi_{fr} i_{ar} \quad (3.7)$$

where the additional subscript r denotes the rated or nominal values of the corresponding variables. Hence, the normalized version of equation (3.6) is

$$T_{en} = \frac{T_e}{T_{er}} = K\frac{\phi_f i_a}{K\phi_{fr} i_{ar}} = \left(\frac{\phi_f}{\phi_{fr}}\right)\left(\frac{i_a}{i_{ar}}\right) = \phi_{fn} i_{an}, \text{ p.u.} \quad (3.8)$$

where the additional subscript n expresses the variables in normalized terms, commonly known as per unit (p.u.) variables.

Section 3.2 Principles of DC Motor Speed Control

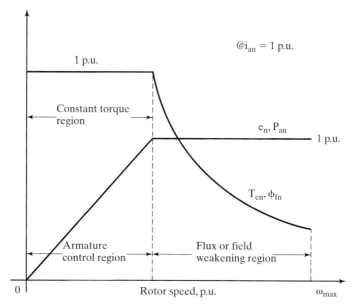

Figure 3.1 Normalized characteristics of a variable-speed dc motor

Normalization eliminates machine constants, compacts the performance equations, and enables the visualization of performance characteristics regardless of machine size on the same scale. Designers and seasoned analysts prefer the p.u. representation because of all these features. The normalized torque, flux, and armature current are

$$T_{en} = \frac{T_e}{T_{er}}, \text{p.u.} \quad (3.9)$$

$$\phi_{fn} = \frac{\phi_f}{\phi_{fr}}, \text{p.u.} \quad (3.10)$$

$$i_{an} = \frac{i_a}{i_{ar}}, \text{p.u.} \quad (3.11)$$

As the armature current is maintained at 1 p.u. in Figure 3.1, the normalized torque becomes

$$T_{en} = \phi_{fn}, \text{p.u.} \quad (3.12)$$

Hence, the normalized electromagnetic torque characteristic coincides with the normalized field flux, as shown in Figure 3.1.

Similarly, the air gap power is,

$$P_{an} = e_n i_{an}, \text{p.u.} \quad (3.13)$$

where e_n is the normalized induced emf.

As i_{an} is set to 1 p.u., the normalized air gap power becomes

$$P_{an} = e_n, \text{p.u.} \quad (3.14)$$

The normalized power output characteristic is similar to the induced emf of the dc motor in the field-weakening and constant-torque regions. The normalized induced emf is the product of the normalized flux and speed. Flux is at 1 p.u. in the armature control region, so the normalized induced emf is equal to the normalized speed. The flux is hyperbolic in the field-weakening region and has an inverse relationship with the speed. Field weakening needs to be discussed in detail here. At rated speed, the motor is delivering rated power with e_n and i_{an} at their rated values. Beyond the rated speed, the field current is decreased to reduce the field flux. This will affect the magnitude of the induced emf and hence the power output. It is very important that the steady-state power output of the machine be kept from exceeding its rated design value, which is 1 p.u. The implication of the air gap power constraint is that the induced emf and field flux are to be coordinated to achieve this objective. The coordination yields the value of field flux as

$$P_{an} = 1 \text{ p.u.} = e_n i_{an} = \phi_{fn} \omega_{mn} i_{an} \tag{3.15}$$

If i_{an} is equal to 1 p.u., then

$$\phi_{fn} \omega_{mn} = 1$$

$$\phi_{fn} = \frac{1}{\omega_{mn}} \tag{3.16}$$

Hence, the normalized induced emf is given as,

$$e_n = \phi_{fn} \omega_{mn} = \frac{1}{\omega_{mn}} \times \omega_{mn} = 1 \text{ p.u.} \tag{3.17}$$

The power output and induced emf are maintained at their rated values in the field-weakening region by programming the field flux to be inversely proportional to the rotor speed. They are shown in Figure 3.1.

Example 3.1

A separately-excited dc motor has the following ratings and constants:
2.625 hp., 120V, 1313 rpm, $R_a = 0.8 \, \Omega$, $R_f = 100 \, \Omega$, $K_b = 0.764$ V.s / rad, $L_a = 0.003$ H, $L_f = 2.2$ H

The dc supply voltage is variable from 0 to 120 V both to the field and armature, independently. Draw the torque–speed characteristics of the dc motor if the armature and field currents are not allowed to exceed their rated values. The rated flux is obtained when the field voltage is 120 V. Assume that the field voltage can be safely taken to a minimum of 12 V only.

Solution (i) Calculation of rated values:

$$\text{Rated speed, } \omega_{mr} = \frac{2\pi N}{60} = \frac{2\pi \times 1313}{60} = 137.56 \text{ rad/sec}$$

$$\text{Rated torque, } T_{er} = \frac{\text{Output power}}{\text{Rated speed}} = \frac{2.625 \times 745.6}{137.56} = 14.23 \text{ N·m}$$

$$\text{Rated armature current, } I_{ar} = \frac{\text{Rated torque}}{K_b} = \frac{14.23}{0.764} = 18.63 \text{ A}$$

Rated field current, $I_{fr} = \dfrac{V_{fr}}{R_f} = \dfrac{120}{100} = 1.2$ A

(ii) Calculation of torque–speed characteristics:

Case (a) Constant-flux/torque region:

$$e_l = V_{max} - I_{ar}R_a = 120 - 18.63 \times 0.8 = 105.1$$

$$\omega_{ml} = \dfrac{e_{ml}}{K_b} = \dfrac{105.1}{0.764} = 137.56 \text{ rad/sec.}$$

$$\omega_{mln} = \dfrac{\omega_{ml}}{\omega_{mr}} = \dfrac{137.56}{137.56} = 1.0 \text{ p.u.}$$

Hence, constant rated torque is available from 0 to 1.0 p.u. speed.

Case (b) Field-weakening region:

For 1 p.u. armature current, the maximum induced emf is

$$e_n = \dfrac{e_l}{e_r} = \dfrac{105.1}{105.1} = 1.0 \text{ p.u.}$$

To maintain this induced emf in the field-weakening region,

$$\phi_{fn} = \dfrac{e_n}{\omega_{mn}} = \dfrac{1.0}{\omega_{mn}} \text{ p.u.}$$

If the range of field variation is known, the maximum speed can be computed as follows:

$$I_{f\,min} = \dfrac{V_{f(min)}}{R_f} = \dfrac{12}{100} = 0.12 \text{ A}$$

1.2 A of field current corresponds to rated field flux and hence 0.12 A corresponds to $0.1\phi_{fr}$, and hence

$$0.1 \text{p.u.} < \phi_{fn} < 1 \text{ p.u.}$$

$$\omega_{max} = \dfrac{1}{0.1} = 10 \text{ p.u.}$$

For various speeds between 1 and 10 p.u., the field flux is evaluated from the equation as

$$\phi_{fn} = \dfrac{1}{\omega_{mn}} \text{ in p.u.}$$

$$T_{en} = \phi_{fn} \text{ for } I_{an} = 1 \text{ p.u.}$$

The torque, power, and flux-vs.-speed plots are shown in Figure 3.2.

Example 3.2

Consider the dc motor given in Example 3.1, and draw the intermittent characteristics if the armature current is allowed to be 300% of rated value.

Solution (i) Constant-flux/torque region

$$I_{max} = 3I_{ar}$$

$$T_{em} = \text{maximum torque} = K_b I_{max} = 0.764 \times 3 \times 18.63 = 42.7 \text{ N·m}.$$

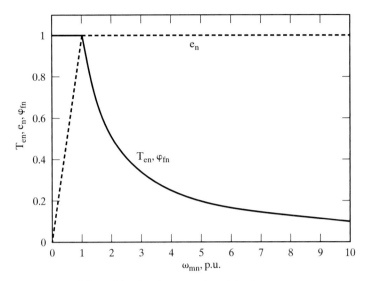

Figure 3.2 Continuous rating of the dc motor

$$T_{en} = \frac{T_{em}}{T_{er}} = \frac{42.7}{14.25} = 3 \text{ p.u.}$$

The maximum induced emf is

$$e_m = V_{max} - I_{max}R_a = 120 - (3 \times 18.63) \times 0.8 = 75.29 \text{ V}$$

Speed corresponding to this induced emf is

$$\omega_{ml} = \frac{e_m}{K_b} = \frac{75.29}{0.764} = 98.54 \text{ rad/sec}$$

$$\omega_{mln} = \frac{98.54}{137.56} = 0.716 \text{ p.u.}$$

Beyond this speed, field weakening is performed.

(ii) Field-weakening region:

$$I_{max} = 3I_{ar}$$
$$e_m = 75.29 \text{ V}$$
$$e_n = \frac{e_m}{105.1} = \frac{75.29}{105.1} = 0.716 \text{ p.u.}$$
$$\omega_{mn} = \frac{e_n}{\phi_{fn}} = \frac{0.716}{\phi_{fn}} \text{ p.u.}$$

The range of the normalized field flux is

$$0.1 < \phi_{fn} < 1$$

The maximum normalized speed is $\omega_{mn} = \dfrac{0.716}{\phi_{fn(min)}} = \dfrac{0.716}{0.1} = 7.16$ p.u.

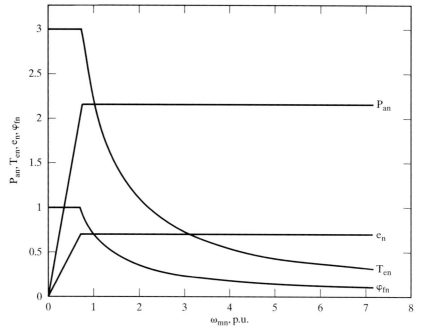

Figure 3.3 Normalized motor characteristics for 3 p.u. armature current

$T_{en} = \phi_{fn}$ per rated current = $3\phi_{fn}$ in the present case.

The intermittent characteristics are drawn from the above equations and are shown in Figure 3.3.

3.2.5 Four-Quadrant Operation

Many applications require controlled starts and stops of the dc motor, such as in robotic actuation. Consider that the machine is operating at a steady speed of ω_m, and it is desired to bring the speed to zero. There are two ways to achieve it:

1. Cut off the armature supply to the machine and let the rotor come to zero speed.
2. The machine can be made to work as a dc generator, thereby the stored kinetic energy can be effectively transferred to the source. This saves energy and brings the machine rapidly to zero speed.

Cutting off supply produces a haphazard speed response; the second method provides a controlled braking of the dc machine. To make the dc machine operating in the motoring mode go to the generating mode, all that needs to be done is to reverse the armature current flow in the dc machine. First, the armature current drawn from the source has to be brought to zero; then, a current in the opposite direction has to be built. Zeroing the current is achieved by making the source dc voltage zero or, better, by making it negative. After this, the armature current is built in the opposite direction by making the source voltage smaller than the induced

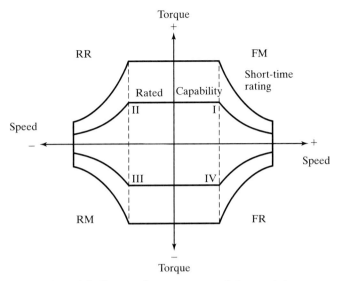

Figure 3.4 Four-quadrant torque–speed characteristics

emf. As the speed reduces, note that the induced emf decreases, necessitating a continued corresponding reduction in the source voltage to keep the armature current constant. The power flows from the machine armature to the dc source. This mode of operation is termed *regenerative braking*. The braking is accomplished by regeneration that implies that a negative torque is generated in the machine as opposed to the positive motoring torque. Hence, a mirror reflection of the speed–torque characteristics, shown in Figure 3.1, is required on the IV quadrant for regeneration. The first and fourth quadrants are for one direction of rotation, say, forward.

Some applications, such as a feed drive in machine tools, require operation in both directions of rotation. In that case, the III quadrant signifies the reverse motoring and II quadrant, the reverse regeneration mode. A motor drive capable of operating in both directions of rotation and of producing both motoring and regeneration is referred to as a four-quadrant variable-speed drive. The torque–speed characteristics of such a four-quadrant dc motor drive are shown in Figure 3.4. This contains two characteristics, one for rated operating condition and the other for short-time or intermittent operation. The short-time characteristic is used for acceleration and deceleration of the machine; it normally encompasses 50 to 100% greater than the rated torque for dc machines. The four-quadrant operation and its relationship to speed, torque, and power output are summarized in Table 3.1.

Figure 3.5 illustrates the speed and torque variation from a point P_1 to Q_1 and Q_1 to P_2 of the dc machine. On receiving the command to go from P_1 to Q_1, the torque is changed to negative by regenerating the machine, as shown by the trajectory P_1M_1. This regeneration torque, along with the load torque, produces a decelerating torque. The torque is maintained at the permitted maximum levels both in the field-weakening and the constant-flux regions. As the machine decelerates, as shown by the trajectory M_1M_2, it will reach zero speed, and keeping the torque at a negative maximum will drive the motor in the reverse direction along the trajectory M_2M_3. Once the desired speed ω_{m2} is reached, the torque is adjusted to equal the

Section 3.2 Principles of DC Motor Speed Control

TABLE 3.1 Four-quadrant dc motor drive characteristics

Function	Quadrant	Speed	Torque	Power Output
Forward Motoring (FM)	I	+	+	+
Forward Regeneration (FR)	IV	+	−	−
Reverse Motoring (RM)	III	−	−	+
Reverse Regeneration (RR)	II	−	+	−

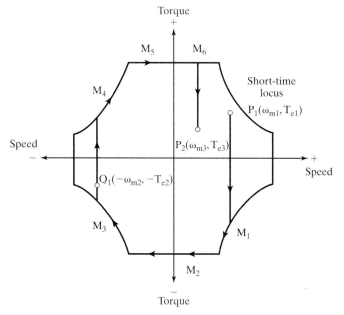

Figure 3.5 Changing the operating points and the use of four quadrants

specified value, $-T_{e2}$, along the trajectory M_3Q_1. Similarly, to change the operating point from Q_1 to P_2, the trajectory shown along $Q_1M_4M_5M_6P_2$ is followed.

From this illustration, it is seen that the use of all quadrants of operation leads to a very responsive motor drive. Contrast this to the *supply cut-off* technique. In such a case, only the load torque contributes to the deceleration, as opposed to the combined machine and load torques in a four-quadrant motor drive.

Converter requirements: The voltage and current requirements for four-quadrant operation of the dc machine are derived as follows. Assuming the field flux is constant, the speed is proportional to the induced emf and hence approximately proportional to the applied voltage to the armature. Also, the electromagnetic torque is proportional to the armature current. Then the speed axis becomes the armature voltage axis and the torque axis is equivalent to the armature current axis. From this observation, the armature voltage and armature current requirements for four-quadrant operation are given in Table 3.2. These requirements, in turn, set the specifications for the electronic converter. For a pump application, the motor needs only unidirectional operation with no regenerative braking. Hence, only first-quadrant operation

TABLE 3.2 Armature voltage and current requirements of a four-quadrant dc motor drive

Operation	Speed	Torque	Voltage	Current	Power Output
FM	+	+	+	+	+
FR	+	−	+	−	−
RM	−	−	−	−	+
RR	−	+	−	+	−

is required, thereby limiting the converter specification to only positive voltage and current variations. Therefore, the power flow is unidirectional from source to load. For a golf cart electric-vehicle propulsion-drive application, a four-quadrant operation is required, with the attendant converter capability to handle power in both directions with bipolar voltage and current requirements. Then the converter is much more complex than that required for a one-quadrant drive.

From a fixed utility ac source, a variable-voltage and variable-current dc output is obtained through two basic methods by using static power converters. The first method uses a controllable rectifier to convert the ac source voltage directly into a variable dc voltage in one single stage of power conversion, using phase-controlled converters. The second method converts the ac source voltage to a fixed dc voltage by a diode bridge rectifier and then converts the fixed dc to variable dc voltage with electronic choppers. The second method involves two-stage power conversion, which is dealt with in Chapter 4; the first method is considered in this chapter.

Thyristor converter: The realization of an ac-to-dc variable-voltage converter by means of the silicon-controlled rectifiers (SCR) known as thyristors is discussed in this subsection. The distinct features of the thyristor are given here, without going into the device physics.

Thyristors are four-element (PNPN), three-junction devices with the terminals of anode (A), cathode (K), and gate (G). A gate current is injected by applying a positive voltage between gate and cathode for turning on the device. The device turns on only if the anode is positive compared to cathode at least by 1 V. After turn-on, the device drop is around 1 V for most of the devices. After turning on, the device acts like a diode. Therefore, to turn off the thyristor, the device has to be reverse biased by making the anode negative with respect to the cathode. This is easily achieved with ac input voltage during its negative half-cycle. Turning off the device goes by the name of *device commutation*. The thyristor, unlike the diode, can hold off conduction even when its anode is positive compared to its cathode, by not triggering the gate.

These features make the thyristors ideal devices for ac-to-variable-dc conversion. Instead of diodes in the diode bridge rectifier, thyristors can be substituted, and, by delaying the conduction from their zero crossings, a part of the ac voltage is rectified for feeding to a load. The dc load volt–sec is reduced from the maximum available volt–sec in the half-cycle ac voltage waveform, which in turn varies the average dc load voltage. The line voltage commutates the devices when it reverses polarity each half cycle and applies a negative voltage across the cathode and anode. This method of commutation is known as natural or line commutation, and converters using this method are termed *line-commutated* converters.

3.3 PHASE-CONTROLLED CONVERTERS

3.3.1 Single-Phase Controlled Converter

A single-phase controlled-bridge converter is shown in Figure 3.6 with its input and output voltage and current waveforms. The load consists of a resistance and an inductance, and the current is assumed to be continuous and constant. The difference between the diode bridge and this thyristor bridge is that conduction can be delayed in the latter beyond positive zero crossing. The delay is introduced in the form of triggering signals to the gates of the thyristors. The delay angle is measured

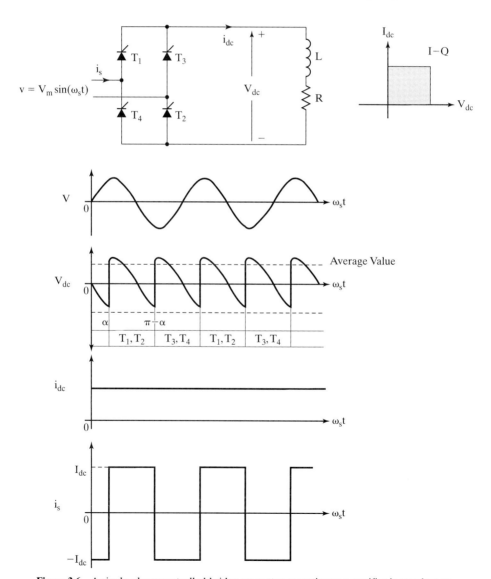

Figure 3.6 A single-phase controlled-bridge converter operating as a rectifier in steady state

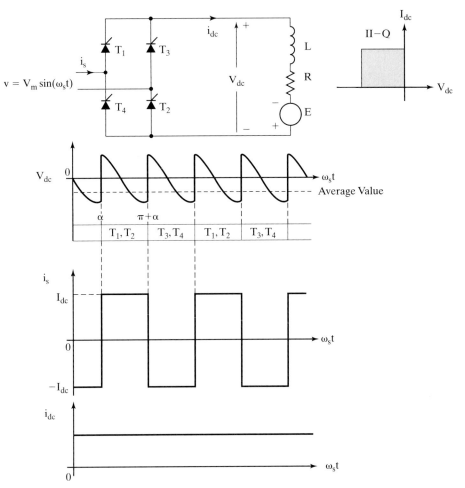

Figure 3.7 Inversion in the steady state of controlled-converter operation

from the zero crossing of the voltage waveform and is generally termed α (in radians) throughout this text. Although the load voltage has both positive and negative volt–seconds, its average is a net positive and is indicated by the horizontal dashed lines denoted as V_{dc}. Assuming continuous load current, this voltage is quantified as

$$V_{dc} = \frac{1}{\pi}\int_\alpha^{\alpha+\pi} V_m \sin(\omega_s t) d(\omega_s t) = \frac{2V_m}{\pi}\cos\alpha \qquad (3.18)$$

where V_m is the crest of the input ac voltage. Increasing the delay angle to greater than 90° produces a negative voltage on average, as is shown in Figure 3.7. Note that the current is in the same direction, thus contributing to a negative power input. Such a feature is known as inversion. It is tacitly assumed here that there is an emf source in the load contributing to the power transfer from the load to source. Such a load is known as an active load.

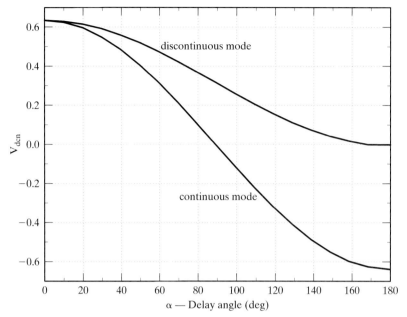

Figure 3.8 Transfer characteristics of a single-phase controlled-bridge rectifier

The load current can be discontinuous; in that case, the average output voltage is derived as

$$V_{dc} = \frac{1}{\pi}\int_{\alpha}^{\alpha+\gamma} V_m \sin(\omega_s t)d(\omega_s t) = \frac{V_m}{\pi}[\cos(\alpha) - \cos(\alpha + \gamma)], \text{ V} \quad (3.19)$$

where γ is the current conduction angle.

Comparing equations (3.18) and (3.19), it is seen that, for certain values of γ, the output voltage for discontinuous conduction can be greater than that for continuous conduction. For example, let

$$\alpha + \gamma = \pi$$
$$\alpha = 30°$$

Hence, the average output voltage for discontinuous current conduction is

$$V_{dc}(\text{dis}) = \frac{V_m}{\pi}[1.866] \quad (3.20)$$

and, for the same triggering angle, the average output voltage for continuous-current conduction is given by

$$V_{dc}(\text{cont}) = \frac{2V_m}{\pi} \times \frac{1.732}{2} = \frac{1.732 V_m}{\pi} \quad (3.21)$$

The transfer characteristics for continuous conduction for an active load and discontinuous conduction for a resistive load are shown in Figure 3.8.

The source inductance delays the current transfer from one pair of conducting thyristors to another set. During this time, the source is short-circuited through the

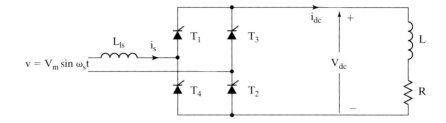

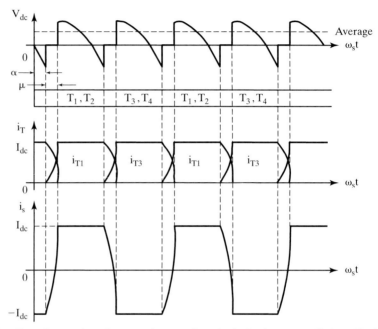

Figure 3.9 The effect of source impedance on the operation of a single-phase controlled rectifier in steady state

source impedance, invariably reducing the load voltage to zero. Hence, the overall effect of source inductance is to reduce the available dc output voltage. Figure 3.9 contains the operational waveforms with source inductance. The source inductance can be introduced by the isolation transformer or by intentionally placed reactors, to reduce the rate of rise of currents in the thyristors. If the source inductance is L_{ls}, the voltage lost due to it is

$$V_x = \frac{1}{\pi}\int_{\alpha}^{\alpha+\mu} V_m \sin(\omega_s t) d(\omega_s t) = \frac{V_m}{\pi}[\cos\alpha - \cos(\alpha + \mu)] \quad (3.22)$$

where μ is the overlap conduction period.

By equating this voltage to the voltage drop in the source reactance, the overlap angle μ is obtained as

$$\mu = \cos^{-1}\left[\cos\alpha - \frac{\pi\omega_s L_{ls} I_{dc}}{V_m}\right] - \alpha \quad (3.23)$$

where I_{dc} is the load current in steady state.

These characteristics are modified when the load includes a counter-emf, as in the case of a dc machine. There is an additional feature in discontinuous operation, with the induced emf appearing across the output of the converter during zero-current intervals, but, if the source emf is instantaneously greater than the back emf, then the conduction starts but will not end immediately when the source emf becomes less than the induced emf, because of the energy stored in the machine inductance and in the external inductance connected in series to the armature of the machine. Therefore, conduction will be prolonged until the energy in these inductances is depleted.

The three-phase controlled full-bridge converter is similar in operation to the single-phase controlled rectifier. Three-phase converters are widely used for both dc and ac motor control. The emphasis is placed in this text on the three-phase converter-controlled dc motor drive. The following section contains the principle of operation, design features of the control circuit, and the characteristics of this converter.

3.3.2 Three-Phase Controlled Converter

A three-phase thyristor-controlled converter is shown in Figure 3.10, and its voltage and current waveforms in the rectifier mode of operation are shown in Figure 3.11. The current is assumed to be continuous for the present. At a given instant, two thyristors are conducting. Assuming that the voltage between phases a and b is maximum, then the thyristors T_1 and T_6 are conducting. The next line voltage to get more positive than ab is ac. At that time, the triggering signal for T_6 will be disabled and that of T_2 will be enabled. Note that the anode of T_2 is more negative than the cathode of T_6, because line voltage ac is greater than the line voltage ab. That will turn off T_6 and transfer the current from T_6 to T_2. The delay in current transfer from T_6 to T_2 is dependent on the source inductance. During this current transfer, T_1, T_6, and T_2 are all conducting, and the load voltage is the average of the line voltages ab and ac. This phenomenon is the commutation overlap, which results in a reduction in the load voltage. The load current will remain the same during commutation of T_6. The current in T_6 declines by the same proportion as current in T_2 rises. It is to be

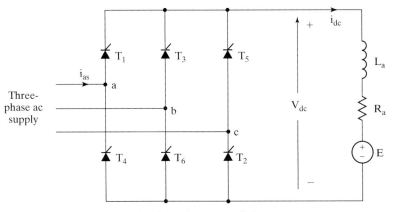

Figure 3.10 Three-phase controlled converter

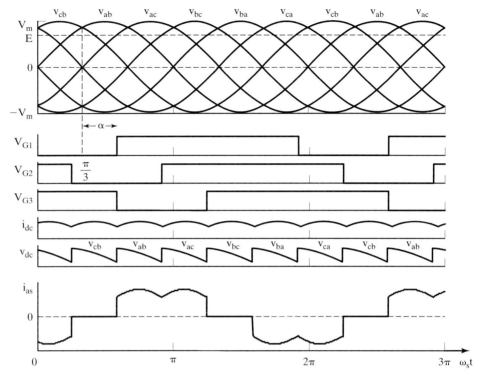

Figure 3.11 Rectification in the three-phase converter in steady state (first-quadrant operation)

observed that the current transfer is effected by the source voltages: voltage *ac* becoming greater than the voltage *ab*, resulting in the reverse biasing of T_6 and forward biasing of T_2. Similarly, it could be seen that the firing/gating sequence is $T_1T_2T_3T_4T_5T_6$ and so on. Also, each of these gating signals is spaced by sixty electrical degrees. The thyristors require small reactors in series to limit the rate of rise of currents and snubbers, which are resistors in series with capacitors across the devices, to limit the rate of rise of voltages when the devices are commutated.

A typical inversion mode of operation is shown in Figure 3.12. Note that this corresponds to a second-quadrant operation of the dc motor drive. The transfer characteristic of the three-phase controlled rectifier is derived as

$$V_{dc} = \frac{1}{\pi/3} \int_{\frac{\pi}{3}+\alpha}^{\frac{2\pi}{3}+\alpha} V_{ab} d(\omega_s t) = \frac{3}{\pi} \int_{\frac{\pi}{3}+\alpha}^{\frac{2\pi}{3}+\alpha} V_m \sin(\omega_s t) d(\omega_s t) = \frac{3}{\pi} V_m \cos \alpha \qquad (3.24)$$

The transfer characteristics are very similar to those of the single-phase converter in both the continuous and the discontinuous mode of conduction. The transfer characteristic for the continuous mode of conduction is shown in Figure 3.13. The characteristic is nonlinear. Hence, the use of this converter as a component in a feedback-control system will cause an oscillatory response. This can be explained as follows.

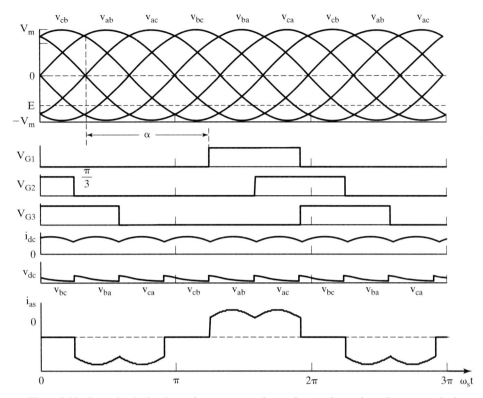

Figure 3.12 Inversion in the three-phase converter in steady state (second-quadrant operation)

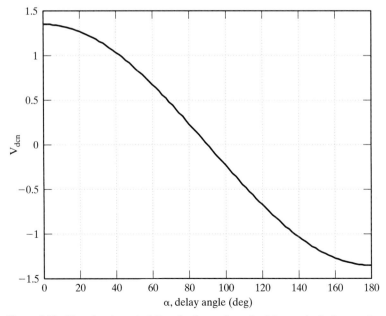

Figure 3.13 Transfer characteristics of a three-phase thyristor-controlled converter

The delay angle will be made a function of speed, current, or position error in a motor-drive system. The error variable is expected to increase or decrease the dc output voltage proportionally. The gain of the converter to a delay angle is not a constant, so it will either overreach or fall short of the required output voltage. This necessitates one more correction in the error signal, causing both time delay and oscillatory response. Such oscillatory responses have been known to create ripple instability in converters.

A control technique to overcome this nonlinear characteristic and the accompanying undesirable dynamic behavior is given in the following. The control input to determine the delay angle is modified to be

$$\alpha = \cos^{-1}\left(\frac{v_c}{V_{cm}}\right) = \cos^{-1}(v_{cn}) \quad (3.25)$$

where v_c is the control input and V_{cm} is the maximum of the absolute value of the control voltage.

Then the dc output voltage is

$$V_{dc} = \frac{3}{\pi}V_m \cos\alpha = \frac{3}{\pi}V_m \cos(\cos^{-1}v_{cn}) = \left[\frac{3}{\pi}V_m\right]v_{cn} = \left[\frac{3}{\pi}\frac{V_m}{V_{cm}}\right]v_c = K_r v_c \quad (3.26)$$

where v_{cn} is the normalized control voltage and K_r is the gain of the converter, defined as

$$K_r = \frac{3}{\pi}\frac{V_m}{V_{cm}} = \frac{3\sqrt{2}V}{\pi V_{cm}} = 1.35\frac{V}{V_{cm}} \quad (3.27)$$

where V is the rms line-to-line voltage.

Then the modified transfer characteristic is linear with a slope of K_r. The control voltage is normalized to keep its magnitude less than or equal to 1, to be able to obtain the inverse cosine of it.

3.3.3 Control Circuit

The control circuit for the three-phase thyristor converter can be realized in many ways. A schematic of a generic implementation is shown in Figure 3.14. The synchronizing signal is obtained from the line voltage between *a* and *c*. The positive–zero crossing of this line voltage forms the starting point for the controller design. The synchronizing signal is multiplexed six times, to have equidistant pulses at 60° intervals. These are decoded to correspond to each gate drive.

The delay angle is obtained from the normalized control voltage v_{cn} through a function generator, so as to make the overall gain of the thyristor converter constant. The delay is incorporated into the synchronized control signal and amplified and fed to the gates of the thyristors. Hence, the maximum limit on the delay angle has to be externally set or commanded. It is essential that sufficient time be given for the thyristor to recover its forward blocking capability. Otherwise, a short of the load and source will occur. Therefore, the maximum limit for the delay angle is usually set in the range of from 150 to 155 degrees. Many adaptive control schemes

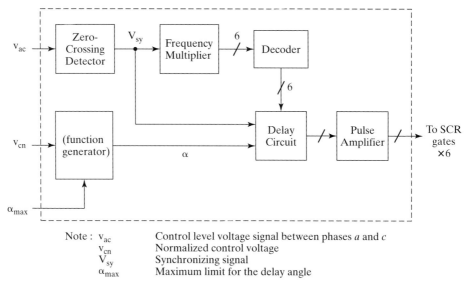

Figure 3.14 Controller schematic for the three-phase converter

adjust this maximum triggering angle as a function of load current. The feature to handle the identification of phase sequence in the ac supply has to be built in, so that the reference synchronization voltage v_{ac} matches with the control circuit.

3.3.4 Control Modeling of the Three-Phase Converter

The converter can be considered as a black box with a certain gain and phase delay for modeling and use in control studies. The gain of the linearized controller-based converter for a maximum control voltage V_{cm} is given in equation (3.27) as

$$K_r = \frac{1.35V}{V_{cm}}, V/V \qquad (3.28)$$

The converter is a sampled-data system. The sampling interval gives an indication of its time delay. Once a thyristor is switched on, its triggering angle cannot be changed. The new triggering delay can be implemented with the succeeding thyristor gating. In the meanwhile, the delay angle can be corrected and will be ready for implementation within 60°, i.e., the angle between two thryistors' gating. Statistically, the delay may be treated as one half of this interval; in time, it is equal to

$$T_r = \frac{60/2}{360} \times (\text{time period of one cycle}) = \frac{1}{12} \times \frac{1}{f_s}, s \qquad (3.29)$$

For a 60-Hz supply-voltage source, note that the time delay is equal to 1.388 ms. The converter is then modeled with its gain and time delay as

$$G_r(s) = K_r e^{-T_r s} \qquad (3.30)$$

and equation (3.30) can also be approximated as a first-order time lag and given as

$$G_r(s) = \frac{K_r}{(1 + sT_r)} \qquad (3.31)$$

For most of the drive-system applications, the model given in equation (3.31) is adequate for phase-controlled converters.

Many low-performance systems have a simple controller with no linearization of its transfer characteristic. The transfer characteristic in such a case is nonlinear. Then, the gain of the converter is obtained as a small-signal gain given by

$$K_r = \frac{\delta V_{dc}}{\delta \alpha} = \frac{\delta}{\delta \alpha}\{1.35\text{ V}\cos\alpha\} = -1.35\text{ V}\sin\alpha \qquad (3.32)$$

The gain is dependent on the operating delay angle denoted by α_0. The converter delay is modeled as an exponential function in Laplace operator s or a first-order lag, describing the transfer function of the converter as in equation (3.31).

3.3.5 Current Source

The key to the control of the machine is to control precisely the electromagnetic torque. This control is achieved in the separately-excited dc machine by controlling its armature current, but the phase-controlled converter provides only a variable voltage output. To make it a controllable current source, a closed-loop control of the dc link current, which in this case is the armature current, is resorted to. A current source can be realized with the phase-controlled converter by incorporating a current feedback loop, as shown in Figure 3.15. Consider a resistive and inductive load combination. The reference current is enforced on the load by comparing it with the actual current in the load. The error current is amplified by a proportional-plus-integral controller and its output is limited so that the control signal will be constrained to be within the maximum triggering angle, α_{max}. The control signal is processed to correspond to linearized operation by the inverse cosine function, and the signals are processed through gate power amplifiers and the converter. Assume that the current reference is a step function and the converter is at rest to start with. The current error will be maximum, which would correspond to minimum triggering angle, thus providing a large voltage across the dc link, i.e., load. This will build up the current in the load, and, when it exceeds the reference value, the current error will reduce from maximum positive to zero. This will enable the triggering angle to be close to $\pi/2$ in this present case, where the dc link voltage, v_{dc}, is not allowed to go below zero. The dc link is constrained to be positive, because the passive load cannot provide for regeneration, and only an active source such as a dc motor will provide the induced emf with the appropriate polarity so that energy can be transferred from the load to the source. The dynamic response of this system with a high proportional gain is shown in Figure 3.16 in normalized units.

The maximum control voltage is 0.7 V, and the control voltage is normalized with respect to this for realizing a linearized controller. The step command of the reference current produces a maximum control voltage, and, correspondingly, the triggering angle is driven to zero to produce the maximum voltage across the dc link. The voltage across the resistor, v_R, starts to increase and that across the induc-

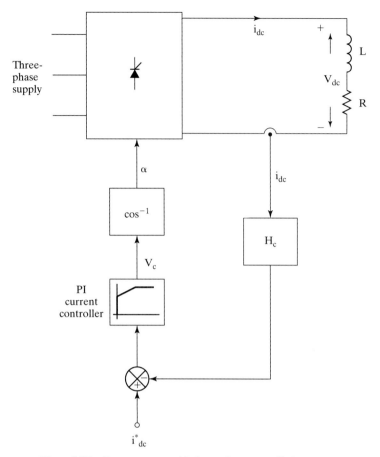

Figure 3.15 Current source with three-phase controlled converter

tance, v_L, begins to decrease. When the current exceeds the reference, the control voltage goes to zero and the triggering angle reaches 90°, to provide an average voltage of zero across the dc link. The current is maintained around the reference value, on average, with a dither, and that dithering is a function of the current controller gains and the load time constant. When the reference current goes to zero, the control voltage becomes zero, and the triggering angle goes to 90° permanently. That forces the current to decay to zero, and the entire voltage of the dc link is borne by the inductance, as shown in Figure 3.16. This type of current source is realized in the 100,000-A range at low voltages for electrolysis in metal processing plants and in a few-thousand-A range at voltages of 600 V to 4,000 V for dc and ac motor drives. The design of the current controller is treated later in this chapter.

3.3.6 Half-Controlled Converter

The converter under study hitherto is a fully controlled bridge converter. Low-power applications can make do with a half-controlled converter, shown in Figure 3.17. The lower half of the bridge has diodes in place of thyristors, thus reducing the cost of

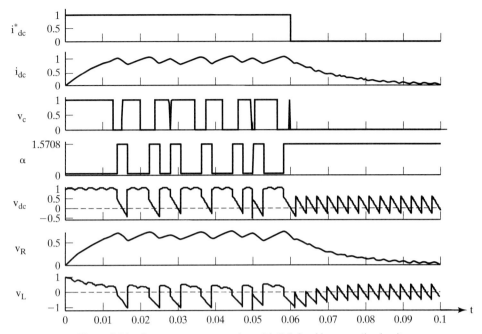

Figure 3.16 Current source operation with R-L load in normalized units

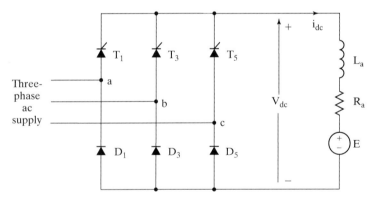

Figure 3.17 Half-controlled three-phase converter

the converter. The control circuit is simplified also. Such a configuration is also possible with a single-phase converter. The half-controlled converters are employed up to 125 hp rating in practice. Note that this converter has only first-quadrant operational capability.

3.3.7 Converters with Freewheeling

A diode is connected across the load as shown in Figure 3.18. The reversal of voltage is not possible now, and hence the converter operates only in the first quadrant, delivering a positive voltage and current. Even though the converter is limited in its capability, the current conduction interval is prolonged by the energy stored in the

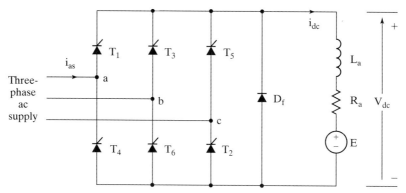

Figure 3.18 Three-phase controlled converter with freewheeling

load inductance. The current continuity has the positive effects of reducing the current ripples in the motor and hence the torque ripples. The waveforms of this freewheeling circuit are shown in Figure 3.19.

There is an alternative way to obtain freewheeling without using a diode across the load. The thyristors of a phase leg can be triggered together and hence can short the load. This, in turn, requires modification of the control circuitry. The configuration utilizing the thyristors has the advantage of minimizing the cost of the converter and making optimal use of the power devices.

3.3.8 Converter Configuration for a Four-Quadrant DC Motor Drive

So far, the converters considered possess one- or two-quadrant operational capability. Figure 3.20 shows the converter configuration for a four-quadrant dc motor drive. It consists of dual three-phase controlled-bridge converters in parallel, with their output polarities reversed. Converter 1 gives both positive and negative dc output voltages with a positive current output, catering to the first- and second-quadrant operation of the dc motor drive, respectively. This converter is hereafter referred to as the forward converter. Similarly, converter 2 delivers negative current to the motor with positive and negative dc output voltages. Such operation encompasses the fourth- and third-quadrant performance of the dc motor drive, respectively. Because of it, this converter is hereafter referred to as a reverse converter. Transitioning from forward to reverse operation in the motor and corresponding changes in the converters require special logic control circuits to avoid short-circuiting the ac supply and for safe operation. The phase-controlled converter has two quadrants of operation: quadrants I and II, or quadrants III and IV, depending on the connection of the converter relative to the machine armature. Note that quadrants I and IV or quadrants II and III are more useful in many unidirectional motion control applications than the ones provided by the converter. That will enable both the motoring and generating/braking torque production contributing to faster acceleration and deceleration in one direction of rotation. The control strategy for a four-quadrant dc motor drive is discussed later in this chapter.

60 Chapter 3 Phase-Controlled DC Motor Drives

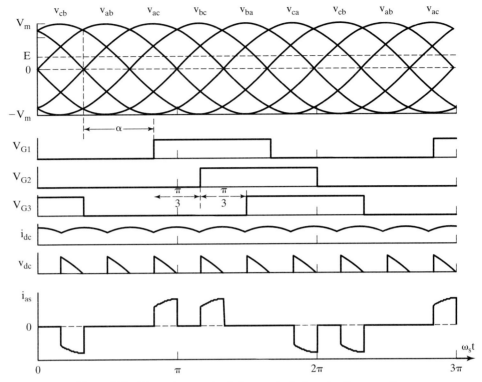

Figure 3.19 Three-phase controlled converter in freewheeling operation

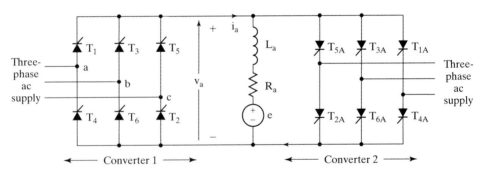

Figure 3.20 Dual three-phase thyristor converter for four-quadrant drive

3.4 STEADY-STATE ANALYSIS OF THE THREE-PHASE CONVERTER-CONTROLLED DC MOTOR DRIVE

3.4.1 Average Analysis

A separately-excited dc motor is fed from a three-phase converter and is operated in one rotational direction, say, in the first and fourth quadrants of torque–speed characteristics. The steady-state performance of this motor drive is described in this section. The steady-state performance, when combined with the load characteristics, provides the basis for evaluating the suitability of the motor drive for the given application. The steady-state performance is developed by assuming that the average values only are considered. Indirectly, it is implied that the average current produces an average torque, which, in combination with load torque, determines the average speed. In that process, the quasi-transients are neglected.

Then the armature voltage equation for the motor in steady state is

$$v_a = R_a i_a + e \tag{3.33}$$

and, in terms of average values (by denoting the variables in capital letters or with a subscript 'av'),

$$V_a = R_a I_a + K\Phi_f \omega_{mav} \tag{3.34}$$

Average electromagnetic torque is given by

$$T_{av} = K\Phi_f I_a \tag{3.35}$$

and, from a previous derivation, the average dc link voltage is

$$V_a = 1.35 V \cos \alpha \tag{3.36}$$

where V is the rms line-to-line ac voltage in a three-phase system.

Then the electromagnetic torque is expressed in terms of delay angle and speed, from equations (3.34) to (3.36), as

$$T_{av} = K\Phi_f \left\{ \frac{V_a - K\Phi_f \omega_{mav}}{R_a} \right\} = K\Phi_f \left\{ \frac{1.35 V \cos \alpha - K\Phi_f \omega_{mav}}{R_a} \right\} \tag{3.37}$$

The equation (3.37) is normalized by dividing the average torque by the rated torque:

$$T_{en} = \frac{T_{av}}{T_{er}} = \frac{T_{av}}{K\Phi_{fr} I_{ar}} = \frac{K\Phi_f \{1.35 V \cos \alpha - K\Phi_f \omega_{mav}\}}{K\Phi_{fr} I_{ar} R_a} = \frac{[1.35 V \cos \alpha - K\Phi_f \omega_{mav}]}{I_{ar} R_a} \Phi_{fn} \tag{3.38}$$

Dividing the numerator and denominator of equation (3.38) by the rated motor voltage, V_r, leads to

$$T_{en} = \frac{\left[1.35 \dfrac{V}{V_r} \cos \alpha - \dfrac{K\Phi_f \omega_{mav}}{V_r}\right]}{\dfrac{I_{ar} R_a}{V_r}} \Phi_{fn} \tag{3.39}$$

and, noting that

$$V_r = K\Phi_{fr} \omega_{mr} \tag{3.40}$$

$$R_{an} = \frac{R_a I_{ar}}{V_r}, \text{ p.u.} \tag{3.41}$$

the normalized electromagnetic torque is given by

$$T_{en} = \frac{[1.35 V_n \cos \alpha - \Phi_{fn} \omega_{mn}]}{R_{an}} \Phi_{fn}, \text{ p.u.} \tag{3.42}$$

where

$$\Phi_{fn} = \frac{\Phi_f}{\Phi_{fr}}, \text{p.u.} \qquad (3.43)$$

$$\omega_{mn} = \frac{\omega_{mav}}{\omega_{mr}}, \text{p.u.} \qquad (3.44)$$

and

$$V_n = \frac{V}{V_r}, \text{p.u.} \qquad (3.45)$$

The normalized equation (3.42) deserves careful scrutiny for use in steady-state performance computation. Positive or motoring torque is produced when the numerator of (3.42) is positive, i.e.,

$$1.35 V_n \cos\alpha - \Phi_{fn}\omega_{mn} > 0 \qquad (3.46)$$

or

$$\cos\alpha > \frac{\Phi_{fn}\omega_{mn}}{1.35 V_n} \qquad (3.47)$$

If $\cos\alpha$ is less than the right-hand side of (3.47), then there is no torque generation in the machine. For some positive values of α, the numerator can become negative, but that will not produce regeneration, since there will be no current flow from the machine to the source with only one converter. The induced emf of the machine will be greater than the applied voltage, thus blocking the conduction of thyristors. If an antiparallel converter is available and is capable of conducting current in the reverse direction to the motoring operation, then regeneration is achieved by decreasing the applied voltage compared with the value of the induced emf. That enables the machine to generate current from the difference between its induced emf and the applied voltage. This step results in power flow from the machine to the ac source.

The electromagnetic torque equation (3.42) expressed as a function of normalized flux, speed, and triggering angle delay has design use. The controller for the converter requires a certain precision in its triggering angle delay. Its resolution can be found by, for instance, finding the minimum and maximum triggering delay angles from the torque requirements at the minimum and maximum speeds of operation by using this expression. The range of triggering delay angle and the finesse with which it needs to be controlled is given by the resolution of speed or torque control.

Example 3.3

Consider a motor drive with $R_{an} = 0.1$ p.u., $\phi_{fn} = 1$ p.u., $V_n = 1.1$ p.u. and extreme load operating points $T_{e1(min)} = 0.1$ p.u., $\omega_{mn(min)} = \omega_{mn1} = 0.1$ p.u., $T_{e2(max)} = 1$ p.u., and $\omega_{mn(max)} = \omega_{mn2} = 1$ p.u.

(i) Find the normalized control voltages to meet these operating points.

Section 3.4 Steady-State Analysis of the Three-Phase Converter-Controlled DC Motor

(ii) Compute the change in control voltages required for a simultaneous change of $\Delta T_{en} = 0.02$ p.u. and $\Delta \omega_{mn} = 0.01$ p.u. for both the extreme operating points. From this, calculate the resolution required for the control voltage.

Solution Assume that the controller is linearized.

$$\therefore \alpha = \cos^{-1}\left\{\frac{V_c}{V_{cm}}\right\} = \cos^{-1}\{V_{cn}\}$$

from which the electromagnetic torque is,

$$T_{en} = \left[\frac{1.35 V_n V_{cn} - \Phi_{fn}\omega_{mn}}{R_{an}}\right]\Phi_{fn}, \text{ p.u.}$$

where V_{cn} is the normalized control voltage for a given steady-state operating point and is obtained as

$$V_{cn} = \frac{T_{en} R_{an} + \phi_{fn}\omega_{mn}}{1.35 V_n}$$

Since $\phi_{fn} = 1$ p.u., the control voltage for minimum torque and speed is

$$V_{cn1} = \frac{T_{en1} R_{an} + \phi_{fn}\omega_{mn1}}{1.35 V_n} = \frac{0.1 * 0.1 + 0.1}{1.35 * 1.1} = 0.074 \text{ p.u.}$$

Similarly for maximum torque and speed, the control voltage is

$$V_{cn2} = \frac{T_{en2} R_{an} + \phi_{fn}\omega_{mn2}}{1.35 V_n} = \frac{1 * 0.1 + 1}{1.35 * 1.1} = 0.74 \text{ p.u.}$$

Incremental control voltage generates incremental torque and speed as

$$V_{cn} + \delta v_{cn} = \frac{R_{an}(T_{en} + \delta T_{en}) + \omega_{mn} + \delta\omega_{mn}}{1.35 V_n}$$

For both changes, $\delta v_{cn} = \dfrac{R_{an}\Delta T_{en} + \delta\omega_{mn}}{1.35 V_n}$

Dividing δV_{cn} by V_{cn} gives an expression in terms of steady-state operating points as

$$\frac{\delta v_{cn}}{V_{cn}} = \frac{R_{an}\delta T_{en} + \delta\omega_{mn}}{R_{an} T_{en} + \omega_{mn}}$$

$\delta T_{en} = 0.02$ p.u., $\delta\omega_{mn} = 0.01$ p.u., $T_{en1} = 0.1$ p.u., $\omega_{mn1} = 0.1$ p.u., $T_{en2} = 1$ p.u., $\omega_{mn2} = 1$ p.u.

For T_{en1}, ω_{mn1}: $\dfrac{\delta v_{cn}}{V_{cn}} = \dfrac{0.1 * 0.02 + 0.01}{0.1 * 0.1 + 0.1} = 0.109$

For T_{en2}, ω_{mn2}: $\dfrac{\delta v_{cn}}{V_{cn}} = \dfrac{0.1 * 0.02 + 0.01}{0.1 * 1 + 1} = 0.0109$

Therefore, the resolution required in control voltage is

$$\delta V_{cn} = 0.109 * V_{cn} = 0.109 * 0.074 = 0.008066 \text{ p.u.}$$

Example 3.4

A separately-excited dc motor has 0.05 p.u. resistance and is fed from a three-phase converter. The normalized voltage and field flux are 1 p.u. Draw the torque–speed characteristics in the first quadrant for constant delay angles of 0, 30, 45, and 60 degrees. Indicate the safe operating region if the maximum torque limit is 2.5 p.u.

Solution

$$T_{en} = \frac{[1.35 V_n \cos \alpha - \Phi_{fn} \omega_{mn}]}{R_{an}} \Phi_{fn}, \text{p.u.}$$

Substituting the given values yields

$$T_{en} = 20[1.35 \cos \alpha - \omega_{mn}], \text{p.u.}$$

The torque–speed characteristics for various angles of delay are shown in Figure 3.21. The safe operating region is shaded in the figure.

3.4.2 Steady-State Solution Including Harmonics

This section considers the steady-state analysis of the dc motor drive with the actual voltage waveforms and not the average values of the input voltages as in the previous section. The advantage of considering the actual current waveforms is to accu-

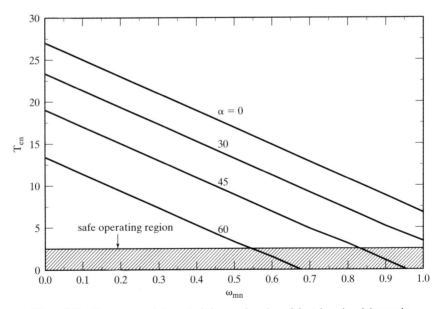

Figure 3.21 Torque–speed characteristics as a function of the triggering delay angle

Section 3.4 Steady-State Analysis of the Three-Phase Converter-Controlled DC Motor

rately predict its electromagnetic torque and hence the magnitude of the torque pulsations. The information on the torque pulsations is necessary for applications such as position servo drives and antenna drives. Additionally, the actual computation of current waveforms assists in the evaluation of harmonic losses and hence of the heating effects on the machine, which will be instrumental in the derating consideration, as explained in a later section.

For this analysis, the motor is considered to be in steady state, i.e., the speed of the machine and the field current are assumed to be constant. In that case, the equation of the motor is sufficiently provided by its electrical part, given as

$$R_a i_a + L_a \frac{di_a}{dt} + K_b \omega_m = v_a \qquad (3.48)$$

where

$$v_a = V_m \sin(\omega_s t + \pi/3 + \alpha), \; 0 < \omega_s t < \pi/3 \qquad (3.49)$$

and for each $\pi/3$ duration, the same is valid. The current may be continuous or discontinuous, depending on the speed, the input line voltage, the triggering angle, and the impedance of the motor. For the sake of simplicity, the commutation effect is neglected in this analysis. It can be easily incorporated for the design calculations without sacrificing the elegance of the present solution. The induced emf is a constant under the assumption of constant speed, and, hence, the solution of the above equation is given by

$$i_a(t) = \left(\frac{V_m}{|Z_a|}\right)\{\sin(\omega_s t + \pi/3 + \alpha - \beta) - \sin(\pi/3 + \alpha - \beta)e^{-t/T_a}\} - \left(\frac{E}{R_a}\right)(1 - e^{-t/T_a}) + i_{ai} e^{-t/T_a} \qquad (3.50)$$

where $\omega_s = 2\pi f_s$, $\beta = \tan^{-1}(\omega_s \cdot L_a/R_a)$ = machine impedance angle, T_a = armature time constant = L_a/R_a, i_{ai} = initial value of current at time $t = 0$, $E = K_b \cdot \omega_m$ = induced emf, V_m = peak value of the line–line input ac voltage, and motor electrical impedance $Z_a = R_a + j\omega_s L_a$

The initial armature current i_{ai} has to be evaluated to obtain the complete solution, and it could be achieved from one other available piece of information. That is that the armature current repeats itself every 60 electrical degrees, similarly to the input voltage. Such a unique situation gives rise to a boundary condition:

$$i_a(\omega_s t) = i_a(\omega_s t + \pi/3) \qquad (3.51)$$

That is, the initial armature current will be equal to the current at the end of 60 electrical degrees. Therefore, evaluating the current at a time corresponding to the 60th electrical degree and equating it to the initial value of the armature current, i_{ai}, as follows, leads to

$$i_a\left(\frac{\pi}{3\omega_s}\right) = i_{ai}$$

$$= \left(\frac{V_m}{|Z_a|}\right)\left\{\sin\left(\frac{2\pi}{3} + \alpha - \beta\right) - \sin\left(\frac{\pi}{3} + \alpha - \beta\right)e^{-\left(\frac{\pi}{3\omega_s T_a}\right)}\right\}$$
$$- \left(\frac{E}{R_a}\right)\left(1 - e^{-\left(\frac{\pi}{3\omega_s T_a}\right)}\right) + i_{ai}e^{-\left(\frac{\pi}{3\omega_s T_a}\right)} \quad (3.52)$$

Rearranging the initial current on one side, it is evaluated as a function of only machine parameters, input voltage, triggering angle, and speed and given as

$$i_{ai} = \frac{\left(\frac{V_m}{|Z_a|}\right)\left\{\sin\left(\frac{2\pi}{3} + \alpha - \beta\right) - \sin\left(\frac{\pi}{3} + \alpha - \beta\right)e^{-\left(\frac{\pi}{3\omega_s T_a}\right)}\right\} - \frac{E}{R_a}\left(1 - e^{-\left(\frac{\pi}{3\omega_s T_a}\right)}\right)}{1 - e^{-\frac{\pi}{3\omega_s T_a}}} \quad (3.53)$$

By using this initial value of the current, the armature current for the complete cycle can be evaluated. The present approach, using the boundary matching condition, is a powerful technique for evaluating the steady state directly, without going through the transients in the solution of the differential equation. This technique is widely applied in the evaluation of steady-state performance in variable-speed dc and ac motor drives and is adopted throughout this text. This technique also gives a closed-form solution that could be used in the analysis and design of the drive systems. Figure 3.22 shows the phase converter output voltage, armature current, and induced emf, using the above solution procedure. Although the input armature voltage contains considerable harmonic, the current has much less harmonic, because of the high harmonic impedance of the machine. The useful electromagnetic torque is

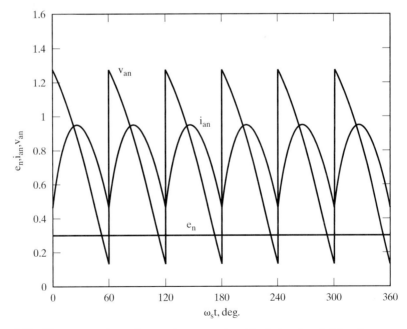

Figure 3.22 The output voltage, the armature current, and the induced emf of the phase converter

Section 3.4 Steady-State Analysis of the Three-Phase Converter-Controlled DC Motor

the average; the harmonic torques contribute to losses only. Note that the electromagnetic torque is very similar to the armature current in shape, with the field current being maintained constant.

3.4.3 Critical Triggering Angle

The triggering angle corresponding to when the armature current is barely continuous is called the critical triggering angle, α_c. It is evaluated by equating the initial armature current, i_{ai}, to zero in equation (3.53). That is given as

$$\alpha_c = \beta + \cos^{-1}\left\{\frac{(E/V_m)}{c_1} \cdot \frac{1}{\cos\beta} \cdot (1 - e^{-(\pi/3\tan\beta)})\right\} - \frac{\pi}{3} + \theta_1 \qquad (3.54)$$

where

$$\left.\begin{array}{l} c_1 = \sqrt{a_1^2 + b_1^2} \\[6pt] a_1 = \dfrac{\sqrt{3}}{2} \\[6pt] b_1 = \dfrac{1}{2} - e^{-\left(\frac{\pi}{3\tan\beta}\right)} \\[6pt] \theta_1 = \tan^{-1}\left(\dfrac{b_1}{a_1}\right) \end{array}\right\} \qquad (3.55)$$

This expression for critical triggering angle is a function of the induced emf, input line-to-line ac voltage, and machine impedance angle, β. The induced emf and the input ac line voltage can be expressed as a single variable, because the input line-to-line voltage is usually a constant, and that leaves the dependence of the critical triggering angle on two variables, E/V_m and β, only. Figure 3.23 shows the critical triggering angle vs. E/V_m for various values of β. It is to be understood that these curves are not machine-specific and could be used to calculate the critical triggering angles for any dc motor-drive system. An increase in the triggering angle beyond its critical value implies that the armature current will become discontinuous.

When one is computing α_c by using equation (3.54), care has to be exercised in the argument of arccosine. If the argument exceeds one, then either E/V_m or β has to be changed to limit it to one. Further, the arccosine term has to be less than the machine impedance angle, β, to be meaningful; accordingly, the computation has to be terminated when α_c becomes less than or equal to zero.

3.4.4 Discontinuous Current Conduction

When the current becomes discontinuous, the voltage across the machine then is the induced emf itself. The steady state under such a condition can be computed from the following relationship:

$$R_a i_a + L_a \frac{di_a}{dt} + E = V_a, \quad 0 < \omega_s t < \omega_s t_x \qquad (3.56)$$

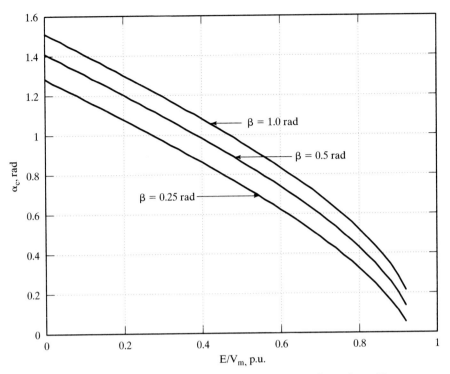

Figure 3.23 The critical triggering angle vs. E/V_m for various values of β

$$i_a = 0, \omega_s t_x < \omega_s t < \pi/3 \quad (3.57)$$
$$v_a = V_m \sin(\omega_s t + \pi/3 + \alpha), 0 < \omega_s t < \omega_s t_x \quad (3.58)$$
$$= E, \omega_s t_x < \omega_s t < \pi/3 \quad (3.59)$$

and $\omega_s t_x = \theta_x$. Time t_x corresponds to the current conduction time, and the waveforms are shown in Figure 3.24. The time t_x can be evaluated from the armature current solution by equating it to zero at time $t = t_x$, as follows:

$$i_a(t_x) = 0 = \frac{V_m}{Z_a}\left[\sin(\omega_s t_x + \pi/3 + \alpha - \beta) - \sin(\pi/3 + \alpha - \beta)e^{-\frac{t_x}{T_a}}\right] - \frac{E}{R_a}\left(1 - e^{-\frac{t_x}{T_a}}\right) \quad (3.60)$$

This equation can be solved iteratively for t_x by using the Newton-Raphson technique. The current conduction time t_x is dependent on the machine parameters, speed (which is contained in the induced emf term), ac input line-to-line voltage, and triggering angle. This triggering angle will be greater than the critical triggering angle evaluated in the earlier section. Because of its dependence on more than two variables, the current conduction time vs. all the variables cannot be contained in a two or three dimensional plot as in the case of the critical triggering angle. It needs to be evaluated for each operating condition. The discontinuous current has a rich content of harmonics compared to the continuous current, resulting in high torque pulsations that might be undesirable for some applications.

Section 3.4   Steady-State Analysis of the Three-Phase Converter-Controlled DC Motor   69

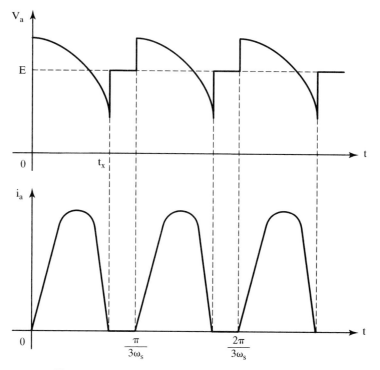

Figure 3.24 Discontinuous current conduction waveforms

Figure 3.25 shows the armature current, the voltage across the armature terminals, and the induced emf for a typical discontinuous conduction. In the discontinuous mode of conduction, note that the linearity of the output voltage with reference voltage will be lost. This has implications in the control contributing to an overall sluggish response of the drive system. The current discontinuity can be overcome by the addition of an external inductor or by suitably designing the dc machine with the required armature inductance to obtain a continuous current. The latter approach is feasible at the system level of planning mostly for new installations. The approach of using an external inductor is the only practical recourse for existing installations; the replacement of the dc machine is an expensive alternative.

Example 3.5

The details and parameters of a separately-excited dc machine are

100 hp, 500 V, 1750 rpm, 153.7 A, R_a = 0.088 Ω, L_a = 0.00183 H, K_b = 2.646 V/(rad/sec)

The machine is supplied from a three-phase controlled converter whose ac input is from a three-phase 415 V, 60 Hz utility supply. Assume that the machine is operating at 100 rpm with a triggering angle delay of 65°. Find the maximum air gap torque ripple at this operating point.

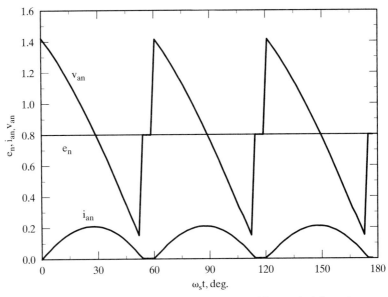

Figure 3.25 The armature current, applied voltage, and induced emf for a typical discontinuous conduction

Solution To find the current ripple, it is essential to determine whether the current is continuous for the given triggering delay by evaluating the critical triggering angle. It is found as follows:

Rotor speed, $\omega_m = \dfrac{2\pi N_r}{60} = \dfrac{2\pi * 100}{60} = 10.48$ rad/sec

Induced emf, $E = K_b\omega_m = 2.646 * 10.48 = 27.7$ V

Peak input voltage, $V_m = \sqrt{2} * 415 = 586.9$ V

Input angular frequency, $\omega_s = 2\pi f_s = 2\pi * 60 = 376.99$ rad/sec

Armature time constant, $T_a = \dfrac{L_a}{R_a} = \dfrac{0.00183}{0.088} = 0.0208$ sec

Machine impedance, $Z_a = \sqrt{R_a^2 + \omega_s^2 L_a^2} = 0.6955\,\Omega$

Machine impedance angle, $\beta = \tan^{-1}\left(\dfrac{\omega_s L_a}{R_a}\right) = 1.444$ rad

The critical triggering angle is

$$\alpha_c = \beta + \cos^{-1}\left\{\dfrac{E/V_m}{c_1} \cdot \dfrac{1}{\cos\beta} \cdot (1 - e^{-(\pi/3\tan\beta)})\right\} - \dfrac{\pi}{3} + \theta_1$$

where

$$a_1 = \dfrac{\sqrt{3}}{2} = 0.866$$

$$b_1 = \dfrac{1}{2} - e^{-\left(\frac{\pi}{3\tan\beta}\right)} = -0.375$$

$$c_1 = \sqrt{a_1^2 + b_1^2} = 0.9437$$

Section 3.5 Two-Quadrant Three-Phase Converter-Controlled DC Motor Drive 71

$$\theta_1 = \tan^{-1}\left(\frac{b_1}{a_1}\right) = -0.4086 \text{ rad}$$

from which the critical angle is obtained as $\alpha_c = 1.5095$ rad $= 86.48°$. The triggering angle α is $65°$, which is less than the critical triggering angle; therefore, the armature current is continuous. Having determined that the drive system is in continuous mode of conduction, we use the relevant equations to calculate the initial current, given by

$$i_{ai} = \frac{\left(\frac{V_m}{|Z_a|}\right)\left\{\sin\left(\frac{2\pi}{3} + \alpha - \beta\right) - \sin\left(\frac{\pi}{3} + \alpha - \beta\right)e^{-\left(\frac{\pi}{3\omega_s T_a}\right)}\right\} - \frac{E}{R_a}\left(1 - e^{-\left(\frac{\pi}{3\omega_s T_a}\right)}\right)}{1 - e^{-\frac{\pi}{3\omega_s T_a}}} = 2308.1 \text{ A}$$

The peak armature current is found by having $\omega_s t = \pi/6$, i.e., at the midpoint of the cycle. This is usually the case, but the operating point can shift it beyond $30°$; therefore, it is necessary to verify graphically or analytically where the maximum current occurs and then substitute that instant to get the peak armature current from the following equation:

$$i_a(t) = \left(\frac{V_m}{|Z_a|}\right)\{\sin(\omega_s t + \pi/3 + \alpha - \beta) - \sin(\pi/3 + \alpha - \beta)e^{-t/T_a}\} - \left(\frac{E}{R_a}\right)(1 - e^{-t/T_a})$$
$$+ i_{ai}e^{-t/T_a} = 2411.5 \text{ A}$$

The armature current ripple magnitude, $\Delta i_a = 2411.5 - 2308.1 = 103.4$ A

The ripple torque magnitude, $\Delta T_e = K_b \Delta i_a = 2.646 * 103.4 = 273.86$ N·m

Average air gap torque, $T_{e(av)} = I_{av} K_b \cong [\{2411.5 + 2308.1\} * 0.5] * 2.646 = 6244$ N·m

Note that the ripple current magnitude is less than 5% and therefore is approximated as a straight line between its minimum and maximum values in each part of its cycle.

Torque ripple as a percent of the operating average torque is

$$\Delta T_{en} = \frac{\Delta T_e}{T_{e(av)}} * 100 = \frac{273.86}{6244} * 100 = 4.4\%$$

3.5 Two-Quadrant Three-Phase Converter-Controlled DC Motor Drive

The control schematic of a two-quadrant converter-controlled separately-excited dc motor drive is shown in Figure 3.26. The motor drive shown is a speed-controlled system. The thyristor bridge converter gets its ac supply through a three-phase transformer and fast-acting ac contactors. The dc output is fed to the armature of the dc motor. The field is separately excited, and the field supply can be kept constant or regulated, depending on the need for the field-weakening mode of operation. The dc motor has a tachogenerator whose output is utilized for closing the speed loop. The motor is driving a load considered to be frictional for this treatment. The output of the tachogenerator is filtered to remove the ripples to provide the signal, ω_{mr}. The speed command ω_r^* is compared to the speed signal to produce a speed error signal. This signal is processed through a proportional-plus-integral (PI) controller to determine the torque command. The torque command is limited, to keep it within the safe

72 Chapter 3 Phase-Controlled DC Motor Drives

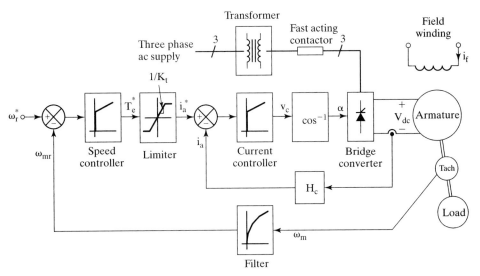

Figure 3.26 Speed-controlled two-quadrant dc motor drive

current limits, and the current command is obtained by proper scaling. The armature current command i_a^* is compared to the actual armature current i_a to have a zero current error. In case there is an error, a PI current controller processes it to alter the control signal v_c. The control signal accordingly modifies the triggering angle α to be sent to the converter for implementation. The implementation of v_c to α in the converter is discussed under control circuit in section 3.3.3.

The inner current loop assures a fast current response and hence also limits the current to a safe preset level. This inner current loop makes the converter a linear current amplifier. The outer speed loop ensures that the actual speed is always equal to the commanded speed and that any transient is overcome within the shortest feasible time without exceeding the motor and converter capability.

The operation of the closed-loop speed-controlled drive is explained from one or two particular instances of speed command. A speed from zero to rated value is commanded, and the motor is assumed to be at standstill. This will generate a large speed error and a torque command and in turn an armature current command. The armature current error will generate the triggering angle to supply a preset maximum dc voltage across the motor terminals. The inner current loop will maintain the current at the level permitted by its commanded value, producing a corresponding torque. As the motor starts running, the torque and current are maintained at their maximum level, thus accelerating the motor rapidly. When the rotor attains the commanded value, the torque command will settle down to a value equal to the sum of the load torque and other motor losses to keep the motor in steady state.

The design of the gain and time constants of the speed and current controllers is of paramount importance in meeting the dynamic specifications of the motor drive. Their systematic designs are considered in the next section.

3.6 TRANSFER FUNCTIONS OF THE SUBSYSTEMS

3.6.1 DC Motor and Load

The dc machine contains an inner loop due to the induced emf. It is not physically seen; it is magnetically coupled. The inner current loop will cross this back-emf loop, creating a complexity in the development of the model. It is shown in Figure 3.27. The interactions of these loops can be decoupled by suitably redrawing the block diagram. The development of such a block diagram for the dc machine is shown in Figure 3.28, step by step. The load is assumed to be proportional to speed and is given as

$$T_l = B_l \omega_m \qquad (3.61)$$

To decouple the inner current loop from the machine-inherent induced-emf loop, it is necessary to split the transfer function between the speed and voltage into two cascade transfer functions, first between speed and armature current and then between armature current and input voltage, represented as

$$\frac{\omega_m(s)}{V_a(s)} = \frac{\omega_m(s)}{I_a(s)} \cdot \frac{I_a(s)}{V_a(s)} \qquad (3.62)$$

where

$$\frac{\omega_m(s)}{I_a(s)} = \frac{K_b}{B_t(1 + sT_m)} \qquad (3.63)$$

$$\frac{I_a(s)}{V_a(s)} = K_1 \frac{(1 + sT_m)}{(1 + sT_1)(1 + sT_2)} \qquad (3.64)$$

$$T_m = \frac{J}{B_t} \qquad (3.65)$$

$$B_t = B_1 + B_l \qquad (3.66)$$

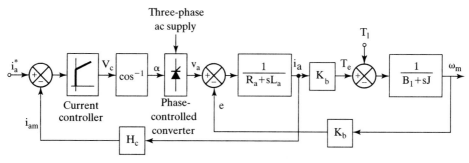

Figure 3.27 DC motor and current-control loop

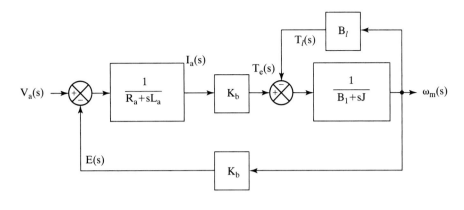

Step 1

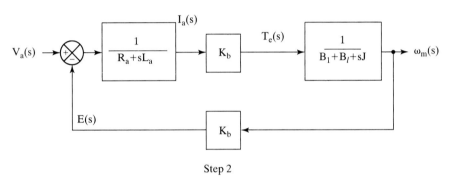

Step 2

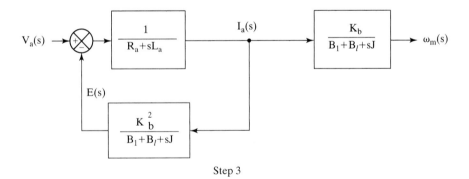

Step 3

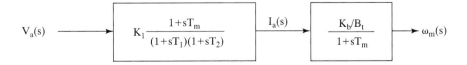

Step 4

Figure 3.28 Step-by-step derivation of a dc machine transfer function

$$-\frac{1}{T_1}, -\frac{1}{T_2} = -\frac{1}{2}\left[\frac{B_t}{J} + \frac{R_a}{L_a}\right] \pm \sqrt{\frac{1}{4}\left(\frac{B_t}{J} + \frac{R_a}{L_a}\right)^2 - \left(\frac{K_b^2 + R_aB_t}{JL_a}\right)} \quad (3.67)$$

$$K_1 = \frac{B_t}{K_b^2 + R_aB_t} \quad (3.68)$$

3.6.2 Converter

The converter after linearization is represented as

$$G_r(s) = \frac{V_a(s)}{v_c(s)} = \frac{K_r}{1 + sT_r} \quad (3.69)$$

The delay time T_r and gain are evaluated and given in section 3.3.4.

3.6.3 Current and Speed Controllers

The current and speed controllers are of proportional-integral type. They are represented as

$$G_c(s) = \frac{K_c(1 + sT_c)}{sT_c} \quad (3.70)$$

$$G_s(s) = \frac{K_s(1 + sT_s)}{sT_s} \quad (3.71)$$

where the subscripts c and s correspond to the current and speed controllers, respectively. The K and T correspond to the gain and time constants of the controllers.

3.6.4 Current Feedback

The gain of the current feedback is H_c. No filtering is required in most cases. In the case of a filtering requirement, a low-pass filter can be included in the analysis. Even then, the time constant of the filter might not be greater than a millisecond.

3.6.5 Speed Feedback

Most high performance systems use a dc tachogenerator, and the filter required is low-pass, with a time constant under 10 ms. The transfer function of the speed feedback filter is

$$G_\omega(s) = \frac{K_\omega}{1 + sT_\omega} \quad (3.72)$$

where K_ω is the gain and T_ω is the time constant.

3.7 DESIGN OF CONTROLLERS

The overall closed-loop system is shown in Figure 3.29. It is seen that the current loop does not contain the inner induced-emf loop. The design of control loops starts from the innermost (fastest) loop and proceeds to the slowest loop, which in this case is the outer speed loop. The reason to proceed from the inner to the outer loop in the design process is that the gain and time constants of only one controller at a time are solved, instead of solving for the gain and time constants of all the controllers simultaneously. Not only is that logical; it also has a practical implication. Note that every motor drive need not be speed-controlled but may be torque-controlled, such as for a traction application. In that case, the current loop is essential and exists regardless of whether the speed loop is going to be closed. Additionally, the performance of the outer loop is dependent on the inner loop; therefore, the tuning of the inner loop has to precede the design and tuning of the outer loop. That way, the dynamics of the inner loop can be simplified and the impact of the outer loop on its performance could be minimized. The design of the current and speed controllers is considered in this section.

3.7.1 Current Controller

The current-control loop is shown in Figure 3.30. The loop gain function is

$$GH_i(s) = \left\{\frac{K_1 K_c K_r H_c}{T_c}\right\} \cdot \frac{(1 + sT_c)(1 + sT_m)}{s(1 + sT_1)(1 + sT_2)(1 + sT_r)} \quad (3.73)$$

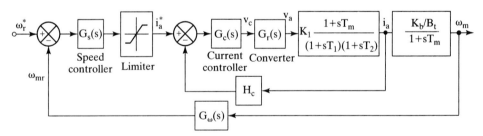

Figure 3.29 Block diagram of the motor drive

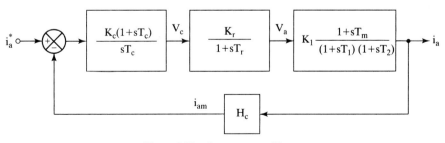

Figure 3.30 Current-control loop

This is a fourth-order system, and simplification is necessary to synthesize a controller without resorting to a computer. Noting that T_m is on the order of a second and in the vicinity of the gain crossover frequency, we see that the following approximation is valid:

$$(1 + sT_m) \cong sT_m \qquad (3.74)$$

which reduces the loop gain function to

$$GH_i(s) \cong \frac{K(1 + sT_c)}{(1 + sT_1)(1 + sT_2)(1 + sT_r)} \qquad (3.75)$$

where

$$K = \frac{K_1 K_c K_r H_c T_m}{T_c} \qquad (3.76)$$

The time constants in the denominator are seen to have the relationship

$$T_r < T_2 < T_1 \qquad (3.77)$$

The equation (3.75) can be reduced to second order, to facilitate a simple controller synthesis, by judiciously selecting

$$T_c = T_2 \qquad (3.78)$$

Then the loop function is

$$GH_i(s) \cong \frac{K}{(1 + sT_1)(1 + sT_r)} \qquad (3.79)$$

The characteristic equation or denominator of the transfer function between the armature current and its command is

$$(1 + sT_1)(1 + sT_r) + K \qquad (3.80)$$

This equation is expressed in standard form as

$$T_1 T_r \left\{ s^2 + s\left(\frac{T_1 + T_r}{T_1 T_r}\right) + \frac{K + 1}{T_1 T_r} \right\} \qquad (3.81)$$

from which the natural frequency and damping ratio are obtained as

$$\omega_n^2 = \frac{K + 1}{T_1 T_r} \qquad (3.82)$$

$$\zeta = \frac{\left(\dfrac{T_1 + T_r}{T_1 T_r}\right)}{2\sqrt{\dfrac{K + 1}{T_1 T_r}}} \qquad (3.83)$$

where ω_n and ζ are the natural frequency and damping ratio, respectively. For good dynamic performance, it is an accepted practice to have a damping ratio of 0.707. Hence, equating the damping ratio to 0.707 in equation (3.83), we get

$$K + 1 = \frac{\left(\dfrac{T_1 + T_r}{T_1 T_r}\right)^2}{\left(\dfrac{2}{T_1 T_r}\right)} \qquad (3.84)$$

Realizing that

$$K \gg 1 \qquad (3.85)$$
$$T_1 \gg T_r \qquad (3.86)$$

tells us that K is approximated as

$$K \cong \frac{T_1^2}{2T_1 T_r} \cong \frac{T_1}{2T_r} \qquad (3.87)$$

By equating (3.76) to (3.87), the current-controller gain is evaluated as

$$K_c = \frac{1}{2} \cdot \frac{T_1 T_c}{T_r} \cdot \left(\frac{1}{K_1 K_r H_c T_m}\right) \qquad (3.88)$$

3.7.2 First-Order Approximation of Inner Current Loop

To design the speed loop, the second-order model of the current loop is replaced with an approximate first-order model. This helps to reduce the order of the overall speed-loop gain function. The current loop is approximated by adding the time delay in the converter block to T_1 of the motor; because of the cancellation of one motor pole by a zero of the current controller, the resulting current loop can be shown in Figure 3.31. The transfer function of the current and its commanded value is

$$\frac{I_a(s)}{I_a^*(s)} = \frac{\dfrac{K_c K_r T_1 T_m}{T_c} \cdot \dfrac{1}{(1 + sT_3)}}{1 + \dfrac{K_1 K_c K_r H_c T_m}{T_c} \cdot \dfrac{1}{(1 + sT_3)}} \qquad (3.89)$$

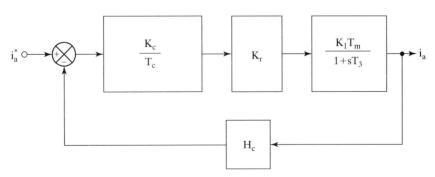

Figure 3.31 Simplified current-control loop

where $T_3 = T_1 + T_r$. The transfer function can be arranged simply as

$$\frac{I_a(s)}{I_a^*(s)} = \frac{K_i}{(1 + sT_i)} \quad (3.90)$$

where

$$T_i = \frac{T_3}{1 + K_{fi}} \quad (3.91)$$

$$K_i = \frac{K_{fi}}{H_c} \cdot \frac{1}{(1 + K_{fi})} \quad (3.92)$$

$$K_{fi} = \frac{K_c K_r K_1 T_m H_c}{T_c} \quad (3.93)$$

The resulting model of the current loop is a first-order system, suitable for use in the design of a speed loop. The gain and delay of the current loop can also be found experimentally in a motor-drive system. That would be more accurate for the speed-controller design.

3.7.3 Speed Controller

The speed loop with the first-order approximation of the current-control loop is shown in Figure 3.32. The loop gain function is

$$GH_s(s) = \left\{ \frac{K_s K_i K_b H_\omega}{B_t T_s} \right\} \cdot \frac{(1 + sT_s)}{s(1 + sT_i)(1 + sT_m)(1 + sT_\omega)} \quad (3.94)$$

This is a fourth-order system. To reduce the order of the system for analytical design of the speed controller, approximation serves. In the vicinity of the gain crossover frequency, the following is valid:

$$(1 + sT_m) \cong sT_m \quad (3.95)$$

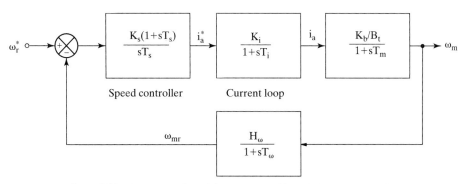

Figure 3.32 Representation of the outer speed loop in the dc motor drive

The next approximation is to build the equivalent time delay of the speed feedback filter and current loop. Their sum is very much less than the integrator time constant, T_s, and hence the equivalent time delay, T_4, can be considered the sum of the two delays, T_i and T_ω. This step is very similar to the equivalent time delay introduced in the simplification of the current-loop transfer function. Hence, the approximate gain function of the speed loop is

$$GH_s(s) \cong K_2 \cdot \frac{K_s}{T_s} \cdot \frac{(1 + sT_s)}{s^2(1 + sT_4)} \tag{3.96}$$

where

$$T_4 = T_i + T_\omega \tag{3.97}$$

$$K_2 = \frac{K_i K_b H_\omega}{B_t T_m} \tag{3.98}$$

The closed-loop transfer function of the speed to its command is

$$\frac{\omega_m(s)}{\omega_r^*(s)} = \frac{1}{H_\omega} \left[\frac{\frac{K_2 K_s}{T_s}(1 + sT_s)}{s^3 T_4 + s^2 + sK_2 K_s + \frac{K_2 K_s}{T_s}} \right] = \frac{1}{H_\omega} \frac{(a_0 + a_1 s)}{(a_0 + a_1 s + a_2 s^2 + a_3 s^3)} \tag{3.99}$$

where

$$a_0 = K_2 K_s / T_s \tag{3.100}$$

$$a_1 = K_2 K_s \tag{3.101}$$

$$a_2 = 1 \tag{3.102}$$

$$a_3 = T_4 \tag{3.103}$$

This transfer function is optimized to have a wider bandwidth and a magnitude of one over a wide frequency range by looking at its frequency response. Its magnitude is given by

$$\left| \frac{\omega_m(j\omega)}{\omega_r^*(j\omega)} \right| = \frac{1}{H_\omega} \sqrt{\frac{a_0^2 + \omega^2 a_1^2}{\{a_0^2 + \omega^2(a_1^2 - 2a_0 a_2) + \omega^4(a_2^2 - 2a_1 a_3) + \omega^6 a_3^2\}}} \tag{3.104}$$

This is optimized by making the coefficients of ω^2 and ω^4 equal zero, to yield the following conditions:

$$a_1^2 = 2a_0 a_2 \tag{3.105}$$

$$a_2^2 = 2a_1 a_3 \tag{3.106}$$

Substituting these conditions in terms of the motor and controller parameters given in (3.100) into (3.103) yields

$$T_s^2 = \frac{2T_s}{K_s K_2} \tag{3.107}$$

resulting in

$$T_s K_s = \frac{2}{K_2} \tag{3.108}$$

Similarly,

$$\frac{T_s^2}{K_s^2 K_2^2} = \frac{2T_s^2 T_4}{K_s K_2} \tag{3.109}$$

which, after simplification, gives the speed-controller gain as

$$K_s = \frac{1}{2K_2 T_4} \tag{3.110}$$

Substituting equation (3.110) into equation (3.108) gives the time constant of the speed controller as

$$T_s = 4T_4 \tag{3.111}$$

Substituting for K_s and T_s into (3.99) gives the closed-loop transfer function of the speed to its command as

$$\frac{\omega_m(s)}{\omega_r^*(s)} = \frac{1}{H_\omega} \left[\frac{1 + 4T_4 s}{1 + 4T_4 s + 8T_4^2 s^2 + 8T_4^3 s^3} \right] \tag{3.112}$$

It is easy to prove that for the open-loop gain function the corner points are $1/4T_4$ and $1/T_4$, with the gain crossover frequency being $1/2T_4$. In the vicinity of the gain crossover frequency, the slope of the magnitude response is -20 dB/decade, which is the most desirable characteristic for good dynamic behavior. Because of its symmetry at the gain crossover frequency, this transfer function is known as a symmetric optimum function. Further, this transfer function has the following features:

(i) Approximate time constant of the system is $4T_4$.
(ii) The step response is given by

$$\omega_r(t) = \frac{1}{H_\omega}(1 + e^{-t/2T_4} - 2e^{-t/4T_4}\cos(\sqrt{3}t/4T_4)) \tag{3.113}$$

with a rise time of $3.1T_4$, a maximum overshoot of 43.4%, and a settling time of $16.5T_4$.

(iii) Since the overshoot is high, it can be reduced by compensating for its cause, i.e., the zero of a pole in the speed command path, as shown in Figure 3.33. The resulting transfer function of the speed to its command is

$$\frac{\omega_m(s)}{\omega_r^*(s)} = \frac{1}{H_\omega} \left[\frac{1}{1 + 4T_4 s + 8T_4^2 s^2 + 8T_4^3 s^3} \right] \tag{3.114}$$

whose step response is

$$\omega_r(t) = \frac{1}{H_\omega}\left(1 - e^{-t/4T_4} - \frac{2}{\sqrt{3}} e^{-t/4T_4}\sin(\sqrt{3}t/4T_4)\right) \tag{3.115}$$

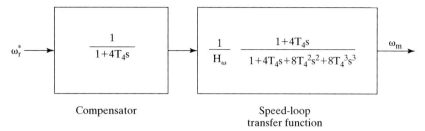

Figure 3.33 Smoothing of the overshoot via a compensator

with a rise time of $7.6T_4$, a maximum overshoot of 8.1%, and a settling time of $13.3T_4$. Even though the rise time has increased, the overshoot has been reduced to approximately 20% of its previous value, and the settling time has come down by 19%.

(iv) The poles of the closed-loop transfer function are

$$s = -\frac{1}{2T_4}; -\frac{1}{4T_4} \pm j\frac{\sqrt{3}}{4T_4} \qquad (3.116)$$

The real parts of the poles are negative, and there are no repeated poles at the origin, so the system is asymptotically stable. Hence, in the symmetric optimum design, the system stability is guaranteed, and there is no need to check for it in the design process. Whether this is true for the original system without approximation will be explored in the following example.

(v) Symmetric optimum eliminates the effects due to the disturbance very rapidly compared to other optimum techniques employed in practical systems, such as linear or modulus optimum. This approach indicates one of the possible methods to synthesize the speed controller. That the judicious choice of approximation is based on the physical constants of the motor, on the converter and transducer gains, and on time delays is to be emphasized here.

That the speed-loop transfer function is expressed in terms of T_4 is significant in that it clearly links the dynamic performance to the speed-feedback and current-loop time constants. That a faster current loop with a smaller speed-filter time constant accelerates the speed response is evident from this. Expressing T_4 in terms of the motor, the converter and transducer gains, and the time delays by using expressions (3.91) and (3.97) yields

$$T_4 = T_i + T_\omega = \frac{T_3}{1 + K_{fi}} + T_\omega = \frac{T_1 + T_r}{1 + K_{fi}} + T_\omega \qquad (3.117)$$

Since $K_{fi} \gg 1$, T_4 is found approximately after substituting for K_{fi} from equation (3.93) in terms of gains and time delays as

$$T_4 \approx \frac{(T_1 + T_r)T_2}{T_m} \cdot \frac{1}{K_1 K_c K_r H_c} + T_\omega \qquad (3.118)$$

This clearly shows the influence of the subsystem parameters on the system dynamics. A clear understanding of this would help the proper selection of the subsystems to obtain the required dynamic performance of the speed-controlled motor-drive system. Further, this derivation demonstrates that the system behavior to a large degree depends on the subsystem parameters rather than only on the current and speed-controller parameters or on the sophistication of their design.

Example 3.6

Design a speed-controlled dc motor drive maintaining the field flux constant. The motor parameters and ratings are as follows:

220 V, 8.3 A, 1470 rpm, $R_a = 4\,\Omega$, $J = 0.0607$ kg$-$m$_2$, $L_a = 0.072$ H, $B_t = 0.0869$ N·m/rad/sec, $K_b = 1.26$ V/rad/sec.

The converter is supplied from 230V, 3-phase ac at 60 Hz. The converter is linear, and its maximum control input voltage is ± 10 V. The tachogenerator has the transfer function $G_\omega(s) = \dfrac{0.065}{(1 + 0.002s)}$. The speed reference voltage has a maximum of 10V. The maximum current permitted in the motor is 20 A.

Solution (i) Converter transfer function:

$$K_r = \frac{1.35\ \text{V}}{V_{cm}} = \frac{1.35 \times 230}{10} = 31.05\ \text{V/V}$$

$$V_{dc}(\max) = 310.5\ \text{V}$$

The rated dc voltage required is 220 V, which corresponds to a control voltage of 7.09 V. The transfer function of the converter is

$$G_r(s) = \frac{31.05}{(1 + 0.00138s)}\ \text{V/V}$$

(ii) Current transducer gain: The maximum safe control voltage is 7.09 V, and this has to correspond to the maximum current error:

$$i_{\max} = 20\ \text{A}$$

$$H_c = \frac{7.09}{I_{\max}} = \frac{7.09}{20} = 0.355\ \text{V/A}$$

(iii) Motor transfer function:

$$K_1 = \frac{B_t}{K_b^2 + R_a B_t} = \frac{0.0869}{1.26^2 + 4 \times 0.0869} = 0.0449$$

$$-\frac{1}{T_1}, -\frac{1}{T_2} = -\frac{1}{2}\left[\frac{B_t}{J} + \frac{R_a}{L_a}\right] \pm \sqrt{\frac{1}{4}\left(\frac{B_t}{J} + \frac{R_a}{L_a}\right)^2 - \left(\frac{K_b^2 + R_a B_t}{JL_a}\right)}$$

$$T_1 = 0.1077\ \text{sec}$$

$$T_2 = 0.0208\ \text{sec}$$

$$T_m = \frac{J}{B_t} = 0.7\ \text{sec}$$

The subsystem transfer functions are

$$\frac{I_a(s)}{V_a(s)} = K_1 \frac{(1 + sT_m)}{(1 + sT_1)(1 + sT_2)} = \frac{0.0449(1 + 0.7s)}{(1 + 0.0208s)(1 + 0.1077s)}$$

$$\frac{\omega_m(s)}{I_a(s)} = \frac{K_b/B_t}{(1 + sT_m)} = \frac{14.5}{(1 + 0.7s)}$$

(iv) Design of current controller:

$$T_c = T_2 = 0.0208 \text{ sec}$$

$$K = \frac{T_1}{2T_r} = \frac{0.1077}{2 \times 0.001388} = 38.8$$

$$K_c = \frac{KT_c}{K_1 H_c K_r T_m} = \frac{38.8 \times 0.0208}{0.0449 \times 0.355 \times 31.05 \times 0.7} = 2.33$$

(v) Current-loop approximation:

$$\frac{I_a(s)}{I_a^*(s)} = \frac{K_i}{(1 + sT_i)}$$

where

$$K_i = \frac{K_{fi}}{H_c} \cdot \frac{1}{(1 + K_{fi})}$$

$$K_{fi} = \frac{K_c K_r K_1 T_m H_c}{T_c} = 38.8$$

$$\therefore K_i = \frac{27.15}{28.09} \cdot \frac{1}{0.355} = 2.75$$

$$T_i = \frac{T_3}{1 + K_{fi}} = \frac{0.109}{1 + 38.8} = 0.0027 \text{ sec}$$

The validity of the approximations is evaluated by plotting the frequency response of the closed-loop current to its command, with and without the approximations. This is shown in Figure 3.34. From this figure, it is evident that the approximations are quite valid in the frequency range of interest.

(vi) Speed-controller design:

$$T_4 = T_i + T_\omega = 0.0027 + 0.002 = 0.0047 \text{ sec}$$

$$K_2 = \frac{K_i K_b H_\omega}{B_t T_m} = \frac{2.75 \times 1.26 \times 0.065}{0.0869 \times 0.7} = 3.70$$

$$K_s = \frac{1}{2K_2 T_4} = \frac{1}{2 \times 3.70 \times 0.0047} = 28.73$$

$$T_s = 4T_4 = 4 \times 0.0047 = 0.0188 = \text{sec}$$

The frequency responses of the speed to its command are shown in Figure 3.35 for cases with and without approximations. That the model reduction with the approximations has given a transfer function very close to the original is obvious from this figure. Further, the

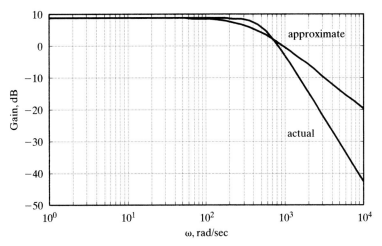

Figure 3.34 Frequency response of the current-transfer functions with and without approximation

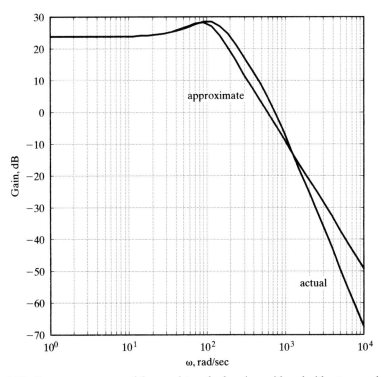

Figure 3.35 Frequency response of the speed-transfer functions with and without approximation

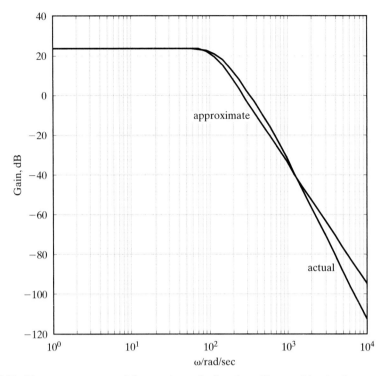

Figure 3.36 Frequency response of the speed-transfer function with smoothing for the cases with and without approximation

smoothing of the overshoot by the cancellation of the zero with a pole at $-1/4T_4$ is shown in Figure 3.36. This figure contains the approximated transfer function of third order for the speed to its command-transfer function and the one without any approximations. Again, the closeness of these two solutions justifies the approximations.

The time responses are important to verify the design of the controllers, and they are shown in Figure 3.37 for the case without smoothing and with smoothing. The case without any approximation is included here for the comparison of all responses.

Example 3.7

Assume motor poles are complex. Develop a design procedure for the current controller.

Solution If motor poles are complex, then the procedure outlined above is not applicable for the design of the current and speed controllers. One alternative is as follows: the current controller is designed by using the symmetric optimum criterion that was applied in the earlier speed controller design. The steps are given below.

Assuming $(1 + sT_m) \cong sT_m$ leads to the following current-loop transfer function:

$$\frac{i_a(s)}{i_a^*(s)} = \frac{K_2 \dfrac{K_c}{T_c}(1 + sT_c)}{b_0 + b_1 s + b_2 s^2 + b_3 s^3}$$

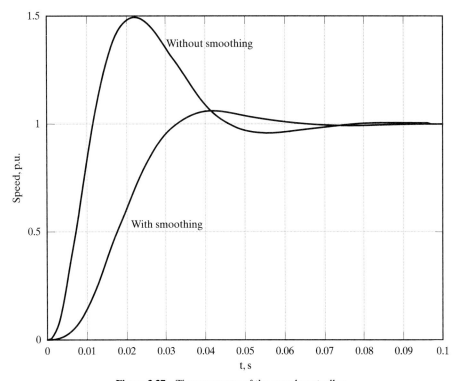

Figure 3.37 Time response of the speed controller

where

$$K_2 = K_1 K_r T_m$$
$$b_0 = 1 + K_2 \frac{K_c}{T_c} H_c$$
$$b_1 = T_1 + T_2 + T_r + K_2 K_c H_c$$
$$b_2 = (T_1 + T_2)T_r + T_1 T_2$$
$$b_3 = T_1 T_2 T_r$$

Applying symmetric-optimum conditions,

$$b_1^2 = 2 b_0 b_2$$
$$(T_1 + T_2 + T_r + K_2 K_c H_c)^2 = 2\left(1 + K_2 \frac{K_c}{T_c} H_c\right)((T_1 + T_2)T_r + T_1 T_2)$$
$$b_2^2 = 2 b_1 b_3$$
$$((T_1 + T_2)T_r + T_1 T_2)^2 = 2(T_1 + T_2 + T_r + K_2 K_c H_c)(T_1 T_2 T_r)$$

but $T_r \ll T_1, T_2$,

$$\therefore (T_1 + T_2)T_r << T_1T_2$$

$$\frac{(T_1T_2)^2}{2(T_1T_2T_r)} = T_1 + T_2 + T_r + K_2K_cH_c$$

$$\frac{T_1T_2}{2T_r} \cong K_2K_cH_c \left[\because T_1 + T_2 + T_r << \frac{T_1T_2}{2T_r} \right]$$

$$\therefore K_c = \frac{T_1T_2}{2T_r} \frac{1}{K_2H_c}.$$

Also,

$$\left(T_1 + T_2 + T_r + \frac{T_1T_2}{2T_r} \right)^2 = 2\left(1 + \frac{T_1T_2}{2T_rT_c} \right)((T_1 + T_2)T_r + T_1T_2)$$

$$\frac{(T_1T_2)^2}{4T_r^2} \cong 2T_1T_2 \left(1 + \frac{T_1T_2}{2T_rT_c} \right)$$

$$\frac{1}{4T_r^2} \cong \frac{1}{T_rT_c}$$

$$T_c \cong \frac{4T_r^2}{T_r} \cong 4T_r.$$

The next step is to obtain the first-order approximation of the current-loop transfer function for the synthesis of the speed controller. Since the time constant T_c is known, the first-order approximation of the current loop is written as

$$\frac{i_a(s)}{i_a^*(s)} \cong \frac{K_i}{1 + sT_i}$$

where the steady-state gain is obtained from the exact transfer function, by setting $s = 0$, as

$$K_i = \frac{\dfrac{K_2K_c}{T_c}}{1 + \dfrac{K_2K_cH_c}{T_c}}$$

and $T_i = T_c$.

From this point, the speed-controller design follows the symmetric-optimum procedure outlined earlier for the case with the real motor poles.

3.8 TWO-QUADRANT DC MOTOR DRIVE WITH FIELD WEAKENING

Flux weakening is exercised in the machine by varying the voltage applied to the field winding. The applied voltage is varied via a single-phase or three-phase half-wave controller rectifier. Because of the large inductance of the field winding, the field current is smooth and hardly has ripple. The converter time lag is negligible compared to the field time constant, and the speed of response is determined primarily by the field time constant. Therefore, there is very little justification for using full-wave converters in motor drives of rating less than 100 kW. A schematic of the two-quadrant dc motor drive with flux weakening is shown in Figure 3.38.

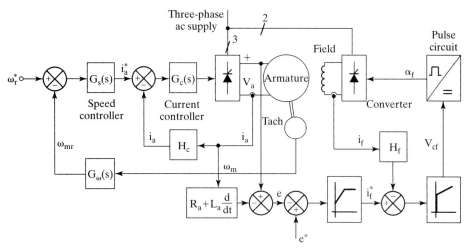

Figure 3.38 A two-quadrant dc motor drive with field weakening

The command field current is determined by the induced-emf error function. The induced-emf error is the difference between the reference induced emf and the estimated machine-induced emf. The machine-induced emf is found by subtracting the resistive and inductive drop from the applied voltage. This induced-emf estimator is machine-parameter sensitive and needs to be adaptive. The induced-emf error is amplified by a PI controller and limited to yield the field-current reference. The field-current reference is enforced by a current-control loop feedback very similar to the armature current-control loop. The outer induced-emf feedback loop enforces a constant induced emf for speeds higher than the base speed. This amounts to

$$e_n = \phi_{fn}\omega_{mn} \qquad (3.119)$$

If the induced emf e_n is kept at rated value, say 1 p.u., then the field flux is inversely proportional to the rotor speed. This condition also enables constant-air gap-power operation.

3.9 FOUR-QUADRANT DC MOTOR DRIVE

A four-quadrant dc motor drive has a set of dual three-phase converters for the power stage. Its control is very similar to that of the two-quadrant dc motor drive. Converters have to be energized depending on the quadrant of operation. Converters 1 and 2 are for forward and reverse directions of rotation of the motor, respectively. The changeover from one converter to another is safely handled by monitoring speed, current-command, and zero-crossing current signals. These signals form the inputs to the selector block, which assigns the pulse-control signals to the appropriate converter. The converters share the same current and speed loops. A control schematic of the four-quadrant dc motor drive is shown in Figure 3.39. The selector block will switch the converters over only when the current in the outgoing

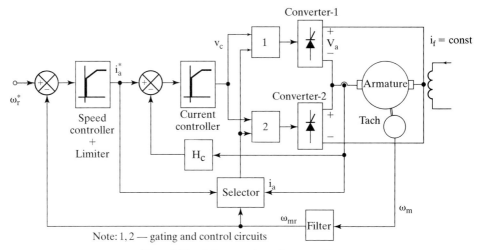

Figure 3.39 A four-quadrant dc motor drive

converter has come to zero. Apart from that, some other conditions have to be satisfied to transfer the control of converters. Assuming that a positive speed command is required for forward running and a negative speed command is required for reverse running, the crossover control for converters is discussed. The machine is running at rated speed and the speed command is changed to reverse rated speed. The current command becomes negative, indicating that the machine has to become regenerative in order to decelerate the motor in the forward direction. Forward regeneration is possible only in quadrant IV, and the converter to provide this quadrant of operation is converter 2. Before the current is reversed, it has to go through zero. That is achieved by increasing the triggering angle of converter 1. When the current is zero, a certain dead time is given to enable the thyristors in converter 1 to recover reverse blocking capability. After this interval, converter 2 is enabled. At this time, the armature current has been forced to zero, but the speed is still positive. The triggering angle of converter 2 is set such that its output voltage equals and opposes the induced emf. Then, slowly decreasing the triggering angle increases the armature current in the opposite direction. By a varying of the triggering angle, the armature current is fully reversed and then is maintained at the reference level. When the motor reaches zero speed, the operation of converter 2 is continued by bringing it into rectification mode, i.e., with α less than $90°$. That will accelerate the motor in the reverse direction until it matches the speed reference.

The function of the selector block is to determine which converter has to be operating. In the previous discussion, as soon as the current command goes negative, the selector block will transfer the control from converter 1 to converter 2 with proper initial triggering angle. If the triggering angle of converter 1 is α_1, then the initial triggering angle, α_2, of converter 2 is $(180° - \alpha_1)$, to match the output voltage of converter 1. If circulating current is not allowed between the converters, zero crossing of the armature current is required to transfer the control from one converter to another. The actual rotor speed is required to determine the quadrant of operation. Based on the rotor speed, the armature current, and its command, the selector block identifies the converter for control and operation. The method discussed here with Figure 3.39 does not allow circulating current between converters 1 and 2. This type of control has a drawback: slower current transfer from one to the other bridge, due to the dead time; but it has the advantage that no additional passive component in the form of an interphase reactor is needed to limit the circulating current.

The other type of crossover control uses a circulating current between the thyristor bridges. It has the advantage of faster current crossover and high dynamic response, but it has the disadvantages of requiring additional interphase reactors and losses associated with them during the operation. The operational details of such an arrangement can be found in references.

3.10 CONVERTER SELECTION AND CHARACTERISTICS

The ratings of the converter and its power switches are derived from the motor and load specifications. Some approximate derivations are given in this section.

Let the maximum current allowed in the motor be I_{max}. The rms value of the current in each device is then based on the fact that it is conducting for 120 electrical degrees in a cycle and that the current is flat. Such an assumption might not be strictly valid under discontinuous conduction, and, anyway, at those points the current will be very much lower than the rated value. The rms value of the current in the power device is

$$I_{rms} = \frac{I_{max}}{\sqrt{3}} = 0.577 I_{max} \qquad (3.120)$$

The voltage rating is the maximum line-to-line voltage of the ac mains,

$$V_t = \sqrt{2} V \qquad (3.121)$$

I_1 is the fundamental rms component of the ac input current, which is derived as

$$I_1 = \frac{1}{\sqrt{2}} \cdot \frac{2\sqrt{3}}{\pi} \cdot I_{max} = \frac{\sqrt{3}\sqrt{2}}{\pi} \cdot I_{max} = 0.78 I_{max} \qquad (3.122)$$

The output power of the converter is given by

$$P_o = V_a I_{max} = \{1.35 V \cos \alpha\} I_{max} = 1.35 V I_{max} \cos \alpha \qquad (3.123)$$

Neglecting losses in the converter, the input power equals the output power. Substituting for I_{max} in terms of the fundamental ac input current from equation (3.122) gives

$$P_i = P_o = 1.35 V I_{max} \cos \alpha = \sqrt{3} V I_1 \cos \alpha \qquad (3.124)$$

This equation gives the real power in a balanced three-phase ac system with a power factor of $\cos \alpha$. Therefore, the angle α gives the power factor angle. Similarly, the reactive power is given by

$$Q_i = \sqrt{3} V I_1 \sin \alpha = 1.35 V I_{max} \sin \alpha \qquad (3.125)$$

92 Chapter 3   Phase-Controlled DC Motor Drives

The input apparent power is

$$P_{VA} = \sqrt{(P_i)^2 + (Q_i)^2} = 1.35 VI_{max} = \sqrt{3} VI_1 \qquad (3.126)$$

Because of the reactive power consumption of the phase-controlled converters, they are expensive to operate where the reactive power is to be paid for. They also generate harmonics, which may be unacceptable to power utilities beyond a certain limit. The above two features are the primary disadvantages of the phase-controlled converter-fed motor drives.

3.11 SIMULATION OF THE ONE-QUADRANT DC MOTOR DRIVE

The equations for various subsystems are derived and then assembled for computer simulation in this section. Key results are discussed. The simulation for either a two- or four-quadrant dc motor drive is very similar to the present development. In the present one-quadrant speed-controlled motor-drive simulation, it is assumed that the field current is constant in the constant torque mode and is varied through a three-phase controlled rectifier for the field-weakening mode of operation to provide constant power over a wide speed range.

3.11.1 The Motor Equations

The motor equations, including that of a simple load modeling, are given below:

$$V_a = R_a i_a + L_a \frac{di_a}{dt} + M i_f \omega_m \qquad (3.127)$$

$$V_f = R_f i_f + L_f \frac{di_f}{dt} \qquad (3.128)$$

$$M i_f i_a - T_l = J \frac{d\omega_m}{dt} + B \omega_m \qquad (3.129)$$

These equations can be rearranged in the following form to facilitate their solution by numerical integration :

$$\frac{di_a}{dt} = -\frac{R_a}{L_a} i_a - \frac{M}{L_a} i_f \omega_m + \frac{V_a}{L_a} \qquad (3.130)$$

$$\frac{di_f}{dt} = -\frac{R_f}{L_f} i_f + \frac{V_f}{L_f} \qquad (3.131)$$

$$\frac{d\omega_m}{dt} = \frac{M}{J} i_f i_a - \frac{B}{J} \omega_m - \frac{T_l}{J} \qquad (3.132)$$

Choose i_a, i_f, and ω_m as state variables and denote them as

$$x_1 = i_a \qquad (3.133)$$

$$x_2 = i_f \qquad (3.134)$$

$$x_3 = \omega_m \qquad (3.135)$$

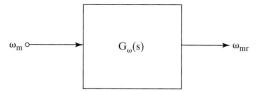

Figure 3.40 Speed-feedback filter

From the motor equations and the above set of definitions, the equations of the motor are written as

$$\dot{x}_1 = -\frac{R_a}{L_a}x_1 - \frac{M}{L_a}x_2 x_3 + \frac{V_a}{L_a} \tag{3.136}$$

$$\dot{x}_2 = -\frac{R_f}{L_f}x_2 + \frac{V_f}{L_f} \tag{3.137}$$

$$\dot{x}_3 = \frac{M}{J}x_1 x_2 - \frac{B}{J}x_3 - \frac{T_l}{J} \tag{3.138}$$

3.11.2 Filter in the Speed-Feedback Loop

Figure 3.40 shows the speed-feedback filter. The transfer function of the filter and tachogenerator can be represented as

$$G_\omega(s) = \frac{\omega_{mr}(s)}{\omega_m(s)} = \frac{H_\omega}{1 + sT_\omega} \tag{3.139}$$

In time domain, this can be rearranged by letting

$$x_4 = \omega_{mr} \tag{3.140}$$

resulting in

$$\dot{x}_4 = \frac{1}{T_\omega}(H_\omega x_3 - x_4) \tag{3.141}$$

The state variable x_4 forms one of the inputs to the speed error/controller block, and that is considered next.

3.11.3 Speed Controller

The speed-controller block diagram is shown in Figure 3.41. The transfer function of the speed controller considered in the present analysis is a proportional-plus-integral controller, given as

$$G_s(s) = K_{ps} + \frac{K_{is}}{s} \tag{3.142}$$

The state diagram of this block is shown in Figure 3.42. Letting

$$\dot{x}_5 = \omega_r^* - \omega_{mr} = \omega_r^* - x_4 \tag{3.143}$$

94 Chapter 3 Phase-Controlled DC Motor Drives

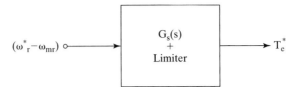

Figure 3.41 Speed-controller block diagram

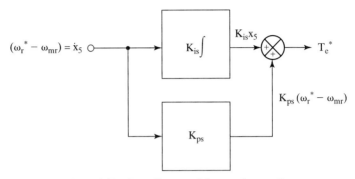

Figure 3.42 State diagram of the speed controller

the torque command signal is derived as

$$T_e^* = K_{ps}(\omega_r^* - x_4) + K_{is}x_5 \qquad (3.144)$$

In order to maintain the drive system in the safe operating region, the torque reference is limited to allowable maximum limits determined by the converter and motor peak capabilities. In this case, let that be $+T_{max}$. This torque reference limit is integrated into the simulation as

$$0 \leq T_e^* \leq + T_{max} \qquad (3.145)$$

3.11.4 Current-Reference Generator

The current reference is derived from the torque reference by using the relationship

$$i_a^* = \frac{T_e^*}{Mi_f} = \frac{K_{ps}}{M} \cdot \frac{\omega_r^*}{x_2} - \frac{K_{ps}}{M} \cdot \frac{x_4}{x_2} + \frac{K_{is}}{M} \cdot \frac{x_5}{x_2} \qquad (3.146)$$

3.11.5 Current Controller

The current controller is of a proportional-plus-integral type with limiter. The transfer function of the current controller is

$$G_i(s) = K_{pi} + \frac{K_{ii}}{s} \qquad (3.147)$$

Similar to the speed-controller equations, the current-controller equations are

$$\dot{x}_6 = (i_a^* - H_c i_a) = (i_a^* - H_c x_1) \qquad (3.148)$$

and

$$v_c = K_{pi}(i_a^* - H_c x_1) + K_{ii} x_6 \qquad (3.149)$$

This control voltage v_c has to be limited to a value V_{cm} corresponding to I_{max}, given in the present case as

$$V_{cm} \equiv H_c I_{max} \qquad (3.150)$$

where I_{max} is the maximum allowable current in the motor and converter. Hence, the limiter is prescribed as

$$0 \leq v_c \leq V_{cm} \qquad (3.151)$$

The lower limit for v_c is made zero: the current in the phase-controlled rectifier cannot be reversed. In the case of a four-quadrant dc motor drive, that limit can be $-V_{cm}$.

Linearizing controller: The linearization of the output-voltage to input-control signal is accomplished by the following:

$$\alpha = \cos^{-1}(v_c/V_{cm}) \qquad (3.152)$$

It could be seen that, when control-signal command goes maximum for maximum positive current error, the result is the triggering angle, α, of zero. That provides the maximum voltage to the armature of the dc motor, building up the armature current and reducing the current error.

Bridge converter: The bridge converter for continuous current can be modeled with its delay as

$$V_a = \sqrt{2} V \sin(\omega_s t + \pi/3 + \alpha) \qquad (3.153)$$

Note that, for every 60 degrees, $\omega_s t$ rolls over, and α is updated at the beginning of each 60 electrical degrees. This can be incorporated into the simulation as

$$\left. \begin{array}{l} \omega_s = 2\pi f_s \\ \omega_s t > \dfrac{\pi}{3}, \omega_s t = 0 \\ \text{When } \omega_s t = 0, \text{ update } \alpha \\ \text{and maintain the same } \alpha \\ \text{until } \omega_s t = \pi/3 \end{array} \right\} \qquad (3.154)$$

3.11.6 Flowchart for Simulation

A flowchart for the digital computer simulation of the one-quadrant dc drive is shown in Figure 3.43. The phase-controlled converter modeling for variations of α greater than 60 degrees has to be considered; accordingly, the algorithm has to be embedded in the software. For precise prediction of SCR and motor-voltage

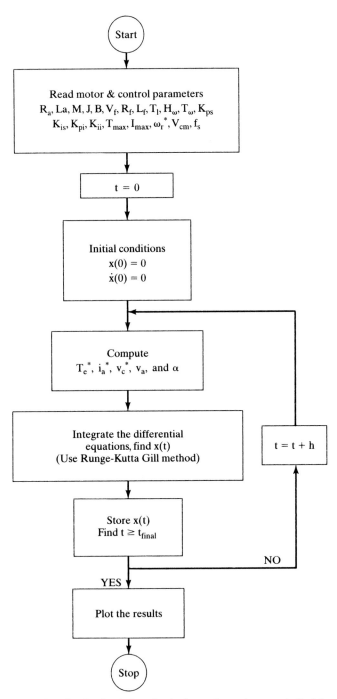

Figure 3.43 Flowchart for the simulation of a single-quadrant phase-controlled dc motor drive

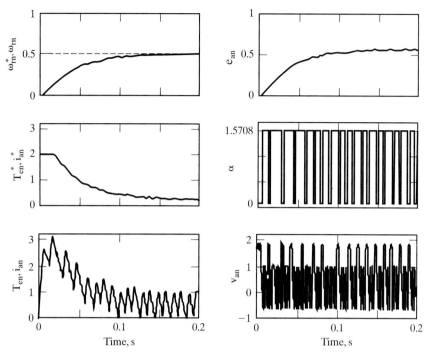

Figure 3.44 Simulation results of armature-controlled dc motor drive with 3-phase SCR converters

waveforms to validate the design of the converter, the use of the right algorithm is required.

In the case of discontinuous armature current, note that a nonzero armature current exists for the conduction duration, which will be less than 60 electrical degrees. For the remaining part of the cycle of 60 electrical degrees duration, the armature current is zero; accordingly, the algorithm has to be modified.

3.11.7 Simulation Results

A set of typical simulation results for a step change in speed reference from standstill to 0.5 p.u with rated field current is shown in Figure 3.44. The speed error is maximum at the start, resulting in a maximum-torque reference, which is limited to 2 p.u. As the speed approaches the set value, the torque reference decreases. The average armature current settles to 0.25 p.u., to generate the electromagnetic torque to counter the load torque, which is set at 0.25 p.u. in this simulation. That happens when the rotor speed drops below the set speed by a small measure. The triggering angle delay is switched to the extremes because of the high proportional gain in the current controller. The speed oscillations are due to high proportional gain of the speed controller. The induced emf follows the rotor speed as the field flux is maintained constant in this simulation.

3.12 HARMONICS AND ASSOCIATED PROBLEMS

The ac input supplies pure sinusoidal voltages and has no harmonic contents in it. It stands to reason that there will be no harmonic currents from the source for a linear load. The SCR control in the converter causes the load to have harmonic voltages besides the dc component. The harmonic voltages produce harmonic currents, which reflect onto the ac-source side. Since the ac source does not contain harmonic sources, the harmonic currents and voltages can be attributed to the phase-controlled converter. Therefore, the phase-controlled converter can be considered a harmonic generator. These harmonics result in heating and torque pulsations in the motor and cause resonance in the power system network. The latter is treated in this section.

3.12.1 Harmonic Resonance

The interaction of the power factor-improving capacitor bank connected to the input of the ac source or the input filters with the phase-controlled dc motor drive provides an ideal setting for resonance, because the circuit is an R, L, and C network. This section contains the resonance condition for such a network and the derivation of the natural system frequency of the power system. An illustrative case study and methods to avoid resonance are outlined.

The input currents to the phase-controlled converter are assumed to be ideally rectangular blocks of alternate polarity and of duration equal to 120 electrical degrees. These waveforms, when resolved into Fourier series, have a number of harmonics besides the dominant fundamental. The fifth and seventh are 1/5th and 1/7th of the strength of the fundamental, respectively. Most of the power-system inputs are installed with large capacitor banks for power-factor correction. The capacitor bank whose reactance is X_c comes in parallel to the impedance of the network including loads $R + jX_1$, as shown in Figure 3.45. That the circuit will resonate under the condition that $X_1 = X_c$ is seen from the equivalent impedance,

$$Z_{eq} = -\frac{jX_c(R + jX_1)}{R + j(X_1 - X_c)} \tag{3.155}$$

When it is in resonance, the equivalent impedance is

$$Z_{eq} = -\frac{jX_c(R + jX_1)}{R} = -\left(1 + \frac{jX_1}{R}\right)jX_c \tag{3.156}$$

If $X_1/R \gg 1$, then the equivalent impedance of the network is approximated as

$$Z_{eq} = \frac{X_1}{R}X_c \tag{3.157}$$

The nth harmonic equivalent impedance is then

$$Z_{eqn} = \left(\frac{nX_1}{R}\right)\left(\frac{X_c}{n}\right) = \frac{X_1}{R}X_c = Z_{eq} \tag{3.158}$$

Section 3.12   Harmonics and Associated Problems   99

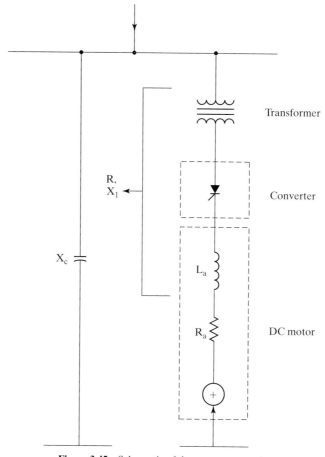

Figure 3.45 Schematic of the system network

That harmonic equivalent impedance does not change from the fundamental is to be noted in this network. Moreover, note the importance of the factor X_1/R in the calculation of the equivalent impedance of the network. By the same approach, the harmonic capacitor current I_{cn} and harmonic capacitor voltage V_{cn} for the nth harmonic are derived in terms of the input current to the power system network, I_n, as

$$I_{cn} \approx j\left(\frac{X_{1n}}{R}\right)I_n \tag{3.159}$$

and

$$V_{cn} = \left(\frac{X_{1n}}{R}\right)X_{cn}I_n = \left(\frac{nX_1}{R}\right)\left(\frac{X_c}{n}\right)I_n \approx \left(\frac{X_1}{R}\right)X_c I_n \tag{3.160}$$

It is seen from the above that the ratio X_1/R is an important factor in such power-system calculations as the equivalent impedance, capacitor current, and capacitor voltage. Note that the reactance of the power-system network changes dynamically,

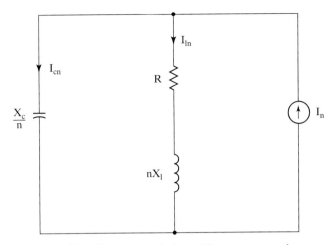

Figure 3.46 Harmonic equivalent of the system network

depending on the loads in the power system. The ratio X_1/R is calculated from the various impedances of the power-system network, such as source, transformer, transmission and distribution lines, and electrical motors and generators. The ratio X_1/R is calculated for the supply frequency. The harmonic currents generated by the phase-controlled converter, I_n, are calculated from the spectrum of the ac currents input to the phase-controlled converter as a function of the triggering angle and load conditions. Alternatively, the harmonic currents can be determined from the output harmonic voltages and harmonic impedances of the load. The harmonic currents are generated by the converter and hence are presented to the network as current sources, as is shown in Figure 3.46. The effect of resonance in this network is illustrated with the following example:

Line-to-line voltage, $V_1 = 13.8$ kV
Capacitive reactance, $X_c = 8.96\ \Omega$
X_1/R of the power system $= 16$
Fifth harmonic current, $I_5 = 60$ A
When $X_{15} = X_{c5}$,

$$I_{c5} = \left(1 + \frac{jX_{15}}{R}\right)(I_5) = \left(\frac{X_{15}}{R}\right)(I_5) = (5)\left(\frac{X_1}{R}\right)(I_5) = (5)(16)(60) = 4800\ \text{A}$$

The voltage across the capacitor bank is

$$V_c = \sqrt{V_{c1}^2 + V_{c5}^2} \qquad (3.161)$$

where V_{c1} is the fundamental-phase voltage and V_{c5} is the phase voltage due to the fifth harmonic current, given by

$$V_{c1} = \frac{V_1}{\sqrt{3}} = \frac{13.8 \times 1000}{\sqrt{3}} = 7967\ \text{V}$$

$$V_{c5} = X_{c5} \times I_{c5} = \left(\frac{X_c}{5}\right)I_{c5} = \frac{8.96 \times 4800}{5} = 8601\ \text{V}$$

Hence,

$$V_c = \sqrt{7967^2 + 8601^2} = 11{,}724 \text{ V per phase}$$

which results in an overvoltage of 47%. This could result in exceeding the voltage and current rating of the capacitor bank. The higher currents could also exceed the circuit breaker rating. Both of these effects can lead to catastrophic failures in the power system, and such a case is cited in reference [15]. The resonance is indicated by the manifold increase in the harmonic currents flowing into the capacitor bank and the resulting overvoltage. Whether resonance might occur can be determined by finding the natural frequency of the power system and whether it coincides with any of the harmonic frequencies generated by the converter. Coincidence of the natural frequency with a harmonic frequency will result in the resonance. To evaluate such a condition, the natural frequency of the power system is determined as follows.

$$\frac{\text{short-circuit volt—ampere rating}}{\text{capacitor volt—ampere rating}} = \frac{\sqrt{3} \cdot V \cdot I_{sc}}{\sqrt{3} \cdot V \cdot I_c} \qquad (3.162)$$

Neglecting the short-circuit resistance, and considering only the short-circuit reactance, X_{sc}, the ratio is given by

$$\frac{\text{short-circuit VA}}{\text{capacitor VA}} = \frac{V \cdot (V/X_{sc})}{V \cdot (V/X_c)} = \frac{X_c}{X_{sc}} = \frac{1}{(2\pi f_s)^2 \cdot L_{sc} \cdot C} \qquad (3.163)$$

Let $f_o^2 = 1/((2\pi)^2 \cdot L_{sc} \cdot C)$, where L_{sc} is the short-circuit inductance and C is the capacitance of the capacitor bank, and substitute this relationship into the ratio of short-circuit-to-capacitor VA rating:

$$\frac{\text{short-circuit VA of the plant}}{\text{capacitor VA capacity}} = \frac{f_o^2}{f_s^2} = f_n^2 \qquad (3.164)$$

where f_n is the natural frequency of the power-system plant expressed in multiples of supply frequency, f_s. The short-circuit rating of the plant is a variable depending on the topology of the network and system loads. To calculate the short-circuit rating of the plant, refer to any standard text on power-system analysis.

The harmonic resonance can be eliminated by attenuating the harmonic currents at their source of generation by installing LC filters selectively tuned to absorb the fifth and other harmonics of concern. Alternatively, inductors can be added in series with the capacitor banks to minimize the current inrush by changing their impedance. It is usual to install LC filters at the input to the converters. These LC filters should not affect the normal functioning of the rest of the power systems, as predetermined by the analysis of the network. This type of solution to the harmonic resonance is ideal for retrofit applications where the addition of new loads has changed the natural frequency. There are other solutions to the harmonics and their resonance. They are of two broad kinds.

Instead of using a six-pulse converter, it is possible to use a twelve-pulse converter, thereby eliminating the fifth and seventh harmonics completely. It is reminded that the most dominant harmonic creating resonance-related problem in the installations is usually the fifth, very rarely the seventh. By eliminating these harmonics, the basic problem of resonance is tackled. The twelve-pulse converter is made up of two six-pulse converters in parallel. The converter sets can be fed through an isolation transformer with two secondaries, one of which is delta-connected, the other star-connected, to provide the displacement in the voltages and hence the steps in the input current drawn from the ac supply. This form of solution to the harmonics is passive and uses a time-honored reliable converter topology with line commutation. Even though the cost of the system will go up (two converter sets, isolation transformer with two secondaries, etc.), there will be considerable reduction in the filter size, and this converter configuration is ideal for high power. The twelve-pulse controlled converter is considered in the following subsection in some detail.

The other kind of harmonic control involves the switching control of the converters. This requires either forced commutation of the SCR switches or self-commutation by using transistors, GTOs, IGBTs, MOSFETs, etc. By a switching of the supply input in a selective manner, the dominant harmonics are selectively eliminated. This is an active solution, but it will involve additional cost compared to the line-commutated converters. Perhaps with time the cost of the self-commutating switches will come down further, to make this option economically attractive for future installations.

3.12.2 Twelve-Pulse Converter for DC Motor Drives

Two six-pulse phase-controlled converters deriving their ac inputs from a set of star (wye) delta-connected secondaries whose primary is delta-connected constitute a twelve-pulse converter. The power schematic is shown in Figure 3.47. The output of the star-connected secondary voltages lags the delta-connected secondary voltages by 30 degrees. The converter outputs are connected through an interphase reactor. This reactor limits the circulating current between the bridge converters and further serves the purpose of paralleling the two outputs of different magnitude provided by the bridge converters. The output voltage is the average of the two bridge converters' outputs.

The control of the bridges is very similar to the control of a six-pulse phase-controlled converter. The operation of the twelve-pulse converter is explained by using Figure 3.48. When v_{a1b1} is positive and 60 degrees from its positive–zero crossing, devices 11 and 12 can be turned on. With a triggering delay angle α, the output is v_1. This output, v_1, is 60 electrical degrees in duration following the line voltage, v_{a1b1}, and, hence, has six pulses per cycle. The output v_2 is very similar to v_1 but is phase-displaced 30 electrical degrees by the phase shift of their ac input voltages. The interphase reactor averages these two output voltages to provide the resultant load voltage v_a, indicated by dashed lines in Figure 3.48. The load voltage contains a twelfth harmonic. The input ac current contains zero 5th and 7th harmonic, as there is no reflected sixth-harmonic current in the load. Hence the minimum higher harmonic present in the ac input is the eleventh. The average output voltage of this twelve-pulse converter for continuous load current is

$$v_a = \frac{v_1 + v_2}{2} \tag{3.165}$$

where the average bridge converter outputs are

$$v_1 = v_2 = 1.35 V \cos \alpha \tag{3.166}$$

Section 3.12　Harmonics and Associated Problems　103

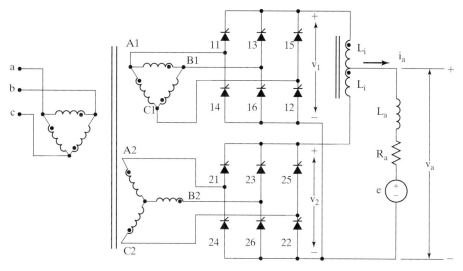

Figure 3.47 Twelve-pulse converter for dc drive

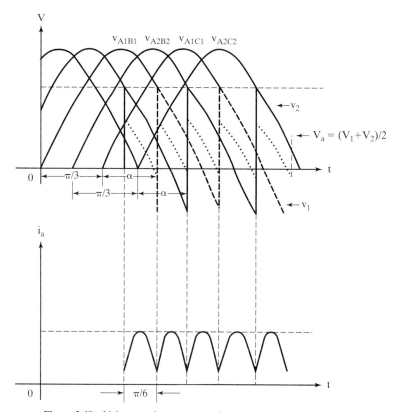

Figure 3.48 Voltage and current waveforms of the 12-pulse converter

104 Chapter 3 Phase-Controlled DC Motor Drives

Hence,

$$v_a = 1.35V \cos \alpha \qquad (3.167)$$

The twelve-pulse converter eliminates the lower harmonics (< eleventh) but does not affect the power factor. This implies that, at low-speed operation of the motor requiring low dc output voltage, the triggering angle α will be close to 90 degrees. This will result in low power factor, and it will be dependent on the motor speed, very much as is the six-pulse converter-controlled dc motor. The improvement in power factor requires additional switching of the bridge converter to control the reactive power, and that is described in the next section.

3.12.3 Selective Harmonic Elimination and Power Factor Improvement by Switching

Increasing the number of switchings in a cycle eliminates the harmonics lower than the switching frequency. For instance, a six-pulse converter has no harmonics below 6 at the output and 5 at the input. Simultaneously, the phase of the input current with respect to its phase voltage can be controlled, resulting in power-factor control and obtaining a desirable operation near or at unity power factor. Such a converter circuit has to have self-commutating switches, i.e., switches capable of turning off regardless of the polarity of the voltage and current. Consider a converter with GTOs, shown in Figure 3.49 with input filter network. The filter contains an inductor L_f and a capacitor C_f per phase. The source has an inductance L_s per phase. Since turn-off is achieved by gate control of the GTO, the switching constraint of the six-pulse SCR converter is removed in this arrangement. The switches can be gated if the anodes are positive with respect to the cathodes over and above their conduction voltage drops. Suppose T_1 and T_2 are conducting and are to be turned off. If the armature current is continuous, a path for its continued flow needs to be ensured. Otherwise, a large voltage due to the inductive drop will arise and appear across the switches, resulting in their destruction. It is averted by gating T_3 and T_6 and shorting

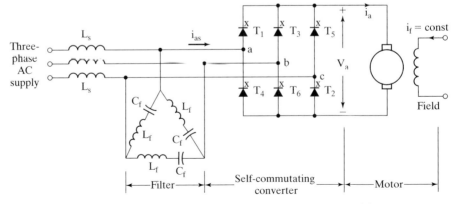

Figure 3.49 Self-commutating converter-controlled dc motor drive

the armature. Due to the delays in turning off T_1 and turning on T_3, an instant may occur with no path for the current, resulting in a large voltage across the devices. That is avoided by overlapping the control signals of T_1 and T_3. As for the ac side, the filter provides a path for the converter input current during device turn-off. During turn-on of a pair of switches, the voltage across the filter will appear across the load, resulting in a large inrush current. This may very well exceed the critical rate of rise of the current of the switches. To prevent such an effect, the filter is provided with the inductor L_f.

The output voltage is controlled by the conduction duration of the power devices. This is in contrast to phase control, in which the output voltage is varied by moving the 60-degree segment of voltage to give the desired average output voltage. Such a movement of the voltage segment results in a movement of current and hence in the phase displacement of the current and a poor power factor. In the present arrangement, the voltage and current are not moved, resulting in zero phase shift of the current. This is illustrated with Figure 3.50. The control of the voltage starts at 60 degrees from the positive–zero crossing of the line voltages, as in the case of the phase control. Each GTO is used for 120 degrees, during which the device *on* and *off* times are varied to provide the desired average voltage to the machine armature. The dc output voltage contains approximately the first harmonic at switching frequency, and the ripple current is minimized by appropriately choosing the maximum available switching frequency consistent with the switch characteristics and loss. The reduction of harmonic armature currents means that a dc reactor might not be necessary in this converter arrangement, hence, none is shown in Figure 3.49. The phase current, i_{as}, consists of the on-time currents of the armature only.

The control of output voltage requires consideration of the turn-on and turn-off times of the power switches. There is a minimum duration of conduction required for the switch, below which the switching losses would dominate the quantum of power transfer from the source to load or vice versa. Hence, low-speed operation requiring low voltages will have problems with pulse-width control. In that region, then, it is necessary to adopt phase control. This comes at the expense of low power factor and high harmonics. The encouraging factor in this mode of operation is that it is required only for less than 10% of the speed range in many applications, beyond which pulse-width control can be resorted to. The control is identical for regeneration mode for both the pulse-width and phase control of this converter.

Care must be exercised in the selection of filter component ratings and snubber ratings for the system, to avoid a resonance due to the coincidence of the switching frequency with the natural frequency of the circuit or the resonant frequency of the filter.

These converter configurations are in practice at 600-kVA ratings for some elevator applications. The ratings of these converters will increase in the future, as the harmonic reduction and power-factor improvement required by the utilities for energy savings become a key economic factor in the day-to-day operation of industries. To reduce the cost of these converter arrangements, the lower half of the bridge can have SCR switches; only the upper half of the bridge need have self-commutating switches.

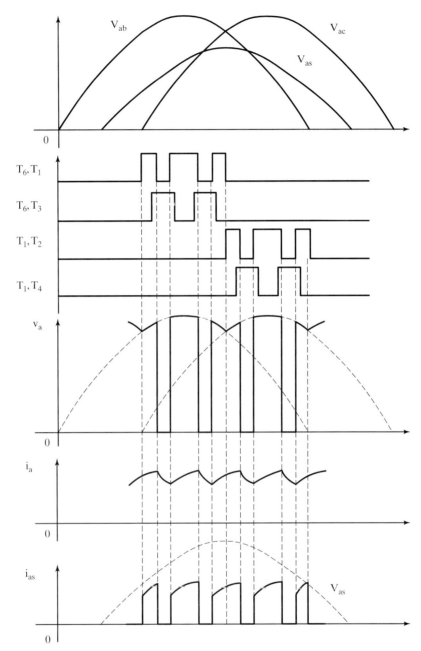

Figure 3.50 Waveforms of the self-commutating converter-controlled dc motor drive

3.13 SIXTH-HARMONIC TORQUE

The previous section dealt with the input-side harmonics of the phase-controlled converter and the associated problems. This section presents the effect of the harmonics on the output side of the six-pulse phase-controlled converter, i.e., the input harmonics to the dc motor. The dominant harmonic in the six-pulse converter is the sixth harmonic, and only that one is considered here. Compared to the effects of the sixth harmonic, the effects of other harmonics are smaller, because the harmonic impedances are very high for higher-order harmonics and thus result in the attenuation of harmonic currents. The approach for calculating the effects of the other harmonics can follow along the same lines as the calculation for the effects of sixth harmonic developed in the following. The effect of the dominant harmonic is to produce a torque whose average is zero and which is therefore useless from the point of view of driving a load. It also creates losses in the form of heating by increasing the rms value of the input armature current. The harmonic torque can contribute to speed ripples at low speed if the mechanical impedance is not adequate to filter it. It can also excite some of the resonant frequencies of the load and disable the smooth functioning of the motor-drive system. It is essential to evaluate the magnitude of the harmonic torque and its effect on the armature heating for application considerations. For both the continuous and discontinuous current-conduction modes, the relevant equations are derived; some are used from the previous section on steady-state analysis.

3.13.1 Continuous Current Conduction Mode

The armature voltage and current waveforms corresponding to Figure 3.22 (given in section 3.4.2) are considered for harmonic analysis. The armature voltage can be resolved into Fourier series as

$$v_a = \sum_{n=0}^{\infty} (a_n \sin n\theta + b_n \cos n\theta) + a_0 \tag{3.168}$$

where

$$a_0 = V_a = 1.35 V \cos \alpha \tag{3.169}$$

$$a_n = \frac{3V_m}{\pi} \left[\frac{\sin \overline{n-1}\theta}{n-1} - \frac{\sin \overline{n+1}\theta}{n+1} \right]_{\frac{n}{3}+\alpha}^{\frac{2\pi}{3}+\alpha} \tag{3.170}$$

$$b_n = \frac{3V_m}{\pi} \left[\frac{\cos \overline{n-1}\theta}{n-1} - \frac{\cos \overline{n+1}\theta}{n+1} \right]_{\frac{n}{3}+\alpha}^{\frac{2\pi}{3}+\alpha} \tag{3.171}$$

where n is the harmonic number and from which the sixth harmonic components are

$$a_6 = \frac{3\sqrt{2}V}{\pi}\left[\frac{\sin 7\alpha}{7} - \frac{\sin 5\alpha}{5}\right] \tag{3.172}$$

$$b_6 = \frac{3\sqrt{2}V}{\pi}\left[\frac{\cos 7\alpha}{7} - \frac{\cos 5\alpha}{5}\right] \tag{3.173}$$

The peak value of the sixth-harmonic armature voltage is

$$V_{6p} = \sqrt{a_6^2 + b_6^2} \qquad (3.174)$$

which, upon substitution of (3.172) and (3.173), yields

$$V_{6p} = \frac{3\sqrt{2}V}{\pi} \cdot \frac{1}{35}\sqrt{74 - 70\cos 2\alpha} \qquad (3.175)$$

This peak sixth-harmonic armature voltage is maximum at a triggering angle of 90 degrees and is given as

$$V_{6p} = 0.463\ V \qquad (3.176)$$

The sixth-harmonic current and electromagnetic torque are derived as

$$I_6 = \frac{V_{6p}}{Z_6} \qquad (3.177)$$

$$T_{e6} = K_b I_6 \qquad (3.178)$$

where the sixth-harmonic impedance is given by

$$Z_6 = R_a + j6\omega_s L_a \cong j6\omega_s = 6X_a \qquad (3.179)$$

The peak sixth-harmonic torque expressed in p.u. is

$$T_{e6n} = \frac{T_{e6}}{T_{er}} = \frac{K_b \dfrac{3\sqrt{2}V}{\pi} \cdot \dfrac{1}{35}\sqrt{74 - 70\cos 2\alpha}}{6X_a K_b I_{ar}} = 0.00643 \frac{V/V_r}{X_a I_{ar}/V_r}\sqrt{74 - 70\cos 2\alpha}$$

$$= 0.00643 \frac{V_n}{X_{an}}\sqrt{74 - 70\cos 2\alpha} \qquad (3.180)$$

where V_n is the normalized voltage and X_{an} is the normalized armature reactance given by

$$V_n = \frac{V}{V_r}\ \text{p.u.} \qquad (3.181)$$

$$X_{an} = \frac{I_{ar} X_a}{V_r}\ \text{p.u.} \qquad (3.182)$$

Note that the sixth-harmonic torque is a function of the triggering angle, which would change with the induced emf and armature current. Hence, the triggering angle corresponding to the induced emf and armature current has to be evaluated to find the sixth-harmonic torque. For a large dc machine with a base voltage of 500 V, base current of 2650 A, 0.027 p.u. resistance, 0.2 p.u. inductance, emf constant of 8V/rad/sec, and input rms line-to-line voltage of 560 V, the peak sixth harmonic is computed; it is shown in Figure 3.51. The torque pulsations at low speeds are higher in magnitude than those at high speeds, because the triggering angle will become higher with decreasing speeds, thus contributing to increasing sixth-harmonic voltage. The fact that the torque pulsation can be as high as 0.42 p.u. at zero speed can make it unsuitable for high-precision applications. If need be, the pulsations can be

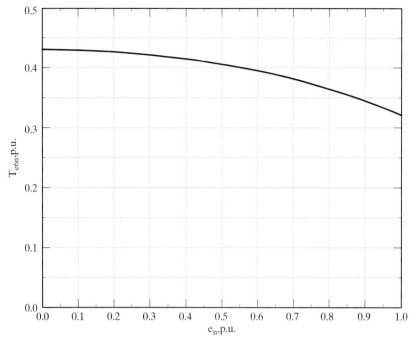

Figure 3.51 Normalized peak sixth-harmonic torque vs. induced emf

reduced by increasing the harmonic impedance with an external reactor. The other solution for decreasing the harmonic voltage is dealt with in the next chapter: using a chopper control.

Calculation of armature resistive loss: The rms armature current is found from

$$I_{rms} = \sqrt{I_{av}^2 + I_{6p}^2/2} \qquad (3.183)$$

where the average dc current is obtained from

$$I_{av} = \frac{1.35V \cos \alpha - E}{R_a} \qquad (3.184)$$

The ratio of armature resistive loss for phase controlled input to a pure dc input is

$$\frac{I_{av}^2 + \dfrac{I_6^2}{2}}{I_{av}^2} = 1 + \frac{1}{2}\left(\frac{I_{6p}}{I_{av}}\right)^2 \qquad (3.185)$$

where

$$\frac{I_{6p}}{I_{av}} = \frac{V_6/6X_a}{(1.35V \cos \alpha - E)/R_a} \qquad (3.186)$$

By substituting for V_6, the ratio of currents is

$$\frac{I_{6p}}{I_{av}} = \frac{0.00643}{\tan\beta} \cdot \frac{\sqrt{74 - 70\cos 2\alpha}}{1.35\cos\alpha - E/V} \quad (3.187)$$

The increase in armature resistive loss due to the sixth-harmonic component is

$$\frac{1}{2}\left(\frac{I_{6p}}{I_{av}}\right)^2 = 0.2067 \times 10^{-4}\left[\frac{74 - 70\cos 2\alpha}{(1.35\cos\alpha - E/V)^2 \cdot \tan^2\beta}\right] \quad (3.188)$$

This is difficult to interpret when the average current becomes zero; hence, an alternative formulation is given below. The normalized armature resistive loss is

$$P_{cn} = \frac{P_c}{P_{cr}} = \frac{I_{av}^2 + I_{6p}^2/2}{I_{ar}^2} = \frac{\left(\frac{1.35V\cos\alpha - E}{R_a}\right)^2 + \frac{0.00643^2}{2} \cdot \frac{V^2}{X_a^2}(74 - 70\cos 2\alpha)}{(V_b/Z_b)^2}$$

$$= \left(\frac{1.35V_n\cos\alpha - E/V_b}{R_{an}}\right)^2 + \frac{0.2067 \times 10^{-4}V_n^2}{X_{an}^2}(74 - 70\cos 2\alpha) \quad (3.189)$$

and this equation is valid for all

$$\alpha \leq \cos^{-1}\left(\frac{E}{1.35V}\right), \text{rad} \quad (3.190)$$

If $V_r = V$, then $V_n = 1$, and hence the normalized armature resistive loss is given by

$$P_{cn} = \left(\frac{1.35V_n\cos\alpha - e_n}{R_{an}}\right)^2 + \frac{0.2067 \times 10^{-4}}{X_{an}^2}(74 - 70\cos 2\alpha), \text{ p.u.} \quad (3.191)$$

where the normalized induced emf and normalized resistance and reactance are

$$e_n = \frac{E}{V}, \text{ p.u.} \quad (3.192)$$

$$R_{an} = R_a/Z_b, \text{ p.u.} \quad (3.193)$$

$$X_{an} = X_a/Z_b, \text{ p.u.} \quad (3.194)$$

For the same machine given in the previous T_{e6n} calculations, the normalized armature resistive loss against e_n is calculated with the above procedure, as shown in Figure 3.52. The average armature current is assumed to be 1 p.u. The copper losses have increased by approximately 9% at zero speed, and at 1 p.u. speed they have increased only by 5%. For continued operation, this increase in losses has to be compensated for by derating the machine, and note this is not negligible.

3.13.2 Discontinuous Current-Conduction Mode

The discontinuous current-conduction mode has been dealt with in the steady-state analysis section. The normalized average and sixth-harmonic voltages are found, from which the respective currents are derived as

$$I_{avn} = (V_{an} - E_n) \cdot \frac{\theta_x}{R_{an} \cdot \pi/3} \quad (3.195)$$

Section 3.13 Sixth-Harmonic Torque 111

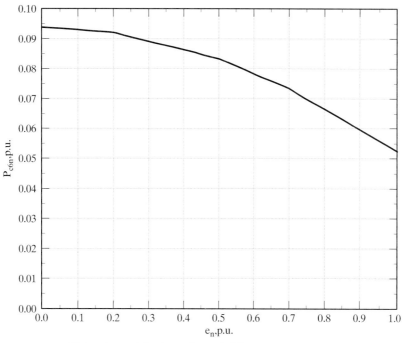

(i) Armature resistive losses due to sixth-harmonic current

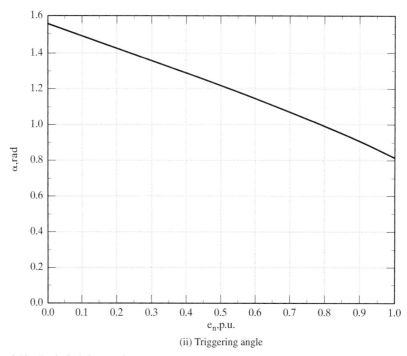

(ii) Triggering angle

Figure 3.52 Peak sixth-harmonic armature resistive losses and triggering angle in normalized units, as a function of normalized induced emf for continuous conduction

where the normalized dc input voltage is

$$V_{an} = \sqrt{2}V_n\sqrt{2 - 2\cos\theta_x} \cdot \frac{\cos(\pi/3 + \alpha - \theta_1)}{\theta_x} \quad (3.196)$$

$$\theta_1 = \tan^{-1}\left(\frac{\sin\theta_x}{1 - \cos\theta_x}\right) \quad (3.197)$$

Note that the average normalized motor input voltage is for the conduction duration, since only it contributes to the current. The current is averaged over the period of $\pi/3$ instead of θ_x. The sixth-harmonic current has a peak given by

$$I_{6pn} = \frac{V_n}{X_{an}} \cdot \left[\frac{1}{\pi\sqrt{2}}(a_1 + a_2) - 2\sqrt{a_1 \cdot a_2}\cos(2\alpha + 2\cdot\pi/3 + \theta_3 - \theta_2)\right] \quad (3.198)$$

where

$$a_1 = \frac{2 - 2\cos 7\theta_x}{49} \quad (3.199)$$

$$a_2 = \frac{2 - 2\cos 5\theta_x}{25} \quad (3.200)$$

$$\theta_2 = \tan^{-1}\left(\frac{\sin 5\theta_x}{\cos 5\theta_x - 1}\right) \quad (3.201)$$

$$\theta_3 = \tan^{-1}\left(\frac{\sin 7\theta_x}{\cos 7\theta_x - 1}\right) \quad (3.202)$$

The triggering angle should be greater than its critical value, derived earlier, to ensure that the operating point is in the discontinuous current-conduction mode. The peak normalized sixth-harmonic torque is equal to the normalized peak sixth-harmonic current, and the resistive loss due to it in the armature is given in normalized units as

$$P_{e6n} = \frac{I_{6pn}^2}{2}, \text{ p.u.} \quad (3.203)$$

The algorithm for the computation of the sixth-harmonic torque is as follows: (i) Find the impedance angle β and the critical triggering angle α_c. (ii) Choose an α greater than α_c that will give the desired average armature current for the assumed normalized induced emf. (iii) Find the conduction angle, θ_x, using equation (3.60), by Newton–Raphson technique. (iv) The sixth-harmonic current and the corresponding armature losses are evaluated from the expression derived above. For the same machine parameters as in the continuous-conduction mode and an average armature current of 1.15 p.u., the triggering angle (α) and conduction angle (γ) vs. normalized induced emf are shown in Figure 3.53. The peak normalized sixth-harmonic torque and its corresponding loss vs. the normalized induced emf are shown in Figure 3.54. The maximum harmonic torque occurs at

Section 3.13　Sixth-Harmonic Torque　113

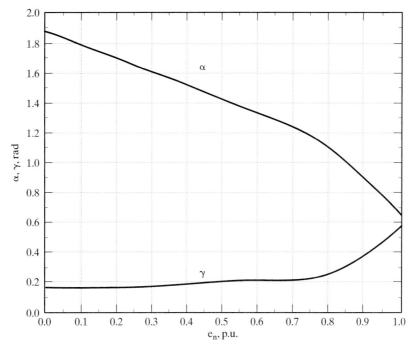

Figure 3.53　Triggering angle and conduction angle vs. induced emf for discontinuous conduction

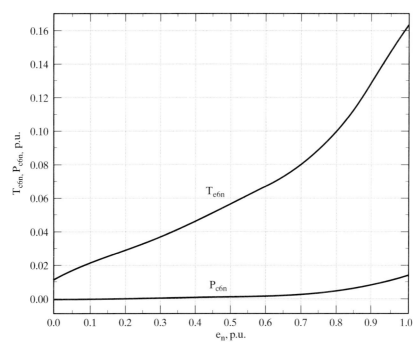

Figure 3.54　Normalized peak sixth-harmonic torque and armature resistive loss vs. induced emf in discontinuous conduction

1 p.u. induced emf or at 1 p.u. speed and has a minimum value at zero speed. This is contrary to the continuous current-conduction mode, where the maximum sixth-harmonic torque occurs at zero speed and the minimum occurs at 1 p.u. speed. The maximum sixth-harmonic torque is approximately 0.16 p.u. for the example considered; that is very high for many high-performance applications. The harmonic currents and torques are lower in the discontinuous current-conduction mode compared to the continuous current-conduction mode, for this machine.

3.14 APPLICATION CONSIDERATIONS

The torque–speed characteristics in steady-state and intermittent operation provide the power and speed ratings of the dc motor. Torque–speed characteristics translate into voltage and current ratings of the motor and converter. The controller design coordinates the steady-state with the dynamic requirements of the motor drive. There are a number of factors to be considered in the selection of the motor, converter, and controller. Some of these factors are listed in the following, from reference [19]:

1. Motor
   - Power, hp.
   - Service factor
   - Armature and field voltage
   - Armature and field current
   - Efficiency
   - Preferred direction of rotation
   - Rated speed
   - Maximum speed
   - Motor friction and inertia
   - Load friction and inertia
   - Type of enclosure
   - Class of insulation
   - Cooling arrangements
   - Safe temperature rise
   - Tachometer characteristics
   - Encoder details
   - Brush resistance, life span, and maintenance
   - Mounting details
   - Gears for speed reduction
   - Protection features
2. Converter
   - Input ac voltage, output dc voltage
   - RMS current rating
   - Maximum current rating (short-time rating)
   - Efficiency vs. load
   - Cooling
   - Safe thermal limits

- Protection features
  - Overvoltage
  - Overcurrent
- Contactors
- Enclosure
3. Controller
   - Features such as regeneration
   - Reversal of speed
   - Steady-state accuracy or error
   - Dynamic responses (overshoot, bandwidth, rise time, etc.)
   - Flexibility to change controller gains
   - Metering and data logging
   - Protection coordination of the motor and converter
   - Interface capability to other processes
   - Loss of field supply, mains supply, etc.

The motor might have to be derated to account for the additional losses produced by the harmonic currents. This derating has to be a function of load and of the type of converter used.

3.15 APPLICATIONS

Because the phase-controlled dc motor drive is one of the earliest and most reliable variable-speed drives, it has found a large number of industrial applications. An example of its application is described here to impart a sense of reality to the theory and analysis described above.

A schematic for a flying shear is shown in Figure 3.55. Flying shears are used in the metal-rolling industry for cutting materials to desired lengths to suit customer needs. A flying shear consists of two blades rotating in opposite directions from their initial positions of c and c' at a velocity of V_b m/s. The material to be cut moves in at a velocity of V_m m/s. The blades are driven by the shafts of two separately-excited dc motor drives connected in series and controlled from a dual set of

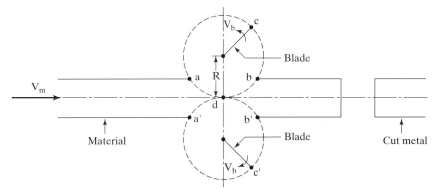

Figure 3.55 Flying-shear system schematic

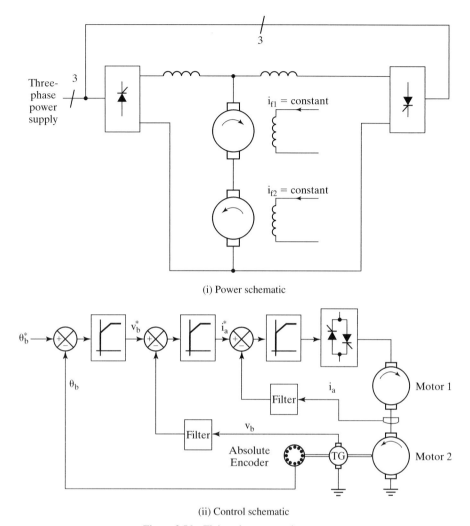

Figure 3.56 Flying-shear control system

three-phase controlled-rectifier bridges, as shown in the power schematic of Figure 3.56. When the material is to be cut, the blades are accelerated until they come to positions a and a'. The acceleration is made to be a function of the velocity of the material inflow, V_m, of the acceleration of the blades, V_b, and of the positions of the blades and the material. A slight deceleration is initiated at this point as the shearing starts and the blades reach point d. At position d, the front end of the material has been cut off. In order not to be in the way obstructing the flow of the material, the blades are accelerated till they reach points b and b', and then they are decelerated to slide into their original positions c and c'. The drive systems controlling the blades have to be highly responsive and are position-controlled drive systems. The key variables of the system are shown in Figure 3.57 for a cycle of operation. The blade velocity exceeds 20 m/s nominally for a short period, and at the time of the metal cutting it equals the velocity of the material. The armature-current commands for the dc

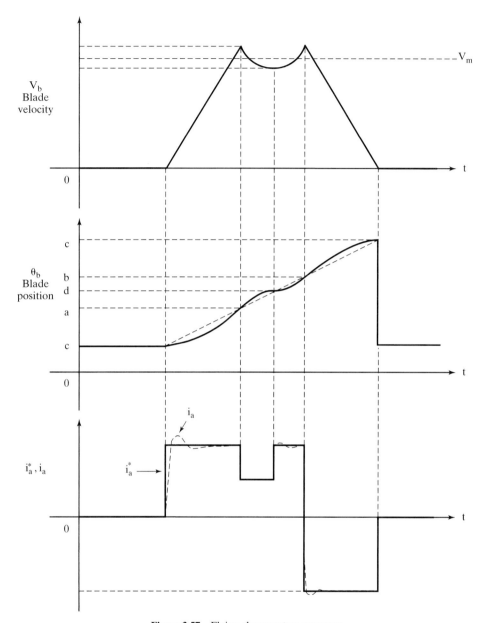

Figure 3.57 Flying-shear system response

motors are at positive and negative maxima during acceleration and deceleration, respectively. The actual motor currents have to follow their commands closely for precise operation of the flying shear. This requires a fast-acting current loop, i.e., a very high current-loop bandwidth for the system. For example, the flying-shear drive system might consist of two 1250-kW motors of 600 V. The rated current of this system is approximately 2200 A. Current reversals and ramping at this magnitude within a few milliseconds, say 50 to 100 ms, are desirable for a fast-acting flying-shear system.

3.16 PARAMETER SENSITIVITY

A change in temperature increases the armature resistance of the dc machine. In the closed-loop drive, this will not have an effect on such output variables as torque and speed. It can increase the armature resistive loss; for example, a temperature rise of 150°C varies the resistance and loss by more than 70% from its nominal value at ambient temperature. Therefore, reconsideration of the thermal limits of the machine and of derating is essential.

In an open-loop motor drive, the changes in armature resistance will decrease the rotor speed. Assuming that the load torque is a constant amounts to a constant armature current; hence, the induced emf decreases for a given input voltage, because the resistance drop has increased.

Decreasing induced emf means a reduction in rotor speed. Note that the ambient conditions and changing conditions are denoted by the subscripts 0 and c, respectively.

$$\omega_{m0} = \frac{e_0}{K_b} = \frac{V_a - i_{a0}R_{a0}}{K_b} \quad (3.204)$$

$$\omega_{mc} = \frac{e_c}{K_b} = \frac{V_a - i_{a0}R_{ac}}{K_b} \quad (3.205)$$

The change in speed due to parameter sensitivity is

$$\frac{\omega_{mc}}{\omega_{m0}} = \frac{V_a - i_{a0}R_{ac}}{V_a - i_{a0}R_{a0}} \quad (3.206)$$

The changing variables can be written as a sum of ambient or normal values and the incremental changes:

$$\omega_{mc} = \omega_{m0} + \delta\omega_m \quad (3.207)$$

$$R_{ac} = R_a + \delta R_a \quad (3.208)$$

When $\delta\omega_m$ is the change in speed due to δR_a change in armature resistance, then by substituting equation (3.207) and (3.208) into (3.206) we get

$$1 + \frac{\delta\omega_m}{\omega_{m0}} = 1 - \frac{i_{a0}\delta R_a}{V_a - i_{a0}R_a} \quad (3.209)$$

from which the incremental change in speed is expressed as

$$\frac{\delta\omega_m}{\omega_{m0}} = -\frac{i_{a0}\delta R_a}{e_0} = -\frac{i_{a0}\delta R_a}{K_b\omega_{m0}} \quad (3.210)$$

Hence,

$$\delta\omega_m = -\left[\frac{i_{a0}}{K_b}\right]\delta R_a \quad (3.211)$$

Likewise, the effect due to saturation can be computed. In the case of saturation, the field flux will change and hence so will the emf constant K_b.

The dynamic performance is affected along with the changing armature resistance and saturation level. That is bound to affect the transient response if the controllers are fixed.

3.17 RESEARCH STATUS

Current research is focused mainly on the following aspects of the dc motor drive:

(i) Adaptive controllers
(ii) Improvement of power factor and reduction of harmonics

Adaptive controllers have become necessary to meet changing load conditions without affecting dynamic performance. This is evident, for example, in the variation of inertia as a robot's arm is gradually moved. A number of processes need to avoid torsional resonance, and they can be controlled by changing either the controller configuration or the controller gains. New control algorithms using fuzzy, neural, and other nonlinear techniques are on the rise; most of them are implemented with microprocessors.

With devices such as GTOs, transistors, and MOSFETs, it is possible to resort to pulse-width modulation of the input ac voltage to eliminate the harmonics selectively. That would keep the input current more nearly sinusoidal and the power factor close to unity. Even though such switching techniques increase the complexity of control and the switching losses, they are worth investigating with the devices and controllers now becoming cost-effective.

3.18 SUGGESTED READINGS

1. David L. Duff and Allan Ludbrook, "Reversing thyristor armature dual converter with logic crossover control," *IEEE Trans. Industry and General Application*, vol. IGA-1, no. 3, pp. 216–222, 1965.
2. T. Krishnan and B. Ramaswami, "A fast-response dc motor speed control system," *IEEE Trans. Industry Applications*, vol. IA-10, no. 5, pp. 643–651, Sept./Oct. 1974.
3. T. Krishnan and B. Ramaswami, "Speed control of dc motor using thyristor dual converter," *IEEE Trans. Industrial Electronics and Control Instrumentation,* vol. IECI-23, no. 4, pp. 391–399, Nov. 1976.
4. Geza Joos and T.H. Barton, "Four-quadrant dc variable speed drives—design considerations," *Proc. IEEE*, vol. 63, no. 12, pp. 1660–1668, Dec. 1975.
5. J. K. Haggerty, J. T. Maynard, and L. A. Koenig, "Application factors for thyristor converter dc motor drives," *IEEE Trans. Industry General Applications*, vol. IGA-7, pp. 718–728, Nov./Dec. 1971.
6. J. P. Sucena-Paiva, R. Hernandez, and L.L. Freries, "Stability study of controlled rectifiers using a new discrete model," *Proc. Institute of Electrical Engineers*, London, vol. 119, pp. 1285–1293, Sept. 1972.
7. C. E. Robinson, "Redesign of dc motors for applications with thyristor power supplies," *IEEE Trans. Industry and General Applications*, vol. IGA-4, pp. 508–514, Sept./Oct. 1968.

8. Leland A. Schlabach, "Conduction limits of a three-phase controlled converter in inversion," *IEEE Trans. Industry Applications*, vol. IA-22, no. 2, pp. 298–303, March/April 1986.
9. B. R. Pelly, *Thyristor Phase-Controlled Converters and Cycloconverter*, New York, Wiley, 1971.
10. S. B. Dewan and W. G. Dunford, "Improved power factor operation of a three-phase rectifier bridge through modified gating," *Conference Record, IEEE-IAS Annual Meeting*, pp. 830–837, Oct. 1980.
11. B. Ilango, R. Krishnan, R. Subramanian, and R. Sadasivam, "Firing circuit for three-phase thyristor bridge rectifier," *IEEE Trans. Industrial Electronics and Control Instrumentation*, vol. IECI-25, no. 1, pp. 45–49, Feb. 1979.
12. Remy Simard and V. Rajagopalan, "Economical equidistant pulse firing scheme for thyristorized dc drives," *IEEE Trans. Industrial Electronics and Control Instrumentation*, vol. IECI-22, no. 3, pp. 425–429, Aug. 1975.
13. J. S. Wade Jr. and L. G. Aya, "Design for simultaneous pulse triggering of SCRs in three phase bridge configuration," *IEEE Trans. Industrial Electronics and Control Instrumentation*, vol. IECI-18, no. 3, pp. 104–106, Aug. 1971.
14. N. Peric and I. Petrovic, "Flying shear control system," *IEEE Trans. on Industry Applications*, vol. 26, no. 6, pp. 1049–1056, Nov./Dec. 1990.
15. Guy Lemieux, "Power system harmonic resonance—A documentary case," *IEEE Trans. on Industry Applications*, vol. 26, no. 3, pp. 483–488, May/June 1990.
16. Friedrich Frohr and Fritz Orttenburger, *Introduction to Electronic Control Engineering*, Wiley Eastern Limited, Madras, 1988.
17. Leland A. Schlabach, "Analysis of discontinuous current in a 12-pulse thyristor dc motor drive," *IEEE Transactions on Industry Applications*, vol. 27, no. 6, pp. 1048–1054, Nov./Dec. 1991.
18. Hiromi Inaba, Seiya Shima, Akiteru Ueda, Takeki Ando, Toshiaki Kurosawa and Yoshio Sakai, "A new speed control system for dc motors using GTO converter and its application to elevators," *IEEE Transactions on Industry Applications*, vol. 21, no. 2, pp. 391–397, March/April 1985.
19. Geza Joos, "Variable Speed Drives," Short-term Course Lecture Notes, Center for Professional Development, NJ, Nov. 1987.

3.19 DISCUSSION QUESTIONS

1. The ratio between the field and armature time constants is of the order of 10 to 1000 in the separately-excited dc machines. Will it be of the same order in series field dc machines?
2. Single-phase converters are avoided in high-performance applications. Why?
3. Three-phase converters are better than single-phase converters when harmonics are considered. Will this advantage be of any use in motor drives?
4. Three-phase converters are faster in response than single-phase converters. Compare their speed of response.
5. The converters require a filter to tackle the harmonics. Which one of the following is preferable: A filter on the input side or one at the output side?
6. A freewheeling diode reduces the harmonics in the output current. This is true for triggering angles above a certain value. What is that limiting triggering angle?

7. Freewheeling can be accomplished without using a diode across the load by using the thyristors in the phase legs. Such a technique requires a modification of the triggering signals. Discuss the conceptual aspects of the modification.
8. Is a microprocessor necessary for implementing the modification in question 7?
9. Steady-state analysis of the dc motor drive assumes that there is no speed ripple. Is that justifiable?
10. Precise control of α is required for accurate control of a dc motor. How can this be ensured in practice?
11. A dc motor drive with an inner current loop alone is a torque amplifier. Is there any use for such a drive?
12. The speed-controlled dc drive uses actual speed for feedback control. In the absence of a tachogenerator, what would be done to close the speed loop?
13. Current feedback can be obtained by sensing the ac-side currents and rectifying them to find the armature current. What is the advantage of this method of current sensing over direct sensing of armature current?
14. Supply transformers are part of the dc motor-drive systems. Are they essential for all ranges of the motor drive?
15. Even though a dc motor-drive system is a sixth-order sampled-data control system, a number of approximations have been made to synthesize the controllers. Why not design the controllers without resorting to any approximations?
16. Can state feedback control be used in the design of current and speed controllers in the dc motor drive? Discuss the implementation of such a control scheme.
17. For speed-controller design, the fourth-order inner current loop has been approximated into a first-order transfer function. Discuss the merits and demerits of the approximation.
18. How critical is the current loop design in a two-/four-quadrant dc motor drive?
19. Discuss the design of the control and gating circuits in the four-quadrant dc motor drive.
20. The power switches have been assumed to be ideal, i.e., they have no forward voltage drops and losses. A dc motor drive is driven from a 12-V ac main. Can the switches be considered ideal now for the analysis of this dc motor drive?
21. Discuss the efficiency of the three-phase fully-controlled converter-fed dc motor drive.
22. "The phase-controlled converter is a harmonic generator." Justify this comment.
23. Load changes have a profound effect on the dynamic response of the dc motor drive. What are the techniques to counter the load sensitivity on the performance of the dc motor drive?

3.20 EXERCISE PROBLEMS

1. Derive the normalized steady-state performance equations of a series-excited dc motor drive.
2. A 100-hp open drip-proof dc motor rated at 500 V and 1750 rpm has the following parameters at rated field current.

   $I_a = 153.71$ A $\quad R_a = 0.088$ Ω
   $L_a = 1.83$ mH $\quad K_b = 2.646$ V/rad/sec

   Draw its torque, induced emf, power, and field flux vs. speed in normalized units for rated armature current and for an intermittent operation at 1.2 p.u. armature current for the speed range of 0 to 2 p.u.

3. The dc motor given in Problem 2 is operated from a fully-controlled three-phase converter fed from a 460-V, 60-Hz, 3-phase ac main. Calculate the triggering angle when the machine is delivering rated torque at rated speed. The armature current is assumed to be continuous.

4. Assuming the current is continuous, draw α vs. speed, maintaining the load torque at rated value, for Problem 3, for a speed range of 0 to 1 p.u.

5. A separately-excited dc motor is controlled from a three-phase full-wave converter fed from a 460-V, 3-phase, 60-Hz ac supply. The dc motor details are as follows: 250 hp, 500 V, 1250 rpm, $R_a = 0.052\ \Omega$, $L_a = 2$ mH.
    (i) Find the rated current and K_b when the field is maintained at rated value.
    (ii) Draw the torque–speed characteristics as a function of triggering angle, α.

6. The motor drive given in Example 3.1 is driving a load proportional to the square of the speed. Draw the torque–speed characteristics of the drive system and α vs. speed characteristics. The combined motor and load mechanical constants are as follows:
    $B_t = 0.06$ N·m/rad/sec, $J = 5$ kg–m²

7. Determine the electrical and mechanical time constants of the dc motor in Problems 5 and 6, respectively.

8. The converter in one case is linearized and in the other case is not modified. The control voltage is ±10 V max in both cases. Design the current controller gains (PI) for the motor drive given in Problem 5. Consider $B_t = 0.06$ N·m/rad/sec and $J = 5$ kg–m². Evaluate the time and frequency responses of the current loop for the two cases.

9. An innermost unity feedback voltage loop is introduced in Problem 8. The voltage controller is of proportional type, with gain K_v. Derive the current-loop transfer function, and explain the impact of the voltage loop on the performance of the current loop.

10. A two-quadrant dc motor drive is fed from a three-phase controlled converter operating on a 230-V, 60-Hz ac main. The motor details are as follows:

    230 V, 40 hp, 1500 rpm,

    $R_a = 0.066\ \Omega$ $L_a = 6.5$ mH $J = 25$ kg–m² $B_1 = 0.4$ N·m/rad/sec
    $H_c = 0.05$ V/A $H_\omega = 0.576$ V/rad/sec $T_\omega = 0.002$ sec $K_b = 1.33$ V/rad/sec

    The load is frictional. (i) Calculate the controller constants by using the method developed in the book. (ii) Evaluate the stability of the motor drive, using the complete model. (iii) Draw the Bode plots for the complete and simplified transfer functions between speed and its commanded value. (iv) Comment on the discrepancy in the results. (v) Select the ratings of the converter power switches. The maximum armature current is not to exceed 1.2 p.u.

11. The dc motor drive given in Example 3.5 has its inner current loop disabled and also its current controller removed. Compare its speed performance to the drive given in Example 3.5. In the present case, the speed controller constants will be different and hence will have to be recalculated.

12. Prove that, when $T_c = T_1$, the damping coefficient and natural frequency of the current-loop transfer function remain equal to the case for $T_c = T_2$. Assume T_r is very, very small compared to T_1 and T_2.

13. Design gating and control circuitry for a half-controlled three-phase converter to drive a dc motor. Analyze the dc output voltage into harmonics and plot the dominant harmonic as a function of triggering angle.

14. A fully-controlled three-phase converter is driving a separately-excited dc motor in open loop. Plot its torque vs. speed characteristics as a function of triggering angle both in the continuous and in the discontinuous modes of operation. Use the motor parameters given in Problem 10.

15. The flying-shear control system described in the application section is to be designed with separately-excited dc motor drives. The power schematic of the drive system is shown in Figure 3.56(i); the control schematic is shown in Figure 3.56(ii). The system details are: Rated power = 1500 kW per motor, Rated voltage = 600 V per motor, Rated current = 2650 A per motor, Rated speed = 600 rpm, Armature resistance, R_a = 0.003645 Ω per motor, Armature inductance, L_a = 0.0001 H per motor, Emf constant, K_b = 9.364 V/rad/sec, Circulating current reactors, L_c = 0.001 H per motor, Total moment of inertia, J = 1500 Kg-m^2, Gear ratio = 7, Radius of blade path = 0.75 m.

Determine the controller gains and current-loop bandwidth if the input voltages supplied to the converter through an isolation transformer are 1000 V, 3-phase, 60 Hz. Assume 10 V equals 5 m/s and 10 V equals 2700 A for the control signals and that the metal input velocity is 4.2 m/s. Write a computer program to simulate the drive system and show the key variables as a function of time.

Refer to the paper 14 cited for relevant equations of the flying-shear control system.

16. Write a user-friendly computer program in a language of your choice to simulate the phase-controlled rectifier-fed separately-excited dc motor drive. The specifications of the drive system are as follows: The motor parameters are 150 hp, 220 V, 1500 rpm, base torque = 711.7 N·m, R_a = 0.05 Ω, L_a = 0.002 H, MI_f = K_b = 1.6 V/rad/sec, J = 0.16 kg − m^2, B = 0.01 N·m/rad/sec. The converter parameters are Supply input: 3-phase, 230 V, 60 Hz; Controller parameters: V_{cm} = 10 V, H_ω = 0.06364 V/rad/sec, T_ω = 0.002, H_c = 0.01124 V/A, K_{pi} = 100, K_{ii} = 1.5, K_{ps} = 10, K_{is} = 1, T_{max} = 2 × Base torque, T_l = 0 N·m. Plot the speed command, rotor speed, torque command, torque, current, current command, triggering angle, and applied voltage vs. time.

CHAPTER 4

Chopper-Controlled DC Motor Drive

4.1 INTRODUCTION

In the case of single stage ac to dc power conversion phase-controlled converters described in Chapter 3 are used to drive the dc machines. Whenever the source is a constant-voltage dc, such as a battery or diode-bridge rectified ac supply, a different type of converter is required to convert the fixed voltage into a variable-voltage/variable-current source for the speed control of the dc motor drive. The variable dc voltage is controlled by chopping the input voltage by varying the on- and off-times of a converter, and the type of converter capable of such a function is known as a chopper.

The principle of operation of a four-quadrant chopper is explained in this chapter. The steady-state and dynamic analysis of the chopper-controlled dc motor drive and its performance characteristics are derived and evaluated. Some illustrative examples are included.

4.2 PRINCIPLE OF OPERATION OF THE CHOPPER

A schematic diagram of the chopper is shown in Figure 4.1. The control voltage to it's gate is v_c. The chopper is on for a time t_{on}, and its off time is t_{off}. Its frequency of operation is

$$f_c = \frac{1}{(t_{on} + t_{off})} = \frac{1}{T} \tag{4.1}$$

and its duty cycle is defined as

$$d = \frac{t_{on}}{T} \tag{4.2}$$

The output voltage across the load during the on-time of the switch is equal to the difference between the source voltage V_s and the voltage drop across the power

Section 4.2   Principle of Operation of the Chopper   125

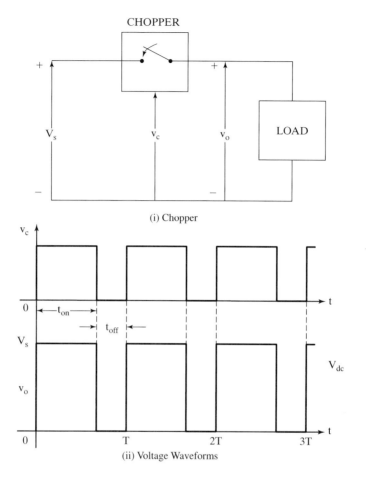

Figure 4.1 Chopper schematic and its waveforms

switch. Assuming that the switch is ideal, with zero voltage drop, the average output voltage V_{dc} is given as

$$V_{dc} = \frac{t_{on}}{T} V_s = dV_s \qquad (4.3)$$

where V_s is the source voltage.

Varying the duty cycle changes the output voltage. Note that the output voltage follows the control voltage, as shown in Figure 4.1, signifying that the chopper is a voltage amplifier. The duty cycle d can be changed in two ways:

(i) By keeping the switching/chopping frequency constant and varying the on-time, to get a changing duty cycle.

(ii) Keeping the on-time constant and varying the chopping frequency, to obtain various values of the duty cycle.

A constant switching frequency has the advantages of predetermined switching losses of the chopper, enabling optimal design of the cooling for the power circuit, and predetermined harmonic contents, leading to an optimal input filter. Both of these advantages are lost by varying the switching frequency of the chopper; hence, this technique for chopper control is not prevalent in practice.

4.3 FOUR-QUADRANT CHOPPER CIRCUIT

A four-quadrant chopper with transistor switches is shown in Figure 4.2. Each transistor has a freewheeling diode across it and a snubber circuit to limit the rate of rise of the voltage. The snubber circuit is not shown in the figure.

The load consists of a resistance, an inductance, and an induced emf. The source is dc, and a capacitor is connected across it to maintain a constant voltage. The base drive circuits of the transistors are isolated, and they reproduce and amplify the control signals at the output. For the sake of simplicity, it is assumed that the switches are ideal and hence, the base drive signals can be used to draw the load voltage.

4.3.1 First-Quadrant Operation

First-quadrant operation corresponds to a positive output voltage and current. This is obtained by triggering T_1 and T_2 together, as is shown in Figure 4.3; then the load voltage is equal to the source voltage. To obtain zero load voltage, either T_1 or T_2 can be turned off. Assume that T_1 is turned off; then the current will decrease in the power switch and inductance. As the current tries to decrease in the inductance, it will have a voltage induced across it in proportion to the rate of fall of current with a polarity opposite to the load-induced emf, thus forward-biasing diode D_4. D_4 provides the path for armature current continuity during this time. Because of this, the circuit configuration changes as shown in Figure 4.4. The load is short-circuited, reducing its voltage to zero. The current and voltage waveforms for continuous and

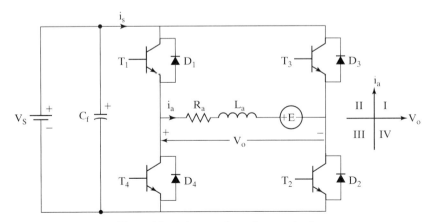

Figure 4.2 A four-quadrant chopper circuit

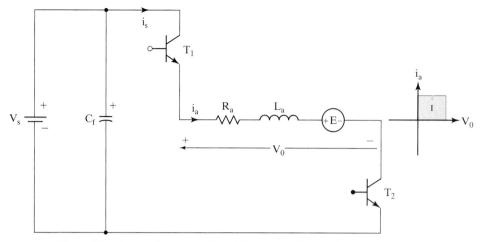

Figure 4.3 First-quadrant operation with positive voltage and current in the load

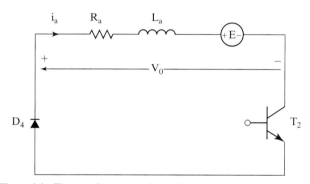

Figure 4.4 First-quadrant operation with zero voltage across the load

discontinuous current conduction are shown in Figure 4.5. Note that, in the discontinuous current-conduction mode, the induced emf of the load appears across the load when the current is zero. The load voltage, therefore, is a stepped waveform. The operation discussed here corresponds to motoring in the clockwise direction, or *forward motoring*. It can be observed that the average output voltage will vary from 0 to V_s; the duty cycle can be varied only from 0 to 1.

The output voltage can also be varied by another switching strategy. Armature current is assumed continuous. Instead of providing zero voltage during turn-off time to the load, consider that T1 and T2 are simultaneously turned off, to enable conduction by diodes D3 and D4. The voltage applied across the load then is equal to the negative source voltage, resulting in a reduction of the average output voltage. The disadvantages of this switching strategy are as follows:

(i) Switching losses double, because two power devices are turned off instead of one only.
(ii) The rate of change of voltage across the load is twice that of the other strategy. If the load is a dc machine, then it has the deleterious effect of causing

128 Chapter 4 Chopper-Controlled DC Motor Drives

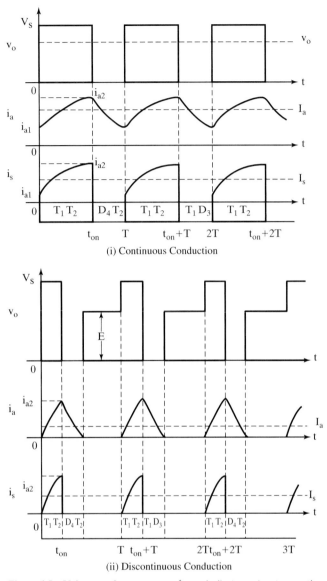

Figure 4.5 Voltage and current waveforms in first-quadrant operation

higher dielectric losses in the insulation and therefore reduced life. Note that the dielectric is a capacitor with a resistor in series.

(iii) The rate of change of load current is high, contributing to vibration of the armature in the case of the dc machine.

(iv) Since a part of the energy is being circulated between the load and source in every switching cycle, the switching harmonic current is high, resulting in additional losses in the load and in the cables connecting the source and converter.

Therefore, this switching strategy is not considered any further in this chapter.

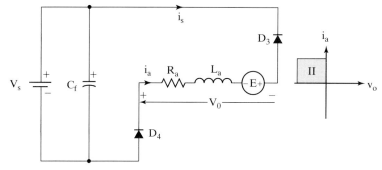

Figure 4.6 Second-quadrant operation, with negative load voltage and positive current

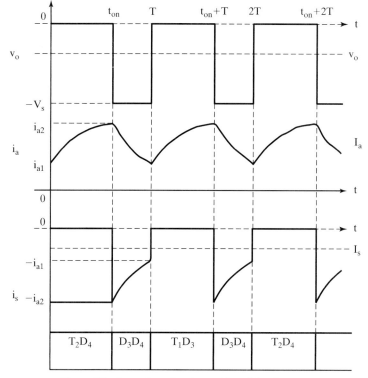

Figure 4.7 Second-quadrant operation of the chopper

4.3.2 Second-Quadrant Operation

Second-quadrant operation corresponds to a positive current with a negative voltage across the load terminals. Assume that the load's emf is negative. Consider that T_1 or T_2 is conducting at a given time. The conducting transistor is turned off. The current in the inductive load has to continue to flow until the energy in it is depleted to zero. Hence, the diodes D_3 and D_4 will take over, maintaining the load current in the same direction, but the load voltage is negative in the new circuit configuration, as is shown in Figure 4.6. The voltage and current waveforms are shown in Figure 4.7. When diodes D_3 and D_4 are conducting, the source receives power from the load. If the source cannot absorb this power, provision has to be made to consume the power. In that case, the overcharge on the filter capacitor is periodically dumped into a resistor connected across the source by controlling the on-time of a transistor in series with a resistor. This form of recovering energy from the load is known as regenerative braking and is common in low-HP motor drives, where the saving in energy might not be considerable or cost-effective. When the current in the load is decreasing, T_2 is turned on. This allows the short-circuiting of the load through T_2 and D_4, resulting in an increase in the load current. Turning off T_2 results in a pulse of current flowing into the source via D_3 and D_4. This operation allows the priming up of the current and a building up of the energy in the inductor from the load's emf, thus enabling the transfer of energy from the load to the source. Note that it is possible to transfer energy from load to source even when E is lower in magnitude than V_s. This particular operational feature is sometimes referred to as *boost operation* in dc-to-dc power supplies. Priming up the load current can also be achieved alternatively, by using T_1 instead of T_2.

4.3.3 Third-Quadrant Operation

Third-quadrant operation provides the load with negative current and voltage. A negative emf source, $-E$, is assumed in the load. Switching on T_3 and T_4 increases the current in the load, and turning off one of the transistors short-circuits the load, decreasing the load current. That way, the load current can be controlled within the externally set limits. The circuit configurations for the switching instants are shown in Figure 4.8. The voltage and current waveforms under continuous and discontinuous

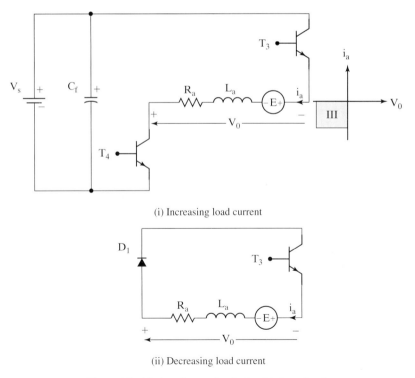

Figure 4.8 Modes of operation in the third quadrant

current-conduction modes are shown in Figure 4.9. Note the similarity between first- and third-quadrant operation.

4.3.4 Fourth-Quadrant Operation

Fourth-quadrant operation corresponds to a positive voltage and a negative current in the load. A positive load-emf source E is assumed. To send energy to the dc source from the load, note that the armature current has to be established to flow

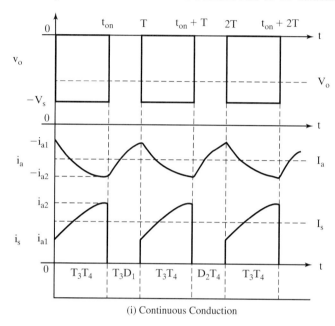

(i) Continuous Conduction

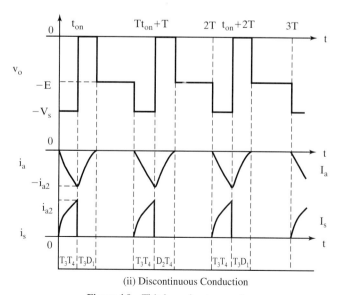

(ii) Discontinuous Conduction

Figure 4.9 Third-quadrant operation

132 Chapter 4 Chopper-Controlled DC Motor Drives

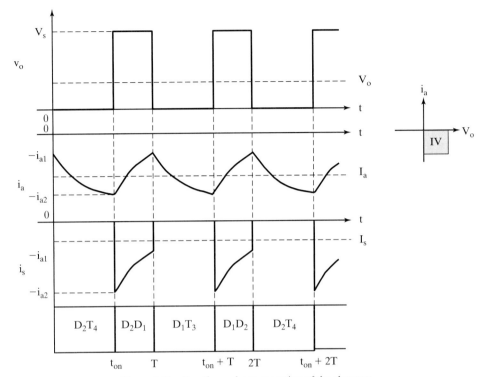

Figure 4.10 Fourth-quadrant operation of the chopper

from the right side to the left side as seen in Figure 4.2. By the convention adopted in this book, that direction of current is negative. Assume that the machine has been operating in quadrant I with a positive current in the armature. When a brake command is received, the torque and armature current command goes negative. The armature current can be driven negative from its positive value through zero. Opening T_1 and T_2 will enable D_3 and D_4 to allow current via the source, reducing the current magnitude rapidly to zero. To establish a negative current, T_4 is turned on. That will short-circuit the load, making the emf source build a current through T_4 and D_2. When the current has reached a desired peak, T_4 is turned off. That forces D_1 to become forward-biased and to carry the load current to the dc input source via D_2 and the load. When the current falls below a lower limit, T_4 is again turned on, to build up the current for subsequent transfer to the source. The voltage and current waveforms are shown in Figure 4.10. The average voltage across the load is positive, and the average load current is negative, indicating that power is transferred from the load to the source. The source power is the product of average source current and average source voltage, and it is negative, as is shown in Figure 4.10.

4.4 CHOPPER FOR INVERSION

The chopper can be viewed as a single-phase inverter because of its ability to work in all the four quadrants, handling leading or lagging reactive loads in series with an emf source. The output fundamental frequency is determined by the rate at which the forward-to-reverse operation is performed in the chopper. The chopper is the building block for a multiphase inverter.

4.5 CHOPPER WITH OTHER POWER DEVICES

Choppers are realized with MOSFETs, IGBTs, GTOs, or SCRs, depending upon the power level required. The MOSFET and transistor choppers are used at power levels up to 50 kW. Beyond that, IGBTs, GTOs, and SCRs are used for the power switches. Excepting SCR choppers, all choppers are self-commutating and hence have a minimum number of power switches and auxiliary components. In the case of SCR choppers, commutating circuits have to be incorporated for each main SCR. Such a circuit is described in Chapter 7.

4.6 MODEL OF THE CHOPPER

The chopper is modeled as a first-order lag with a gain of K_{ch}. The time delay corresponds to the statistical average conduction time, which can vary from zero to T. The transfer function is then

$$G_r(s) = \frac{K_r}{1 + \frac{sT}{2}} \qquad (4.4)$$

where $K_r = V_s/V_{cm}$, V_s is the source voltage, and V_{cm} is the maximum control voltage.

Increasing the chopping frequency decreases the delay time, and hence the transfer function becomes a simple gain.

4.7 INPUT TO THE CHOPPER

The input to the chopper is either a battery or a rectified ac supply. The rectified ac is the prevalent form of input. The ac input is rectified through a diode bridge, and its output is filtered to keep the dc voltage a constant. The use of the diode bridge has the advantage of near-unity power factor, thus overcoming one of the serious disadvantages of the phase-controlled converter. It has a disadvantage: it cannot transfer power from the dc link into ac mains. In that case, the regenerative energy has to be dumped in the braking resistor, as shown in Figure 4.11, or a phase-controlled converter is connected antiparallel to the diode bridge to handle the regenerative energy, as shown in Figure 4.12. In the latter configuration, the phase-controlled converter can have a smaller rating than other power converters, because the rms value of the regenerative current will be small—the duration of regeneration is only a fraction of the motoring time for most loads.

The regenerating converter has to be operated at triggering angles greater than 90°, to reverse the output dc voltage across the converter to match the polarity of V_s. The phase-controlled converter is enabled only when V_s is greater than the

134 Chapter 4 Chopper-Controlled DC Motor Drives

allowable magnitude ΔV over and above the nominal dc voltage obtained from the ac source. It is usual to set ΔV to be 15 or 20% of V. In such a case, there is a need for a step-up transformer in the path of the phase-controlled converter, to match the dc link voltage V_s. The phase converter is disabled when V_s is slightly greater than 1.35 V, where V is the line-to-line rms voltage, to prevent energy flow from source to dc link and from dc link to source via the phase-controlled converter.

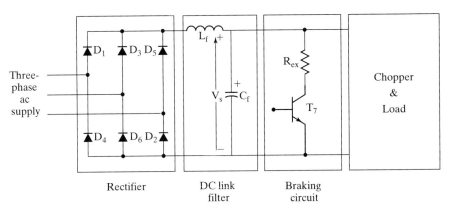

Figure 4.11 Front-end of the chopper circuit

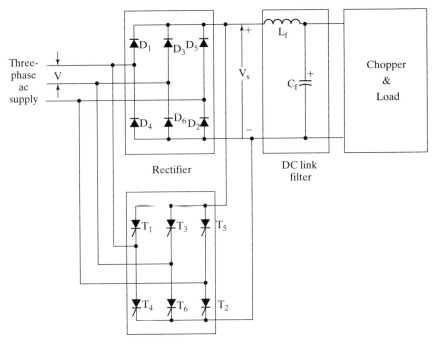

Figure 4.12 Chopper with regeneration capability

4.8 OTHER CHOPPER CIRCUITS

Not all applications demand four-quadrant operation. The power circuit is therefore simplified to accommodate only the necessary operation. Power devices are reduced for one- and two-quadrant drives, resulting in economy of the power converter. Apart from the chopper circuit shown in Figure 4.2 and its variations for reduced quadrants of operation, shown in Figure 4.13, a number of unique

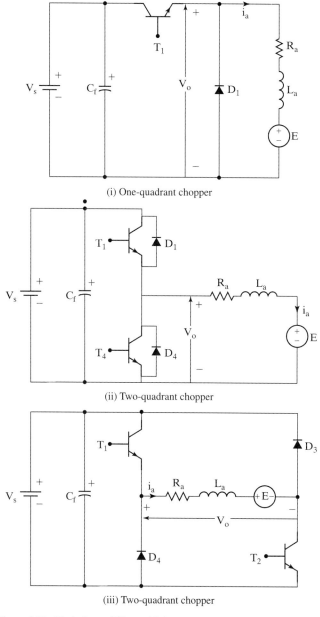

Figure 4.13 Variations of Figure 4.2 for one- and two-quadrant operation

chopper circuits are used. *Morgan* and *Jones* circuits are some of the commonly used choppers for dc motor speed control. They differ in their circuit topologies and commutation of the current from the circuit discussed in this chapter but they all have identical transfer characteristics. Considering this aspect, these circuits are not described and interested readers are referred to power-electronics textbooks.

4.9 STEADY-STATE ANALYSIS OF CHOPPER-CONTROLLED DC MOTOR DRIVE

The steady-state performance of the chopper-controlled dc motor drive is obtained either with average values, by neglecting harmonics, or including harmonics. The justification for using average values is that the average torque is the useful torque that is transmitted to the load. The torque components due to the current harmonics produce an average torque of zero over one cycle of switching. They do not contribute to useful power production. Further, they result in increased armature losses, because the harmonic currents increase the effective motor current. From an output point of view, neglecting harmonics and using only average values gives easier steady-state computation. The analysis by this method is known as *analysis by averaging*.

When losses, maximum steady-state current, and precise electromagnetic torque are required to fully analyze and design the drive system for an application, the true current waveforms in steady state need to be computed. Therefore, the harmonics cannot be excluded in the steady-state computation. A computationally efficient, analytical closed-form expression is obtained by the novel technique of boundary-matching conditions. This method is referred to as *instantaneous steady-state computation*.

It is assumed that the rotor speed is constant and the field is separately excited. For the following analysis, the field flux is maintained at rated value. For any other value of the field flux, the derivations need to be changed only with regard to the induced-emf term.

4.9.1 Analysis by Averaging

The average armature current is

$$I_{av} = \frac{V_{dc} - E}{R_a} \tag{4.5}$$

where

$$V_{dc} = dV_s \tag{4.6}$$

The electromagnetic torque is

$$T_{av} = K_b I_{av} \tag{4.7}$$

Section 4.9   Steady-State Analysis of Chopper-Controlled DC Motor Drive   137

The torque, written in terms of duty cycle and speed from equations (4.5), (4.6) and (4.7), is

$$T_{av} = \frac{K_b(dV_s - K_b\omega_m)}{R_a} \text{ (N·m)} \quad (4.8)$$

The electromagnetic torque is normalized by dividing it by the base torque, T_b, and by dividing both its numerator and denominator by the base voltage, V_b. The normalized torque is obtained by simplifying the expressions and substituting for V_b in terms of the base speed, ω, and emf constant:

$$T_{en} = \frac{T_{av}/V_b}{T_b/V_b} = \frac{K_b(dV_s - K_b\omega_m)/V_b}{K_bI_bR_a/V_b} = \frac{dV_n - \omega_{mn}}{R_{an}}, \text{ p.u.} \quad (4.9)$$

where

$$R_{an} = \frac{I_bR_a}{V_b}, \text{ p.u.} \quad (4.10)$$

$$V_n = \frac{V_s}{V_b}, \text{ p.u.} \quad (4.11)$$

$$\omega_{mn} = \frac{\omega_m}{\omega_b}, \text{ p.u.} \quad (4.12)$$

R_{an}, V_n, and ω_{mn} are normalized resistance, voltage, and speed, respectively. By assigning the product of normalized torque and p.u. resistance on the y axis, a set of normalized performance curves is drawn for various values of duty cycles, normalized voltage, and speed, as is shown in Figure 4.14. From the characteristics and the given duty cycle and voltage, the torque can be evaluated for a given speed, if the normalized resistance is known.

4.9.2 Instantaneous Steady-State Computation

The instantaneous steady-state armature current and electromagnetic torque including harmonics are evaluated in this section, both for continuous and discontinuous current conduction, by boundary-matching conditions. The relevant waveforms are shown in Figure 4.15. For each of the current-conduction modes, the performance is evaluated separately.

4.9.3 Continuous Current Conduction

The relevant electrical equations of the motor for on and off times are as follows:

$$V_s = E + R_a i_a + L_a \frac{di_a}{dt}, \quad 0 \leq t \leq dT \quad (4.13)$$

$$0 = E + R_a i_a + L_a \frac{di_a}{dt}, \quad dT \leq t \leq T \quad (4.14)$$

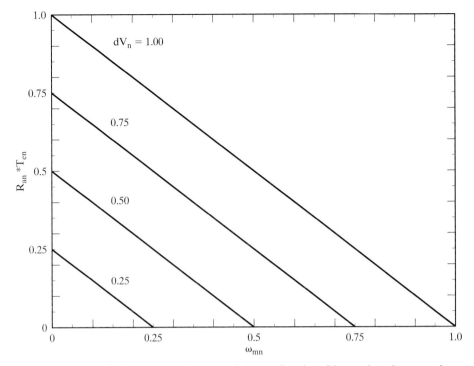

Figure 4.14 Normalized torque–speed characteristics as a function of duty cycle and source voltage

The solution of equation (4.13) is

$$i_a(t) = \frac{V_s - E}{R_a}(1 - e^{-\frac{t}{T_a}}) + I_{a0}e^{-\frac{t}{T_a}}, \quad 0 < t < dT \quad (4.15)$$

where

$$T_a = \text{Armature time constant} = \frac{L_a}{R_a} \quad (4.16)$$

Similarly, the solution of equation (4.14) is

$$i_a(t) = -\frac{E}{R_a}(1 - e^{-\frac{t^1}{T_a}}) + I_{a1}e^{-\frac{t^1}{T_a}}, \quad dT \leq t \leq dT \quad (4.17)$$

and

$$t^1 = t - dT \quad (4.18)$$

From Figure 4.15, it is seen that

$$i_a(t) = i_a(t + T) \quad (4.19)$$

By using this boundary condition, I_{a0} and I_{a1} are evaluated as

Section 4.9 Steady-State Analysis of Chopper-Controlled DC Motor Drive

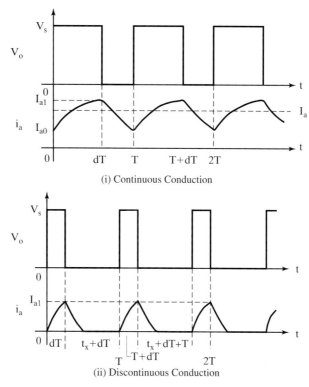

Figure 4.15 Applied voltage and armature current in a chopper-controlled dc motor drive

$$I_{a1} = \frac{V_s(1 - e^{-dT/T_a})}{R_a(1 - e^{-T/T_a})} - \frac{E}{R_a} \quad (4.20)$$

$$I_{a0} = \frac{V_s(e^{dT/T_a} - 1)}{R_a(e^{T/T_a} - 1)} - \frac{E}{R_a} \quad (4.21)$$

Having evaluated I_{a0} and I_{a1}, you can use equations (4.15) and (4.17) to evaluate the instantaneous armature current in steady state.

The limiting or minimum value of duty cycle for continuous current is evaluated by equating I_{a0} to zero. This value is termed the *critical duty cycle*, d_c, and is given by

$$d_c = \left(\frac{T_a}{T}\right) \log_e \left[1 + \frac{E}{V_s}(e^{\frac{T}{T_a}} - 1)\right] \quad (4.22)$$

Duty cycles lower than d_c will produce discontinuous current in the motor. Note that this critical value is dependent on the ratio between chopping time period and armature time constant and also on the ratio between induced emf and source voltage.

4.9.4 Discontinuous Current Conduction

The relevant equations for discontinuous current-conduction mode are obvious from Figure 4.15;

$$V_s = E + R_a i_a + L_a \frac{di_a}{dt}, \qquad 0 < t < dT \qquad (4.23)$$

$$0 = E + R_a i_a + L_a \frac{di_a}{dt}, \qquad dT < t < (t_x + dT) \qquad (4.24)$$

with

$$i_a(t_x + dT) = 0 \qquad (4.25)$$
$$i_a(0) = 0 \qquad (4.26)$$

Hence,

$$I_{a1} = \frac{V_s - E}{R_a}(1 - e^{-dT/T_a}) \qquad (4.27)$$

$$i_a(t_x + dT) = -\frac{E}{R_a}(1 - e^{-\frac{t_x}{T_a}}) + I_{a1} e^{-\frac{t_x}{T_a}} \qquad (4.28)$$

This equation is equal to zero, by the constraint given in equation (4.25), and, from that, t_x is evaluated as

$$t_x = T_a \log_e \left[1 + \frac{I_{a1} R_a}{E} \right] \qquad (4.29)$$

The solution for the armature current in three time segments is

$$i_a(t) = \frac{V_s - E}{R_a}(1 - e^{-t/T_a}), \qquad 0 < t < dT \qquad (4.30)$$

$$i_a(t) = I_{a1} e^{-\frac{(t-dT)}{T_a}} - \frac{E}{R_a}\left(1 - e^{-\frac{(t-dT)}{T_a}}\right), \qquad dT < t < t_x + dT \qquad (4.31)$$

$$i_a(t) = 0, \qquad (t_x + dT) < t < T \qquad (4.32)$$

The steady-state performance is calculated by using equations (4.30) to (4.32).

Example 4.1

A dc motor is driven from a chopper with a source voltage of 24V dc and at a frequency of 1 kHz. Determine the variation in duty cycle required to have a speed variation of 0 to 1 p.u. delivering a constant 2 p.u. load. The motor details are as follows:

1 hp, 10 V, 2500 rpm, 78.5 % efficiency, $R_a = 0.01\ \Omega$, $L_a = 0.002$ H, $K_b = 0.03819$ V/rad/sec

The chopper is one-quadrant, and the on-state drop voltage across the device is assumed to be 1 V regardless of the current variation.

Solution (i) Calculation of rated and normalized values

$$V_b = 10 \text{ V}$$

$$V_n = \frac{V_s}{V_b} = \frac{24 - 1}{10} = 2.3 \text{ p.u.}$$

$$\omega_{mr} = \frac{2500 \times 2\pi}{60} = 261.79 \text{ rad/sec}$$

$$I_{ar} = \frac{\text{Output}}{\text{Voltage} \times \text{Efficiency}} = \frac{1 \times 746}{10 \times 0.785} = 95 \text{ A} = I_b$$

$$R_{an} = \frac{I_b R_a}{V_b} = \frac{95 \times 0.001}{10} = 0.095 \text{ p.u.}$$

$$T_{en} = 2 \text{ p.u.}$$

(ii) Calculation of duty cycle

The minimum and maximum duty cycles occur at 0 and 1 p.u. speed, respectively, and at 2 p.u. load. From equation (4.9),

$$d = \frac{T_{en} R_{an} + \omega_{mn}}{V_n}$$

$$d_{min} = \frac{2 \times 0.095 + 0}{2.3} = 0.0826$$

$$d_{max} = \frac{2 \times 0.095 + 1}{2.3} = 0.517$$

The range of duty cycle variation required, then, is

$$0.0826 \leq d \leq 0.517$$

Example 4.2

The critical duty cycle can be changed by varying either the electrical time constant or the chopping frequency in the chopper. Draw a set of curves showing the effect of these variations on the critical duty cycle for various values of E/V_s.

Solution

$$d_c = \left(\frac{T_a}{T}\right) \log_e \left[1 + \frac{E}{V_s}\left(e^{\frac{T}{T_a}} - 1\right)\right]$$

In terms of chopping frequency,

$$d_c = f_c T_a \log_e \left[1 + \frac{E}{V_s}\left(e^{\frac{1}{T_a f_c}} - 1\right)\right]$$

Assigning various values of E/V_s and varying $f_c T_a$ would yield a set of critical duty cycles. The graph between d_c and $f_c T_a$ for varying values of E/V_s is shown in Figure 4.16. The maximum value of $f_c T_a$ is chosen to be 10.

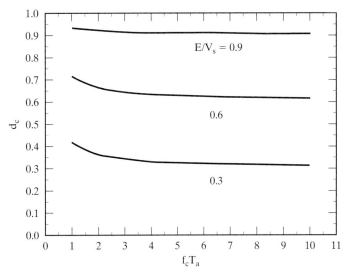

Figure 4.16 Critical duty cycles vs. the product of chopping frequency and electrical time constant of the dc motor, as a function of induced emf and source voltage

---

Example 4.3

A 200-hp, 230-V, 500-rpm separately-excited dc motor is controlled by a chopper. The chopper is connected to a bridge-diode rectifier supplied from a 230-V, 3-ϕ, 60-Hz ac main. The motor chopper details are as follows:

$$R_a = 0.04\ \Omega,\ L_a = 0.0015\ H,\ K_b = 4.172\ V/rad/sec,\ f_c = 2\ kHz.$$

The motor is running at 300 rpm with 55% duty cycle in the chopper. Determine the average current from steady-state current waveform and the electromagnetic torque produced in the motor. Compare these results with those obtained by averaging.

Solution The critical duty cycle is evaluated to determine the current continuity at the given duty cycle of 0.55,

$$d_c = \left(\frac{T_a}{T}\right)\log_e\left[1 + \frac{E}{V_s}(e^{\frac{T}{T_a}} - 1)\right]$$

$$T_a = \frac{0.0015}{0.04} = 0.0375\ s$$

$$T = \frac{1}{f_c} = \frac{1}{2 \times 10^3} = 0.5\ ms$$

$$\frac{T_a}{T} = 75$$

$$V_s = 1.35\ V\cos\alpha = 1.35 \times 230 \times \cos 0° = 310.5\ V$$

$$E = K_b\omega_m = 4.172 \times \frac{2\pi \times 300}{60} = 131.1\ V$$

$$d_c = 75\log_e\left[1 + \frac{131.1}{310.5}(e^{\frac{1}{75}} - 1)\right] = 0.423$$

The given value of d is greater than the critical duty cycle; hence, the armature current is continuous.

$$I_{a0} = \frac{V_s(e^{dT/T_a} - 1)}{R_a(e^{T/T_a} - 1)} - \frac{E}{R_a} = \frac{310.5(e^{0.55/75} - 1)}{0.04(e^{1/75} - 1)} - \frac{131.1}{0.04} = 979 \text{ A}$$

$$I_{a1} = \frac{V_s(1 - e^{-dT/T_a})}{R_a(1 - e^{-T/T_a})} - \frac{E}{R_a} = \frac{310.5(1 - e^{-0.55/75})}{0.04(1 - e^{-1/75})} - \frac{131.1}{0.04} = 1004.7 \text{ A}$$

The average current is

$$I_{av} = \frac{1}{T}\left[\int_0^{dT}\left(\frac{V_s - E}{R_a}(1 - e^{-t/T_a}) + I_{a0}e^{-t/T_a}\right)dt + \int_0^{(1-d)T}\left(-\frac{E}{R_a}(1 - e^{-t/T_a}) + I_{a1}e^{-t/T_a}\right)dt\right]$$

$$= \frac{1}{T}\begin{bmatrix} \dfrac{V_s - E}{R_a}\{dT + T_a(e^{-dT/T_a} - 1)\} + I_{a0}T_a(1 - e^{-dT/T_a}) \\ - \dfrac{E}{R_a}\{(1-d)T - T_a + T_a e^{-(1-d)T/T_a}\} + I_{a1}T_a(1 - e^{-(1-d)T/T_a}) \end{bmatrix} = 991.8 \text{ A}$$

$$T_{av} = K_b I_{av} = 4.172 \times 991.8 = 4137.7 \text{ N·m}$$

Steady state by averaging

$$I_{av} = \frac{(dV_s - K_b\omega_m)}{R_a} = \frac{0.55 \times 310.5 - 4.172 \times 31.42}{0.04} = 991.88 \text{ A}$$

$$T_{av} = K_b I_{av} = 4138.1 \text{ N·m}$$

There is hardly any significant difference in the results by these two methods.

4.10 RATING OF THE DEVICES

The chopper shown in Figure 4.2 is considered, for illustration. The armature current is assumed to be continuous, with no ripples. If I_{max} is the maximum allowable current in the dc machine, the rms value of the power switch is dependent on this value and its duty cycle. The duty cycle of one device is slightly more than the duty cycle of the chopper, because one of the power switches continues to carry current during freewheeling while the other is turned off. In order to equal the current loading of the switches, the power switch carrying the freewheeling current is turned off in the next cycle and the previously inactive switch is allowed to carry the freewheeling current alternately. Accordingly, the current waveform of one power switch and the diode is as shown in Figure 4.17. Note that the diodes are, like their respective power switches, alternately carrying the freewheeling current, and only motoring action is considered for the calculation.

The rms value of the power switch current and average diode current are given by,

$$I_t = \sqrt{\frac{I_{max}^2}{2T}(T + dT)} = \sqrt{\frac{1+d}{2}} \cdot I_{max} \qquad (4.33)$$

$$I_d = \left(\frac{1-d}{2}\right) \cdot I_{max} \qquad (4.34)$$

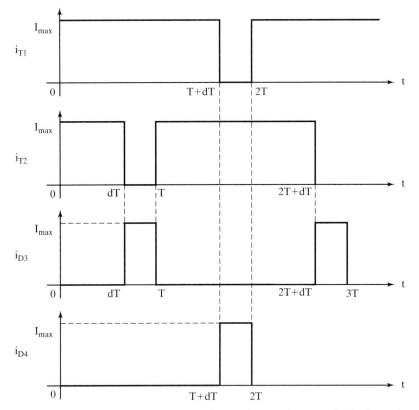

Figure 4.17 Transistor and diode currents in the chopper for motoring operation in the continuous-conduction mode

where the subscripts d and T refer to the diode and power switch, respectively. The average diode and rms power switch currents vs. duty cycle for continuous conduction are shown in Figure 4.18, for design use.

The minimum voltage rating for both the devices is,

$$V_T = V_d = V_s \tag{4.35}$$

Whenever regeneration occurs in the motor drive, the current in the freewheeling diodes would change, and, depending on the frequency and duration of regeneration, the diode currents have to be recalculated. To optimize the chopper rating in comparison to the motor and load demands, the operating conditions have to be known beforehand. The extreme operating conditions would then prevail on the design and hence on the final rating of the chopper.

4.11 PULSATING TORQUES

The armature current has ac components. These ac components or harmonics produce corresponding pulsating torques. The average of the harmonic torques is zero,

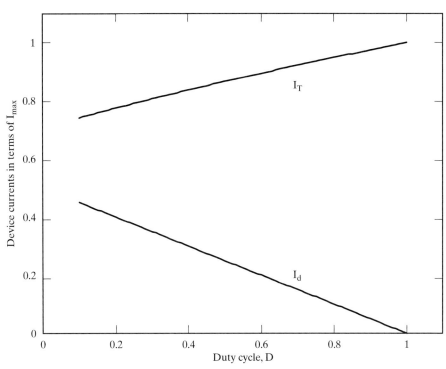

Figure 4.18 Device currents vs. duty cycle for continuous-conduction mode

and they do not contribute to useful torque and power. Some high-performance applications, such as machine tools and robots, require the pulsating torque to be a minimum so as not to degrade the process and the products. In that case, an estimation of the pulsating torques is in order.

The pulsating currents are evaluated from the harmonic voltages and the harmonic armature impedances of the dc machine. The applied voltage shown in Figure 4.10 is resolved into Fourier components as

$$v_a(t) = V_a + \sum_{n=1}^{\infty} A_n \sin(n\omega_c t + \theta_n) \tag{4.36}$$

where

$$V_a = \frac{dT}{T} \cdot V_s = dV_s, \quad V \tag{4.37}$$

$$\omega_c = 2\pi f_c = \frac{2\pi}{T}, \quad \text{rad/sec} \tag{4.38}$$

$$A_n = \frac{2V_s}{n\pi} \sin\frac{n\omega_c dT}{2}, \quad V \tag{4.39}$$

$$\theta_n = \frac{\pi}{2} - \frac{n\omega_c dT}{2}, \quad \text{rad} \tag{4.40}$$

and n is the order of the harmonic. The armature current is expressed as

$$i_a(t) = I_{av} + \sum_{n=1}^{\infty} \frac{A_n}{|Z_n|} \sin(n\omega_c t + \theta_n - \phi_n) \qquad (4.41)$$

where

$$I_{av} = \frac{V_a - E}{R_a} \qquad (4.42)$$

$$Z_n = R_a + jn\omega_c L_a \qquad (4.43)$$

$$\phi_n = \cos^{-1}\left\{\frac{R_a}{\sqrt{R_a^2 + n^2\omega_c^2 L_a^2}}\right\} \qquad (4.44)$$

and Z_{an} is the harmonic impedance of the armature. The input power is

$$P_i = v_a(t) i_a(t)$$

$$= \left\{ \begin{array}{l} V_a I_{av} + I_{av} \sum_{n=1}^{\infty} A_n \sin(n\omega_c t + \theta_n) + V_a \sum_{n=1}^{\infty} \frac{A_n}{|Z_n|} \sin(n\omega_c t + \theta_n - \phi_n) \\ + \sum_{n=1}^{\infty} \left(\frac{A_n^2}{|Z_n|} \sin(n\omega_c t + \theta_n)\sin(n\omega_c t + \theta_n - \phi_n)\right) \end{array} \right\} \qquad (4.45)$$

The right-hand side of equation (4.45) can be simplified further, because the following pulsating terms reduce to an average of zero.

$$I_{av} \sum_{n=1}^{\infty} A_n \sin(n\omega_c t + \theta_n) = 0 \qquad (4.46)$$

$$V_a \sum_{n=1}^{\infty} \frac{A_n}{|Z_n|} \sin(n\omega_c t + \theta_n - \phi_n) = 0 \qquad (4.47)$$

The other terms tend to zero, as is shown below.

$$\sum_{n=1}^{\infty} \left(\frac{A_n^2}{|Z_n|} \sin(n\omega_c t + \theta_n)\sin(n\omega_c t + \theta_n - \phi_n)\right)$$

$$= \sum_{n=1}^{\infty} \left(\frac{A_n^2}{|Z_n|} \cdot \frac{1}{2}\{\cos\phi_n - \cos(2n\omega_c t + 2\theta_n - \phi_n)\}\right) \qquad (4.48)$$

The average of the double-frequency term is zero, and the power factor is nearly zero because

$$n^2\omega_c^2 L_a^2 \gg R_a \qquad (4.49)$$

Hence, the expression (4.48) is almost equal to zero. The average input power becomes a constant quantity and is expressed as

$$P_{av} = V_a I_{av} \qquad (4.50)$$

Power calculation can use the average values, and there is no need to resort to harmonic analysis, as is shown by the above derivation. The fundamental-harmonic peak pulsating torque then is

$$T_{e1} = K_b i_{a1} \qquad (4.51)$$

and the fundamental-harmonic armature current is given by

$$i_{a1} = \frac{2V_s}{\pi\sqrt{R_a^2 + \omega_c^2 L_a^2}} \sin\left(\frac{\omega_c dT}{2}\right) \quad (4.52)$$

The fundamental armature current can be alternately expressed in terms of duty cycle as

$$i_{a1} = \frac{2V_s}{\pi\sqrt{R_a^2 + \omega_c^2 L_a^2}} \sin(\pi d) \quad (4.53)$$

When the duty cycle is 50%, the fundamental armature current, and hence the pulsating torque, is maximum. For a duty cycle of 100%, there are no pulsating-torque components. The instantaneous electromagnetic torque is the sum of the dc and harmonic torques, written as

$$T_e(t) = T_{av} + T_{eh} \quad (4.54)$$

where h is the harmonic order. The harmonic torques, T_{eh}, do not contribute to the load: their averages are zero.

The fundamental pulsating torque is expressed as a fraction of average torque, to examine the impact of duty cycle on the pulsating torque. It is facilitated by the following development.

$$\frac{T_{e1}}{T_{av}} = \frac{K_b i_{a1}}{K_b I_{av}} = \frac{2V_s \sin(\pi d)/(\pi\sqrt{R_a^2 + \omega_c^2 L_a^2})}{(dV_s - E)/R_a} = \frac{2R_a}{\pi\sqrt{R_a^2 + \omega_c^2 L_a^2}} \cdot \frac{V_s}{(dV_s - E)} \sin(\pi d) \quad (4.55)$$

$$= \left(\frac{2}{\pi}\cos\phi_1\right) \frac{\sin(\pi d)}{\left(d - \frac{E}{V_s}\right)} = k_1 \cdot \frac{\sin(\pi d)}{\left(d - \frac{E}{V_s}\right)}$$

where ϕ_1 is the fundamental power-factor angle, which can be extracted from

$$\cos\phi_1 = \frac{R_a}{\sqrt{R_a^2 + \omega_c^2 L_a^2}} \quad (4.56)$$

and

$$k_1 = \frac{2}{\pi}\cos\phi_1 \quad (4.57)$$

A set of normalized curves is shown in Figure 4.19 to indicate the influence of duty cycle and of the ratio between the induced emf and the source voltage on the magnitude of the pulsating torque.

By noting that the fundamental is the predominant component among the ac components, the rms value of the armature current is approximated as

$$I_{rms} = \sqrt{I_{av}^2 + I_1^2} \quad (4.58)$$

where I_1 is the fundamental rms current, given by

$$I_1 = \frac{i_{a1}}{\sqrt{2}} = \frac{\sqrt{2}V_s}{\pi\sqrt{R_a^2 + \omega_c^2 L_a^2}} \sin(\pi d) \quad (4.59)$$

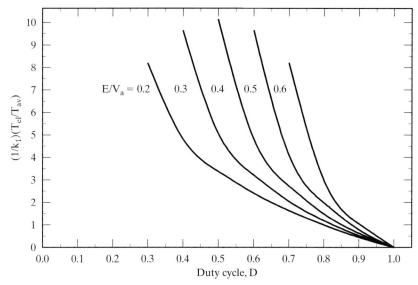

Figure 4.19 Normalized fundamental torque pulsation vs. duty cycle as a function of the ratio between induced emf and source voltage

The armature resistive losses are,

$$P_c = I_{rms}^2 R_a \tag{4.60}$$

This indicates that the thermal capability of the motor is degraded by the additional losses produced by the harmonic currents.

Minimization of the dominant-harmonic torque is of importance in many applications, mainly in positioning of machine tool drives. The key to mitigation of the harmonic torque is revealed by the expression for harmonic current given by equation (4.53). Given a fixed voltage source, there are only two variables that could be utilized to reduce harmonic current: the chopping frequency, and the machine inductance. Note that d cannot be used; it is a variable dependent on the speed and load. The chopping frequency is limited by the selection of power device, by its switching losses, and by other factors, such as electromagnetic compatibility. The advantage of varying the carrier frequency is that it is machine-independent, and hence the solution is contained within the chopper. This may not be feasible, as in the case of large (>100 hp) motor drives, but it is possible to increase the armature inductance of the machine during the design of the motor or to include an external inductor to increase the effective inductance in the armature path. The latter solution is the only practical approach in retrofit applications.

Example 4.4

A separately-excited dc motor is controlled by a chopper whose input dc voltage is 180 V. This motor is considered for low-speed applications requiring less than 2% pulsating torque at 300 rpm. (i) Evaluate its suitability for that application. (ii) If it is found unsuitable, what is the chopping frequency that will bring the pulsating torque to the specification? (iii) Alternatively, a series inductor in the armature can be introduced to meet the specification. Determine the value of that inductor. The motor and chopper data are as follows:

3 hp, 120 V, 1500 rpm, $R_a = 0.8\ \Omega$, $L_a = 0.003$ H, $K_b = 0.764$ V/rad/sec, $f_c = 500$ Hz

Solution Rated torque,

$$T_{er} = \frac{3 \times 745.6}{2\pi \times 1500/60} = 14.25\ \text{N}\cdot\text{m}$$

Maximum pulsating torque permitted $= 0.02 \times T_{er} = 0.02 \times 14.25 = 0.285$ N·m

To find the harmonic currents, it is necessary to know the duty cycle. That is approximately determined by the averaging-analysis technique, by assuming that the motor delivers rated torque at 300 rpm.

$$I_b = \frac{T_{er}}{K_b} = \frac{14.25}{0.764} = 18.65\ \text{A}$$

$$V_a = E + I_{av}R_a = K_b\omega_m + I_{av}R_a = 0.764 \times \frac{2\pi \times 300}{60} + 18.65 \times 0.8 = 38.91\ \text{V}$$

$$d = \frac{V_a}{V_s} = \frac{38.91}{180} = 0.216$$

(i) The fundamental pulsating torque is assumed to be predominant for this analysis.

$$i_{a1} = \frac{2V_s}{\pi\sqrt{R_a^2 + \omega_c^2 L_a^2}} \sin(\pi d) = \frac{2 \times 180}{\pi\sqrt{0.8^2 + (2\pi \times 500 \times 0.003)^2}} \sin(0.216\pi) = 7.6\ \text{A}.$$

$$T_{e1} = K_b i_{a1} = 0.764 \times 7.6 = 5.8\ \text{N}\cdot\text{m}$$

This pulsating torque exceeds the specification, and, hence, in the present condition, the drive is unsuitable for use.

(ii) The fundamental current to produce 2% pulsating torque is

$$i_{a1(spec)} = \frac{T_{e1(spec)}}{K_b} = \frac{0.285}{0.764} = 0.373\ \text{A}$$

$$i_{a1(spec.)} = \frac{2V_s}{\pi\sqrt{R_a^2 + \omega_c^2 L_a^2}} \sin(\pi d)$$

from which the angular switching frequency to meet the fundamental current specification is obtained as

$$\omega_{c1} = \sqrt{\frac{4V_s^2 \sin^2(\pi d)}{\pi^2 L_a^2 i_{a1(spec.)}^2} - \frac{R_a^2}{L_a^2}} = \sqrt{(4 \times 180^2 \sin^2(0.216\pi)/(\pi^2(0.003)^2(0.373)^2) - \left(\frac{0.8}{0.003}\right)^2}$$

$$= 64{,}278\ \text{rad/s}$$

$$f_{c1} = \frac{\omega_{c1}}{2\pi} = 10.23\ \text{kHz}$$

150 Chapter 4 Chopper-Controlled DC Motor Drives

Note that f_{c1} is the chopping frequency in Hz, which decreases the pulsating torque to the specification.

(iii) Let L_{ex} be the inductor introduced in the armature circuit. Then its value is

$$L_{ex} = \sqrt{\frac{4V_s^2\sin^2(\pi d)}{\pi^2\omega_c^2 I_{al(spec)}^2} - \frac{R_a^2}{\omega_c^2}} - L_a$$

$$= \sqrt{(4 \times 180^2 \sin^2(0.216\pi))/(\pi^2(2\pi \times 500)^2(0.373)^2) - \left(\frac{0.8}{2\pi \times 500}\right)^2} - 0.003 = 71.5 \text{ mH}$$

Example 4.5

Calculate (i) the maximum harmonic resistive loss and (ii) the derating of the motor drive given in Example 4.4. The motor is operated with a base current of 18.65 A, which is inclusive of the fundamental-harmonic current. Consider only the dominant-harmonic component, to simplify the calculation.

Solution

$$I_{rms}^2 = I_b^2 = I_{av}^2 + I_1^2$$

By dividing by the square of the base current, the equation is expressed in terms of the normalized currents as

$$I_{avn}^2 + I_{1n}^2 = 1 \text{ p.u.}$$

where

$$I_{1n} = \frac{\sqrt{2}}{\pi} \frac{V_s}{\sqrt{R_a^2 + \omega_c^2 L_a^2}} \sin(\pi d) \frac{1}{I_b} = \frac{\sqrt{2}}{\pi} \frac{V_{sn}}{Z_{an}} \sin(\pi d)$$

where

$$V_{sn} = \frac{V_s}{V_b}; Z_{an} = \frac{Z_a}{Z_b} = \frac{\sqrt{R_a^2 + \omega_c^2 L_a^2}}{Z_b}$$

$$Z_b = \frac{V_b}{I_b} = \frac{120}{18.65} = 6.43 \text{ }\Omega$$

$$V_{sn} = \frac{180}{120} = 1.5 \text{ p.u.}$$

$$Z_a = \sqrt{0.8^2 + (2\pi * 500 * 0.003)^2} = 9.42 \text{ }\Omega$$

$$Z_{an} = \frac{9.42}{6.43} = 1.464 \text{ p.u.}$$

For a duty cycle of 0.5, the dominant-harmonic current is maximum and is given as

$$I_{1n} = \frac{\sqrt{2}}{\pi} \frac{1.5}{1.464} = 0.46 \text{ p.u.}$$

(i) The dominant-harmonic armature resistive loss is

$$P_{1n} = I_{1n}^2 R_{an} = 0.46^2 \frac{0.8}{6.43} = 0.02628 \text{ p.u.}$$

Section 4.12   Closed-Loop Operation   151

(ii) For equality of losses in the machine with pure and chopped-current operation, the average current in the machine with chopped-current operation is derived as

$$I_{avn} = \sqrt{1 - I_{1n}^2} = \sqrt{1 - .46^2} = 0.887 \text{ p.u.}$$

which translates into an average electromagnetic torque of 0.887 p.u., resulting in 11.3% derating of the torque and hence of output power.

4.12 CLOSED-LOOP OPERATION

4.12.1 Speed-Controlled Drive System

The speed-controlled dc-motor chopper drive is very similar to the phase-controlled dc-motor drive in its outer speed-control loop. The inner current loop and its control are distinctly different from those of the phase-controlled dc motor drive. This difference is due to the particular characteristics of the chopper power stage. The current loop and speed loop are examined, and their characteristics are explained, in this section. The closed-loop speed-controlled separately-excited dc motor drive is shown in Figure 4.20 for analysis, but the drive system control strategy is equally applicable to a series motor drive.

4.12.2 Current Control Loop

With inner current loop alone, the motor drive system is a torque amplifier. The commanded value of current is compared to the actual armature current, and its error is processed through a current controller. The output of the current controller, in conjunction with other constraints, determines the base drive signals of the chopper switches. The current controller can be either of the following types:

(i) Pulse-Width-Modulation (PWM) controller
(ii) Hysteresis controller

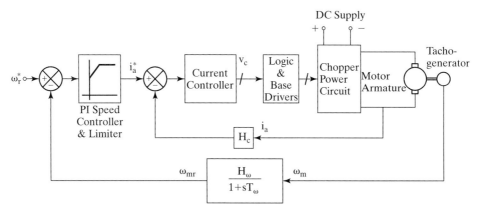

Figure 4.20   Speed-controlled dc-motor chopper drive

152 Chapter 4 Chopper-Controlled DC Motor Drives

The selection of the current controller affects its transient response (and hence the overall speed loop bandwidth indirectly). These two controllers are described in the following sections.

4.12.3 Pulse-Width-Modulated Current Controller

The current error is fed into a controller, which could be proportional (P), proportional plus integral (PI), or proportional, integral, and differential (PID). The most commonly used controller among them is the PI controller. The current error is amplified through this controller and emerges as a control voltage, v_c. It is required to generate a proportional armature voltage from the fixed source through a chopper operation. Therefore, the control voltage is equivalent to the duty cycle of the chopper. Its realization is as follows. The control voltage is compared with a ramp signal to generate the on- and off-times, as shown in Figure 4.21. *On* signal is produced if the control voltage is greater than the ramp (carrier) signal; *off* signal is generated when the control signal is less than the ramp signal.

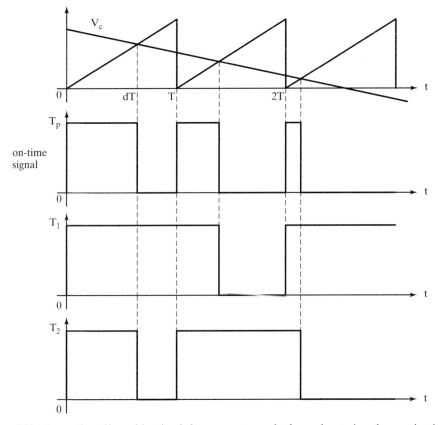

Figure 4.21 Generation of base-drive signals from current error for forward motoring when one is using the chopper shown in Figure 4.2

This logic amounts to the fact that the duration for which the control signal exceeds the ramp signal determines the duty cycle of the chopper. The on- and off-time signals are combined with other control features, such as interlock, minimum on- and off-times, and quadrant selection. The interlock feature prevents the turning on of the transistor (top/bottom) in the same leg before the other transistor (bottom/top) is turned off completely. This is ensured by giving a time delay between the turn-off instant of one device and the turn-on instant of the other device in the same phase leg. Simultaneous conduction of the top and bottom devices in the same leg results in a short circuit of the dc source; it is known as shoot-through failure in the literature.

Figure 4.21 corresponds to the forward motoring operation in the four-quadrant chopper shown in Figure 4.2. When the motor drive is to operate in the third and fourth quadrants, the armature current reverses. This calls for a change in the current-control circuitry. A block diagram of the current controller is shown in Figure 4.22, including all the constraints pertaining to the operational quadrant. The speed and current polarities, along with that of the control voltage, determine the quadrant and hence the appropriate gating signals. The on-time is determined by comparing the ramp signal with the absolute value of the control voltage. The current-error signal, which determines the control voltage v_c, is rectified to find the intersection point between the carrier ramp and v_c. A unidirectional carrier-ramp signal can be used when the control voltage is also unidirectional, but the control voltage will be negative when the current error becomes negative. It happens for various cases, such as reducing the reference during transient operation and changing the polarity of the reference to go from quadrant one to three or four. Taking the polarity of the control voltage and

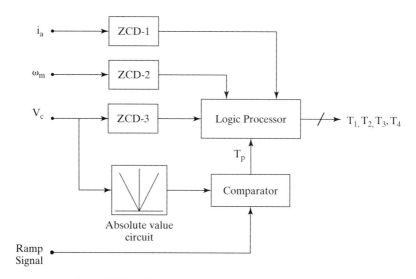

Note : ZCD — Zero Crossing Detector for polarity detection

Figure 4.22 PWM current-controller implementation with ramp carrier signal

combining it with the polarities of the current and speed gives the operational quadrant. The chopper *on* and *off* pulses generated with the intersection of rectified v_c and carrier ramp will then be combined with the quadrant-selector signals of speed, current, and control-voltage polarities to generate the base-drive signals to the chopper switching devices. An illustration is given in the drive-system simulation section.

Instead of a ramp signal for carrier waveform, a unidirectional sawtooth waveform could be used. It is advantageous in that it has symmetry between the rising and falling sides of the waveform, unlike the ramp signal. Its principle of operation is explained in the following.

The voltage applied to the load is varied within one cycle of the carrier signal. This is illustrated in Figure 4.23. The switching logic is summarized as follows:

$$i_a^* - i_a \geq \text{carrier frequency saw tooth waveform magnitude}, T_p = 1, v_a = V_s \quad (4.61)$$

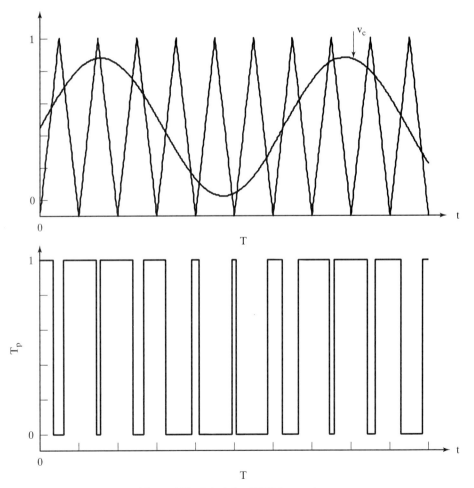

Figure 4.23 Principle of PWM operation

$i_a^* - i_a <$ carrier frequency saw tooth waveform magnitude, $T_p = 0$, $v_a = 0$ \quad (4.62)

For a fast response, the current error is amplified so that a small current error would activate the chopper control. The PWM controller has the advantage of smaller output ripple current for a given switching frequency, compared to the hysteresis current controller described later.

The pulses generated from the PWM controller are substituted for T_p in the block diagram shown in Figure 4.22. They are then processed for quadrant selection, interlock, and safety features, and appropriate base-drive signals are generated for application to the chopper circuit.

4.12.4 Hysteresis-Current Controller

The PWM current controller acts once a cycle, controlling the duty cycle of the chopper. The chopper then is a variable voltage source with average current control. Instantaneous current control is not exercised in the PWM current controller. In between two consecutive switchings, the current can exceed the maximum limit; if the PWM controller is sampled and held once a switching cycle, then the current is controlled on an average but not on an instantaneous basis. The hysteresis controller overcomes such a drawback by converting a voltage source into a fast-acting current source. The current is controlled within a narrow band of excursion from its desired value in the hysteresis controller. The hysteresis window determines the allowable or preset deviation of current, Δi. Commanded current and actual current are shown in Figure 4.24 with the hysteresis windows. The voltage applied to the load is determined by the following logic:

$$i_a \leq i_a^* - \Delta i, \quad \text{set } v_a = V_s \quad (4.63)$$

$$i_a \geq i_a^* + \Delta i, \quad \text{reset } v_a = 0 \quad (4.64)$$

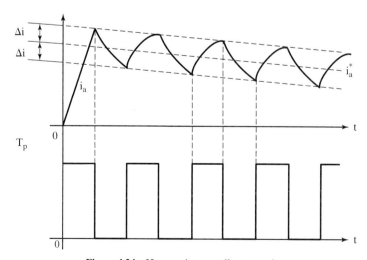

Figure 4.24 Hysteresis-controller operation

156 Chapter 4 Chopper-Controlled DC Motor Drives

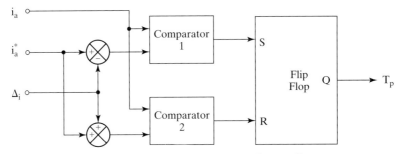

Figure 4.25 Realization of hysteresis controller

TABLE 4.1 Comparison of current controllers

Characteristics	Current Controllers	
	Hysteresis	PWM
Switching Frequency	Varying	Fixed at Carrier Frequency
Speed of Response	Fastest	Fast
Ripple Current	Adjustable	Fixed
Filter Size	Dependent on Δi	Usually small
Switching Losses	Usually high	Low

The realization of this logic is shown in Figure 4.25. The window, Δi, can either be externally set as a constant or be made a fraction of armature current, by proper programming. The chopping frequency is a varying quantity, unlike the constant frequency in the PWM controller. This has the disadvantage of higher switching losses in the devices with increased switching frequency.

The T_p pulses issued from the hysteresis controller replace the block consisting of the comparator output in the Figure 4.22. All other features remain the same for the implementation of the hysteresis controller. This controller provides the fastest response, by means of its instantaneous action. A qualitative comparison of the PWM and hysteresis controllers is summarized in Table 4.1.

4.12.5 Modeling of Current Controllers

The current-error amplifier is modeled as a gain and is given by

$$G_c(s) = K_c \tag{4.65}$$

The chopper is modeled as a first-order lag, with a gain given by

$$G_r(s) = \frac{K_r}{\left(1 + \frac{sT}{2}\right)} \tag{4.66}$$

The PWM current controller has a delay of half the time period of the carrier waveform, and its gain is that of the chopper. Hence, its transfer function, including that of the chopper, is

$$G_c G_r(s) = \frac{K_c K_r}{\left(1 + \frac{sT}{2}\right)} \quad (4.67)$$

where K_c is the gain of the PWM current controller, K_r is the gain of the chopper, and the time constant T is given by,

$$T = \frac{1}{\text{carrier frequency}} = \frac{1}{f_c} \quad (4.68)$$

The gain of the PWM current controller is dependent on the gain of the current-error amplifier. For all practical purposes, the PWM current control loop can be modeled as a unity-gain block if the delay due to the carrier frequency is negligible.

The hysteresis controller has instantaneous response; hence, the current loop is approximated as a simple gain of unity.

4.12.6 Design of Current Controller

The current loop is not easily approximated into a first-order transfer function, unlike the case of the phase-controlled-rectifier drive system. The chopping frequency is considered to be high enough that the time constant of the converter is very much smaller than the electrical time constants of the dc motor. That leads to the converter model given by the product of the converter and current-controller gains. Then the closed-loop current-transfer function is written as

$$\frac{i_a(s)}{i_a^*(s)} = K_c K_r K_1 \frac{(1 + sT_m)}{(1 + sT_1)(1 + sT_2) + H_c K_r K_c K_1 (1 + sT_m)} \quad (4.69)$$

where $K_1 = \dfrac{B_t}{K_b^2 + R_a B_t}$ and K_c and K_r are the current-controller and chopper gains, respectively. The chopper gain is derived as

$$K_r = \frac{V_s}{V_{cm}} \quad (4.70)$$

where V_s is the dc link voltage and V_{cm} is the maximum control voltage.

The gain of the current controller is not chosen on the basis of the damping ratio, because the poles are most likely to be real ones. Lower the gain of the current controller; the poles will be far removed from the zero. The higher the value of the gain, the closer will one pole move to the zero, leading to the approximate cancellation of the zero. The other pole will be far away from the origin and will contribute to the fast response of the current loop. Consider the worked-out Example 3.4 from the previous chapter. The values of various constants are:

$$K_1 = 0.049 \quad H_c = 0.355 \quad T_m = 0.7 \quad T_1 = 0.1077$$
$$T_2 = 0.0208 \quad V_s = 285 \text{ V} \quad V_{cm} = 10\text{V} \quad K_r = 28.5 \text{ V}$$

158 Chapter 4 Chopper-Controlled DC Motor Drives

The zero of the closed current-loop transfer function is at −1.42. Note that this is not affected by the current controller. The closed-loop poles for current-controller gains of 0.1, 1, and 10 are given below, along with their steady-state gains, in the following table.

K_c	Poles	Steady-state gain
0.1	−7.18, −65.11	21.33
1.0	−3.24, −203.45	213.38
10.0	−1.66, −1549.0	2580.30

It is seen that, as the gain increases, one of the poles moves closer to the zero at $-\dfrac{1}{T_m}$, enabling cancellation and a better dynamic response.

In high-performance motor-drive systems, it is usual to have a PI current controller instead of the simple proportional controller illustrated in this section. The PI controller provides zero steady-state current error, whereas the proportional controller will have a steady-state error. In the case of the PI current controller, the design procedure for the phase-controlled dc motor-drive system can be applied here without any changes. The interested reader can refer to Chapter 3 for further details on the design of the PI current controller.

4.12.7 Design of Speed Controller by the Symmetric-Optimum Method

The block diagram for the speed-controlled drive system with the substitution of the current-loop transfer function is shown in Figure 4.26. Assuming that the time constant of the speed filter is negligible, the speed-loop transfer function is derived from Figure 4.26 as

$$\frac{\omega_m(s)}{\omega_r^*(s)} = \frac{1}{H_\omega} \cdot \frac{a_0(1 + sT_s)}{a_0 + a_1 s + a_2 s^2 + a_3 s^3} \tag{4.71}$$

where

$$a_0 = K_5 \frac{K_s}{T_3} \tag{4.72}$$

$$a_1 = 1 + H_c K_r K_c K_1 + K_5 K_s \tag{4.73}$$

$$a_2 = T_1 + T_2 + H_c K_r K_c K_1 T_m \tag{4.74}$$

$$a_3 = T_1 T_2 \tag{4.75}$$

$$K_5 = K_b \frac{H_\omega K_r K_c K_1}{B_t} \tag{4.76}$$

This is very similar to the equation derived in Chapter 3, from which the following symmetric optimum conditions are imposed:

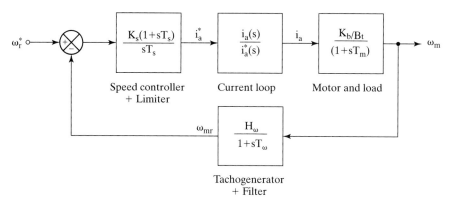

Figure 4.26 Simplified speed-controlled dc motor drive fed from a chopper with hysteresis-current control

$$a_1^2 = 2a_0 a_2 \tag{4.77}$$

$$a_2^2 = 2a_1 a_3 \tag{4.78}$$

Conditions given by equations (4.77) and (4.78) result in speed-controller time and gain constants given by

$$K_s = \frac{1}{K_5} \cdot \left\{ \frac{a_2^2}{2a_3} - (1 + H_c K_r K_c K_1) \right\} \tag{4.79}$$

$$T_s = \frac{2K_5 K_s a_2}{a_1^2} \tag{4.80}$$

For the same example, the speed-controller gain and time constants for various gains of the current controller are given next, together with the closed-loop poles and zeros and the steady-state gains of the speed-loop transfer function.

Current controller gain	Speed controller gain	Speed controller time constant	Steady-state gain	Zero	Poles
0.1	106.7	0.045	1628	−22.22	−18.3 ± j31, −36.9
1.0	102.9	0.0188	38,000	−53.2	−52.7 ± j88.9, −105.1
10.0	597.0	0.0026	1.6×10^7	−384.6	−393 ± j666.8, −792.4

Increasing current-controller gain has drastically reduced the speed-loop time constant without appreciably affecting the damping ratio of the closed-loop speed-control system. Because of the large integral gain in the speed controller, its output will saturate in time. An anti-windup circuit is necessary to overcome the saturation in this controller design and thus to keep the speed controller responsive. The anti-windup circuit can be realized in many ways. One of the implementations is shown in Figure 4.27.

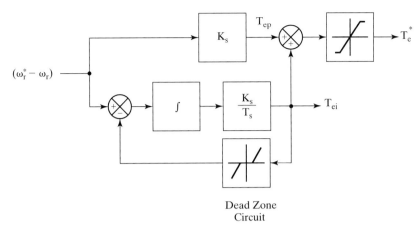

Figure 4.27 Anti-windup circuit with PI speed controller

The saturation due to the integral action alone is countered in this implementation. It is achieved by a negative-feedback control of the integral-controller output through a dead-zone circuit. This dead-zone circuit produces an output only when the integral controller output exceeds a preset absolute maximum, i.e., when the controller output saturates. This feedback is subtracted from the speed error and the resulting signal constitutes the input to the integrator. When the integral-controller output saturates, the input to the integral controller is reduced. This action results in the reduction of the integral controller's output, thus pulling the integral controller from saturation and making the controller very responsive. If there is no saturation of the integral-controller output, then the feedback is zero (because of the dead-zone circuit); hence, the anti-windup circuit is inactive in this implementation. The outputs of the integral and the proportional controllers are summed, then limited, to generate the torque reference signal. By keeping the outputs of the proportional and integral controllers separate, their individual tuning and the beneficial effect of the high proportional gain are maintained.

4.13 DYNAMIC SIMULATION OF THE SPEED-CONTROLLED DC MOTOR DRIVE

A dynamic simulation is recommended before a prototype or an actual drive system is built and integrated, to verify its capability to meet the specifications. Such a simulation needs to consider all the motor-drive elements, with nonlinearities. The transfer-function approach could become invalid, because of the nonlinear current loop. This drawback is overcome with the time-domain model developed below. The speed-controlled drive shown in Figure 4.20, but modified to include the effects of field excitation variation and with a PWM or a hysteresis controller, is considered for the simulation.

Section 4.13 Dynamic Simulation of the Speed-Controlled DC Motor Drive

4.13.1 Motor Equations

The dc motor equations, including its field, are

$$V_a = R_a i_a + L_a \frac{di_a}{dt} + K\Phi_f \omega_m \tag{4.81}$$

$$V_f = R_f i_f + L_f \frac{di_f}{dt} \tag{4.82}$$

$$T_e - T_l = J \frac{d\omega_m}{dt} + B_1 \omega_m \tag{4.83}$$

$$T_e = K\Phi_f i_a \tag{4.84}$$

where

$$K\Phi_f = M i_f \tag{4.85}$$

and M is the mutual inductance between the armature and field windings. The state variables are defined as

$$x_1 = i_a \tag{4.86}$$

$$x_2 = \omega_m \tag{4.87}$$

$$x_3 = i_f \tag{4.88}$$

The motor equations in terms of the state variables are

$$\dot{x}_1 = -\frac{R_a}{L_a} x_1 - \frac{M}{L_a} x_2 x_3 + \frac{1}{L_a} V_a \tag{4.89}$$

$$\dot{x}_2 = \frac{M}{J} x_1 x_3 - \frac{B_1}{J} x_2 - \frac{T_l}{J} \tag{4.90}$$

$$\dot{x}_3 = -\frac{R_f}{L_f} x_3 + \frac{1}{L_f} V_f \tag{4.91}$$

4.13.2 Speed Feedback

The tachogenerator and the filter are combined in the transfer function as

$$\frac{\omega_{mr}(s)}{\omega_m(s)} = \frac{H_\omega}{1 + sT_\omega} \tag{4.92}$$

The state diagram of equation (4.92) is shown in Figure 4.28, where

$$x_4 = \omega_{mr} \tag{4.93}$$

and the state equation is obtained from the state diagram as

$$\dot{x}_4 = \frac{1}{T_\omega}(H_\omega x_2 - x_4) \tag{4.94}$$

162 Chapter 4 Chopper-Controlled DC Motor Drives

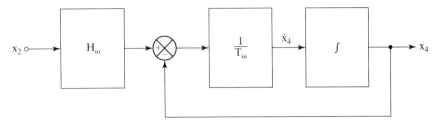

Figure 4.28 State diagram of the speed feedback

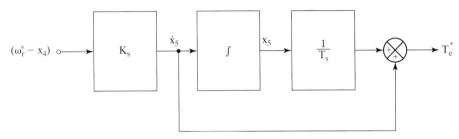

Figure 4.29 State diagram of the speed controller

4.13.3 Speed Controller

The input to the speed controller is the speed error, and the controller is of a PI type given by

$$\frac{T_e^*(s)}{(\omega_r^* - \omega_{mr})} = \frac{K_s(1 + sT_s)}{sT_s} \qquad (4.95)$$

The state diagram of the speed controller is shown in Figure 4.29, where

$$y_1 = (\omega_r^* - \omega_{mr}) \qquad (4.96)$$

The state equation and the torque equation are

$$\dot{x}_5 = K_s y_1 = K_s(\omega_r^* - \omega_{mr}) = K_s(\omega_r^* - x_4) \qquad (4.97)$$

$$T_e^* = \frac{x_5}{T_s} + \dot{x}_5 = \frac{x_5}{T_s} + K_s y_1 = \frac{x_5}{T_s} + K_s(\omega_r^* - x_4) = -K_s x_4 + \frac{1}{T_s} x_5 + K_s \omega_r^* \quad (4.98)$$

There is a limit on the maximum torque command, which is included as a constraint as

$$-T_e(\text{max}) \leq T_e^* \leq +T_e(\text{max}) \qquad (4.99)$$

4.13.4 Command Current Generator

The current command is calculated from the torque command and is given as

$$i_a^* = \frac{T_e^*}{Mi_f} = \frac{T_e^*}{M} \cdot \frac{1}{x_3} \qquad (4.100)$$

Section 4.13 Dynamic Simulation of the Speed-Controlled DC Motor Drive

Substituting for T_e^* from equation (4.98) into (4.100) gives the current command as

$$i_a^* = \left(-\frac{K_s}{M}\right) \cdot \frac{x_4}{x_3} + \left(\frac{1}{T_s M}\right) \cdot \frac{x_5}{x_3} + \frac{K_s}{M} \cdot \frac{\omega_r^*}{x_3} \quad (4.101)$$

4.13.5 Current Controller

The current error is

$$i_{er} = i_a^* - i_a \quad (4.102)$$

which, in terms of the motor and controller parameters, is

$$i_{er} = \left(-\frac{K_s}{M}\right) \cdot \frac{x_4}{x_3} + \left(\frac{1}{T_s M}\right) \cdot \frac{x_5}{x_3} + \left(\frac{K_s}{M} \cdot \omega_r^*\right) \cdot \frac{1}{x_3} - x_1 \quad (4.103)$$

The current-controller logic is illustrated for hysteresis control. It is given as

$$i_{er} > \Delta I, \quad \text{set } V_a = V_s \quad (4.104)$$
$$i_{er} < -\Delta I, \quad \text{reset } V_a = 0 \quad (4.105)$$

Note that the control logic considers only the forward-motoring mode. Suitable logic has to be incorporated for the rest of the quadrants' operation. For PWM control, the logic given by equations (4.61) and (4.62) is used.

4.13.6 System Simulation

The anti-windup controller for the speed controller can be modelled and can be incorporated in place of the PI speed controller. For the example illustrated in the following pages, the anti-windup circuit is not incorporated with the PI speed controller. Equations (4.89), (4.90), (4.91), (4.94), and (4.97) constitute the state equations; equations (4.100), (4.104), and (4.105) are to incorporate the limits and current controller. The solution of the state equations is achieved by numerical integration, with ω_r^* serving as the only input into the block diagram. A flowchart for the dynamic simulation is given in Figure 4.30.

Typical dynamic performance of a speed-controlled chopper-fed dc motor drive is shown in Figure 4.31. The current controller is of hysteresis type, and a 0.25-p.u. steady-state load torque is applied. 0.5 p.u. step speed is commanded, with a maximum torque limit of 2 p.u. resulting in the armature current limit of 2 p.u. The current window is 0.1 p.u., sufficient to cause low switching frequency for this drive system. The references are shown with dotted lines. The torque response is obtained in 1.3 ms, and the speed response is uniform and without any noticeable oscillations. Note that, as the rotor speed reaches the commanded speed, the speed error (and hence the electromagnetic torque command) decreases. During this time, to reduce the electromagnetic torque, only zero voltage is applied across the armature terminals. When the armature current goes below its reference value, then the source voltage is applied to the armature.

164 Chapter 4 Chopper-Controlled DC Motor Drives

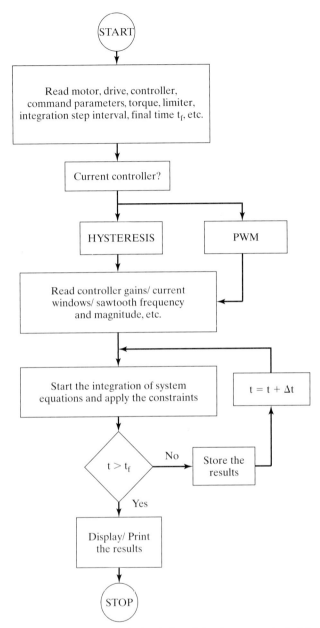

Figure 4.30 Flowchart for the dynamic simulation of the chopper-controlled dc motor drive

Section 4.13 Dynamic Simulation of the Speed-Controlled DC Motor Drive 165

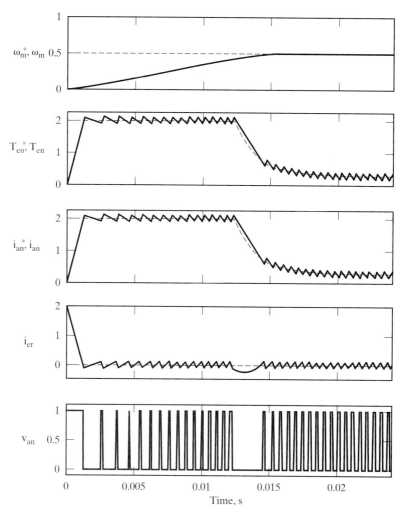

Figure 4.31 Dynamic performance of a one-quadrant chopper-controlled separately-excited dc motor drive for a step command in speed reference, in normalized units

Figure 4.32 shows the performance of a four-quadrant drive system with hysteresis-current controller, for a load torque of 0.25 p.u.. The reference speed is step-commanded, so the speed error reaches a maximum of 2 p.u., as in the first-quadrant operation described in Figure 4.31. When the speed reference goes to zero and then to negative speed, the torque command follows, but the armature current is still positive. To generate a negative torque, the armature current has to be reversed, i.e, in this case to negative value. The only way the current can be reversed is by taking it through zero value. To reduce it to zero, the armature current is made to charge the dc link by opening the transistors, thus enabling the diodes across the other pair of nonconducting transistors to conduct. This corresponds to part of the

166　Chapter 4　Chopper-Controlled DC Motor Drives

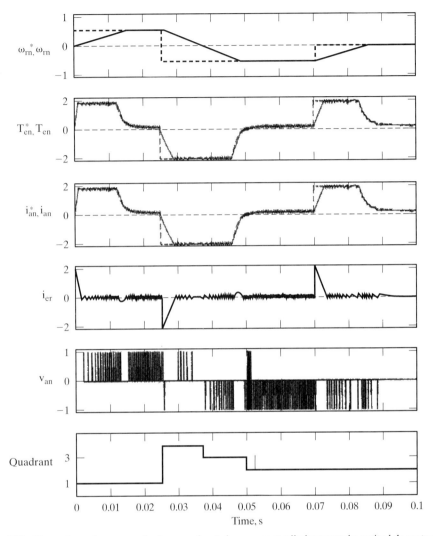

Figure 4.32 Dynamic performance of a four-quadrant chopper-controlled separately-excited dc motor drive for a step command in bipolar speed reference, in normalized units

second-quadrant operation, as shown in Figure 4.7, with the diodes D_3D_4 conducting. When the current is driven to zero, note that the speed, and hence the induced emf, are still positive. The current is reversed with the aid of the induced emf by switching T_4 only, as in to the fourth-quadrant operation shown in Figure 4.10. The induced emf enables a negative current through D_2, the machine armature, and T_4. The current rises fast; when it exceeds the current command by the hysteresis window, switch T_4 is turned off, enabling D_1 to conduct along with D_2. The armature current charges the dc source, and the voltage applied across the machine is $+V_s$ during this instant. If the current falls below the command by its hysteresis window, then the armature is shorted with T_4 again, to let the current build up. This priming and charging cycle continues until the speed reaches zero. The drive algorithm for the part discussed is briefly summarized as follows.

ω_r^*	T_e^*	ω_r	i_a^*	i_a	Condition
+	+	+	+	+	If $\omega_r < \omega_r^*$, $i_{er} \geq \Delta i$, set $V_a = V_s$
					$i_{er} \leq -\Delta i$, reset $V_a = 0$
					(Quadrant I)
+	−	+	−	+	If $\omega_r > \omega_r^*$, $V_a = -V_s$ until $i_a \to 0$
					(Quadrant I)
+	−	+	−	−	If $\omega_r > \omega_r^*$, $i_{er} \geq \Delta i$, set $V_a = V_s$
					$i_{er} \leq -\Delta i$, reset $V_a = 0$
					(Quadrant IV)

To reverse the rotor speed, the negative torque generation has to be continued. When the stored energy has been depleted in the machine as it reaches standstill, energy has to be applied to the machine, with the armature receiving a negative current to generate the negative torque. This is achieved by the switching of T_3 and T_4. From now on, it corresponds to third- and second-quadrant operations for the machine to operate in the reverse direction at the desired speed and to bring it to zero speed, respectively. For the operation in the reverse direction, the input armature voltage conditions can be derived as in the table given above. The quadrants of operation are also plotted in the figure, to appreciate the correlation of speed, torque, voltage, and currents in the motor drive system.

4.14 APPLICATION

Forklift trucks are used in material-handling applications, such as for loading, unloading, and transportation of materials and finished goods in warehouses and manufacturing plants. The forklift trucks are operated by rechargeable battery-driven dc series motor-drive systems. Such electrically operated equipment is desired, to comply with zero pollution in closed workplaces and to minimize fire hazards. The dc series motor for a typical forklift [6] is rated at 10 hp, 32 Vdc, 230 A, 3600 rpm and is totally enclosed and fan-cooled. It operates with a duty cycle of 20%. Four-quadrant operation is desired, to regenerate and charge the batteries during braking, and both directions of rotation are required, for moving in the forward and reverse directions. The torque is increased by having a gear box between the motor and the forklift drive train. A schematic diagram of the forklift control is shown in Figure 4.33. The presence of an operator means that the forklift has no automatic closed-loop speed control; the operator provides the speed feedback and control by continuous variation of speed reference.

Chopper-controlled dc motor drives are also used in conveyors, hoists, elevators, golf carts, people carriers in airport lobbies, and some variable-speed hand tools.

168 Chapter 4 Chopper-Controlled DC Motor Drives

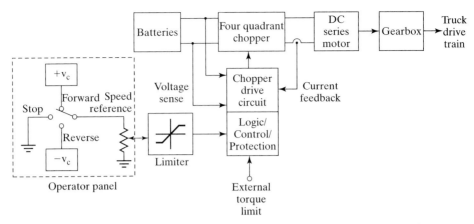

Figure 4.33 Block-diagram control schematic of a battery-fed chopper-controlled dc series motor-drive system

Example 4.6

A separately-excited chopper-controlled dc motor drive is considered for a paper winder in a winding–unwinding controller. This application uses essentially constant horsepower because the tension of the strip decreases as the build-up increases on the roll; hence, the winder can be speeded up. The maximum increase in speed is determined by the output power of the drive system. The motor ratings and parameters are as follows: 200 kW, 600 V, 2400 rpm, $R_a = 0.05\ \Omega, L_a = 0.005\ H, K_b = 2.32\ V/rad/sec, B_l = 0.05\ N\cdot m/rad/sec, J = 100\ K\ g\text{-}m^2, R_f = 30\ \Omega, L_f = 20\ H$, Input = $460\ V \pm 10\%, 3\ ph, 60\ Hz \pm 3\ Hz$.

Design the chopper power circuit with a current capacity of 2 p.u. for short duration and dc link voltage ripple factor of 1% for dominant harmonic. (The ripple factor is the ratio between the rms ac ripple voltage and the average dc voltage.)

Solution

$$\omega_b = \frac{2\pi \times 2400}{60} = 251.33\ \text{rad/sec}$$

$$T_b = \frac{P_b}{\omega_b} = \frac{200 \times 1000}{251.33} = 795.77\ \text{N}\cdot\text{m}$$

$$I_b = \frac{T_b}{K_b} = \frac{795.77}{2.32} = 343.01\ \text{A}$$

$$I_{max} = 2 \times I_b = 686.02\ \text{A}$$

Consider the power circuit shown in Figure 4.11.

(i) DC Diode-Bridge Rectifier

Considering the maximum ac input voltage,

$$\text{Minimum peak inverse voltage} = \sqrt{2}(V + 0.1V) = \sqrt{2} \times 460 \times 1.1 = 715.6\ V$$

$$\text{RMS current per diode (approximate)} = I_{max}\sqrt{\frac{1}{3}} = (680)\sqrt{\frac{1}{3}} = 392.6\ A$$

(ii) DC Link Filter

The dominant input harmonic to the dc link voltage is the sixth. Its rms value is

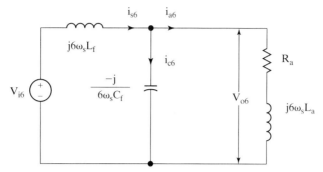

Figure 4.34 Harmonic-equivalent circuit of the dc motor drive

$$V_{i6} = \frac{1}{\sqrt{2}}\left(\frac{3\sqrt{2}V}{\pi} \cdot \frac{2}{35}\right)$$

from equation (3.175), which when substituted for nominal line-to-line voltage of 460 V, gives

$$V_{i6} = 25.1 \text{ V}$$

Given ripple factor $= 0.01 = \dfrac{V_{o6}}{V_{dc}}$,

where V_{o6} is the output voltage ripple across the capacitor and V_{dc} is the average dc voltage,

$$V_{o6} = 0.01 V_{dc} = 0.01 \times 1.35 \times V = 0.01 \times 1.35 \times 460 = 6.21 \text{ V}$$

Hence, the ratio of sixth-harmonic output voltage to input voltage is given as

$$\frac{V_{o6}}{V_{i6}} = \frac{6.21}{25.1} = 0.247$$

The sixth-harmonic equivalent circuit for the filter and motor load is shown in Figure 4.34.

From the figure, it is seen that, to minimize the harmonic current to the load, the capacitive reactance has to be approximately 10 times smaller than the load impedance:

$$\sqrt{R_a^2 + (6\omega_s L_a)^2} \gg \frac{1}{6\omega_s C_f} \cong \frac{10}{6\omega_s C_f}$$

from which C_f is evaluated as

$$C_f = \frac{10}{6\omega_s \sqrt{R_a^2 + (6\omega_s L_a)^2}} = \frac{10}{(6 \times 358.14)\sqrt{(0.05)^2 + (6 \times 358.14 \times 0.005)^2}} = 433 \text{ }\mu F$$

where ω_s corresponds to the lowest line frequency, i.e.,

$$2\pi(f_s - 3) = 2\pi(60 - 3) = 358.14 \text{ rad/sec}$$

The load impedance is 10 times the capacitive reactance; for the purpose of finding the ratio between the capacitive voltage and the input source voltage, then, the load impedance can be treated as an open circuit. This gives rise to

$$\frac{V_{o6}}{V_{i6}} = \frac{\dfrac{1}{n\omega_s C_f}}{n\omega_s L_f - \dfrac{1}{n\omega_s C_f}} = \frac{1}{n^2 \omega_s^2 L_f C_f - 1} = 0.247$$

from which $L_f C_f$ is obtained as

$$L_f C_f = \frac{\frac{1}{0.247} + 1}{n^2 \omega_s^2} = 1.09 \times 10^{-6}$$

Hence, the filter inductance is obtained, by using the previously calculated C_f, as

$$L_f = \frac{1.09 \times 10^{-6}}{C_f} = \frac{1.09 \times 10^{-6}}{433 \times 10^{-6}} = 2.52 \text{ mH}$$

(iii) Chopper Device Ratings

$$I_T = I_{max} = 686 \text{ A}$$

$$V_D = V_T = \sqrt{2}(V + 0.1V) = 715.6 \text{ V}$$

(iv) Device Selection
IGBT.

A single device per switch is sufficient in both these categories.

4.15 SUGGESTED READINGS

1. P. W. Franklin, "Theory of dc motor controlled by power pulses, Part I—Motor operation," *IEEE Trans. on Power Apparatus and Systems*, vol. PAS-91, pp. 249–255, Jan./Feb. 1972.
2. P. W. Franklin, "Theory of dc motor controlled by power pulses, Part II—Braking methods, commutation and additional losses," *IEEE Trans. on Power Apparatus and Systems*, vol. PAS-91, pp. 256–262, Jan./Feb. 1972.
3. Nisit K. De, S. Sinha, and A. K. Chattophadya, "Microcomputer as a programmable controller for state feedback control of a dc motor employing thyristor amplifier," *Conf. Record, IEEE–IAS Annual Meeting*, pp. 586–592, Oct. 1984.
4. S. R. Doradla and N. V. P. R. Durga Prasad, "Open-loop and closed-loop performance of an ac-dc PWM converter controlled separately excited dc motor drive," *Conf. Record, IEEE–IAS Annual meeting*, pp. 411–418, Oct. 1985.
5. H. Irie, T. Hirasa, and K. Taniguchi, "Speed control of dc motor driven by integrated voltage control method of chopper," *Conf. Record, IEEE IAS Annual Meeting*, pp. 405–410, Oct. 1985.
6. Report Number: DOE/BP 34906—, "Adjustable Speed Drive Applications Guidebook," 1990.
7. S. B. Dewan and A. Mirbod, "Microprocessor based optimum control for four quadrant chopper," *Conf. Record, IEEE–IAS Annual meeting*, pp. 646–652, Oct. 1980.
8. G. A. Perdikaris, "Computer control of a dc motor," *Conf. Record, IEEE – IAS Annual meeting*, pp. 502–507, Oct. 1980.
9. N. A. Loric, A. S. Sedra, and S. B. Dewan, "A phase-locked dc servo system for contouring numerical control," *Conf. Record, IEEE–IAS Annual Meeting*, pp. 1203–1207, Oct. 1981.

4.16 DISCUSSION QUESTIONS

1. Why is the duty cycle usually changed by varying the on-time rather than the chopping frequency?
2. Very small duty cycles near zero are not possible to realize in practice. What are the constraints in such cases?
3. Similarly, a duty cycle of one is not feasible in the chopper. What is the reason?
4. Draw the gating pulses for each of the four quadrants of operation of the chopper-controlled dc motor drive.
5. Is a sine-wave output current possible from a chopper circuit?
6. Is a sine-wave voltage output possible from a chopper circuit?
7. The chopper can be switched at very high frequencies. What are the factors that limit high-frequency operation?
8. A 5-hp dc motor is to be driven from a chopper for two applications: a robot, and a winder. Discuss the power device to be selected for each of the applications.
9. Compare a four-quadrant chopper and a phase-controlled converter fed from a 3-phase ac supply with regard to the number of power devices, speed of response, control complexity, and harmonics generated.
10. A two-quadrant drive to operate in FM and FR is required. Which one has to be recommended between Figures 4.13(ii) and (iii)?
11. The averaging and the instantaneous steady-state computation techniques give identical results in the continuous-conduction mode. Is this true for the discontinuous current mode, too?
12. Ripple current is minimized by either increasing the chopper frequency or including an inductance in series with the armature of the dc motor. Discuss the merits and demerits of each alternative. Are the alternatives constrained by the size of the motor drive?
13. Discuss the control circuit design for a two-quadrant chopper circuit.
14. Discuss the impact of the choice of the current controllers on the dynamic performance of the dc motor-drive system.
15. Design a digital controller for a hysteresis-current controller.
16. Design an analog version of the PWM current controller. Use commercially available integrated chips and operational amplifiers in the design.
17. Which one of the current controllers produces the highest switching losses and motor copper losses?
18. Is an innermost voltage loop necessary for fast response of the dc motor drive?
19. The hysteresis window, Δi, can be made a constant or a variable. Discuss the merits and demerits of both alternatives.
20. Likewise, the PWM carrier frequency can be either a constant or a variable. Discuss the merits and demerits of both alternatives.
21. The window in the hysteresis controller and the carrier frequency in the PWM controller are made to be variable quantities. What are they to be dependent upon for variation?
22. Is the parameter sensitivity of the open-loop chopper-controlled dc motor drive different from that of the phase-controlled dc motor drive?
23. Time-domain simulation can eliminate the errors in the final construction and testing of a chopper-controlled dc motor drive. Discuss a method of including the front-end uncontrolled rectifier and filter in the simulation.

24. The time-domain dynamic model derived and developed in this chapter has to be very slightly modified to treat either a series or a cumulatively-excited dc motor drive. What are the changes to be introduced?

25. The transfer-function model is true for both small- and large-signal analysis of the separately-excited dc motor drive. A chopper-controlled dc series motor drive has to be considered for its speed-controller design. Will the same block diagram be adequate for the task at hand?

4.17 EXERCISE PROBLEMS

1. A chopper is driving a separately-excited dc motor whose details are given in Example 4.1. The load torque is frictional and delivers rated torque at rated speed. Calculate and plot the duty cycle vs. speed. The maximum electromagnetic torque allowed in the motor is 2 p.u. The friction coefficient of the motor is 0.002 N·m/rad/sec.

2. Consider the dc motor chopper details given in Example 4.2. Compute the torque–speed characteristics for a duty cycle of 0.4, using the averaging and instantaneous steady state computation techniques. Do the methods compare well in the discontinuous current-conduction mode?

3. Compare the switch ratings of the two-quadrant choppers shown in Figures 4.13(ii) and (iii).

4. A chopper-controlled dc series motor drive is intended for traction application. Calculate its torque-speed characteristics for various duty cycles. The motor details are as follows: 100 hp, 500 V, 1500 rpm, $R_a + R_f = 0.01\ \Omega$, $L_a + L_{se} = 0.012$ H, M = 0.1 H, J = 3 Kg·m², $B_1 = 0.1$ N·m/rad/sec. The chopper has an input source voltage of 650 V and operates at 600 Hz.

5. The motor chopper given in Example 4.4 is used in a position servo application. It is required that the pulsating torque be less than 1% of the rated torque. At zero speed, the motor drive is producing a maximum of 3 p.u. torque. Determine (i) the increase in switching frequency and (ii) the value of series inductance to keep the pulsating torque within the specification.

6. The above problem has an outermost position loop. The position controller is of PI type. Determine the overall-position transfer function and find its bandwidth for $K_p = 1$ and $T_p = 0.2$ sec, if the system is stable. Assume that the speed controller is of proportional type, with a gain of 100. The current controller is a hysteresis controller.

7. A chopper-controlled dc motor has the following parameters:

    6.3 A, 200 V, 1000 rpm, $R_a = 4\ \Omega$, $L_a = 0.018$ H, $K_b = 1.86$ V/rad/sec, J = 0.1 kg·m², $B_1 = 0.0162$ N·m/rad/sec, $f_c = 500$ Hz, $V_s = 285$ V

    Determine the following:
    (i) Torque–speed characteristics for duty cycles of 0.2, 0.4, 0.6, and 0.8 in the forward-motoring mode;
    (ii) The average currents at 500 rpm, using averaging and instantaneous steady-state evaluation techniques; (Assume a suitable duty cycle.)
    (iii) Critical duty cycle vs. speed, with and without an external inductance of 20 mH. (Draw it.)

8. Determine the speed-controller gain and time constant for the following separately-excited chopper-controlled dc motor drive:

250 hp, 500 V, 1250 rpm, 92% Efficiency, $R_a = 0.052\ \Omega$, $L_a = 1$ mH, $K_b = 3.65$ V/rad/sec, $J = 5$ kg·m^2, $B_1 = 0.2$ N·m/rad/sec, $V_s = 648$ V.

The current controller has a PWM strategy with a carrier frequency of 2 kHz. Calculate also the speed response for a speed command of 0 to 0.8 p.u., when the load torque is maintained at 0.5 p.u. {Hint: Calculate the speed controller gain and time constants, assuming $B_1 = 0$.}

CHAPTER 5

Polyphase Induction Machines

5.1 INTRODUCTION

A brief introduction to the theory and principle of operation of induction machines is given in this chapter. The notations are introduced, and consistent use of them in the text is adhered to. The principle of operation of the induction motor is developed to derive the steady-state equivalent circuit and performance calculations. They are required to evaluate the steady-state response of variable-speed induction-motor drives in the succeeding chapters. The dynamic simulation is one of the key steps in the validation of the design process of the motor-drive systems, eliminating inadvertent design mistakes and the resulting errors in the prototype construction and testing and hence the need for dynamic models of the induction machine. The dynamic model of the induction machine in direct-, quadrature-, and zero-sequence axes (known as dqo axes) is derived from fundamentals. This is generally considered difficult and is avoided by many practicing engineers. Hence, care is taken to maintain simplicity in the derivation, while physical insight is introduced. The dqo model uses two windings for each of the stator and rotor of the induction machine. The transformations used in the derivation of various dynamic models are based on simple trigonometric relationships obtained as projections on a set of axes. The dynamic model is derived with the frames of observation rotating at an arbitrary speed. The most useful models in stationary, rotor, and synchronous reference frames are obtained as particular cases of the arbitrary reference-frames model. The dynamic model is used to obtain transient responses, small-signal equations, and a multitude of transfer functions, all of which are useful in the study of converter-fed induction-motor drives. Space-phasor approach has further simplified the polyphase induction machine model to one equivalent stator and one rotor winding, thereby evoking a powerful similarity to the dc machine to correspond with its armature and field windings. The space-phasor modeling of the induction motor is introduced as a simple extension of dqo models. The merits and demerits of the space-phasor representation are described. There is a small but significant following among international engineers for this model. This fact necessitates an understanding of this model to follow their research work and patents using the space-phasor model. A number of worked examples are given to illustrate the key concepts and methods in the analysis of induction machines.

5.2 CONSTRUCTION AND PRINCIPLE OF OPERATION

5.2.1 Machine Construction

This section outlines some salient aspects of the machine construction but does not go into design aspects. Design details may be found in many of the standard texts, some of which are cited in the references.

Magnetic Part: The stator and rotor of the induction machine are made up of magnetic steel laminations of thickness varying from 0.0185 to 0.035 inch (0.47 to 0.875 mm), machine-punched with slots at the inner periphery for the stator and at the outer periphery for the rotor. These slots can be partially closed or fully open in the stator laminations to adjust the leakage inductances of the stator windings. In the rotor laminations, the slots can be partially open or fully closed. The rotor slots can be very deep compared to the width and might contain from two to five compartments to hold rotor bars of different shapes placed in them. In such a case, the rotor is known as a *deep bar* rotor. Depending on the number of parallel copper bars placed in a slot, the machine is referred to as a double-, triple-, or multiple-cage induction motor. The multiple-cage rotor is intended to maximize the electromagnetic torque during starting and to minimize the rotor copper losses during steady-state operation. The stator laminations are aligned and stacked in a fixture and pressed by heavy presses of capacity varying from 40 to 80 tons, to pack the laminations very closely and to remove the air gap between them. With these steps, the stator magnetic part of the machine is ready for insertion of windings.

The rotor windings are skewed by one half or a full slot pitch from one end to the other, to minimize or to completely cancel some of the time harmonics. To accommodate such skewing, the rotor laminations are assembled with a skew in a jig and fixture and then pressed to make the rotor magnetic block.

Stator and Rotor Windings: Consider a three-phase induction machine having three windings each on its stator and rotor. The phase windings are displaced in space from each other by 120 electrical degrees, where

$$\text{Electrical degrees} = (\text{Pairs of poles}) \times \text{Mechanical degrees} \tag{5.1}$$

The three-phase rotor windings are short-circuited either within the rotor or outside of the rotor, with or without external resistances connected to them. If the rotor windings are connected through slip rings mounted on the shaft and adjacent to the rotor of the induction motor to provide external access to the rotor windings, the machine is referred to as a *slip-ring induction* motor. Alternately, the rotor windings can be bars of copper or aluminum, with two end rings attached to short-circuit the bars on the rotor itself, thus making the rotor very compact. Such a construction is known as a *squirrel-cage induction* motor.

The numbers of slots in the stator and rotor are unequal to avoid harmonic crawling torques. The windings are distributed in the slots across the periphery of the stator and rotor. The windings can have different progressions, such as concentric or lap windings. In concentric windings, the windings are centered around slots within a pole pitch. In the case of lap windings, a coil is spread over a fixed pitch angle, say, 180 electrical degrees, and connected to the coil in the adjacent slot, and so on. With a coil pitch of 180 electrical degrees, the induced emf of the coil side under the north pole will be equal but opposite to the induced emf in the coil side under the south pole. The induced emf in the coil is equal to the sum of the induced emfs in both the coil sides, which will be twice the voltage induced in one coil side. The windings need not have a pitch of 180 electrical degrees and might have less than that, to eliminate some fixed number of harmonics. This case is known as short-chorded winding. Short chording reduces the resultant voltage, because the coils are not displaced by 180 degrees but by less, with the outcome that their phasor sum is less than their algebraic sum. Such an effect is included in the pitch factor, c_p.

The slots have a phase shift, both to allow for mechanical integrity and to control the short-chording angle, so the induced emf in the adjacent coils of a phase will have a phase displacement. When the emfs in these coils of a phase are summed up, the resultant induced emf is the sum of the phasor voltages induced in these coils. A reduction in the voltage results because of the phasor addition compared to their algebraic sum. The factor to account for this aspect is known as the distribution factor, c_d.

The resultant voltage in a phase winding is reduced both by the pitch and distribution factors. Therefore, the winding factor, k_w, which reflects the effectiveness of the winding, is given as

$$k_w = c_d c_p \tag{5.2}$$

Induced Emf: With this understanding, the induced emf in a stator phase winding is derived from the first principles. The mutual flux linkages are distributed sinusoidally. The induced emf is equal in magnitude to the rate of change of mutual flux linkages, which in turn is equal to the product of the effective number of turns in the winding and the mutual flux. From this, the induced emf in a phase winding is derived as

$$e_{as} = -\frac{d\lambda}{dt} = -\frac{d}{dt}\{(k_{w1}T_1)\Phi_m \sin(2\pi f_s t)\} = -2\pi f_s k_{w1} T_1 \Phi_m \cos(2\pi f_s t) \tag{5.3}$$

where f_s is the supply frequency in Hz, Φ_m is the peak value of the mutual flux, T_1 is the total number of turns in phase *a*, and k_{w1} is the stator winding factor. The rms value of the induced emf is given by

$$E_{as} = \frac{|e_{as}|}{\sqrt{2}} = 4.44 k_{w1} T_1 \Phi_m f_s \tag{5.4}$$

The expression for the induced emf in the rotor phase windings is very similar to equation (5.4) if appropriate winding factor, number of turns, and frequency for the induced emf in the rotor are inserted into equation (5.4). More on this is deferred for the present.

Winding Method: Two types of winding methods are common in induction machines. They are known as *random-* and *form-wound* windings. They are briefly described in the following.

(i) **Random-Wound Windings:** The coils are placed in the slots and separated from the magnetic steel with an insulation paper, such as a mica sheet. Each coil in a slot contains a number of circular but stranded enameled wires which are wound on a former. This type of winding is referred to as random-wound. They are used for low-voltage (<600 V) motors. The disadvantages of this method of winding the coil are: (i) The adjacent round wires in the worst case can be the first and the last turn in the coil. Because of this, the turn-to-turn voltage can be maximum in such a case and will equal the full coil voltage but not equal the sequential turn-to-turn voltage, which is only a small fraction of the full coil voltage. (ii) They are likely to have considerable air pockets between the round wires. These air pockets form capacitors between strands of wire. With the application of repetitive voltages at high frequency with high rate of change of voltages from the inverter, a discharge current flows into the air capacitor. This is known as partial discharge. These partial discharges cause insulation failures.

Random-wound machines are economical, have low losses, and hence have a higher efficiency and tend to run cooler. Random-wound machines use a semiclosed slot, which reduces the flux density in the teeth, resulting in lower core losses—as much as 20 to 30%. Some methods have very recently been suggested to reduce the partial-discharge possibility of random-wound machines. The methods recommend using (i) heavier insulation; (ii) a wind-in-place insertion method, making the wire placement sequential; (iii) extra strategically placed insulation within a phase of the motor; (iv) extra insulating sleeves on the turns nearer to the line leads.

(ii) **Form-Wound Windings:** Higher-voltage (>600 V) induction machines are usually form-wound, meaning that each wire of rectangular cross section is placed in sequence with a strand of insulation in between them and then the bundle is wrapped with mica ground wall insulation, over which an armor covering is applied to keep all the wire arrangement firmly in place. The form-wound coils are placed in a fully open slot, as it is not possible to use the semi-closed slot for this arrangement, with the consequence that the machine will have higher core losses. This is because the tooth flux density increases: there is a sizable reduction of its cross sectional area compared to a semi-closed slotted tooth. On the other hand, form-wound windings have a much higher resistance to partial discharges, as their turn-to-turn voltage is minimal and a heavier insulation build between them also helps mitigate them. The disadvantage of this winding is that it is relatively expensive compared to random windings.

Rotor Construction: Two methods of rotor construction are used for induction machines. They are *fabrication* and *die casting* of the rotor. Fabricated rotor construction is possible for both aluminum and copper bar rotors, but aluminum fabricated rotors are hardly ever used. The reason is that a fabricated aluminum rotor is expensive, whereas a die-cast aluminum rotor is inexpensive. The fabricated copper-bar rotor is used in large machines where the aluminum die-cast rotor is not available or in high-inertia loads demanding frequent starts, such as in crushers and shredders. Frequent line-supply starts of the induction machine result in higher inrush currents that are multiple times the rated currents and hence produce more losses and forces capable of dislodging the rotor bars.

The aluminum die-cast rotor construction is used in applications having lower load inertia than recommended by National Electrical Manufacturers Association (NEMA) and not required to meet stall condition or very high starting torques. This rotor type is used prevalently and covers as much as 90% of applications.

Insulation: The stator and rotor with windings on the magnetic circuit are immersed in varnish and heat-treated for drying. The insulation sheets between the slots and coils and on the enameled wires and between turns in the coil consist of insulation materials of different classes, known as A, B, F, and H. The choice depends on the maximum temperature rise permissible for each class. The NEMA MG1 and American National Standards Institution CR50.41 specifications for allowable stator temperature rise for various insulation classes are given in Table 5.1. Service factor is the ratio between the steady-state maximum power output capability and the rated power output of the machine. Because of the uncertainty in sizing some loads, service factors greater than 1 are recommended in field applications. Note that a service factor of 1.15 means that the machine can generate 15% more power than its nameplate rating. A majority of induction machines have class F insulation; motors for servo-drive applications usually need class H insulation. Underutilized machines in nonrigorous applications with practically no overloads and infrequent starts can use lower-class insulations. Higher winding temperatures usually result in transmission of heat to bearings, resulting in failures and frequent replacements. Therefore, higher operating winding temperature is not preferred, and that is the reason for a majority of motors to have an insulation class of F or lower.

Rotor Shaft: The rotor shaft is usually made of forged steel for higher speeds (> 3,600 rpm) and has to conform to sizes recommended for power levels by NEMA or other applications-specific industrial standards such as American Petroleum Institute (API) standard 541.

TABLE 5.1 Standards for stator temperature rise

Motor Rating & Voltage	Method of Temperature Measurement	Maximum Winding Temperature (°C) for various classes of insulation						
		Service Factor = 1				Service Factor = 1.15		
		A	B	F	H	A	B	F
All	Resistance	60	80	105	125	70	90	115
≤ 1500 hp	Embedded Detectors	70	90	115	140	80	100	125
> 1500 hp, ≤ 700 V	Embedded Detectors	65	85	110	135	75	95	120
> 1500 hp, > 700 V	Embedded Detectors	50	80	105	125	70	90	115

Enclosure: Windings are inserted in the stator laminations and the stator laminations are fitted inside of a nonmagnetic steel housing. Two end covers will be attached to the steel housing with part of the bearings attached to them. They then will be assembled with through bolts after inserting the rotor with shaft. The nonmagnetic steel housing is intended for protecting the stator and rotor assembly from environmental factors (such as rain, water, snow, insects, birds, and rodents making an ingress into the machine), for providing an intermediary medium for exchange of heat between the machine windings and ambient, and for providing mechanical strength and ease of assembly. Commonly used enclosures are open drip-proof (ODP), weather-protected types I and II (WPI, WPII), totally enclosed air-to-air cooled (TEAAC), totally enclosed fan cooled (TEFC), and totally enclosed water-to-air cooled (TEWAC) types. These provide varying degrees of protection from environment and come with cost differentials among them. The enclosures are briefly summarized in the following.

ODP: It will resist the entrance of water that falls at an angle less than 15° from the vertical. It is widely used, but it is not suited for an outdoor application or a dirt-filled environment.

WPI: It is an ODP with the additional screens or louvers to prevent the entrance of objects with diameter greater than 0.75" (1·gcm). It is not ideally suited for harsh outdoor application or persistent pollution with airborne materials or abrasive dust. Examples are in textiles manufacturing and mines.

WPII: This is suitable for outdoor applications and can prevent windblown rain or snow. The heavier particles in the air are caught with blow-throughs which are 90° bends creating a low velocity area where the heavier particles drop out before they reach inner electrical parts of the machine. These can also be provided with filters to eliminate particles greater than 10 microns in size. Periodic cleaning and replacement of filters becomes necessary in this situation but comes with the advantage of longer life for the motor.

Totally Enclosed Motors: These enclosures offer the greatest protection against dirt and environment. TEAAC and TEWAC have higher power density compared to TEFC machines. These enclosures are expensive compared to other types. TEAAC has heat pipes, also known as cooling tubes or heat exchangers, that increase the overall volume of the machine with their mounting on top of the machine and also carry the disadvantage of periodic maintenance of the cooling tubes, which could clog. TEWAC has the disadvantage of requiring a clean water supply and auxiliary pump for the operation of its cooling system. It has the advantage of being practically independent of local ambient conditions.

Industrial-grade motors that are not operated continuously should be equipped with a drain, a breather, and a space heater to prevent condensation of moist air due to the breathing in of air that takes place with all types of enclosures.

Bearings: With the proper enclosure chosen for an application, the bearing selection and maintenance is the next step. Between the shaft and the end covers of the stator structure, one bearing on each end is mounted. Two kinds of bearings are available: sleeve and antifriction bearings. Antifriction bearings are cheaper than sleeve bearings and require lubrication (grease) at periodic intervals, whereas the sleeve bearings need to be checked on a daily basis for oil level and vibration. Further, sleeve bearings cannot take side or vertically upward loads. Assembly and disassembly are much easier with antifriction than with sleeve bearings. Sleeve bearings are not good with underload or overload, whereas the anti friction bearings have a moderate tolerance for both the situations. All of these facts make antifriction bearings a popular choice in a majority of applications.

Rotor Balancing: Before final assembly, the rotor is dynamically balanced so that no eccentricities of any kind are presented to the bearings. The balancing is achieved either by removing some magnetic iron material in smaller motors or by adding a magnetically and electrically inert material compound in larger machines.

For detailed information on the construction details of the machine, refer to the papers and standard texts cited in the reference section of this chapter or to a multitude of textbooks on electrical machines.

5.2.2 Principle of Operation

It is well known that, when a set of three-phase currents displaced in time from each other by angular intervals of 120° is injected into a stator having a set of three-phase windings displaced in space by 120° electrical, a rotating magnetic field is produced. This rotating magnetic field has a uniform strength and travels at an angular speed equal to its stator frequency. It is assumed that the rotor is at standstill. The rotating magnetic field in the stator induces electromagnetic forces in the rotor windings. As the rotor windings are short-circuited, currents start circulating in them, producing a reaction. As known from Lenz's law, the reaction is to counter the source of the rotor currents, i.e., the induced emfs in the rotor and, in turn, the rotating magnetic field itself. The induced emfs will be countered if the difference in the speed of the rotating magnetic field and the rotor becomes zero. The only way to achieve it is for the rotor to run in the same direction as that of the stator magnetic field and catch up with it eventually. When the differential speed between the rotor and magnetic field in the stator becomes zero, there is zero emf, and hence zero rotor currents resulting in zero torque production in the motor. Depending on the shaft load, the rotor will settle down to a speed, ω_r, always less than the speed of the rotating magnetic field, called the *synchronous speed* of the machine, ω_s. The speed differential is known as the *slip speed*, ω_{sl}. The elementary relationships between slip speed, rotor speed, and stator frequency are given below.

Synchronous speed is given as

$$\omega_s = 2\pi f_s, \text{ rad/sec} \tag{5.5}$$

where f_s is the supply frequency.

If ω_m is the mechanical rotor speed, slip speed is

$$\omega_{sl} = \omega_s - \omega_r = \omega_s - \frac{P}{2}\omega_m, \text{ rad/sec} \tag{5.6}$$

where P is the number of poles.

The differential speed between the stator magnetic field and rotor windings is slip speed, and that is responsible for the frequency of the induced emfs in the rotor and hence their currents. Therefore, the rotor currents are at slip frequency, which can be obtained from the angular slip speed by dividing it by 2π. The slip is defined as

$$s = \frac{\omega_{sl}}{\omega_s} \tag{5.7}$$

Note that slip is nondimensional. It is proven later that this is one of the most important variables in the control and operation of the induction machines. Combining equations (5.6) and (5.7), the rotor electrical speed is given as

$$\omega_r = \omega_s(1 - s), \text{rad/sec} \tag{5.8}$$

From this, the rotor speed in rpm, denoted by n_r, is expressed as

$$n_r = n_s(1 - s)(\text{rpm}) \tag{5.9}$$

where n_s is the synchronous speed or the speed of the stator magnetic field in rpm, given by

$$n_s = \frac{120 f_s}{P} \text{rpm} \tag{5.10}$$

5.3 INDUCTION-MOTOR EQUIVALENT CIRCUIT

The equivalent circuit of the induction motor is very similar to that for a transformer. Although the rotor currents are at slip frequency, the rotor is incorporated into the circuit in a simple way. Recognizing the fact that the stator and rotor windings have resistances and leakage inductances and that the mutual inductance for modeling the mutual flux links the stator and rotor windings, Figure 5.1 gives the elementary equivalent circuit. The parameters are stator resistance per phase, R_s, rotor resistance per phase, R_{rr}, mutual inductance, L_m, stator leakage inductance

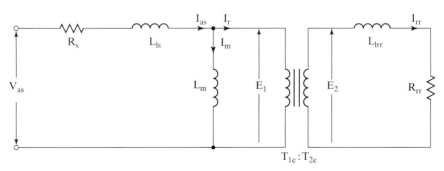

Figure 5.1 Elementary equivalent circuit for the induction motor

per phase, L_{ls}, rotor leakage inductance per phase, L_{lrr}, stator turns per phase, T_1, rotor turns per phase, T_2, induced emf in the stator per phase, E_1, and induced emf in the rotor per phase, E_2. Effective stator or rotor turns per phase is equal to the product of the number of turns per phase and the winding factor of the stator and rotor and, respectively, is denoted as k_{w1} and k_{w2}. The winding factor is the product of pitch and distribution factors that account for the specific winding characteristics. The evaluation of these factors can be found in any standard textbook on electrical machines.

The relationship between the induced emfs is

$$\frac{E_2}{E_1} = s\frac{T_{2e}}{T_{1e}} = \frac{s}{a} \quad (5.11)$$

where T_{1e} and T_{2e} are the effective stator and rotor turns per phase, and the turns ratio, a, between the stator-to-rotor effective turns per phase is given as

$$a = \frac{T_{1e}}{T_{2e}} = \frac{k_{w1}T_1}{k_{w2}T_2} = \text{turns ratio} \quad (5.12)$$

The rotor current I_{rr}, then, is

$$I_{rr} = \frac{E_2}{R_{rr} + j\omega_{sl}L_{lrr}} = \frac{E_2}{R_{rr} + js\omega_s L_{lrr}} \quad (5.13)$$

Substituting for E_2 from equation (5.11) into equation (5.13), the rotor current is

$$I_{rr} = \frac{E_1}{\frac{aR_{rr}}{s} + j\omega_s(aL_{lrr})} = \frac{E_1/a}{\frac{R_{rr}}{s} + j\omega_s L_{lrr}} \quad (5.14)$$

Equation (5.14) is incorporated into the equivalent circuit as shown in Figure 5.2. Note that both the rotor and stator uniformly have the same frequency, which is that of the stator in equation (5.14). The rotor current reflected into the stator is denoted as I_r in Figure 5.2 and is given in terms of the rotor current I_{rr} as

$$I_r = \frac{I_{rr}}{a} \quad (5.15)$$

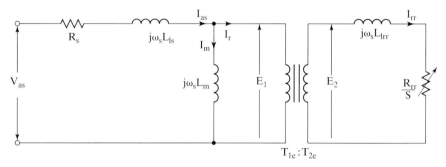

Figure 5.2 Equivalent circuit of the induction motor

Note that the slip does not enter in this; the ratio between the stator and rotor induced emfs viewed at stator frequency does not contain it, and rotor impedance absorbs this slip. Substitution of equation (5.14) into equation (5.13) yields

$$I_r = \frac{E_1}{\dfrac{(a^2 R_{rr})}{s} + j\omega_s(a^2 L_{rr})} = \frac{E_1}{\dfrac{R_r}{s} + j\omega_s L_{lr}} \tag{5.16}$$

where R_r and L_{lr} are the stator-referred rotor resistance and leakage inductance, respectively, given as

$$R_r = a^2 R_{rr}$$
$$L_{lr} = a^2 L_{lrr} \tag{5.17}$$

As the fictitious rotor and stator at the air gap have the same induced emf, E_1, the physical isolation can be removed to get a connected circuit. The final equivalent circuit, referred to the stator, is shown in Figure 5.3. Magnetization is accounted for by the magnetizing branch of the equivalent circuit, consisting of the magnetizing inductance that is lossless and hence cannot represent core losses. An equivalent resistance can represent the core losses. This core-loss resistance is in parallel to the magnetizing inductance, because the core losses are dependent on the flux and hence proportional to the flux linkages and the resulting air gap voltage E_1. The self-inductances of the stator phase winding, L_s, and of the rotor phase winding referred to the stator, L_r, are obtained as the sum of the magnetizing inductance and respective leakage inductances:

$$L_s = L_m + L_{ls} \tag{5.18}$$
$$L_r = L_m + L_{lr} \tag{5.19}$$

The corresponding reactances are obtained by multiplying the inductances by the stator angular frequency. A simple phasor diagram of the induction motor is drawn by using the equivalent circuit given in Figure 5.3. The basis is provided by the following relationships, contained in equations from (5.20) to (5.25).

When the machine is energized with no load on the rotor, then the rotor circuit is open-circuited, because the slip is zero. The stator current drawn during this condition, known as the no-load current, contributes to the magnetization of the machine

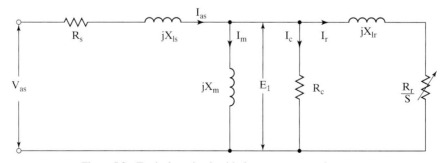

Figure 5.3 Equivalent circuit with the rotor at stator frequency

184 Chapter 5 Polyphase Induction Machines

and resulting core losses. Then this no-load current is viewed as the sum of the magnetizing and core-loss components of the current and accordingly is written as

$$I_o = I_m + I_c \tag{5.20}$$

where I_o is the no-load phase current, I_m is the magnetizing current, and I_c is the core-loss current. The magnetizing current in terms of the air gap voltage and the magnetizing reactance is written as

$$I_m = \frac{E_1}{jX_m} \tag{5.21}$$

where E_1 is the air gap voltage and X_m is the magnetizing reactance. Similarly, the core-loss component of the stator current is written as

$$I_c = \frac{E_1}{R_c} \tag{5.22}$$

where R_c is the resistance to account for the core losses. The rotor phase current is given by

$$I_r = \frac{E_1}{\frac{R_r}{s} + jX_{lr}} \tag{5.23}$$

where I_r is the rotor phase current. The stator phase current then is

$$I_{as} = I_r + I_o \tag{5.24}$$

In terms of the induced emf, stator current, and stator parameters, the applied stator voltage is expressed as the sum of the induced emf and stator impedance voltage drop and is given by

$$V_{as} = E_1 + (R_s + jX_{ls})I_{as} \tag{5.25}$$

where V_{as} is the stator phase voltage.

The phasor diagram is shown in Figure 5.4. The angle ϕ between the stator input voltage and the stator current is the power factor angle of the stator, and λ_m is the mutual flux. The steady-state performance is taken up next by using the equivalent circuit.

5.4 STEADY-STATE PERFORMANCE EQUATIONS OF THE INDUCTION MOTOR

The key variables in the machine are the air gap power, mechanical and shaft output power, and electromagnetic torque. These are derived from the equivalent circuit of the induction machine as follows. The real power transmitted from the stator, P_i, to the air gap, P_a, is the difference between total input power to the stator windings and copper losses in the stator and is given as

$$P_a = P_i - 3I_s^2 R_s \quad (W) \tag{5.26}$$

Section 5.4 Steady-State Performance Equations of the Induction Motor

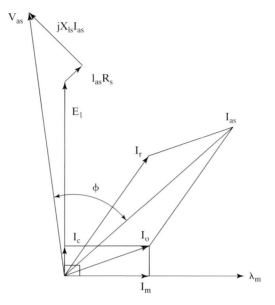

Figure 5.4 Simplified phasor diagram of the induction motor

Neglecting the core losses, the air gap power is equal to the total power dissipated in R_r/s in the three phases of the machine; there is no other element to consume power in the rotor equivalent circuit. It is given as

$$P_a = 3I_r^2 \frac{R_r}{s} \tag{5.27}$$

which could be written alternatively as

$$P_a = 3I_r^2 R_r + 3I_r^2 R_r \frac{(1-s)}{s} \tag{5.28}$$

The $I_r^2 R_r$ term is recognized as the rotor copper loss, and hence the remainder, $I_r^2 R_r \frac{(1-s)}{s}$, gives the power converted into mechanical form. The rotor copper losses are equal to the product of the slip and air gap power from equation (5.27), and this is referred to as slip power. The common term of three in equations (5.27) and (5.28) accounts for the number of phases in the machines, which throughout this text is taken as three. The mechanical power output, P_m, is obtained as

$$P_m = 3I_r^2 R_r \frac{(1-s)}{s} \quad (W) \tag{5.29}$$

Alternately, in terms of the electromagnetic torque and rotor speed, the mechanical power output is equal to their product:

$$P_m = T_e \omega_m \tag{5.30}$$

where T_e is the internal or electromagnetic torque, derived from equations (5.29) and (5.30) as

$$T_e = \frac{3 I_r^2 R_r (1-s)}{s \omega_m} \tag{5.31}$$

Substituting for the rotor speed in terms of the slip and stator frequency, given by

$$\omega_m = \frac{\omega_r}{P/2} = \frac{\omega_s (1-s)}{P/2} \tag{5.32}$$

into (5.29), the electromagnetic or air gap torque is obtained as

$$T_e = 3 \left(\frac{P}{2} \right) \frac{I_r^2 R_r}{s \omega_s} \tag{5.33}$$

To obtain the shaft output power of the machine, P_s, the windage and friction losses of the rotor, denoted as P_{fw}, have to be subtracted from the mechanical output power of the machine, symbolically given as follows:

$$P_s = P_m - P_{fw} \tag{5.34}$$

The friction and windage losses are two distinct and separate losses; they are proportional to the speed and the square of the speed, respectively, therefore they have to be represented as a function of speed for evaluation of the variable-speed performance of the induction motor. There are also losses due to stray magnetic fields in the machine; they are covered by the term *stray load* losses. The stray-load losses vary from 0.25 to 0.5 percent of the rated machine output. The stray-load losses are obtained from the measurements on the machine under load from the remainder of the difference between the input power and the sum of the known losses such as the stator and rotor copper and core losses, friction and windage losses, and power output. Note that the stray-load losses have not been accounted for in the equivalent circuit of the machine. Various analytical formulae and empirical relationships are in use, but a precise prediction of the stray losses is very difficult.

Example 5.1

(i) Find the efficiency of an induction motor operating at full load. The machine details are given in the following:

2000 hp, 2300 V, 3 phase, Star connected, 4 pole, 60 Hz, Full load slip = 0.03746

$R_s = 0.02\ \Omega$; $R_r = 0.12\ \Omega$; $R_c = 451.2\ \Omega$; $X_m = 50\ \Omega$; $X_{ls} = X_{lr} = 0.32\ \Omega$

(ii) The line power factor needs to be improved to unity by installing capacitors at the input terminals of the induction motor. Calculate the per-phase capacitance required to obtain a line power factor of unity.

Section 5.4 Steady-State Performance Equations of the Induction Motor

Solution (i) The equivalent circuit of the induction motor is utilized to solve the problem. First, calculate the equivalent input impedance of the induction motor. This is achieved by finding the equivalent impedance between the magnetizing and core-loss resistance, and then the combined impedance of this with the rotor impedance is found, which, when added to the stator impedance, gives the equivalent impedance of the induction machine.

Magnetizing-branch equivalent impedance, $Z_o = \dfrac{jX_m R_c}{R_c + jX_m} = 5.47 + j49.39 \, \Omega$

Rotor impedance, $Z_r = \dfrac{R_r}{s} + jX_{lr} = \dfrac{0.12}{0.03447} + j0.32 \, \Omega$

Equivalent rotor and magnetizing impedance, $Z_{eq} = \dfrac{Z_o Z_r}{Z_o + Z_r} = 3.13 + j0.51 \, \Omega$

Motor equivalent impedance, $Z_{im} = R_s + jX_{ls} + Z_{eq} = 3.15 + j0.83 \, \Omega$

Phase input voltage, $V_{as} = \dfrac{V_{ll}}{\sqrt{3}} = \dfrac{2300}{\sqrt{3}} = 1327.9 \, V$

Stator current, $I_s = \dfrac{V_{as}}{Z_{im}} = 394.2 - j104.1 \, A$

Rotor current, $I_r = \dfrac{Z_{eq}}{Z_{im}} I_s = 393.87 - j78.07 \, A$

No-load current, $I_o = I_s - I_r = 0.33 - j26.03 \, A$

Core-loss current, $I_c = \dfrac{Z_o}{R_c} I_o = 2.85 - j0.28 \, A$

Rotor angular speed,
$\omega_m = (1 - s) \dfrac{\omega_s}{P/2} = (1 - s) \dfrac{2\pi f_s}{P/2} = 0.03746 \dfrac{2 * \pi * 60}{4/2} = 181.43 \, rad/sec \, (mech)$

Air-gap torque, $T_e = 3 \dfrac{P}{2} |I_r|^2 \dfrac{R_r}{s \omega_s} = 8220.1 \, N \cdot m$

Mechanical power, $P = \omega_m T_e = 1491.2 \, kW$
Shaft power output, $P_m = P_s - P_m = 1491.2 - 0 = 1491.2 \, kW$
Stator resistive losses, $P_{sc} = 3|I_s|^2 R_s = 3 * 407.74^2 * 0.2 = 9.975 \, kW$
Rotor resistive losses, $P_{rc} = 3|I_r|^2 R_r = 3 * 401.53^2 * 0.12 = 58.04 \, kW$
Core losses, $P_{co} = 3|I_c|^2 R_c * 2.865^2 * 451.2 = 11.11 \, (kW)$
Input power, $P_i = P_m + P_{sce} + P_{rc} + P_{co} = 1491.2 + 9.975 + 58.04 + 11.11 = 1570.5 \, kW$

% Efficiency, $\eta = \dfrac{P_s}{P_i} 100 = \dfrac{1491.2}{1570.5} 100 = 94.96\%$

(ii) The principle of power-factor improvement with capacitor installation at the machine stator terminals is based on the capacitor's drawing a leading reactive current from the supply to cancel the lagging reactive current drawn by the induction machine. In order for the line power factor to be unity, the reactive component of the line current must be zero. The reactive line current is the sum of the capacitor and induction machine reactive currents. Therefore, the capacitive reactive current (I_{cap}) has to be equal in magnitude but opposite in direction to the machine lagging reactive current, but the machine reactive current is the imaginary part of the stator current and is given by

$$I_{cap} + \text{imag}(I_s) = 0$$

$$\text{Hence,} \; I_{cap} = -\text{imag}(I_s) = -(-j104.1) = j104.1 \; A$$

This current is controlled by the capacitors installed at the input, and therefore the capacitor required is

$$C = \frac{I_{cap}}{j\omega_s V_{as}} = \frac{j104.1}{j377 * 1327.9} = 20.792 \; \mu F$$

5.5 STEADY-STATE PERFORMANCE

A flowchart for the evaluation of the steady-state performance of the motor is given in Figure 5.5. The relevant equations for use are from (5.22) to (5.33). A set of sample torque-vs.-slip characteristics for constant input voltage is shown in Figures 5.6(a) and (b).

The torque-vs.-slip characteristics are shown for slip varying from 0 to 1. The slip is chosen in place of rotor speed because it is nondimensional and so is applicable to any motor frequency. Near the synchronous speed, i.e., at low slips, the torque is linear and is proportional to slip; beyond the maximum torque (also known as breakdown torque), the torque is approximately inversely proportional to slip, as is seen from the Figure 5.6(a). At standstill, the slip equals unity, and, at this operating point, the torque produced is known as standstill torque. To accelerate a load, this standstill torque has to be greater than the load torque. It is preferable to operate near low slips to have higher efficiency. This is due to the fact that the rotor copper losses are directly proportional to slip and are equal to the slip power, and, hence, at low slips, the rotor copper losses are small.

The positive-slope region of the torque–slip characteristics provides stable operation. Consider the machine operating at 1 p.u. with a low slip, and let the load torque be increased to 1.5 p.u. The rotor slows down and thereby develops a larger slip, which increases the electromagnetic torque capable of meeting the load torque. The new steady state is reached at 1.5 p.u. torque after a transient period with oscillations in torque. If the operating point at 1 p.u. torque at a slip of 0.85 is considered, the load torque disturbance will lead to increasing slip, resulting in less and less torque generation, thereby diverging more and more from the new load torque leading to a final pullout of the machine and reaching standstill. As this discussion is addressed to the steady-state characteristics of the machine, the two regions are named statically stable and unstable, as indicated in Figure 5.6(a).

Figure 5.6(b) shows the torque vs. slip for a wide range of slip, from −2 to 2, with the stator supplied with rated voltages and frequency. Consider the motor as spinning in the direction opposite to that of a phase sequence *abc*. Assume that a set of stator voltages with a phase sequence *abc* is applied at supply frequency. This creates a stator flux linkage counter to the direction of rotor speed, resulting in a braking action. This also creates a slip greater than one: the rotor speed is negative with

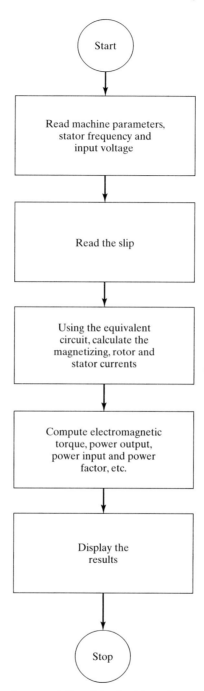

Figure 5.5 Flowchart for the computation of steady-state performance of the induction motor

190　Chapter 5　Polyphase Induction Machines

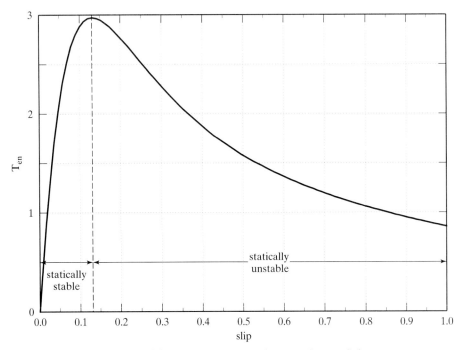

Figure 5.6　(a) Induction motor speed–torque characteristics

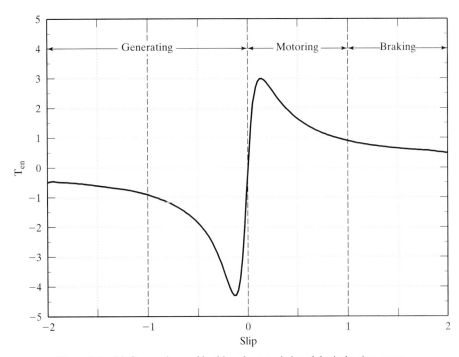

Figure 5.6　(b) Generation and braking characteristics of the induction motor

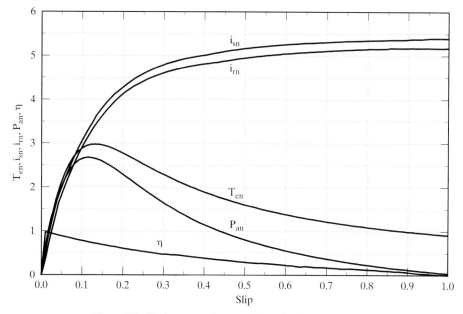

Figure 5.7 Performance characteristics of an induction motor

respect to synchronous speed. This braking action brings rotor speed to standstill in time. Consider the rotor electrical speed to be greater than the synchronous speed, resulting in a negative slip. A negative slip changes the generation of positive (motoring) to negative (generating) torque as the induced emf phase is reversed. Hence, for negative slip, the torque-vs.-slip characteristic is similar to the motoring characteristic discussed earlier, except that the breakdown torque is much higher with negative-slip operation. This is due to the fact that the mutual flux linkages are strengthened by the generator action of the induction machine. The reversal of rotor current reduces the motor impedance voltage drop, resulting in a boost of magnetizing current and hence in an increase of mutual flux linkages and torque.

The performance characteristics are shown in Figure 5.7 for rated voltage and frequency. The stator current at standstill on full voltage is 5.35 p.u. The efficiency is approximately equal to $(1 - s)$. Even though there is torque generated at standstill, the power output is zero (because of zero rotor speed). For variable-speed applications, the range of interest for the slip is 0 to 0.13 p.u., i.e., in the positive slope region of the torque-vs.-slip characteristics. It is seen that the efficiency is high in this region, as explained elsewhere.

Starting Torque: The starting torque is obtained from the torque expression by substituting for $s = 1$. Neglecting the magnetizing current, the electromagnetic torque at starting is approximately

$$T_e \cong \frac{3}{\omega_s} \cdot \frac{P}{2} \cdot \frac{V_{as}^2 \cdot R_r}{(R_s + R_r)^2 + (X_{ls} + X_{lr})^2} \qquad (5.35)$$

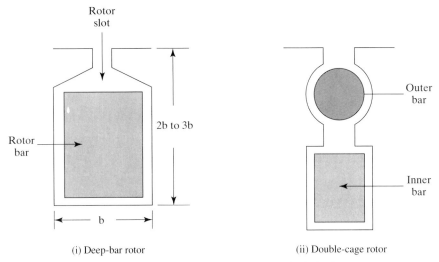

(i) Deep-bar rotor (ii) Double-cage rotor

Figure 5.8 Deep-bar and double-cage rotors for the induction motor

This torque can be augmented for motors started directly with full line voltages (what are commonly termed *line start* motors), by increasing the rotor resistance. That increases the numerator proportionally and the denominator only slightly in equation (5.35), contributing to a higher torque. Rotor resistance is increased through the connection of external resistors in the slip-ring rotor. In the case of the squirrel-cage rotor, it is realized through either deep-bar or double-cage rotors, shown in Figure 5.8. Note that, at standstill, the rotor currents are at supply frequency, and therefore the rotor conductors have considerable skin effect, resulting in a higher resistance. Further, the larger slot height, 2 to 3 times the slot width, results in higher leakage flux, which prevents the flow of current in the inner part of the rotor bars, further decreasing the cross-sectional area of the rotor bar for current conduction and resulting in increased rotor resistance. When the machine is operating near synchronous speed, the rotor current frequency is very low, and hence the skin effect becomes negligible, resulting in an even flow of current in the rotor bar. That reduces the rotor resistance and hence provides an efficient operation of the motor in steady state. By similar reasoning, it could be seen that the outer bar becomes effective during starting and the inner bar predominates during steady-state operation in the double-cage rotor shown in Figure 5.8(ii). Note that the outer bars have smaller area of cross-section compared to the inner bars.

Induction Motor Classifications: The National Electrical Manufacturers Association (NEMA Class A, B, C, and D) has classified induction motors based on the torque–slip characteristics. Typical characteristics of a NEMA B and NEMA D induction motor are shown in Figure 5.9. Class A machine characteristic is similar to Class B machine. NEMA Class A and B are for general-purpose applications, such as fans and pumps. Class C has lower peak torque but has a higher starting torque than class B. NEMA Class C is for driving compressor pumps. NEMA Class D is to provide high starting torque and a wide statically stable speed range of operation, with the disadvantage of low efficiency compared to other types.

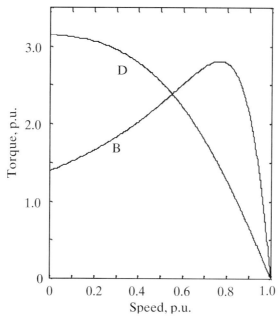

Figure 5.9 Typical speed–torque curves for squirrel-cage induction motors with NEMA Design classification B and D.

From the previous discussion on starting torques, it is seen that various types of motors are realized by means of different rotor slot and bar constructions.

5.6 MEASUREMENT OF MOTOR PARAMETERS

The measurement of motor parameters is based on the equivalent circuit of the induction motor. The various tests to measure the motor parameters are described, and the calculation of the parameters is considered.

5.6.1 Stator Resistance

With the rotor at standstill, the stator phase resistance is measured by applying a dc voltage and the resulting current. While this procedure gives only the dc resistance at a certain temperature, the ac resistance has to be calculated by considering the wire size, the stator frequency, and the operating temperature.

5.6.2 No-Load Test

The induction motor is driven at synchronous speed by another motor, preferably a dc motor. Then the stator is energized by applying rated voltage at rated frequency. The input power per phase, P_1, is measured. The corresponding equivalent circuit for this condition is shown in Figure 5.10. The slip is zero, so the rotor circuit is open. For the present, the stator impedance can be neglected; it is small compared to the

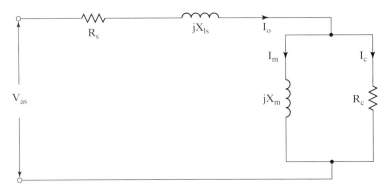

Figure 5.10 Equivalent circuit of the induction motor at no load

effective impedance of the magnetizing and core-loss branches. The mutual inductance and the core-loss resistance are calculated as follows:

The no-load power factor is given by

$$\cos \phi_o = \frac{P_1}{V_{as} I_o} \tag{5.36}$$

from which the magnetizing current is calculated as

$$I_m = I_o \sin \phi_o \tag{5.37}$$

and the core-loss current is given by

$$I_c = I_o \cos \phi_o \tag{5.38}$$

The magnetizing inductance is computed from

$$L_m = \frac{V_{as}}{2\pi f_s I_m} \tag{5.39}$$

and the core-loss resistance is given by

$$R_c = \frac{V_{as}}{I_c} \tag{5.40}$$

For most applications, such an approximate procedure for the evaluation of L_m and R_c is sufficient.

5.6.3 Locked-Rotor Test

The rotor of the induction motor is locked to keep it at standstill and a set of low three-phase voltages is applied to circulate rated stator currents. The input power per phase, P_{sc}, is measured along with the input voltage and stator current. The slip is unity for the locked-rotor condition and hence the circuit resembles that of a secondary-shorted transformer. The equivalent circuit for this situation is shown in Figure 5.11. The magnetizing branch impedance is very high compared to the rotor impedance, $(R_r + jX_{lr})$, and hence the approximate effective impedance is only the smaller of the two, which is the rotor impedance in this case. The parameters are calculated as follows.

Section 5.6 Measurement of Motor Parameters 195

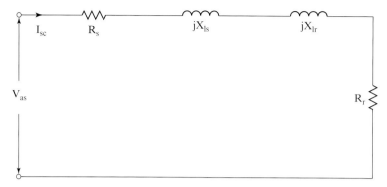

Figure 5.11 Equivalent circuit of the induction motor at standstill

The short-circuit power factor obtained from the equivalent circuit is

$$\cos \phi_{sc} = \frac{P_{sc}}{V_{sc} I_{sc}} \tag{5.41}$$

and the short-circuit impedance is given by

$$Z_{sc} = \frac{V_{sc}}{I_{sc}} \tag{5.42}$$

from which the rotor resistance and total leakage reactance are computed as

$$R_r = Z_{sc} \cos \phi_{sc} - R_s \tag{5.43}$$
$$X_{eq} = Z_{sc} \sin \phi_{sc} \tag{5.44}$$

where the total leakage reactance per phase, X_{eq}, is the sum of the stator and referred-rotor leakage reactances, given as

$$X_{eq} = X_{ls} + X_{lr} \tag{5.45}$$

It is usual to assume that the stator leakage inductance is equal to the rotor leakage inductance. For precise calculation, it is a good practice to consult the manufacturer's data sheet. Some rules of thumb are available as to their relationship for various types of motors.

Example 5.2

The no-load and locked-rotor test results for a three-phase, star-connected, 60-Hz, 2000-hp induction machine with a stator phase resistance of 0.02 Ω are as follows:

Test	Input line to line voltage, V	Line current, A	Three-phase input power, kW
No load	2300	26.55	11.617
Locked rotor	462.68	407.75	319.22

Find the machine equivalent-circuit parameters.

Solution (i) From the no-load test results:

Input power, $P_1 = 11.617/3 = 3.872$ kW/phase

Power factor, $\cos \phi_o = \dfrac{P_1}{V_{as}I_o} = \dfrac{11.617 * 10^3}{\sqrt{3} * 2300 * 26.55} = 0.1098$

Magnetizing current, $I_m = I_o \sin \phi_o = 26.39$ (A)

Core-loss branch current, $I_c = I_o \cos \phi_o = 2.9216$ (A)

Neglecting stator impedance, the following are calculated:

Magnetizing inductance, $L_m = \dfrac{V_{as}}{2\pi f_s I_m} = \dfrac{2300/\sqrt{3}}{2\pi * 60 * 26.39} = 0.1335$ H

Core-loss resistance, $R_c = \dfrac{V_{as}}{I_c} = \dfrac{2300/\sqrt{3}}{2.916} = 455.37$ (Ω)

(ii) From the locked-rotor test results:

Power factor, $\cos \phi_{sc} = \dfrac{P_{sc}}{V_{sc}I_{sc}} = \dfrac{(319.22/3) * 10^3}{(462.68/\sqrt{3}) * 407.75} = 0.2137$

Phase angle, $\phi_{sc} = 1.355$ rad

Locked-rotor impedance, $Z_{sc} = \dfrac{V_{sc}}{I_{sc}} = \dfrac{462.68/\sqrt{3}}{407.75} = \dfrac{267.13}{407.75} = 0.6551\,\Omega$

Stator-referred rotor resistance per phase, $R_r = Z_{sc} \cos \phi_{sc} - R_s = 0.12\,\Omega$

Total leakage reactance, $X_{eq} = Z_{sc} \sin \phi_{sc} = 0.64\,\Omega$

The per-phase stator and rotor leakage reactances are half of the total leakage reactance; hence,

$$X_{ls} = X_{lr} = 0.32\,\Omega$$

and the leakage inductances are

$$L_{ls} = L_{lr} = 0.8488 \text{ mH}$$

(iii) Further refinement of the equivalent-circuit parameters: The stator impedance was neglected in the calculation of magnetizing inductance and core-loss resistance because, at that time, the stator leakage inductance was not available. Now, having obtained it from the locked rotor test, we can use the no-load equivalent circuit to recalculate the magnetizing inductance and core-loss resistance; their final values are obtained as 0.1326 H and 451.2 Ω, respectively.

5.7 DYNAMIC MODELING OF INDUCTION MACHINES

The steady-state model and equivalent circuit developed in earlier sections are useful for studying the performance of the machine in steady state. This implies that all electrical transients are neglected during load changes and stator frequency variations. Such variations arise in applications involving variable-speed drives. The variable-speed drives are converter-fed from finite sources, unlike the utility sources, due to the limitations of the switch ratings and filter sizes. This results in their incapability to supply large transient power. Hence, we need to evaluate the dynamics of converter-fed variable-speed drives to assess the adequacy of the converter switches and the converters for a given motor and their interaction to determine the excursions of currents and torque in the converter and motor. The dynamic model considers the instantaneous effects of varying voltages/currents, stator frequency, and torque disturbance. The dynamic model of the induction motor is derived by using a two-phase motor in direct and quadrature axes. This approach is desirable because of the conceptual simplicity obtained with two sets of windings, one on the stator and the other on the rotor. The equivalence between the three-phase and two-phase machine models is derived from simple observation, and this approach is suitable for extending it to model an n-phase machine by means of a two-phase machine. The concept of power invariance is introduced: the power must be equal in the three-phase machine and its equivalent two-phase model. The required transformation in voltages, currents, or flux linkages is derived in a generalized way. The reference frames are chosen to be arbitrary, and particular cases, such as stationary, rotor, and synchronous reference frames, are simple instances of the general case. Derivations for electromagnetic torque involving the currents and flux linkages are given. The differential equations describing the induction motor are nonlinear. For stability and controller design studies, it is important to linearize the machine equations around a steady-state operating point to obtain small-signal equations. Such a small-signal model is derived in this chapter. Algorithms and flowchart are given to compute the eigenvalues, transfer functions, and frequency responses of the induction motor. Various stages in the dynamic modeling of the three-phase induction machine are shown in Figure 5.12. For the sake of completeness, the space-phasor model is derived from the dynamic model in direct and quadrature axes. The space-phasor model powerfully evokes the similarity and equivalence between the induction machines and dc machines from the modeling and control points of view. Examples are included in this section to illustrate the essential concepts.

5.7.1 Real-Time Model of a Two-Phase Induction Machine

The following assumptions are made to derive the dynamic model:

(i) uniform air gap;
(ii) balanced rotor and stator windings, with sinusoidally distributed mmf;
(iii) inductance vs. rotor position is sinusoidal; and
(iv) saturation and parameter changes are neglected.

A two-phase induction machine with stator and rotor windings is shown in Figure 5.13. The windings are displaced in space by 90 electrical degrees, and the rotor winding, α, is at an angle θ_r from the stator d axis winding. It is assumed that the d axis is leading the q axis for clockwise direction of rotation of the rotor. If the clockwise phase sequence is dq, the rotating magnetic field will be revolving at the angular speed of the supply frequency but counter to the phase sequence of the stator supply.

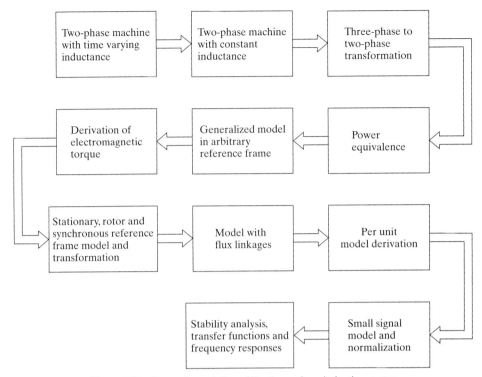

Figure 5.12 Dynamic modeling of the three-phase induction motor

Therefore, the rotor is pulled in the direction of the rotating magnetic field—that is, counter clockwise, in this case. The currents and voltages of the stator and rotor windings are marked in Figure 5.13. The number of turns per phase in the stator and rotor, respectively are T_1 and T_2. A pair of poles is assumed for this figure, but it is applicable with slight modification for any number of pairs of poles if it is drawn in terms of electrical degrees. Note that θ_r is the electrical rotor position at any instant, obtained by multiplying the mechanical rotor position by pairs of electrical poles. The terminal voltages of the stator and rotor windings can be expressed as the sum of the voltage drops in resistances and rates of change of flux linkages, which are the products of currents and inductances. The equations are as follows:

$$v_{qs} = R_q i_{qs} + p(L_{qq} i_{qs}) + p(L_{qd} i_{ds}) + p(L_{q\alpha} i_\alpha) + p(L_{q\beta} i_\beta) \quad (5.46)$$

$$v_{ds} = p(L_{dq} i_{qs}) + R_d i_{ds} + p(L_{dd} i_{ds}) + p(L_{d\alpha} i_\alpha) + p(L_{d\beta} i_\beta) \quad (5.47)$$

$$v_\alpha = p(L_{\alpha q} i_{qs}) + p(L_{\alpha d} i_{ds}) + R_\alpha i_\alpha + p(L_{\alpha\alpha} i_\alpha) + p(L_{\alpha\beta} i_\beta) \quad (5.48)$$

$$v_\beta = p(L_{\beta q} i_{qs}) + p(L_{\beta d} i_{ds}) + p(L_{\beta\alpha} i_\alpha) + R_\beta i_\beta + p(L_{\beta\beta} i_\beta) \quad (5.49)$$

where p is the differential operator d/dt, and the various inductances are explained as follows. $v_{qs}, v_{ds}, v_\alpha, v_\beta$ are the terminal voltages of the stator q axis, d axis, and rotor α and β windings, respectively. i_{qs} and i_{ds} are the stator q axis and d axis currents, respectively. i_α and i_β are the rotor α and β windings currents, respectively. $L_{qq}, L_{dd}, L_{\alpha\alpha}$, and $L_{\beta\beta}$ are the stator q and d axes winding and rotor α and β winding self-inductances, respectively.

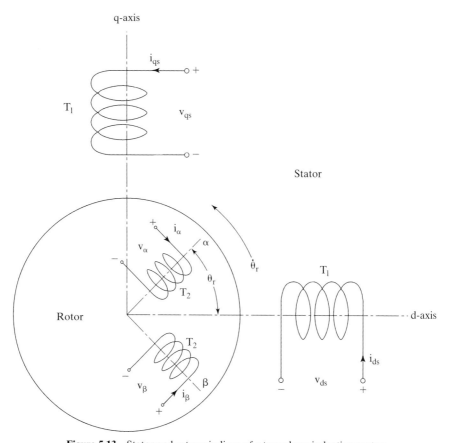

Figure 5.13 Stator and rotor windings of a two-phase induction motor

The mutual inductances between any two windings are denoted by L with two subscripts, the first subscript denoting the winding at which the emf is measured due to the current in the other winding, indicated by the second subscript. For example, L_{qd} is the mutual inductance between the q and d axes windings due to a current in the d axis winding. Under the assumption of uniform air gap, the self-inductances are independent of angular positions; hence, they are constants:

$$L_{\alpha\alpha} = L_{\beta\beta} = L_{rr} \tag{5.50}$$

$$L_{dd} = L_{qq} = L_s \tag{5.51}$$

The mutual inductances between the stator windings and between the rotor windings are zero, because the flux set up by a current in one winding will not link with the other winding displaced in space by 90 degrees. This leads to the following simplifications.

$$L_{\alpha\beta} = L_{\beta\alpha} = 0 \tag{5.52}$$

$$L_{dq} = L_{qd} = 0 \tag{5.53}$$

The mutual inductances between the stator and rotor windings are a function of the rotor position, θ_r, and they are assumed to be sinusoidal functions because of the assumption of sinusoidal mmf distribution in the windings. Symmetry in windings and construction causes the mutual inductances between one stator and one rotor winding to be the same whether they are viewed from the stator or the rotor:

$$L_{\alpha d} = L_{d\alpha} = L_{sr} \cos \theta_r \tag{5.54}$$

$$L_{\beta d} = L_{d\beta} = L_{sr} \sin \theta_r \tag{5.55}$$

$$L_{\alpha q} = L_{q\alpha} = L_{sr} \sin \theta_r \tag{5.56}$$

$$L_{\beta q} = L_{q\beta} = -L_{sr} \cos \theta_r \tag{5.57}$$

where L_{sr} is the peak value of the mutual inductance between a stator and a rotor winding. The last equation has a negative term, because a positive current in β winding produces a negative flux linkage in the q axis winding, and vice versa. Substitution of equations from (5.50) to (5.57) into equations from (5.46) to (5.49) results in a system of differential equations with time-varying inductances. The resulting equations are as follows:

$$v_{qs} = (R_s + L_s p)i_{qs} + L_{sr} p(i_\alpha \sin \theta_r) - L_{sr} p(i_\beta \cos \theta_r) \tag{5.58}$$

$$v_{ds} = (R_s + L_s p)i_{ds} + L_{sr} p(i_\alpha \cos \theta_r) + L_{sr} p(i_\beta \sin \theta_r) \tag{5.59}$$

$$v_\alpha = L_{sr} p(i_{qs} \sin \theta_r) + L_{sr} p(i_{ds} \cos \theta_r) + (R_{rr} + L_{rr} p)i_\alpha \tag{5.60}$$

$$v_\beta = -L_{sr} p(i_{qs} \cos \theta_r) + L_{sr} p(i_{ds} \sin \theta_r) + (R_{rr} + L_{rr} p)i_\beta \tag{5.61}$$

where

$$R_s = R_q = R_d \tag{5.62}$$

$$R_{rr} = R_\alpha = R_\beta \tag{5.63}$$

The solution of these equations is time-consuming, because of their dependence on the product of the instantaneous rotor-position-dependent cosinusoidal functions and winding currents, and an elegant set of equations leading to a simple solution procedure is necessary. Transformations performing such a step are discussed subsequently.

5.7.2 Transformation to Obtain Constant Matrices

The transformation to obtain constant inductances is achieved by replacing the actual with a fictitious rotor on the q and d axes, as shown in Figure 5.14. In that process, the fictitious rotor will have the same number of turns for each phase as the actual rotor phase windings and should produce the same mmf. That leads to a cancellation of the number of turns on both sides of that equation, resulting in a relationship of the actual currents to fictitious rotor currents i_{qrr} and i_{drr}. Then the fictitious rotor currents i_{qrr} and i_{drr} are equal to the sum of the projections of i_α and i_β on the q and d axis, respectively, as given below:

$$\begin{bmatrix} i_{drr} \\ i_{qrr} \end{bmatrix} = \begin{bmatrix} \cos \theta_r & \sin \theta_r \\ \sin \theta_r & -\cos \theta_r \end{bmatrix} \begin{bmatrix} i_\alpha \\ i_\beta \end{bmatrix} \tag{5.64}$$

Section 5.7   Dynamic Modeling of Induction Machines   201

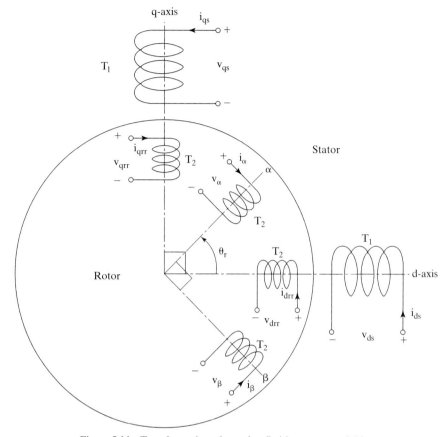

Figure 5.14   Transformation of actual to fictitious rotor variables

Equation (5.64) is written compactly as

$$i_{dqrr} = [T_{\alpha\beta}] i_{\alpha\beta} \tag{5.65}$$

where

$$i_{dqrr} = [i_{drr} \quad i_{qrr}]^t \tag{5.66}$$
$$i_{\alpha\beta} = [i_\alpha \quad i_\beta]^t \tag{5.67}$$

and

$$T_{\alpha\beta} = \begin{bmatrix} \cos\theta_r & \sin\theta_r \\ \sin\theta_r & -\cos\theta_r \end{bmatrix} \tag{5.68}$$

This transformation is valid for voltages, currents, and flux-linkages in a machine. That the transformation from α, β axes to d, q axes and vice versa is elegant is due to the fact that

$$T_{\alpha\beta} = T_{\alpha\beta}^{-1} \tag{5.69}$$

This matrix is both orthogonal and symmetric. Applying this transformation to the α and β rotor-winding currents and rotor voltages in equations (5.60) and (5.61), the following matrix equation is obtained:

$$\begin{bmatrix} v_{qs} \\ v_{ds} \\ v_{qrr} \\ v_{drr} \end{bmatrix} = \begin{bmatrix} R_s + L_s p & 0 & L_{sr} p & 0 \\ 0 & R_s + L_s p & 0 & L_{sr} p \\ L_{sr} p & -L_{sr} \dot{\theta}_r & R_{rr} + L_{rr} p & -L_{rr} \dot{\theta}_r \\ L_{sr} \dot{\theta}_r & L_{sr} p & L_{rr} \dot{\theta}_r & R_{rr} + L_{rr} p \end{bmatrix} \begin{bmatrix} i_{qs} \\ i_{ds} \\ i_{qrr} \\ i_{drr} \end{bmatrix} \quad (5.70)$$

where $\dot{\theta}_r$ is the time derivative of θ_r. The rotor equations need to be referred to the stator, as in the case of the transformer-equivalent circuit. This step removes the physical isolation and facilitates the corresponding stator and rotor d and q axes windings in becoming physically connected. The steps involved in referring these rotor parameters and variables to the stator are as follows:

$$R_r = a^2 R_{rr} \quad (5.71)$$

$$L_r = a^2 L_{rr} \quad (5.72)$$

$$i_{qr} = \frac{i_{qrr}}{a} \quad (5.73)$$

$$i_{dr} = \frac{i_{drr}}{a} \quad (5.74)$$

$$v_{qr} = a v_{qrr} \quad (5.75)$$

$$v_{dr} = a v_{drr} \quad (5.76)$$

where

$$a = \frac{\text{Stator effective turns per phase}}{\text{Rotor effective turns per phase}} = \frac{k_{w1} T_1}{k_{w2} T_2} \quad (5.77)$$

Note that the magnetizing and mutual inductances are,

$$L_m \propto T_1^2 \quad (5.78)$$

$$L_{sr} \propto T_1 T_2 \quad (5.79)$$

From equations (5.77) and (5.79), the magnetizing inductance of the stator is derived as

$$L_m = a L_{sr} \quad (5.80)$$

Substituting equations from (5.71) to (5.80) into (5.70), the machine equations referred to the stator are obtained as

$$\begin{bmatrix} v_{qs} \\ v_{ds} \\ v_{qr} \\ v_{dr} \end{bmatrix} = \begin{bmatrix} R_s + L_s p & 0 & L_m p & 0 \\ 0 & R_s + L_s p & 0 & L_m p \\ L_m p & -L_m \dot{\theta}_r & R_r + L_r p & -L_r \dot{\theta}_r \\ L_m \dot{\theta}_r & L_m p & L_r \dot{\theta}_r & R_r + L_r p \end{bmatrix} \begin{bmatrix} i_{qs} \\ i_{ds} \\ i_{qr} \\ i_{dr} \end{bmatrix} \quad (5.81)$$

This equation is in the form where the voltage vector is equal to the product of the impedance matrix and current vector. Note that the impedance matrix has constant inductance terms and is no longer dependent on the rotor position. Some of the impedance matrix elements are dependent on the rotor speed, and only when they are constant, as in steady state, does the system of equations become linear. In the case of varying rotor speed and if its variation is dependent on the currents, then the system of equations becomes nonlinear. It is derived later that the electromagnetic torque, as a function of winding currents and rotor speed, is determined by the electromagnetic and load torques along with load parameters such as inertia and friction. In that case, it can be seen that the induction-machine system is nonlinear.

5.7.3 Three-Phase to Two-Phase Transformation

The model that has been developed so far is for a two-phase machine. Three-phase induction machines are common; two-phase machines are rarely used in industrial applications. A dynamic model for the three-phase induction machine can be derived from the two-phase machine if the equivalence between three and two phases is established. The equivalence is based on the equality of the mmf produced in the two-phase and three-phase windings and equal current magnitudes. Figure 5.15 shows the three-phase and two-phase windings. Assuming that each of the three-phase windings has T_1 turns per phase and equal current magnitudes, the two-phase windings will have $3T_1/2$ turns per phase for mmf equality. The d and q axes mmfs are found by resolving the mmfs of the three phases along the d and q axes.

The common term, the number of turns in the winding, is canceled on either side of the equations, leaving the current equalities. The q axis is assumed to be lagging the a axis by θ_c. The relationship between dqo and abc currents is as follows:

$$\begin{bmatrix} i_{qs} \\ i_{ds} \\ i_0 \end{bmatrix} = \frac{2}{3} \begin{bmatrix} \cos\theta_c & \cos\left(\theta_c - \frac{2\pi}{3}\right) & \cos\left(\theta_c + \frac{2\pi}{3}\right) \\ \sin\theta_c & \sin\left(\theta_c - \frac{2\pi}{3}\right) & \sin\left(\theta_c + \frac{2\pi}{3}\right) \\ \frac{1}{2} & \frac{1}{2} & \frac{1}{2} \end{bmatrix} \begin{bmatrix} i_{as} \\ i_{bs} \\ i_{cs} \end{bmatrix} \quad (5.82)$$

The current i_0 represents the imbalances in the a, b, and c phase currents and can be recognized as the zero-sequence component of the current. Equation (5.79) can be expressed in a compact form by

$$i_{qdo} = [T_{abc}] i_{abc} \quad (5.83)$$

where

$$i_{qdo} = [i_{qs} \ i_{ds} \ i_0]^t \quad (5.84)$$
$$i_{abc} = [i_{as} \ i_{bs} \ i_{cs}]^t \quad (5.85)$$

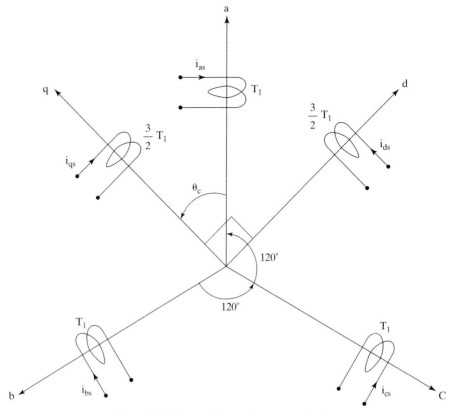

Figure 5.15 Two- and three-phase stator windings

and the transformation from *abc* to *qdo* variables is

$$[T_{abc}] = \frac{2}{3}\begin{bmatrix} \cos\theta_c & \cos\left(\theta_c - \frac{2\pi}{3}\right) & \cos\left(\theta_c + \frac{2\pi}{3}\right) \\ \sin\theta_c & \sin\left(\theta_c - \frac{2\pi}{3}\right) & \sin\left(\theta_c + \frac{2\pi}{3}\right) \\ \frac{1}{2} & \frac{1}{2} & \frac{1}{2} \end{bmatrix} \quad (5.86)$$

The zero-sequence current, i_o, does not produce a resultant magnetic field. The transformation from two-phase currents to three-phase currents can be obtained as

$$i_{abc} = [T_{abc}]^{-1} i_{qdo} \quad (5.87)$$

where

$$[T_{abc}]^{-1} = \begin{bmatrix} \cos\theta_c & \sin\theta_c & 1 \\ \cos\left(\theta_c - \frac{2\pi}{3}\right) & \sin\left(\theta_c - \frac{2\pi}{3}\right) & 1 \\ \cos\left(\theta_c + \frac{2\pi}{3}\right) & \sin\left(\theta_c + \frac{2\pi}{3}\right) & 1 \end{bmatrix} \quad (5.88)$$

This transformation could also be thought of as a transformation from three (*abc*) axes to three new (*qdo*) axes; for uniqueness of the transformation from one set of axes to another set of axes, including unbalances in the *abc* variables requires three variables such as the *dqo*. The reason for this is that it is easy to convert from three *abc* variables to two *qd* variables if the *abc* variables have an inherent relationship among themselves, such as the equal-phase displacement and magnitude. Therefore, in such a case, there are only two independent variables in *abc*; the third is a dependent variable, obtained as the negative sum of the other two variables. Hence a *qd*-to-*abc* transformation is unique under that circumstance. When the *abc* variables have no such inherent relationship, then there are three distinct and independent variables; hence, the third variable cannot be recovered from the knowledge of the other two variables only. It also means that they are not recoverable from two variables *qd* but require another variable, such as the zero-sequence component, to recover the *abc* variables from the *dqo* variables.

Under balanced conditions only, there are four system equations, as given in (5.81). Under unbalanced conditions, note that two more system equations, one for the stator zero-sequence voltage and the other for the rotor zero-sequence voltage, emerge. They are given as [8]

$$v_{os} = R_s + L_{ls}\, pi_{os} \tag{5.89}$$

$$v_{or} = R_r + L_{lr}\, pi_{or} \tag{5.90}$$

where in the variables the first subscript, *o*, denotes the zero-sequence component and the second subscript denotes the stator and rotor by *s* and *r*, respectively. They could be derived from the stator and rotor inductance matrices in the *abc* frames, then converted into *dqo* frames by using the transformation derived above. It is interesting to observe that only leakage inductances and phase resistances influence the zero-sequence voltages and currents, unlike in the *dq* component variables, which are influenced by the self and mutual inductances and phase resistances.

It is usual to align the *q* axis with the phase *a* winding; this implies that the *qd* frames are fixed to the stator. The model is known as Stanley's model or the *stator reference frames* model. In that case, $\theta_c = 0$, and the transformation from *abc* to *qdo* variables is given as

$$T^s_{abc} = \frac{2}{3}\begin{bmatrix} 1 & -\frac{1}{2} & -\frac{1}{2} \\ 0 & -\frac{\sqrt{3}}{2} & \frac{\sqrt{3}}{2} \\ \frac{1}{2} & \frac{1}{2} & \frac{1}{2} \end{bmatrix} \tag{5.91}$$

In a balanced three-phase machine, the sum of the three-phase currents is zero and is given as

$$i_{as} + i_{bs} + i_{cs} = 0 \tag{5.92}$$

leading to a zero-sequence current of zero value:

$$i_o = \frac{1}{3}(i_{as} + i_{bs} + i_{cs}) = 0 \tag{5.93}$$

With equations (5.86) and (5.88), the equivalence between the two-phase and three-phase induction machines is established. It is instructive to know that the transformation derived is applicable to currents, voltages, and flux-linkages.

Example 5.3

An induction motor has the following parameters:

5 hp, 200 V, 3-phase, 60 Hz, 4-pole, star-connected; $R_s = 0.277 \, \Omega$
$R_r = 0.183 \, \Omega$ $L_m = 0.0538 \, H$ $L_s = 0.0553 \, H$ $L_r = .056 \, H$
Effective stator to rotor turns ratio, a = 3

The motor is supplied with its rated and balanced voltages. Find the q and d axes steady-state voltages and currents and phase currents I_{qrr}, I_{drr}, I_α, and I_β when the rotor is locked. Use the stator-reference-frames model of the induction machine.

Solution   The applied phase voltages are as follows:

$$v_{as} = \frac{200}{\sqrt{3}} \times \sqrt{2} \sin \omega_s t = 163.3 \sin \omega_s t$$

$$v_{bs} = 163.3 \sin\left(\omega_s t - \frac{2\pi}{3}\right)$$

$$v_{cs} = 163.3 \sin\left(\omega_s t + \frac{2\pi}{3}\right)$$

The d and q axes voltages are

$$\begin{bmatrix} v_{qs} \\ v_{ds} \\ v_o \end{bmatrix} = T^s_{abc} \begin{bmatrix} v_{as} \\ v_{bs} \\ v_{cs} \end{bmatrix}$$

Hence, $v_{qs} = \frac{2}{3}\left[v_{as} - \frac{1}{2}(v_{bs} + v_{cs})\right]$

For a balanced three-phase input,

$$v_{as} + v_{bs} + v_{cs} = 0$$

Substituting for v_{bs} and v_{cs} in terms of v_{as} yields

$$v_{qs} = \frac{2}{3}\left[\frac{3}{2} v_{as}\right] = v_{as}$$

Similarly,

$$v_{ds} = \frac{1}{\sqrt{3}}(v_{cs} - v_{bs})$$

and $v_o = 0$

$$v_{qs} = v_{as} = 163.3 \sin \omega_s t = 163.3 \angle 0° = 163.3 \, V$$

Section 5.7 Dynamic Modeling of Induction Machines

$$v_{ds} = \frac{1}{\sqrt{3}}(v_{cs} - v_{bs}) = 163.3 \cos \omega_s t = 163.3 \angle 90° = j163.3 \text{ V}$$

The rotor is locked; hence,

$$\dot{\theta}_r = 0$$

For steady state evaluation,

$$p = j\omega_s = j2\pi f_s = j2\pi \cdot 60 = j377 \text{ rad/sec}$$

The system equations in steady state are

$$\begin{bmatrix} v_{qs} \\ v_{ds} \\ 0 \\ 0 \end{bmatrix} = \begin{bmatrix} R_s + j\omega_s L_s & 0 & j\omega_s L_m & 0 \\ 0 & R_s + j\omega_s L_s & 0 & j\omega_s L_m \\ j\omega_s L_m & 0 & R_r + j\omega_s L_r & 0 \\ 0 & j\omega_s L_m & 0 & R_r + j\omega_s L_r \end{bmatrix} \begin{bmatrix} i_{qs} \\ i_{ds} \\ i_{qr} \\ i_{dr} \end{bmatrix}$$

Note that the rotor windings are short-circuited, and hence rotor voltages are zero. The numerical values for the parameters and variables are substituted to solve for the currents. The currents are

$$i_{qs} = 35.37 - j108.18 = 113.81 \angle -71.90°$$
$$i_{ds} = 108.18 + j35.37 = 113.81 \angle 18.10°$$
$$i_{qr} = -34.88 + j103.63 = 109.34 \angle 108.6°$$
$$i_{dr} = -103.63 - j34.88 = 109.34 \angle -161.4°$$

Note that the stator and rotor currents are displaced by 90° among themselves, as expected in a two-phase machine. The zero-sequence currents are zero, because zero-sequence voltages are nonexistent with balanced supply voltages.

The phase currents are

$$\begin{bmatrix} i_{as} \\ i_{bs} \\ i_{cs} \end{bmatrix} = \begin{bmatrix} 1 & 0 & 1 \\ -1/2 & -\sqrt{3}/2 & 1 \\ -1/2 & \sqrt{3}/2 & 1 \end{bmatrix} \begin{bmatrix} i_{qs} \\ i_{ds} \\ i_o \end{bmatrix} = \begin{bmatrix} 113.8 \angle -71.9° \\ 113.8 \angle 168.1° \\ 113.8 \angle 48.1° \end{bmatrix}$$

The various rotor currents are

$$i_{qrr} = a i_{qr} = 328.02 \angle 108.6°$$
$$i_{drr} = a i_{dr} = 328.02 \angle -161.4°$$

The α and β currents, assuming $\theta_r = 0$, are

$$\begin{bmatrix} i_\alpha \\ i_\beta \end{bmatrix} = \begin{bmatrix} \cos\theta_r & \sin\theta_r \\ \sin\theta_r & -\cos\theta_r \end{bmatrix} \begin{bmatrix} i_{drr} \\ i_{qrr} \end{bmatrix} = \begin{bmatrix} 1 & 0 \\ 0 & -1 \end{bmatrix} \begin{bmatrix} i_{drr} \\ i_{qrr} \end{bmatrix} = \begin{bmatrix} 328.02 \angle -161.4° \\ -328.02 \angle 108.6° \end{bmatrix}$$

Example 5.4

Derive the steady-state equivalent circuit from the dynamic equations of the induction motor.

Solution   The steady-state equivalent circuit of the induction motor can be derived by substituting for the d and q axes voltages in the system equations:

$$v_{as} = V_m \sin \omega_s t$$

$$v_{bs} = V_m \sin\left(\omega_s t - \frac{2\pi}{3}\right)$$

$$v_{cs} = V_m \sin\left(\omega_s t + \frac{2\pi}{3}\right)$$

Hence,

$$\begin{bmatrix} v_{qs} \\ v_{ds} \\ v_o \end{bmatrix} = [T_{abc}] \begin{bmatrix} v_{as} \\ v_{bs} \\ v_{cs} \end{bmatrix} = \begin{bmatrix} V_m \sin \omega_s t \\ V_m \cos \omega_s t \\ 0 \end{bmatrix} = \begin{bmatrix} V_m \angle 0° \\ V_m \angle 90° \\ 0 \end{bmatrix}$$

In steady state,

$$p = j\omega_s$$

$$v_{qr} = v_{dr} = 0$$

Substituting these into the system equations yields

$$\begin{bmatrix} V_m \\ jV_m \\ 0 \\ 0 \end{bmatrix} = \begin{bmatrix} R_s + j\omega_s L_s & 0 & j\omega_s L_m & 0 \\ 0 & R_s + j\omega_s L_s & 0 & j\omega_s L_m \\ j\omega_s L_m & -\omega_r L_m & R_r + j\omega_s L_r & -L_r \omega_r \\ \omega_r L_m & j\omega_s L_m & L_r \omega_r & R_r + j\omega_s L_r \end{bmatrix} \begin{bmatrix} i_{qs} \\ i_{ds} \\ i_{qr} \\ i_{dr} \end{bmatrix}$$

The input voltages are in quadrature, so the currents have to be in quadrature, because the system in steady state is linear, and they can be represented as

$$i_{ds} = j i_{qs}$$
$$i_{dr} = j i_{qr}$$

Substituting these equations into the above equation and considering only one stator and rotor equation with rms values yields

$$V_s = (R_s + j\omega_s L_s) I_s + j\omega_s L_m I_r$$
$$0 = jL_m(\omega_s - \omega_r) I_s + (R_r + j(\omega_s - \omega_r) L_r) I_r$$

where V_s is the stator rms voltage and I_s and I_r are the stator and rotor rms currents, respectively. Rearrange the rotor equation with the aid of

$$\omega_{sl} = \omega_s - \omega_r = s\omega_s$$

The rotor equation then is

$$0 = jL_m \omega_s I_s + \left(\frac{R_r}{s} + j\omega_s L_r\right) I_r$$

The rotor and stator equations, when combined, give the equivalent circuit, with an understanding that the sum of stator and rotor currents gives the magnetizing current. Note that the stator and stator-referred rotor self-inductances are equal to their magnetizing inductance and respective leakage inductances and are given as

$$L_s = L_m + L_{ls}$$
$$L_r = L_m + L_{lr}$$

5.7.4 Power Equivalence

The power input to the three-phase motor has to be equal to the power input to the two-phase machine to have meaningful interpretation in the modeling, analysis, and simulation. Such an identity is derived in this section. The three-phase instantaneous power input is

$$p_i = v_{abc}^t i_{abc} = v_{as}i_{as} + v_{bs}i_{bs} + v_{cs}i_{cs} \tag{5.94}$$

From equation (5.82), the *abc* phase currents and voltages are transformed into their equivalent qd-axes currents and voltages as

$$i_{abc} = [T_{abc}]^{-1} i_{qdo} \tag{5.95}$$

$$v_{abc} = [T_{abc}]^{-1} v_{qdo} \tag{5.96}$$

Substituting equations (5.95) and (5.96) into equation (5.94) gives the power input as

$$p_i = v_{qdo}^t ([T_{abc}]^{-1})^t [T_{abc}]^{-1} i_{qdo} \tag{5.97}$$

Expanding the right-hand side of the equation (5.97) gives the power input in *dqo* variables:

$$p_i = \frac{3}{2}((v_{qs}i_{qs} + v_{ds}i_{ds}) + 2v_o i_o) \tag{5.98}$$

For a balanced three-phase machine, the zero-sequence current does not exist; hence, the power input is compactly represented by

$$p_i = \frac{3}{2}(v_{qs}i_{qs} + v_{ds}i_{ds}) \tag{5.99}$$

The model development has so far kept the d and q axes stationary with respect to the stator. These axes or frames are known as reference frames. A generalized treatment where the speed of the reference frames is arbitrary is derived in the next section. The input power given by equation (5.98) remains valid for all occasions, provided that the voltages and currents correspond to the frames under consideration.

5.7.5 Generalized Model in Arbitrary Reference Frames

Reference frames are very much like observer platforms, in that each of the platforms gives a unique view of the system at hand as well as a dramatic simplification of the system equations. For example, consider that, for the purposes of control, it is desirable to have the system variables as dc quantities, although the actual variables are sinusoidal. This could be accomplished by having a reference frame revolving at the same angular speed as that of the sinusoidal variable. As the reference frames are moving at an angular speed equal to the angular frequency of the sinusoidal supply, say, then the differential speed between them is reduced to zero, resulting in the sinusoid being perceived as a dc signal from the reference frames. Then, by moving to that plane, it becomes easier to develop a small-signal equation out of a nonlinear equation, as the quiescent or operating point is described only by dc values; this then leads to the linearized system around an operating point. Now, it is easier to synthesize a compensator for the system by using standard linear control-system techniques. Likewise, the independent rotor field position determines the induced emf and affects the dynamic system equations of both the wound-rotor and permanent-magnet synchronous machines. Therefore, looking at the entire system from the rotor, i.e., rotating reference frames, the system inductance matrix becomes independent of rotor position, thus leading to the simplification and compactness of the system equations. Such advantages are many from using reference frames. Instead of deriving the transformations for each and every particular reference frame, it is advantageous to derive the general transformation for an arbitrary rotating reference frame. Then any particular reference frame model can be derived by substituting the appropriate frame speed and position in the generalized reference model.

Reference frames rotating at an arbitrary speed are hereafter called arbitrary reference frames. Other reference frames are particular cases of these arbitrary reference frames. The relationship between the stationary reference frames denoted by d and q axes and the arbitrary reference frames denoted by d^c and q^c axes are shown in Figure 5.16. From now on, the three-phase machine is assumed to have balanced windings and balanced inputs, thus making the zero-sequence components be zero and eliminating the zero-sequence equations from further consideration. Note that the zero-sequence equations have to be included only for unbalanced operation of the motor, a situation common with a fault in the machine or converter.

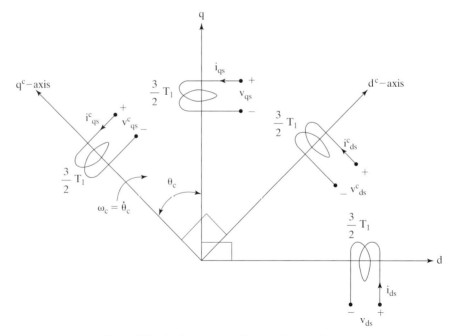

Figure 5.16 Stationary and arbitrary reference frames

Section 5.7 Dynamic Modeling of Induction Machines

Assuming that the windings have equal number of turns on both of the reference frames, the arbitrary reference frame currents are resolved on the d and q axes to find the currents in the stationary reference frames. The relationships between the currents are written as

$$i_{qds} = [T^c] i_{qds}^c \qquad (5.100)$$

where

$$i_{qds} = [i_{qs} \; i_{ds}]^t \qquad (5.101)$$
$$i_{qds}^c = [i_{qs}^c \; i_{ds}^c]^t \qquad (5.102)$$

and

$$T^c = \begin{bmatrix} \cos\theta_c & \sin\theta_c \\ -\sin\theta_c & \cos\theta_c \end{bmatrix} \qquad (5.103)$$

The speed of the arbitrary reference frames is

$$\dot{\theta}_c = \omega_c \qquad (5.104)$$

Similarly, the fictitious rotor currents are transformed into arbitrary frames by using T^c, and they are written as

$$i_{qdr} = [T^c] i_{qdr}^c \qquad (5.105)$$

where

$$i_{qdr} = [i_{qr} \; i_{dr}]^t \qquad (5.106)$$
$$i_{qdr}^c = [i_{qr}^c \; i_{dr}^c]^t \qquad (5.107)$$

Likewise, the voltage relationships are

$$v_{qds} = [T^c] v_{qds}^c \qquad (5.108)$$
$$v_{qdr} = [T^c] v_{qdr}^c \qquad (5.109)$$

where

$$v_{qds} = [v_{qs} \; v_{ds}]^t \qquad (5.110)$$
$$v_{qds}^c = [v_{qs}^c \; v_{ds}^c]^t \qquad (5.111)$$
$$v_{qdr} = [v_{qr} \; v_{dr}]^t \qquad (5.112)$$
$$v_{qdr}^c = [v_{qr}^c \; v_{dr}^c]^t \qquad (5.113)$$

By substituting equations from (5.100) to (5.113) into equation (5.81), the induction-motor model in arbitrary reference frames is obtained. It is given below for use in subsequent sections.

$$\begin{bmatrix} v_{qs}^c \\ v_{ds}^c \\ v_{qr}^c \\ v_{dr}^c \end{bmatrix} = \begin{bmatrix} R_s + L_s p & \omega_c L_s & L_m p & \omega_c L_m \\ -\omega_c L_s & R_s + L_s p & -\omega_c L_m & L_m p \\ L_m p & (\omega_c - \omega_r) L_m & R_r + L_r p & (\omega_c - \omega_r) L_r \\ -(\omega_c - \omega_r) L_m & L_m p & -(\omega_c - \omega_r) L_r & R_r + L_r p \end{bmatrix} \begin{bmatrix} i_{qs}^c \\ i_{ds}^c \\ i_{qr}^c \\ i_{dr}^c \end{bmatrix} \qquad (5.114)$$

where

$$\omega_r = \dot{\theta}_r \qquad (5.115)$$

ω_r is the rotor speed in electrical radians/sec. The relationship between the arbitrary reference frame variables and the *a, b,* and *c* variables is derived by using

$$i^c_{qds} = [T^c]^{-1} i_{qds} \qquad (5.116)$$

By substituting from equation (5.91) for i_{qds} in terms of *a, b,* and *c* phase currents in the stator reference frames, the *qdo* currents in the arbitrary reference frames are obtained as

$$i^c_{qdo} = \begin{bmatrix} [T^c]^{-1} & 0 \\ [0] & 1 \end{bmatrix} [T^s_{abc}][i_{abc}] = [T_{abc}][i_{abc}] \qquad (5.117)$$

where [0] is a 1 × 2 null vector. Note that the zero-sequence currents remain unchanged in the arbitrary reference frames. This transformation is valid for currents, voltages, and flux-linkages for both the stator and the rotor. Particular cases of the reference frames are derived in a later section. The next section contains the derivation of the electromagnetic torque in terms of the current variables in the arbitrary reference frames.

5.7.6 Electromagnetic Torque

The electromagnetic torque is an important output variable that determines such mechanical dynamics of the machine as the rotor position and speed. Therefore, its importance cannot be overstated in any of the simulation studies. It is derived from the machine matrix equation by looking at the input power and its various components, such as resistive losses, mechanical power, rate of change of stored magnetic energy, and reference-frame power. Elementary reasoning leads to the fact that there cannot be a power component due to the introduction of reference frames. Similarly, the rate of change of stored magnetic energy must be zero in steady state. Hence, the output power is the difference between the input power and the resistive losses in steady state. Dynamically, the rate of change of stored magnetic energy need not be zero. Based on these observations, the derivation of the electromagnetic torque is made as follows.

The equation (5.114) can be written as

$$V = [R]i + [L]pi + [G]\omega_r i + [F]\omega_c i \qquad (5.118)$$

where the vectors and matrices are identified by observation. Premultiplying the equation (5.118) by the transpose of the current vector gives the instantaneous input power as,

$$p_i = i^t V = i^t[R]i + i^t[L]pi + i^t[G]\omega_r i + i^t[F]\omega_c i \qquad (5.119)$$

where the [R] matrix consists of resistive elements, the [L] matrix consists of the coefficients of the derivative operator *p*, the [G] matrix has elements that are the coefficients of the electrical rotor speed ω_r, and [F] is the frame matrix in terms of the coefficients of the reference frame speed, ω_c. The term $i^t[R]i$ gives stator and rotor resistive losses. The term $i^t[F]\omega_c i$ is the reference frame power, and upon expansion is found to be identically equal to zero, as it should be, because there cannot be a power associated with a fictitious element introduced for the sake of simplifying the model and analysis. The term $i^t [L] pi$ denotes the rate of change of stored magnetic energy. Therefore, what is left of the power component must be equal to the air gap power, given by the term $i^t [G] \omega_r i$. From the fundamentals, it is known that the air gap power has to be associated with the rotor speed. The air gap power is the product of the mechanical rotor speed and air gap or electromagnetic torque. Hence air gap torque, T_e, is derived from the terms involving the rotor speed, ω_m in mechanical rad/sec, as

$$\omega_m T_e = P_a = i^t [G] i \times \omega_r \quad (5.120)$$

Substituting for ω_r in terms of ω_m leads to electromagnetic torque as

$$T_e = \frac{P}{2} i^t [G] i \quad (5.121)$$

By substituting for $[G]$ in equation (5.121) by observation from (5.114), the electromagnetic torque is obtained as

$$T_e = \frac{3}{2} \frac{P}{2} L_m (i_{qs}^c i_{dr}^c - i_{ds}^c i_{qr}^c) \quad (5.122)$$

The factor $\frac{3}{2}$ is introduced into the right-hand side of equation (5.122) from the power-equivalence condition between the three-phase and two-phase induction motors. The next section considers the frequently used models in various reference frames and their derivation from the generalized induction-motor model in arbitrary reference frames.

5.7.7 Derivation of Commonly Used Induction-Motor Models

Three particular cases of the generalized model of the induction motor in arbitrary reference frames are of general interest:

(i) stator reference frames model;
(ii) rotor reference frames model;
(iii) synchronously rotating reference frames model.

Their derivations and transformation relationships are considered in this section.

5.7.7.1 Stator reference frames model. The speed of the reference frames is that of the stator, which is zero; hence,

$$\omega_c = 0 \quad (5.123)$$

is substituted into equation (5.114). The resulting model is

$$\begin{bmatrix} v_{qs} \\ v_{ds} \\ v_{qr} \\ v_{dr} \end{bmatrix} = \begin{bmatrix} R_s + L_s p & 0 & L_m p & 0 \\ 0 & R_s + L_s p & 0 & L_m p \\ L_m p & -\omega_r L_m & R_r + L_r p & -\omega_r L_r \\ \omega_r L_m & L_m p & \omega_r L_r & R_r + L_r p \end{bmatrix} \begin{bmatrix} i_{qs} \\ i_{ds} \\ i_{qr} \\ i_{dr} \end{bmatrix} \quad (5.124)$$

For convenience, the superscript s is omitted for the stator reference frames model hereafter. The torque equation is

$$T_e = \frac{3}{2}\frac{P}{2} L_m (i_{qs}i_{dr} - i_{ds}i_{qr}) \qquad (5.125)$$

Note that equations (5.124) and (5.81) are identical. The transformation for variables is obtained by substituting $\theta_c = 0$ in $[T_{abc}]$ and will be the same as $[T^s_{abc}]$, defined in equation (5.91). This model is used when stator variables are required to be actual, i.e., the same as in the actual machine stator, and rotor variables can be fictitious. This model allows elegant simulation of stator-controlled induction-motor drives, such as phase-controlled and inverter-controlled induction-motor drives, because the input variables are well defined and could be used to find the stator q and d axes voltages through a set of simple algebraic equations, for a balanced polyphase supply input, given by:

$$v_{qs} = v_{as} \qquad (5.126)$$

$$v_{ds} = \frac{(v_{cs} - v_{bs})}{\sqrt{3}} \qquad (5.127)$$

Such algebraic relationships reduce the number of computations and thus lend themselves to real-time control applications in high-performance variable-speed drives requiring the computation of stator currents, stator flux linkages, and electromagnetic torque for both control and parameter adaptation.

5.7.7.2 Rotor reference frames model.
The speed of the rotor reference frames is

$$\omega_c = \omega_r \qquad (5.128)$$

and the angular position is

$$\theta_c = \theta_r \qquad (5.129)$$

Substituting in the upper subscript r for rotor reference frames and equation (5.128) in the equation (5.114), the induction-motor model in rotor reference frames is obtained. The equations are given by

$$\begin{bmatrix} v^r_{qs} \\ v^r_{ds} \\ v^r_{qr} \\ v^r_{dr} \end{bmatrix} = \begin{bmatrix} R_s + L_s p & \omega_r L_s & L_m p & \omega_r L_m \\ -\omega_r L_s & R_s + L_s p & -\omega_r L_m & L_m p \\ L_m p & 0 & R_r + L_r p & -0 \\ 0 & L_m p & 0 & R_r + L_r p \end{bmatrix} \begin{bmatrix} i^r_{qs} \\ i^r_{ds} \\ i^r_{qr} \\ i^r_{dr} \end{bmatrix} \qquad (5.130)$$

and the electromagnetic torque is

$$T_e = \frac{3}{2}\frac{P}{2} L_m (i^r_{qs} i^r_{dr} - i^r_{ds} i^r_{qr}) \qquad (5.131)$$

The transformation from *abc* to *dqo* variables is obtained by substituting (5.129) into $[T_{abc}]$, defined in (5.86) as

$$[T^r_{abc}] = \frac{2}{3} \begin{bmatrix} \cos\theta_r & \cos\left(\theta_r - \frac{2\pi}{3}\right) & \cos\left(\theta_r + \frac{2\pi}{3}\right) \\ \sin\theta_r & \sin\left(\theta_r - \frac{2\pi}{3}\right) & \sin\left(\theta_r + \frac{2\pi}{3}\right) \\ \frac{1}{2} & \frac{1}{2} & \frac{1}{2} \end{bmatrix} \quad (5.132)$$

The rotor reference frames model is useful where the switching elements and power are controlled on the rotor side. Slip-power recovery scheme is one example where this model will find use in the simulation of the motor-drive system.

5.7.7.3 Synchronously rotating reference frames model.
The speed of the reference frames is

$$\omega_c = \omega_s = \text{Stator supply angular frequency/rad/sec} \quad (5.133)$$

and the instantaneous angular position is

$$\theta_c = \theta_s = \omega_s t \quad (5.134)$$

By substituting (5.134) into (5.114), the induction-motor model in the synchronous reference frames is obtained. By using the superscript e to denote this electrical synchronous reference frame, the model is obtained as

$$\begin{bmatrix} v^e_{qs} \\ v^e_{ds} \\ v^e_{qr} \\ v^e_{dr} \end{bmatrix} = \begin{bmatrix} R_s + L_s p & \omega_s L_s & L_m p & \omega_s L_m \\ -\omega_s L_s & R_s + L_s p & -\omega_s L_m & L_m p \\ L_m p & (\omega_s - \omega_r)L_m & R_r + L_r p & (\omega_s - \omega_r)L_r \\ -(\omega_s - \omega_r)L_m & L_m p & -(\omega_s - \omega_r)L_r & R_r + L_r p \end{bmatrix} \begin{bmatrix} i^e_{qs} \\ i^e_{ds} \\ i^e_{qr} \\ i^e_{dr} \end{bmatrix} \quad (5.135)$$

The electromagnetic torque is,

$$T_e = \frac{3}{2}\frac{P}{2} L_m(i^e_{qs} i^e_{dr} - i^e_{ds} i^e_{qr}) \quad (\text{N·m}) \quad (5.136)$$

The transformation from *abc* to *dqo* variables is found by substituting (5.134) into equation (5.86) and is given as

$$[T^e_{abc}] = \frac{2}{3} \begin{bmatrix} \cos\theta_s & \cos\left(\theta_s - \frac{2\pi}{3}\right) & \cos\left(\theta_s + \frac{2\pi}{3}\right) \\ \sin\theta_s & \sin\left(\theta_s - \frac{2\pi}{3}\right) & \sin\left(\theta_s + \frac{2\pi}{3}\right) \\ \frac{1}{2} & \frac{1}{2} & \frac{1}{2} \end{bmatrix} \quad (5.137)$$

It may be seen that the synchronous reference frames transform the sinusoidal inputs into dc signals. This model is useful where the variables in steady state need to be dc quantities, as in the development of small-signal equations. Some high-performance control schemes use this model to estimate the control inputs; this led to a major breakthrough in induction-motor control, by decoupling the torque and flux channels for control in a manner similar to that for separately-excited dc motor drives. This is dealt with in detail in Chapter 8, on vector control schemes.

Example 5.5

Solve the problem given in Example 5.1 when its speed is 1400 rpm, using the synchronous reference frames model and rotating reference frames model.

Solution (i) Synchronous Reference Frames

$$\begin{bmatrix} v_{qs}^e \\ v_{ds}^e \\ v_o \end{bmatrix} = [T_{abc}^e] \begin{bmatrix} v_{as} \\ v_{bs} \\ v_{cs} \end{bmatrix}$$

where

$$v_{as} = V_m \sin \omega_s t$$
$$v_{bs} = V_m \sin\left(\omega_s t - \frac{2\pi}{3}\right)$$
$$v_{cs} = V_m \sin\left(\omega_s t + \frac{2\pi}{3}\right)$$

and

$$V_m = 163.3 \text{ V}$$

Substituting these, and solving for the d and q axes stator voltages in the synchronous reference frames, yields

$$\begin{bmatrix} v_{qs}^e \\ v_{ds}^e \\ v_o \end{bmatrix} = [T_{abc}^e] \begin{bmatrix} v_{as} \\ v_{bs} \\ v_{cs} \end{bmatrix} = \begin{bmatrix} 0 \\ V_m \\ 0 \end{bmatrix}$$

The d and q axes stator voltages are dc quantities; hence, the responses will be dc quantities, too, because the system is linear. Hence,

$$pi_{qs}^e = pi_{ds}^e = pi_{qr}^e = pi_{dr}^e = 0$$

Substituting these into the system equations gives

$$\begin{bmatrix} 0 \\ V_m \\ 0 \\ 0 \end{bmatrix} = \begin{bmatrix} R_s & \omega_s L_s & 0 & \omega_s L_m \\ -\omega_s L_s & R_s & -\omega_s L_m & 0 \\ 0 & \omega_{sl} L_m & R_r & \omega_{sl} L_r \\ -\omega_{sl} L_m & 0 & -\omega_{sl} L_r & R_r \end{bmatrix} \begin{bmatrix} i_{qs}^e \\ i_{ds}^e \\ i_{qr}^e \\ i_{dr}^e \end{bmatrix}$$

where

$$\omega_{sl} = \omega_s - \omega_r = \text{slip speed}$$

Section 5.7 Dynamic Modeling of Induction Machines

The currents are solved by inverting the impedance matrix and premultiplying with the input-voltage vector. The currents in the synchronous frame are

$$\begin{bmatrix} i_{qs}^e \\ i_{ds}^e \\ i_{qr}^e \\ i_{dr}^e \end{bmatrix} = \begin{bmatrix} -9.27 \text{ A} \\ 14.35 \text{ A} \\ -1.69 \text{ A} \\ -14.60 \text{ A} \end{bmatrix}$$

The electromagnetic torque is given by

$$T_e = \frac{3}{2}\frac{P}{2} L_m (i_{qs}^e i_{dr}^e - i_{ds}^e i_{qr}^e)(\text{N·m}) = 25.758 \text{ N·m}$$

The actual phase currents are obtained by the following procedure:

$$i_{abc} = [T_{abc}]^{-1} i_{qdo}^e$$

Hence,

$$i_{as} = 17.08 \sin(\omega_s t - 0.57)$$
$$i_{bs} = 17.08 \sin(\omega_s t - 2.67)$$
$$i_{cs} = 17.08 \sin(\omega_s t + 1.52)$$

Note that the three-phase currents are balanced and that the power-factor angle is 0.57 radians (32.9°) lagging.

(ii) Rotor Reference Frames

$$\begin{bmatrix} v_{qs}^r \\ v_{ds}^r \\ v_o \end{bmatrix} = [T_{abc}^r] \begin{bmatrix} v_{as} \\ v_{bs} \\ v_{cs} \end{bmatrix}$$

Substituting for the transformation from equation (5.118) gives

$$\begin{bmatrix} v_{qs}^r \\ v_{ds}^r \\ v_o \end{bmatrix} = \begin{bmatrix} V_m \sin \omega_{sl} t \\ V_m \cos \omega_{sl} t \\ 0 \end{bmatrix} = \begin{bmatrix} 0 - jV_m \\ V_m + j0 \\ 0 \end{bmatrix}$$

The stator voltages appear at slip frequency in rotor reference frames; hence, the currents are at slip frequency in steady state. By substituting for $p = j\omega_{sl}$ in the system equations, the following is obtained:

$$\begin{bmatrix} -jV_m \\ V_m \\ 0 \\ 0 \end{bmatrix} = \begin{bmatrix} R_s + j\omega_{sl}L_s & \omega_r L_s & j\omega_{sl}L_m & \omega_r L_m \\ -\omega_r L_s & R_s + j\omega_{sl}L_s & -\omega_r L_m & j\omega_{sl}L_m \\ j\omega_{sl}L_m & 0 & R_r + j\omega_{sl}L_r & 0 \\ 0 & j\omega_{sl}L_m & 0 & R_r + j\omega_{sl}L_r \end{bmatrix} \begin{bmatrix} i_{qs}^r \\ i_{ds}^r \\ i_{qr}^r \\ i_{dr}^r \end{bmatrix}$$

from which the rotor reference frame currents are

$$\begin{bmatrix} i_{qs}^r \\ i_{ds}^r \\ i_{qr}^r \\ i_{dr}^r \end{bmatrix} = \begin{bmatrix} 17.08 \sin(\omega_{sl} t - 0.57) \\ 17.08 \cos(\omega_{sl} t - 0.57) \\ 14.69 \sin(\omega_{sl} t - 3.03) \\ 14.69 \cos(\omega_{sl} t - 3.03) \end{bmatrix}$$

from which, by using the inverse transformation $[T_{abc}^r]^{-1}$, the actual phase stator currents can be evaluated. It can be seen that the stator q and d axes currents are displaced from each other by 90°, and so are the rotor q and d axes currents in the rotor reference frames. They can be written as

$$I_{qs}^r = I_s \angle \gamma; \quad I_{ds}^r = I_s \angle \gamma + 90° = j I_{qs}^r; \quad I_{qr}^r = I_r \angle \eta; \quad I_{dr}^r = I_r \angle \eta + 90° = jI_r \angle \eta = j I_{qr}^r$$

where $I_s = 17.08$ A; $I_r = 14.69$ A; $\gamma = -0.57$ rad; $\eta = -3.03$ rad.

Using the torque expression shows that $T_e = 0$, even though it is incorrect. In steady state, speed is constant, and therefore the air gap power is proportional to the torque; however, the torque is a function of products of two currents, as seen from the expression derived earlier. The currents are complex variables in steady state, so when products are involved, one of them in each product has to be a complex conjugate, as in the case of the evaluation of power in ac circuits. Hence, the torque expression for use in this case is modified as follows:

$$T_e = \frac{3}{2} \frac{P}{2} L_m (I_{qs}^e \overline{I_{qs}^e} - I_{ds}^e \overline{I_{qr}^e}) \quad (\text{N·m}) = 25.758 \text{ N·m}$$

Note the bar on the rotor currents, to indicate that they are complex conjugates of their respective values. This gives the correct value for the torque and matches that from using synchronous reference frames, given in the above. The solution for torque, like that for other output variables, such as the rotor speed and input and output power, is independent of reference frames; it is important to realize that the observation frames cannot affect these machine values. A similar approach is to be used in the stator reference frame model also, for the evaluation of the steady-state electromagnetic torque of the machine: the currents turn out to be complex values.

5.7.8 Equations in Flux Linkages

The dynamic equations of the induction motor in arbitrary reference frames can be represented by using flux linkages as variables. This involves the reduction of a number of variables in the dynamic equations, which greatly facilitates their solution by using analog and hybrid computers. Even when the voltages and currents are discontinuous, the flux linkages are continuous. This gives the advantage of differentiating these variables with numerical stability. In addition, the flux-linkages representation is used in motor drives to highlight the process of the decoupling of the flux and torque channels in the induction and synchronous machines.

The stator and rotor flux linkages in the arbitrary reference frames are defined as

$$\left. \begin{array}{l} \lambda_{qs}^c = L_s i_{qs}^c + L_m i_{qr}^c \\ \lambda_{ds}^c = L_s i_{ds}^c + L_m i_{dr}^c \\ \lambda_{qr}^c = L_r i_{qr}^c + L_m i_{qs}^c \\ \lambda_{dr}^c = L_r i_{dr}^c + L_m i_{ds}^c \end{array} \right\} \quad (5.138)$$

The zero-sequence flux linkages are

$$\left. \begin{array}{l} \lambda_{os} = L_{ls} i_{os} \\ \lambda_{or} = L_{lr} i_{or} \end{array} \right\} \quad (5.139)$$

The q axis stator voltage in the arbitrary reference frame is

$$v_{qs}^c = R_s i_{qs}^c + \omega_c (L_s i_{ds}^c + L_m i_{dr}^c) + L_m p i_{qr}^c + L_s p i_{qs}^c \quad (5.140)$$

Section 5.7 Dynamic Modeling of Induction Machines

Substituting from the defined flux-linkages into the voltage equation yields

$$v_{qs}^c = R_s i_{qs}^c + \omega_c \lambda_{ds}^c + p\lambda_{qs}^c \quad (5.141)$$

Similarly, the stator d axis voltage, the d and q axes rotor voltages, and the zero-sequence voltage equations are derived as

$$v_{ds}^c = R_s i_{ds}^c - \omega_c \lambda_{qs}^c + p\lambda_{ds}^c \quad (5.142)$$
$$v_{os} = R_s i_{os} + p\lambda_{os} \quad (5.143)$$
$$v_{qr}^c = R_r i_{qr}^c + (\omega_c - \omega_r)\lambda_{dr}^c + p\lambda_{qr}^c \quad (5.144)$$
$$v_{dr}^c = R_r i_{dr}^c - (\omega_c - \omega_r)\lambda_{qr}^c + p\lambda_{dr}^c \quad (5.145)$$
$$v_{or} = R_r i_{or} + p\lambda_{or} \quad (5.146)$$

These equations can be represented in equivalent circuits.

Power-system engineers and design engineers use normalized or per unit (p.u.) values for the variables. The normalization of the variables is made via reactances rather than inductances. To facilitate such a step, a modified flux linkage is defined whose unit in volts is

$$\psi_{qs}^c = \omega_b \lambda_{qs}^c = \omega_b (L_s i_{qs}^c + L_m i_{qr}^c) = X_s i_{qs}^c + X_m i_{qr}^c \quad (5.147)$$

where ω_b is the base frequency in rad/sec. Similarly the other modified flux linkages are written as

$$\psi_{ds}^c = X_s i_{ds}^c + X_m i_{dr}^c \quad (5.148)$$
$$\psi_{qr}^c = X_r i_{qr}^c + X_m i_{qs}^c \quad (5.149)$$
$$\psi_{dr}^c = X_r i_{dr}^c + X_m i_{ds}^c \quad (5.150)$$
$$\psi_{os} = X_{ls} i_{os} \quad (5.151)$$
$$\psi_{or} = X_{lr} i_{or} \quad (5.152)$$

Substituting the flux linkages in terms of the modified flux linkages yields

$$\lambda_{qs}^c = \frac{\psi_{qs}^c}{\omega_b}, \lambda_{ds}^c = \frac{\psi_{ds}^c}{\omega_b}, \lambda_{os}^c = \frac{\psi_{os}^c}{\omega_b}, \lambda_{qr}^c = \frac{\psi_{qr}^c}{\omega_b}, \lambda_{dr}^c = \frac{\psi_{dr}^c}{\omega_b}, \lambda_{or}^c = \frac{\psi_{or}^c}{\omega_b} \quad (5.153)$$

and, by substituting equation (5.153) into equations from (5.141) to (5.146), the resulting equations in modified flux linkages are

$$v_{qs}^c = R_s i_{qs}^c + \frac{\omega_c}{\omega_b}\psi_{ds}^c + \frac{p}{\omega_b}\psi_{qs}^c, \; v_{ds}^c = R_s i_{ds}^c - \frac{\omega_c}{\omega_b}\psi_{qs}^c + \frac{p}{\omega_b}\psi_{ds}^c, \; v_{os} = R_s i_{os} + \frac{p}{\omega_b}\psi_{os},$$

$$v_{qr}^c = R_r i_{qr}^c + \frac{(\omega_c - \omega_r)}{\omega_b}\psi_{dr}^c + \frac{p}{\omega_b}\psi_{qr}^c, \; v_{dr}^c = R_r i_{dr}^c - \frac{(\omega_c - \omega_r)}{\omega_b}\psi_{qr}^c + \frac{p}{\omega_b}\psi_{dr}^c,$$

$$v_{or} = R_r i_{or} + \frac{p}{\omega_b}\psi_{or} \quad (5.154)$$

The electromagnetic torque in flux linkages and currents is derived as

$$T_e = \frac{3}{2}\frac{P}{2}L_m(i_{qs}^c i_{dr}^c - i_{ds}^c i_{qr}^c) = \frac{3}{2}\frac{P}{2}(i_{qs}^c(L_m i_{dr}^c) - i_{ds}^c(L_m i_{qr}^c)) \quad (5.155)$$

The rotor currents can be substituted in terms of the stator currents and stator flux linkages from the basic definitions of the flux linkages. From equation (5.138),

$$L_m i_{dr}^c + L_s i_{ds}^c = \lambda_{ds}^c \tag{5.156}$$

Hence,

$$L_m i_{dr}^c = (\lambda_{ds}^c - L_s i_{ds}^c) \tag{5.157}$$

Similarly,

$$L_m i_{qr}^c = (\lambda_{qs}^c - L_s i_{qs}^c) \tag{5.158}$$

Substituting equations (5.157) and (5.158) into equation (5.152) gives the electromagnetic torque in stator flux linkages and stator currents as

$$T_e = \frac{3}{2}\frac{P}{2}(i_{qs}^c(\lambda_{ds}^c - L_s i_{ds}^c) - i_{ds}^c(\lambda_{qs}^c - L_s i_{qs}^c)) = \frac{3}{2}\frac{P}{2}(i_{qs}^c \lambda_{ds}^c - i_{ds}^c \lambda_{qs}^c) \tag{5.159}$$

Alternatively, the electromagnetic torque in terms of modified flux linkages and currents is

$$T_e = \frac{3}{2}\frac{P}{2}\frac{1}{\omega_b}(i_{qs}^c \psi_{ds}^c - i_{ds}^c \psi_{qs}^c) \tag{5.160}$$

The electromagnetic torque can be expressed in only rotor variables and in many other forms. Note that they are all derivable from equation (5.152) by proper substitutions.

5.7.9 Per-Unit Model

The normalized model of the induction motor is derived by defining the base variables both in the *abc* and the *dqo* variables. In the *abc* frames, let the rms values of the rated phase voltage and current form the base quantities, given as

$$\text{Base Power} = P_b = 3V_{b3}I_{b3} \tag{5.161}$$

where V_{b3} and I_{b3} are three-phase base voltage and current, respectively. Selecting the base quantities in *dq* frames denoted by V_b and I_b to be equal to the peak value of the phase voltage and current in *abc* frames, we get

$$V_b = \sqrt{2}V_{b3} \tag{5.162}$$

$$I_b = \sqrt{2}I_{b3} \tag{5.163}$$

Hence, the base power is defined as

$$P_b = 3V_{b3}I_{b3} = 3\frac{V_b}{\sqrt{2}}\frac{I_b}{\sqrt{2}} = \frac{3}{2}V_b I_b \tag{5.164}$$

The arbitrary reference frame model is chosen to illustrate the normalization process. Consider the *q* axis stator voltage to begin with, which is given in the following:

$$v_{qs}^c = R_s i_{qs}^c + \omega_c (L_s i_{ds}^c + L_m i_{dr}^c) + L_m p i_{qr}^c + L_s p i_{qs}^c \tag{5.165}$$

Section 5.7 Dynamic Modeling of Induction Machines

Writing the inductances in terms of reactances at base frequency of ω_b gives

$$v_{qs}^c = R_s i_{qs}^c + \frac{X_s}{\omega_b} p i_{qs}^c + \frac{\omega_c X_s}{\omega_b} i_{ds}^c + \frac{X_m}{\omega_b} p i_{qr}^c + \frac{\omega_c X_m}{\omega_b} i_{dr}^c \qquad (5.166)$$

It is normalized by dividing by the base voltage, V_b:

$$v_{qsn}^c = \frac{v_{qs}^c}{V_b} = \frac{R_s}{V_b} i_{qs}^c + \frac{X_s}{\omega_b V_b} p i_{qs}^c + \frac{\omega_c X_s}{\omega_b V_b} i_{ds}^c + \frac{X_m}{\omega_b V_b} i_{qr}^c + \frac{\omega_c X_m}{\omega_b V_b} i_{dr}^c \qquad (5.167)$$

Substituting for base voltage in terms of base current, I_b, and base impedance, Z_b, as follows,

$$V_b = I_b Z_b \qquad (5.168)$$

into equation (5.167) yields

$$v_{qsn}^c = \left(\frac{R_s}{Z_b}\right)\left(\frac{i_{qs}^c}{I_b}\right) + \frac{1}{\omega_b}\left(\frac{X_s}{Z_b}\right) p\left(\frac{i_{qs}^c}{I_b}\right) + \left(\frac{\omega_c}{\omega_b}\right)\left(\frac{X_s}{Z_b}\right)\left(\frac{i_{ds}^c}{I_b}\right)$$
$$+ \frac{1}{\omega_b}\left(\frac{X_m}{Z_b}\right) p\left(\frac{i_{qr}^c}{I_b}\right) + \left(\frac{\omega_c}{\omega_b}\right)\left(\frac{X_m}{Z_m}\right)\left(\frac{i_{dr}^c}{I_b}\right) \qquad (5.169)$$

By defining the normalized parameters and variables in the following manner,

$$R_{sn} = \frac{R_s}{Z_b}\text{ p.u.}, \; X_{sn} = \frac{X_s}{Z_b}\text{ p.u.}, \; \omega_{mn} = \frac{\omega_r}{\omega_b}\text{ p.u.}, \; i_{dsn}^c = \frac{i_{ds}^c}{I_b}\text{ p.u.}, \; v_{dsn}^c = \frac{v_{ds}^c}{V_b}\text{ p.u.}$$

$$R_{rn} = \frac{R_r}{Z_b}\text{ p.u.}, \; X_{rn} = \frac{X_r}{Z_b}\text{ p.u.}, \; i_{qsn}^c = \frac{i_{qs}^c}{I_b}\text{ p.u.}, \; i_{drn}^c = \frac{i_{dr}^c}{I_b}\text{ p.u.}, \; v_{qrn}^c = \frac{v_{qr}^c}{V_b}\text{ p.u.} \qquad (5.170)$$

$$X_{mn} = \frac{X_m}{Z_b}\text{ p.u.}, \; \omega_{cn} = \frac{\omega_c}{\omega_b}\text{ p.u.}, \; i_{qrn}^c = \frac{i_{qr}^c}{I_b}\text{ p.u.}, \; v_{qsn}^c = \frac{v_{qs}^c}{V_b}\text{ p.u.}, \; v_{drn}^c = \frac{v_{dr}^c}{V_b}\text{ p.u.}$$

and by substituting equations in (5.170) into appropriate places in (5.169), v_{qsn}^c is obtained as,

$$v_{qsn}^c = \left(R_{sn} + \frac{X_{sn}}{\omega_b} p\right) i_{qsn}^c + \omega_{cn} X_{sn} i_{dsn}^c + \frac{X_{mn}}{\omega_b} p i_{qrn}^c + \omega_{cn} X_{mn} i_{drn}^c \qquad (5.171)$$

Similarly, the other equations are normalized and given below:

$$\begin{bmatrix} v_{qsn}^c \\ v_{dsn}^c \\ v_{qrn}^c \\ v_{drn}^c \end{bmatrix} = \begin{bmatrix} R_{sn} + \frac{X_{sn}}{\omega_b} p & \omega_{cn} X_{sn} & \frac{X_{mn}}{\omega_b} p & \omega_{cn} X_{mn} \\ -\omega_{cn} X_{sn} & R_{sn} + \frac{X_{sn}}{\omega_b} p & -\omega_{cn} X_{mn} & \frac{X_{mn}}{\omega_b} p \\ \frac{X_{mn}}{\omega_b} p & (\omega_{cn} - \omega_{rn}) X_{mn} & R_{rn} + \frac{X_{rn}}{\omega_b} p & (\omega_{cn} - \omega_{rn}) X_{mn} \\ -(\omega_{cn} - \omega_{rn}) X_{mn} & \frac{X_{mn}}{\omega_b} p & -(\omega_{cn} - \omega_{rn}) X_{rn} & R_{rn} + \frac{X_{rn}}{\omega_b} p \end{bmatrix} \begin{bmatrix} i_{qsn}^c \\ i_{dsn}^c \\ i_{qrn}^c \\ i_{drn}^c \end{bmatrix}$$

$$(5.172)$$

The electromagnetic torque is,

$$T_e = \frac{3}{2}\frac{P}{2}\frac{1}{\omega_b}(i_{qs}^c \psi_{ds}^c - i_{ds}^c \psi_{qs}^c) \quad (5.173)$$

but the base torque is,

$$T_b = \frac{P_b}{\frac{\omega_b}{P/2}} = \frac{P}{2}\frac{P_b}{\omega_b} \quad (5.174)$$

from which the normalized electromagnetic torque is obtained as

$$T_{en} = \frac{T_e}{T_b} = \frac{\frac{3}{2}\frac{P}{2}\frac{1}{\omega_b}}{\frac{P}{2}\frac{P_b}{\omega_b}}(i_{qs}^c \psi_{ds}^c - i_{ds}^c \psi_{qs}^c) = \frac{3/2}{P_b}(i_{qs}^c \psi_{ds}^c - i_{ds}^c \psi_{qs}^c) = \frac{3/2}{(3/2)V_b I_b}(i_{qs}^c \psi_{ds}^c - i_{ds}^c \psi_{qs}^c)$$

$$= \left(\frac{i_{qs}^c}{I_b}\right)\left(\frac{\psi_{ds}^c}{V_b}\right) - \left(\frac{i_{ds}^c}{I_b}\right)\left(\frac{\psi_{qs}^c}{V_b}\right) = (i_{qsn}^c \psi_{dsn}^c - i_{dsn}^c \psi_{qsn}^c) \text{ p.u.} \quad (5.175)$$

where

$$\left.\begin{array}{l}\psi_{dsn}^c = \dfrac{\psi_{ds}^c}{V_b} \text{ p.u.} \\ \psi_{qsn}^c = \dfrac{\psi_{qs}^c}{V_b} \text{ p.u.}\end{array}\right\} \quad (5.176)$$

Note that the modified flux linkages, ψ_{qs}^c and ψ_{qs}^c, are in volts, and hence ψ_{dsn}^c and ψ_{qsn}^c become dimensionless. The electromechanical dynamic equation is given by

$$T_e = J\frac{d\omega_m}{dt} + T_l + B\omega_m \quad (5.177)$$

where J is the moment of inertia of motor and load, T_l is the load torque, B is the friction coefficient of the load and motor, and ω_m is the mechanical rotor speed. Normalizing this equation yields

$$T_{en} = \frac{T_e}{T_b} = \frac{J\frac{d\omega_m}{dt}}{\left(\frac{P_b \times P/2}{\omega_b}\right)} + \frac{T_l}{T_b} + \frac{B\omega_m}{\left(\frac{P/2 \times P_b}{\omega_b}\right)} = \frac{J\omega_b \omega_b}{(P/2)^2 P_b}\frac{d}{dt}\left(\frac{\omega_r}{\omega_b}\right) + T_{ln} + \frac{B\omega_b \omega_b \omega_r}{(P/2)^2 P_b \omega_b}$$

$$= \frac{J\omega_b^2}{(P/2)^2 P_b}\frac{d\omega_{rn}}{dt} + T_{ln} + \frac{B\omega_b^2}{(P/2)^2 P_b}\omega_{rn} = 2Hp\omega_{rn} + T_{ln} + B_n \omega_{rn} \quad (5.178)$$

where

$$H = \frac{1}{2}\frac{J\omega_b^2}{P_b(P/2)^2} \quad (5.179)$$

is known as inertial constant, and the normalized friction constant is

$$B_n = \frac{B\omega_b^2}{P_b(P/2)^2} \quad (5.180)$$

The normalized equations of the induction machine in the arbitrary reference frames are given by equations (5.172), (5.175), and (5.178).

5.8 DYNAMIC SIMULATION

The dynamic simulation of the induction machine is explained in this section. The equations of the induction machine in arbitrary reference frames in p.u. are cast in the state-space form as

$$P_1 p X_1 + Q_1 X_1 = u_1 \quad (5.181)$$

where

$$P_1 = \begin{bmatrix} \dfrac{X_{sn}}{\omega_b} & 0 & \dfrac{X_{mn}}{\omega_b} & 0 \\ 0 & \dfrac{X_{sn}}{\omega_b} & 0 & \dfrac{X_{mn}}{\omega_b} \\ \dfrac{X_{mn}}{\omega_b} & 0 & \dfrac{X_{rn}}{\omega_b} & 0 \\ 0 & \dfrac{X_{mn}}{\omega_b} & 0 & \dfrac{X_{rn}}{\omega_b} \end{bmatrix} \quad (5.182)$$

$$Q_1 = \begin{bmatrix} R_{sn} & \omega_{cn} X_{sn} & 0 & \omega_{cn} X_{mn} \\ -\omega_{cn} X_{sn} & R_{sn} & -\omega_{cn} X_{mn} & 0 \\ 0 & (\omega_{cn} - \omega_{rn}) X_{mn} & R_{rn} & (\omega_{cn} - \omega_{rn}) X_{rn} \\ -(\omega_{cn} - \omega_{rn}) X_{mn} & 0 & -(\omega_{cn} - \omega_{rn}) X_{rn} & R_{rn} \end{bmatrix} \quad (5.183)$$

$$X_1 = [i_{qsn}^c \ i_{dsn}^c \ i_{qrn}^c \ i_{drn}^c]^t \quad (5.184)$$

$$u_1 = [v_{qsn}^c \ v_{dsn}^c \ v_{qrn}^c \ v_{drn}^c]^t \quad (5.185)$$

The equation (5.178) can be rearranged in the state-space form as follows:

$$pX_1 = P^{-1}(u_1 - Q_1 X_1) \quad (5.186)$$

This can be written as

$$pX_1 = A_1 X_1 + B_1 u_1 \quad (5.187)$$

where $A_1 = -P_1^{-1} Q_1$ and $B_1 = P_1^{-1}$, and they are evaluated to be

$$B_1 = P_1^{-1} = \frac{1}{\Delta} \begin{bmatrix} X_{rn} & 0 & -X_{mn} & 0 \\ 0 & X_{rn} & 0 & -X_{mn} \\ -X_{mn} & 0 & X_{sn} & 0 \\ 0 & -X_{mn} & 0 & X_{sn} \end{bmatrix} \quad (5.188)$$

where

$$\Delta = \left(\frac{X_{sn}X_{rn} - X_{mn}^2}{\omega_b}\right) \quad (5.189)$$

$$A_1 = -\frac{1}{\Delta}\begin{bmatrix} R_{sn}X_{rn} & k_1 + \omega_{rn}X_{mn}^2 & -R_{rn}X_{mn} & \omega_{rn}X_{rn}X_{mn} \\ -k_1 - \omega_{rn}X_{mn}^2 & R_{sn}X_{rn} & -\omega_{rn}X_{rn}X_{mn} & -R_{rn}X_{mn} \\ -R_{sn}X_{mn} & -\omega_{rn}X_{mn}X_{sn} & R_{rn}X_{sn} & k_1 - \omega_{rn}X_{sn} \\ \omega_{rn}X_{mn}X_{sn} & -R_{sn}X_{mn} & -k_1 + \omega_{rn}X_{rn}X_{sn} & R_{rn}X_{sn} \end{bmatrix} \quad (5.190)$$

where

$$k_1 = \omega_{cn}(X_{sn}X_{rn} - X_{mn}^2) \quad (5.191)$$

The electromechanical equation is

$$T_{en} = 2Hp\omega_{rn} + T_{ln} + B_n\omega_{rn} \quad (5.192)$$

where

$$T_{en} = i_{qsn}^c \psi_{dsn}^c - i_{dsn}^c \psi_{qsn}^c \quad (5.193)$$

The modified flux linkages require additional computation; the torque can be conveniently expressed in terms of the normalized currents as

$$T_{en} = X_{mn}(i_{qsn}^c i_{drn}^c - i_{dsn}^c i_{qrn}^c) \quad (5.194)$$

and hence the electromechanical equation can be written as

$$p\omega_{rn} = \frac{X_{mn}}{2H}(i_{qsn}^c i_{drn}^c - i_{dsn}^c i_{qrn}^c) - \frac{T_{ln}}{2H} - \frac{B_n}{2H}\omega_{rn} \quad (5.195)$$

Solving equations (5.187) and (5.195) by numerical integration gives the solutions for currents and speed from which the torque, power output, and losses are calculated. A flowchart describing the steps in the dynamic simulation is shown in Figure 5.17.

Example 5.6

An induction motor has the following constants and ratings:

200 V, 4 pole, 3 phase, 60 Hz, Y connected

$R_s = 0.183\ \Omega, R_r = 0.277\ \Omega, L_m = 0.0538$ H

$L_s = 0.0553$ H, $L_r = 0.056$ H, B = 0

Load torque = $T_1 = 0$ N·m, J = 0.0165 kg-m^2

Base power is 5 hp

The motor is at standstill. A set of balanced three-phase voltages at 70.7% of rated values at 60 Hz is applied. Plot the starting characteristics of speed vs. torque, instantaneous stator and rotor currents in all the three reference frames, and stator flux linkages and flux phasor.

Section 5.8 Dynamic Simulation 225

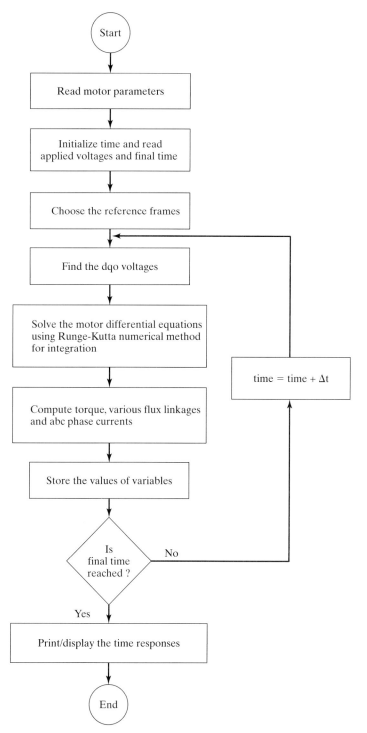

Figure 5.17 Flowchart for dynamic simulation of the induction motor

226 Chapter 5 Polyphase Induction Machines

Solution The plots are shown in Figures 5.18(a), (b) and (c). The machine outputs, such as air gap torque, speed, actual stator and rotor phase currents, and magnitude of stator flux linkages are all the same regardless of reference frames. Also note that the flux linkages are in real units, while other variables are in normalized units. Line start produces higher currents, flux linkages, and dc offsets in them. The torque pulsations are very severe, and repeated line starting could endanger the mechanical integrity of the motor. Higher stator and rotor currents also produce resistive losses that are multiple times the design limit.

5.9 SMALL-SIGNAL EQUATIONS OF THE INDUCTION MACHINE

5.9.1 Derivation

The electrical equations of the induction machine and the electromechanical subsystems given in (5.187), (5.194), and (5.195) combine to give the dynamic equations of the motor–load system. These dynamic equations are nonlinear: some of the terms are the products of two current variables or a current variable and rotor speed. The transient responses or the solution of the dynamic equations are obtained by numerical integration. While any of the standard subroutines can be used for numerical integration, the fourth-order Runge–Kutta method is adequate for the study of induction machines. For the solution, the equations have to be arranged in the state-space form. For controller design with linear control-system design techniques, the nonlinear dynamic equations cannot be directly used. They have to be linearized around an operating point by using perturbation techniques. For small-signal inputs or disturbances, the linearized equations are valid. The ideal model for perturbation to get the linearized model is the one with steady-state operating-state variables as dc values. This is possible only with the synchronously rotating reference frames model of the induction motor. The linearized equations are obtained as follows.

The voltages, currents, torque, stator frequency, and rotor speed in their steady state are designated by an additional subscript with an 'o' in the variables, and the perturbed increments are designated by a δ preceding the variables. Accordingly, the variables in SI units after perturbation are

$$v_{qs}^e = v_{qso}^e + \delta v_{qs}^e \tag{5.196}$$

$$v_{ds}^e = v_{dso}^e + \delta v_{ds}^e \tag{5.197}$$

$$i_{qs}^e = i_{qso}^e + \delta i_{qs}^e \tag{5.198}$$

$$i_{ds}^e = i_{dso}^e + \delta i_{ds}^e \tag{5.199}$$

$$i_{qr}^e = i_{qro}^e + \delta i_{qr}^e \tag{5.200}$$

$$i_{dr}^e = i_{dro}^e + \delta i_{dr}^e \tag{5.201}$$

$$T_e = T_{eo} + \delta T_e \tag{5.202}$$

$$T_l = T_{lo} + \delta T_l \tag{5.203}$$

$$\omega_s = \omega_{so} + \delta \omega_s \tag{5.204}$$

$$\omega_r = \omega_{ro} + \delta \omega_r \tag{5.205}$$

Section 5.9 Small-Signal Equations of the Induction Machine

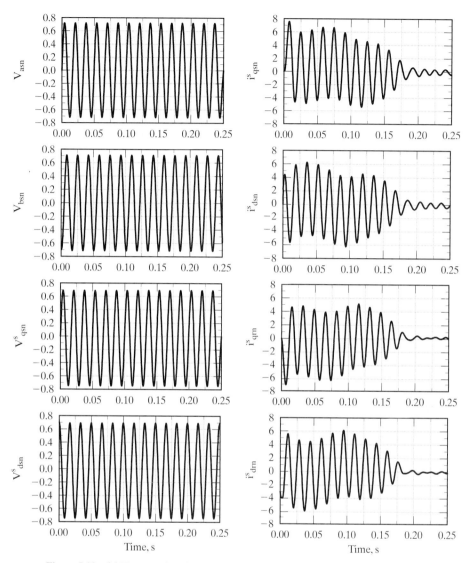

Figure 5.18 (a) Free-acceleration characteristics in stator reference frames (Part I)

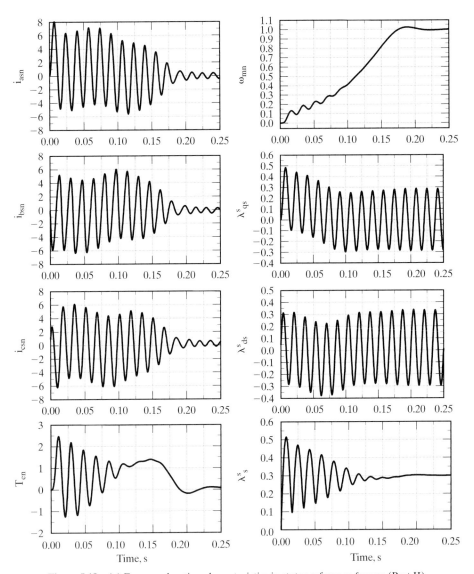

Figure 5.18 (a) Free-acceleration characteristics in stator reference frames (Part II)

Section 5.9 Small-Signal Equations of the Induction Machine 229

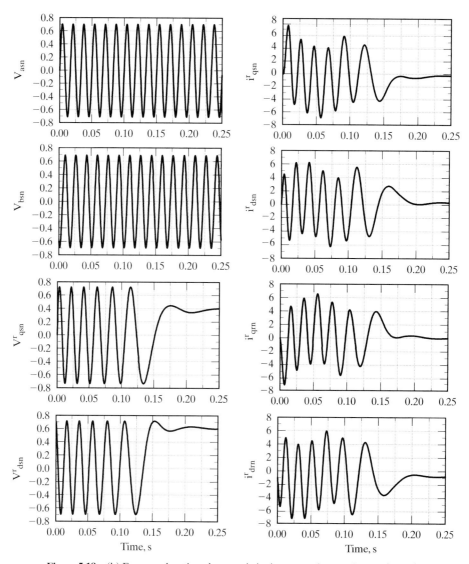

Figure 5.18 (b) Free-acceleration characteristics in rotor reference frames (Part I)

230 Chapter 5 Polyphase Induction Machines

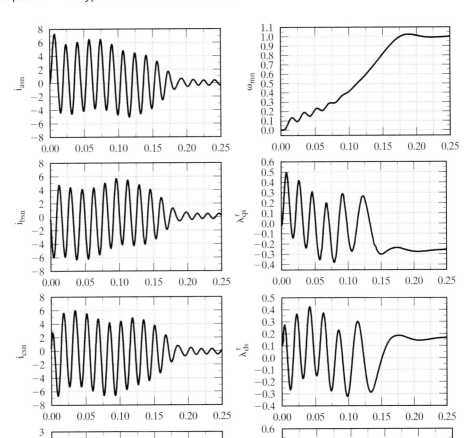

Figure 5.18 (b) Free-acceleration characteristics in rotor reference frames (Part II)

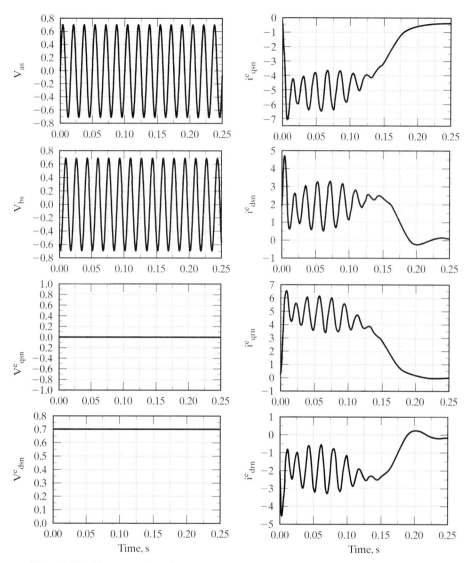

Figure 5.18 (c) Free-acceleration characteristics in synchronous reference frames (Part I)

232 Chapter 5 Polyphase Induction Machines

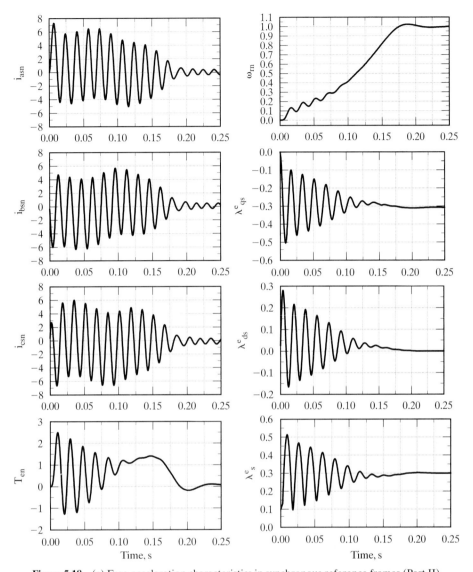

Figure 5.18 (c) Free-acceleration characteristics in synchronous reference frames (Part II)

Section 5.9 Small-Signal Equations of the Induction Machine

where

$$T_{eo} = \frac{3}{2}\frac{P}{2}L_m(i^e_{qso}i^e_{dro} - i^e_{dso}i^e_{qro}) \tag{5.206}$$

By substituting from equations (5.196) to (5.206) into equations (5.114), (5.122), and (5.177), by neglecting the second-order terms, and by canceling the steady-state terms on the right- and left-hand sides of the equations, the small-signal dynamic equations are obtained:

$$\delta v^e_{qs} = (R_s + L_s p)\delta i^e_{qs} + \omega_{so}L_s\delta i^e_{ds} + L_m p\delta i^e_{qr} + \omega_{so}L_m\delta i^e_{dr}$$
$$+ (L_s i^e_{dso} + L_m i^e_{dro})\delta\omega_s \tag{5.207}$$

$$\delta v^e_{ds} = -\omega_{so}L_s\delta i^e_{qs} + (R_s + L_s p)\delta i^e_{ds} - \omega_{so}L_m\delta i^e_{qr} + L_m p\delta i^e_{dr} - (L_s i^e_{qso} + L_m i^e_{qro})\delta\omega_s \tag{5.208}$$

$$\delta v^e_{qr} = L_m p\delta i^e_{qs} + (\omega_{so} - \omega_{ro})L_m\delta i^e_{ds} + (R_r + L_r p)\delta i^e_{qr} + (\omega_{so} - \omega_{ro})L_r\delta i^e_{dr}$$
$$- (L_m i^e_{dso} + L_r i^e_{dro})\delta\omega_r + (L_m i^e_{dso} + L_r i^e_{dro})\delta\omega_s \tag{5.209}$$

$$\delta v^e_{dr} = -(\omega_{so} - \omega_{ro})L_m\delta i^e_{qs} + L_m p\delta i^e_{ds} - (\omega_{so} - \omega_{ro})L_r\delta i^e_{qr} + (R_r + L_r p)\delta i^e_{dr}$$
$$+ (L_m i^e_{qso} + L_r i^e_{qro})\delta\omega_r - (L_m i^e_{qso} + L_r i^e_{qro})\delta\omega_s \tag{5.210}$$

$$Jp\delta\omega_r + B\delta\omega_r = \frac{P}{2}(\delta T_e - \delta T_l) \tag{5.211}$$

$$\delta T_e = \frac{3}{2}\frac{P}{2}L_m(i^e_{qso}\delta i^e_{dr} + i^e_{dro}\delta i^e_{qs} - i^e_{dso}\delta i^e_{qr} - i^e_{qro}\delta i^e_{ds}) \tag{5.212}$$

Combining equations from (5.207) to (5.211) and casting them in state-space form gives

$$pX = AX + B_1U \tag{5.213}$$

where

$$X = [\delta i^e_{qs} \quad \delta i^e_{ds} \quad \delta i^e_{qr} \quad \delta i^e_{dr} \quad \delta\omega_r]^t \tag{5.214}$$

$$U = [\delta v^e_{qs} \quad \delta v^e_{ds} \quad \delta v^e_{qr} \quad \delta v^e_{dr} \quad \delta\omega_s \quad \delta T_l]t \tag{5.215}$$

$$A = P_1^{-1}Q_1 \tag{5.216}$$

$$B_1 = P_1^{-1}R_1 \tag{5.217}$$

$$P_1 = \begin{bmatrix} L_s & 0 & L_m & 0 & 0 \\ 0 & L_s & 0 & L_m & 0 \\ L_m & 0 & L_r & 0 & 0 \\ 0 & L_m & 0 & L_r & 0 \\ 0 & 0 & 0 & 0 & J \end{bmatrix} \tag{5.218}$$

234 Chapter 5 Polyphase Induction Machines

$$Q_1 = \begin{bmatrix} -R_s & -\omega_{so}L_s & 0 & -\omega_{so}L_m & 0 \\ \omega_{so}L_s & -R_s & \omega_{so}L_m & 0 & 0 \\ 0 & -(\omega_{so}-\omega_{ro})L_m & -R_r & -(\omega_{so}-\omega_{ro})L_r & L_m i_{dso}^e + L_r i_{dro}^e \\ (\omega_{so}-\omega_{ro})L_m & 0 & (\omega_{so}-\omega_{ro})L_r & -R_r & -(L_m i_{qso}^e + L_r i_{qro}^e) \\ k_2 i_{dro}^e & -k_2 i_{qro}^e & -k_2 i_{dso}^e & k_2 i_{qso}^e & -B \end{bmatrix}$$

(5.219)

$$k_2 = \frac{3}{2}\left(\frac{P}{2}\right)^2 L_m \tag{5.220}$$

$$R_1 = \begin{bmatrix} 1 & 0 & 0 & 0 & -L_s i_{dso}^e + L_m i_{dro}^e & 0 \\ 0 & 1 & 0 & 0 & L_s i_{qso}^e + L_m i_{qro}^e & 0 \\ 0 & 0 & 1 & 0 & -L_m i_{dso}^e + L_r i_{dro}^e & 0 \\ 0 & 0 & 0 & 1 & L_m i_{qso}^e + L_r i_{qro}^e & 0 \\ 0 & 0 & 0 & 0 & 0 & -\dfrac{P}{2} \end{bmatrix}$$

(5.221)

The output of interest can be a function of the state variables and inputs, expressed as

$$y = CX + DU \tag{5.222}$$

where C and D are now vectors of appropriate dimensions. The system and its output are described by the equations (5.213) and (5.222), respectively.

5.9.2 Normalized Small-Signal Equations

The small-signal equations for the normalized model can be derived as in the procedure adopted for the SI-unit model. The equations thus obtained are presented in this section. The small-signal equations for the torque and voltages are given as follows:

$$\delta T_{en} = X_{mn}(i_{qsno}^e \delta i_{drn}^e + i_{drno}^e \delta i_{qsn}^e - i_{dsno}^e \delta i_{qrn}^e - i_{qrno}^e \delta i_{dsn}^e) \tag{5.223}$$

$$\delta v_{qsn}^e = \left(R_{sn} + \frac{X_{sn}}{\omega_b}p\right)\delta i_{qsn}^e + \omega_{sno} X_{sn} \delta i_{dsn}^e + \frac{X_{mn}}{\omega_b} p \delta i_{qrn}^e + \omega_{sno} X_{mn} \delta i_{drn}^e$$
$$+ (X_{sn} i_{dsno}^e + X_{mn} i_{drno}^e) \delta\omega_{sn} \tag{5.224}$$

$$\delta v_{dsn}^e = -\omega_{sno} X_{sn} \delta i_{qsn}^e + \left(R_{sn} + \frac{X_{sn}}{\omega_b}p\right)\delta i_{dsn}^e - \omega_{sno} X_{mn} \delta i_{qrn}^e + \frac{X_{mn}}{\omega_b} p \delta i_{dn}^e$$
$$- (X_{sn} i_{qsno}^e + X_{mn} i_{qrno}^e) \delta\omega_{sn} \tag{5.225}$$

$$\delta v_{qrn}^e = \frac{X_{mn}}{\omega_b} p \delta i_{qsn}^e + (\omega_{sno} - \omega_{rno}) X_{mn} \delta i_{dsn}^e + \left(R_{rn} + \frac{X_{rn}}{\omega_b}p\right)\delta i_{qrn}^e$$
$$+ (\omega_{sno} - \omega_{rno}) X_{rn} \delta i_{drn}^e - (X_{mn} i_{dsno}^e + X_{rn} i_{drno}^e) \delta\omega_{rn} + (X_{mn} i_{dsno}^e + X_{rn} i_{drno}^e) \delta\omega_{sn} \tag{5.226}$$

Section 5.9   Small-Signal Equations of the Induction Machine

$$\delta v_{drn}^e = -(\omega_{sno} - \omega_{rno})X_{mn}\delta i_{qsn}^e + \frac{X_{mn}}{\omega_b}p\delta i_{dsn}^e - (\omega_{sno} - \omega_{rno})X_{rn}\delta i_{qrn}^e$$
$$+ \left(R_{rn} + \frac{X_{rn}}{\omega_b}p\right)\delta i_{drn}^e + (X_{mn}i_{qsno}^e + X_{rn}i_{qrno}^e)\delta\omega_{rn} - (X_{mn}i_{qsno}^e + X_{rn}i_{qrno}^e)\delta\omega_{sn} \quad (5.227)$$

$$2Hp\delta\omega_{rn} + B_n\delta\omega_{rn} = (\delta T_{en} - \delta T_{ln}) \quad (5.228)$$

Combining the equations from (5.223) to (5.228) and casting them in state-space form gives

$$px = Ax + B_1 u \quad (5.229)$$

where

$$x = [\delta i_{qsn}^e \quad \delta i_{dsn}^e \quad \delta i_{qrn}^e \quad \delta i_{drn}^e \quad \delta\omega_{rn}]^t \quad (5.230)$$

$$u = [\delta v_{qsn}^e \quad \delta v_{dsn}^e \quad \delta v_{qrn}^e \quad \delta v_{drn}^e \quad \delta\omega_{sn} \quad \delta T_{ln}]^t \quad (5.231)$$

$$A = P_1^{-1} Q_1 \quad (5.232)$$

$$B_1 = P_1^{-1} R_1 \quad (5.233)$$

$$P_1 = \begin{bmatrix} \frac{X_{sn}}{\omega_b} & 0 & \frac{X_{mn}}{\omega_b} & 0 & 0 \\ 0 & \frac{X_{sn}}{\omega_b} & 0 & \frac{X_{mn}}{\omega_b} & 0 \\ \frac{X_{mn}}{\omega_b} & 0 & \frac{X_{rn}}{\omega_b} & 0 & 0 \\ 0 & \frac{X_{mn}}{\omega_b} & 0 & \frac{X_{rn}}{\omega_b} & 0 \\ 0 & 0 & 0 & 0 & 2H \end{bmatrix} \quad (5.234)$$

$$Q_1 = \begin{bmatrix} -R_{sn} & -\omega_{sno}X_{sn} & 0 & -\omega_{sno}X_{mn} & 0 \\ \omega_{sno}X_{sn} & -R_{sn} & \omega_{sno}X_{mn} & 0 & 0 \\ 0 & -\omega_{slno}X_{mn} & -R_{rn} & -\omega_{slno}X_{rn} & X_{mn}i_{dsno}^e + X_{rn}i_{drno}^e \\ \omega_{slno}X_{mn} & 0 & \omega_{slno}X_{rn} & -R_{rn} & X_{mn}i_{qsno}^e + X_{rn}i_{qrno}^e \\ X_{mn}i_{drno}^e & -X_{mn}i_{qrno}^e & -X_{mn}i_{dsno}^e & X_{mn}i_{qsno}^e & -B_n \end{bmatrix} \quad (5.235)$$

where

$$\omega_{slno} = \omega_{sno} - \omega_{rno} = \text{slip speed (p.u.)} \quad (5.236)$$

$$R_1 = \begin{bmatrix} 1 & 0 & 0 & 0 & -X_{sn}i_{dsno}^e + X_{mn}i_{drno}^e & 0 \\ 0 & 1 & 0 & 0 & X_{sn}i_{qsno}^e + X_{mn}i_{qrno}^e & 0 \\ 0 & 0 & 1 & 0 & -(X_{mn}i_{dsno}^e + X_{rn}i_{drno}^e) & 0 \\ 0 & 0 & 0 & 1 & X_{mn}i_{qsno}^e + X_{rn}i_{qsno}^e & 0 \\ 0 & 0 & 0 & 0 & 0 & -1 \end{bmatrix} \quad (5.237)$$

The output of interest can be a function of the state variables and inputs, expressed as

$$y = Cx + Du \qquad (5.238)$$

where C and D are now vectors of appropriate dimensions.

The system and its output are described by the equations (5.229) and (5.238), respectively. They can be used to study the control characteristics of the system, as shown in the next section.

5.10 EVALUATION OF CONTROL CHARACTERISTICS OF THE INDUCTION MACHINE

The control characteristics of the induction machine consist of stability and of frequency and time responses. They require the evaluation of various transfer functions. Stability is evaluated by finding the eigenvalues of the system matrix A in the equation (5.213). It can be found by using standard subroutines available in a software library. The evaluation of transfer functions and of frequency and time responses is available in control-system simulation libraries. Simple algorithms to develop the above are given in this section.

5.10.1 Transfer Functions and Frequency Responses

Taking the Laplace transform of equations (5.213) and (5.222) with the assumption of zero initial conditions gives

$$sX(s) = AX(s) + B_1 u(s) \qquad (5.239)$$

$$y(s) = CX(s) + Du(s) \qquad (5.240)$$

By manipulating equations (5.239) and (5.240), the output can be expressed as

$$y(s) = (C(sI - A)^{-1} B_1 + D) u(s) \qquad (5.241)$$

where I is an identity matrix of appropriate dimensions.

The transfer function involves one input, so the input matrix product $B_1 u(s)$ is then written as

$$B_1 u(s) = b_i u_i(s) \qquad (5.242)$$

where b_i is the ith column vector of B matrix and i corresponds to the element number in the input vector. Then, correspondingly,

$$Du(s) = d_i u_i(s) \qquad (5.243)$$

Then the resulting equations are

$$\begin{aligned} sX(s) &= AX(s) + b_i u_i(s) \\ y(s) &= CX(s) + d_i u_i(s) \end{aligned} \qquad (5.244)$$

The evaluation of transfer functions is made simple if the canonical or phase-variable form of the state equation given in equation (5.244) is found. Assume that is effected by the following transformation,

Section 5.10 Evaluation of Control Characteristics of the Induction Machine

$$X = T_p X_p \tag{5.245}$$

Then the state and output equations are transformed to

$$pX_p = A_p X_p + B_p u_i \tag{5.246}$$
$$y = C_p X_p + d_i u_i \tag{5.247}$$

where

$$A_p = T_p^{-1} A T_p \tag{5.248}$$
$$B_p = T_p^{-1} b_i \tag{5.249}$$
$$C_p = C T_p \tag{5.250}$$

These matrices and vectors are of the form

$$A_p = \begin{bmatrix} 0 & 1 & 0 & 0 & 0 \\ 0 & 0 & 1 & 0 & 0 \\ 0 & 0 & 0 & 1 & 0 \\ 0 & 0 & 0 & 0 & 1 \\ -m_1 & -m_2 & -m_3 & -m_4 & -m_5 \end{bmatrix} \tag{5.251}$$

$$B_p = \begin{bmatrix} 0 & 0 & 0 & 0 & 1 \end{bmatrix}^t \tag{5.252}$$
$$C_p = \begin{bmatrix} n_1 & n_2 & n_3 & n_4 & n_5 \end{bmatrix} \tag{5.253}$$

and the transfer function, by observation, is written as

$$\frac{y(s)}{u_j(s)} = \frac{n_1 + n_2 s + n_3 s^2 + n_4 s^3 + n_5 s^4}{m_1 + m_2 s + m_3 s^2 + m_4 s^3 + m_5 s^4 + s^5} + d_i \tag{5.254}$$

The problem lies in finding the transformation matrix, T_p. An algorithm to construct T_p is given below:

$$\left.\begin{aligned} T_p &= \begin{bmatrix} t_1 & t_2 & t_3 & t_4 & t_5 \end{bmatrix} \\ t_5 &= b_i \\ t_{5-k} &= A t_{5-k+1} + m_{5-k+1} b_i\,;\, k = 1, 2, 3, 4 \end{aligned}\right\} \tag{5.255}$$

where, $t_1, \ldots, t_5$ are the column vectors. The last equation needs the coefficients of the characteristic equation and is computed beforehand by using the Leverrier algorithm. The Leverrier algorithm is given in the following.

$$\left.\begin{aligned} m_5 &= -\text{trace}(A);\; H_5 = A + m_5 I \\ m_4 &= -\frac{1}{2}\text{trace}(AH_5);\; H_4 = AH_5 + m_4 I \\ &\;\vdots \\ m_1 &= -\frac{1}{5}\text{trace}(AH_2) \end{aligned}\right\} \tag{5.256}$$

where the trace of a matrix is equal to the sum of its diagonal elements. The frequency response is evaluated from equation (5.241) by substituting $s = j\omega$ wherever s occurs. The magnitude and phase plots can be drawn over the desired frequency range for the evaluation of control properties. A flowchart for the computation of transfer functions and of frequency and time responses is given in Figure 5.19. The computation of time responses is considered next.

5.10.2 Computation of Time Responses

If the state equation given in (5.213) is transformed into diagonal form by the transformation

$$X = T_d Z \qquad (5.257)$$

then the transformed equations are

$$\dot{Z} = T_d^{-1} A T_d Z + T_d^{-1} B_1 U = MZ + HU \qquad (5.258)$$

where M is a diagonal matrix with distinct eigenvalues and

$$M = T_d^{-1} A T_d \qquad (5.259)$$

$$H = T_d^{-1} B \qquad (5.260)$$

Solving for $z_1, \ldots, z_n$ from equation (5.260) gives

$$z_n(t) = \left(\sum_{j=1}^{m} H_{nj} u_j\right) \frac{(-1 + e^{\lambda_n t})}{\lambda_n} \qquad (5.261)$$

where λ_n is the nth eigenvalue and m is the number of inputs. Once vector Z is evaluated via equation (5.263), the output y is obtained from the following:

$$y = CX + DU = CT_d Z + DU \qquad (5.262)$$

Example 5.7

For the induction motor given in Example 5.4 and for the steady-state operating conditions given below, find the transfer function between the rotor speed and load torque and the poles of the system.

$$\omega_{so} = 377 \text{ rad/sec}$$
$$\omega_{ro} = 370 \text{ rad/sec}$$

Solution The steady-state current vector is obtained as follows: The impedance matrix in synchronous reference frame is

$$Z = \begin{bmatrix} R_s & L_s \omega_{so} & 0 & L_m \omega_{so} \\ -L_s \omega_{so} & R_s & -L_m \omega_{slo} & 0 \\ 0 & L_m \omega_{slo} & R_r & L_s \omega_{slo} \\ -L_m \omega_{slo} & 0 & -L_s \omega_{slo} & R_r \end{bmatrix}$$

Section 5.10   Evaluation of Control Characteristics of the Induction Machine   239

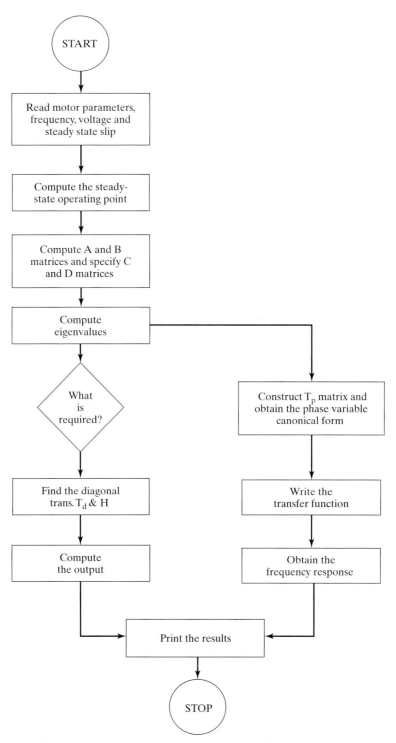

Figure 5.19   Flowchart for the computation of transfer functions and frequency responses

where

$$\omega_{so} = 377 \text{ rad/sec}$$

$$\omega_{slo} = \omega_{so} - \omega_{ro} = 37 - 370 = 7 \text{ rad/sec}$$

The q axis stator voltage in synchronous reference frame is

$$v_{qso}^e = 0$$

and

$$v_{dso}^e = \frac{\sqrt{2}}{\sqrt{3}} V = 163.3 \text{ V}$$

$$v = \begin{bmatrix} 0 \\ 163.3 \\ 0 \\ 0 \end{bmatrix}$$

$$I = [Z^{-1}] \cdot v = \begin{bmatrix} -9.48 \\ 15.15 \\ 1.90 \\ -15.44 \end{bmatrix}, A = \begin{bmatrix} I_{qso}^e \\ I_{dso}^e \\ I_{qro}^e \\ I_{dro}^e \end{bmatrix}$$

The steady-state torque is

$$T_{eo} = \frac{3}{2} \cdot \frac{P}{2} \cdot L_m (I_{qso}^e I_{dro}^e - I_{dso}^e I_{qro}^e)$$

$$= 18.98 \text{ N·m}$$

The next step is to compute the system matrices A and B_1 as follows :

$$A = P_1^{-1} Q$$

$$B_1 = P_1^{-1} R_1$$

where

$$P_1 = \begin{bmatrix} L_s & 0 & L_m & 0 & 0 \\ 0 & L_s & 0 & L_m & 0 \\ L_m & 0 & L_r & 0 & 0 \\ 0 & L_m & 0 & L_r & 0 \\ 0 & 0 & 0 & 0 & J \end{bmatrix}$$

$$R_1 = \begin{bmatrix} 1 & 0 & 0 & 0 & -C_1 & 0 \\ 0 & 1 & 0 & 0 & C_2 & 0 \\ 0 & 0 & 1 & 0 & -C_3 & 0 \\ 0 & 0 & 0 & 1 & C_4 & 0 \\ 0 & 0 & 0 & 0 & 0 & -\frac{P}{2} \end{bmatrix}$$

Section 5.10 Evaluation of Control Characteristics of the Induction Machine

$$Q_1 = \begin{bmatrix} -R_s & -L_s\omega_{so} & 0 & -L_m\omega_{so} & 0 \\ L_s\omega_{so} & -R_s & L_m\omega_{so} & 0 & 0 \\ 0 & -L_m\omega_{slo} & -R_r & -L_r\omega_{slo} & C_3 \\ -L_m\omega_{slo} & 0 & -L_r\omega_{slo} & -R_r & -C_4 \\ K_1 I^e_{dro} & -k_1 I^e_{qro} & -k_1 I^e_{dso} & k_1 I^e_{qso} & -B \end{bmatrix}$$

where

$$C_1 = L_s I^e_{dso} + L_m I^e_{dro} = 0.007$$
$$C_2 = L_s I^e_{qso} + L_m I^e_{qro} = -0.422$$
$$C_3 = L_m I^e_{dso} + L_r I^e_{dro} = -0.05$$
$$C_4 = L_m I^e_{qso} + L_r I^e_{qro} = -0.404$$
$$k_1 = \frac{3}{2}\left(\frac{P}{2}\right)^2 \cdot L_m = 0.323$$

The fifth column in the R_1 matrix is due to the $\delta\omega_s$ term. The output is incremental torque, which is written as the elements of C vector as follows:

$$C = \begin{bmatrix} k_2 I^e_{dro} & -k_2 I^e_{qro} & -k_2 I^e_{dso} & k_2 I^e_{qso} & 0 \end{bmatrix}$$

where

$$k_2 = \frac{3}{2} \cdot \frac{P}{2} \cdot L_m$$

Note D = 0.

The coefficients of the characteristic polynomial are obtained from Leverrier's algorithm by using the A matrix, as:

$$m_1 = 1.132 \times 10^{11}$$
$$m_2 = 2.706 \times 10^9$$
$$m_3 = 1.646 \times 10^7$$
$$m_4 = 1.75 \times 10^5$$
$$m_5 = 253.63$$

To compute the numerator polynomial, the column vector of the B_1 matrix is identified as corresponding to the input which is the sixth column for the load torque input:

$$b_1 = 6^{th} \text{ column of } B_1 = \begin{bmatrix} 0 \\ 0 \\ 0 \\ 0 \\ -119.8 \end{bmatrix}$$

By using A, m_1 to m_5, and b_1, the column vectors t_5 to t_1 are found via the algorithm

$$t_n = b_1 \quad \text{for } n = 5$$
$$t_{n-1} = At_n + m_n b_1 \quad \text{for } n = 5, 4, 3, 2$$

The transformation matrix T_p is

$$T_p = [t_1 \; t_2 \; t_3 \; t_4 \; t_5]$$

$$= \begin{bmatrix} -2.31 \times 10^{10} & 1.39 \times 10^8 & -2.91 \times 10^5 & -1.58 \times 10^3 & 0 \\ 9.25 \times 10^{10} & 1.88 \times 10^9 & 1.62 \times 10^6 & 1.29 \times 10^4 & 0 \\ 2.51 \times 10^{10} & -1.17 \times 10^8 & 3.07 \times 10^5 & 1.63 \times 10^3 & 0 \\ -9.48 \times 10^{10} & -1.93 \times 10^9 & -1.73 \times 10^6 & -1.32 \times 10^4 & 0 \\ -4.57 \times 10^{10} & -1.71 \times 10^9 & -1.90 \times 10^7 & -3.03 \times 10^4 & -119.8 \end{bmatrix}$$

and

$$A_p = T_p^{-1} \cdot A \cdot T_p = \begin{bmatrix} 0 & 1 & 0 & 0 & 0 \\ 0 & 0 & 1 & 0 & 0 \\ 0 & 0 & 0 & 1 & 0 \\ 0 & 0 & 0 & 0 & 1 \\ -1.13 \times 10^{11} & -2.7 \times 10^9 & -1.65 \times 10^7 & -1.75 \times 10^5 & -253.63 \end{bmatrix}$$

$$C_p = C \cdot T_p = [1.13 \times 10^{11} \;\; 2.32 \times 10^9 \;\; 2.12 \times 10^6 \;\; 1.62 \times 10^4 \;\; 0] = [n_1 \; n_2 \; n_3 \; n_4 \; n_5]$$

The transfer function then is

$$\frac{\delta T_e(s)}{\delta T_1(s)} = \frac{n_1 + n_2 s + n_3 s^2 + n_4 s^4 + n_5 s^5}{m_1 + m_2 s + m_3 s^2 + m_4 s^4 + m_5 s^4 + s^5}$$

Compute $B_p = T_p^{-1} \cdot b_1$ to check the calculations.

$$B_p = \begin{bmatrix} 0 \\ 0 \\ 0 \\ 0 \\ 1 \end{bmatrix}$$

proving the correctness of the solution.

5.11 SPACE-PHASOR MODEL

5.11.1 Principle

The stator and rotor flux-linkage phasors are the resultant stator and rotor flux linkages and are found by taking the vector sum of the respective d and q components of the flux linkages. Note that the flux-linkage phasor describes its spatial distribution. Instead of using two axes such as the d and q for a balanced polyphase machine, the flux-linkage phasors can be thought of as being produced by equivalent single-phase stator and rotor windings. Such a representation has many advantages: (i) the system of equations could be compact and be reduced from four to two; (ii) the system reduces to a two-winding system like the dc machine—hence the apparent similarity of them in control to obtain a decoupled independent flux and torque control as in the dc machine; (iii) a clear conceptualization of the dynamics of the machine—it is easier to visualize the interaction of two windings rather than four windings, resulting in an in-depth understanding of the dynamic process used in developing high-performance control strategies; (iv) a meaningful interpretation of the eigenvalues of the system, regardless of the reference frames considered in a study; and (v) easier analytical solution of dynamic transients of the key machine variables, involving only the solution of two differential equations with complex coefficients. Such an analytical solution improves the understanding of the machine behavior in terms of machine parameters, leading to the formulation of the machine design requirements for variable-speed applications. Such a space-vector model is derived in this section.

5.11.2 DQ Flux-Linkages Model Derivation

The induction-machine model, in terms of the flux linkages described in section 5.7.8, is made use of in this section. The equations can be cast in the normalized form by finding the currents in terms of the flux linkages from equation (5.138):

$$\left.\begin{array}{l} i_{qs}^c = \dfrac{1}{\Delta_1}(L_r \lambda_{qs}^c - L_m \lambda_{qr}^c) \\[4pt] i_{ds}^c = \dfrac{1}{\Delta_1}(L_r \lambda_{ds}^c - L_m \lambda_{dr}^c) \\[4pt] i_{qr}^c = \dfrac{1}{\Delta_1}(-L_m \lambda_{qs}^c + L_s \lambda_{qr}^c) \\[4pt] i_{dr}^c = \dfrac{1}{\Delta_1}(-L_m \lambda_{ds}^c + L_s \lambda_{dr}^c) \end{array}\right\} \quad (5.263)$$

$$\text{where } \Delta_1 = (L_s L_r - L_m^2) \quad (5.264)$$

Then, substituting for the currents in the flux-linkage equations given in (5.141), (5.142), (5.144), and (5.145), the model in arbitrary reference frames in normalized units is derived as

$$\frac{d}{d\tau}\begin{bmatrix}\lambda_{qsn}^c \\ \lambda_{dsn}^c \\ \lambda_{qm}^c \\ \lambda_{drn}^c\end{bmatrix} = \begin{bmatrix} -\dfrac{1}{\tau_s'} & -\omega_{cn} & \dfrac{k_r}{\tau_s'} & 0 \\[6pt] \omega_{cn} & -\dfrac{1}{\tau_s'} & 0 & \dfrac{k_r}{\tau_s'} \\[6pt] \dfrac{k_s}{\tau_r'} & 0 & -\dfrac{1}{\tau_r'} & (\omega_{rn}-\omega_{cn}) \\[6pt] 0 & \dfrac{k_s}{\tau_r'} & (\omega_{cn}-\omega_{rn}) & -\dfrac{1}{\tau_r'} \end{bmatrix}\begin{bmatrix}\lambda_{qsn}^c \\ \lambda_{dsn}^c \\ \lambda_{qm}^c \\ \lambda_{drn}^c\end{bmatrix} + \begin{bmatrix}1 & 0 \\ 0 & 1 \\ 0 & 0 \\ 0 & 0\end{bmatrix}\begin{bmatrix}v_{qsn}^c \\ v_{dsn}^c\end{bmatrix} \quad (5.265)$$

where $k_s = L_m/L_s =$ stator coupling factor, $k_r = L_m/L_r =$ rotor coupling factor, $\sigma = 1 - L_m^2/(L_s L_r) = 1 - k_s k_r =$ leakage coefficient, $\tau_s = L_s/R_s =$ stator time constant, $\tau_r = L_r/R_r =$ rotor time constant, $\tau_s' = \sigma \tau_s =$ transient stator time constant, $\tau_r' = \sigma \tau_r =$ transient rotor time constant, $\tau = \omega_b t =$ normalized time. To gain further insight, various time constants and normalized time are introduced into this

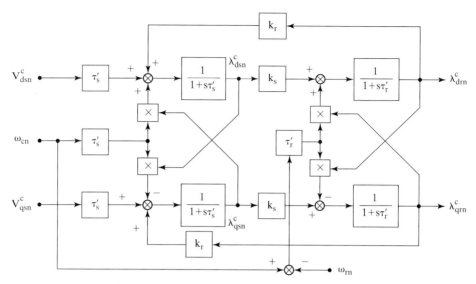

Figure 5.20 Signal-flow graph of the induction machine

model. The signal-flow graph of the system is shown in Figure 5.20. This does not give any additional advantage in analyzing the system and therefore is not considered any further.

The eigenvalues of the system as the rotor speed is increased from zero to 0.5 p.u. are shown in Figure 5.21 for a machine with normalized parameters of R_{sn} = 0.0446 p.u., R_{rn} = 0.054 p.u., L_{mn} = 2.89 p.u., L_{sn} = 3.005 p.u., and L_{rn} = 3.13 p.u and in stator reference frames. The root loci on the left represent the two rotor eigenvalues; the other represents the stator eigenvalues. Their distinction comes from the fact that the rotor eigenvalues increase in their imaginary values as the rotor speed increases. As the stator does not experience the physical rotation, its eigenfrequencies are more or less stationary. The stator and rotor eigenvalues are complex conjugates; thus, it is sufficient to consider one of each to obtain a full picture of the dynamic characteristics of the machine. Note that by combining the two rotor and two stator equations into a resultant rotor and stator equation, the order of the system is reduced from four to two, resulting in the root loci containing one of the rotor and stator eigenvalues. That is precisely what the space-phasor model attempts to capture, with the attendant advantages, as shown in the succeeding sections. The real values denoting the transient time constants associated with the stator and rotor are to be noted; that they approach close to each other in value at higher speeds is also seen clearly.

5.11.3 Root Loci of the DQ Axes Based Induction Machine Model

Figure 5.22 shows the loci of the eigenvalues when the reference frames are fixed arbitrarily at a speed of 0.5 p.u. instead of at zero reference frames speed as in the Figure 5.21. The eigenfrequencies (the imaginary part of the eigenvalues) are shifted

Section 5.11   Space Phasor Model   245

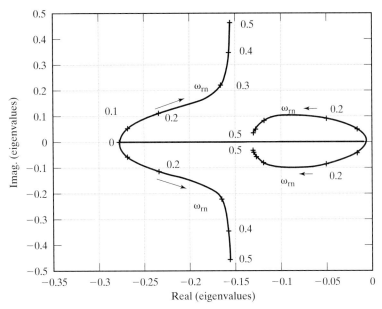

Figure 5.21 Loci of the eigenvalues in stator reference frames with rotor speed varying from 0 to 0.5 p.u.

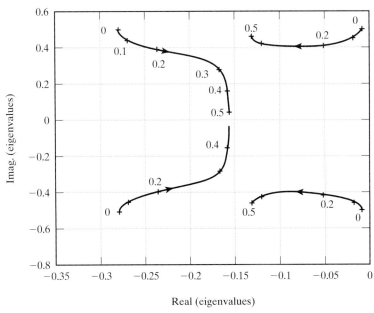

Figure 5.22 Loci of the eigenvalues in arbitrary reference frames with a velocity of 0.5 p.u. and rotor speed varying from 0 to 0.5 p.u.

by the reference-frames speed, i.e., 0.5 p.u. in both the stator and rotor root loci. The negative eigenfrequencies are added to the reference-frames speed, whereas the positive eigenfrequencies are subtracted from the reference-frames speed, as becomes clearly discernible in this figure. In effect, the root loci computed for one frame are convertible to another reference frame by simply adding and subtracting the reference-frames speed from their eigenfrequencies. This has great significance in that it demonstrates the interrelationship of the machine eigenvalues in various reference frames.

Consider that every eigenvalue contributes to a response term of exponential nature, with one, due to the real part, indicating the damping and the other, due to the eigenfrequency term (i.e., imaginary value of the root), in the form of a phase. The reference frames merely displace the eigenfrequencies so that the phase term of the response cancels the phase rotation added to the input variables, resulting in the elimination of the contribution from the reference-frames speed term in the responses, but by having intact the real parts regardless of the reference frames, the damping is preserved uniformly in the responses in all of the reference frames.

An alternative approach is to look at the induction machine with a set of two windings, one in the stator and the other on the rotor, using a space-phasor approach.

5.11.4 Space-Phasor Model Derivation

Let the space phasor of the stator flux linkages in arbitrary reference frames be given by

$$\lambda_{sn}^c = \lambda_{qsn}^c - j\lambda_{dsn}^c \tag{5.266}$$

This definition is valid for currents, voltages, and flux linkages in the machine for both the stator and rotor variables. Such a relationship can alternatively be expressed in complex phasor form as

$$\lambda_{sn}^c = e^{-j\omega_{cn}t}\left\{\frac{2}{3}(\lambda_{asn} + e^{j2\pi/3}\lambda_{bsn} + e^{j4\pi/3}\lambda_{csn})\right\} \tag{5.267}$$

where the first exponential term gives the reference-frame rotation at an arbitrary speed of ω_{cn}. Note that the operation in the parentheses denotes the resolution of the three phases from abc axes to the q and d axes, with the q axis aligned to phase A. The resolution from abc to qdo axes is achieved by projecting the abc axes variables onto the q and d axes through simple trigonometric relationships developed in section 5.7.3. Further, the expression inside the parentheses, on expansion, results in

$$\frac{2}{3}(\lambda_{asn} + e^{j2\pi/3}\lambda_{bsn} + e^{j4\pi/3}\lambda_{csn}) = \frac{2}{3}\left\{\lambda_{asn} - 0.5(\lambda_{bsn} + \lambda_{csn}) - j\frac{\sqrt{3}}{2}(\lambda_{csn} - \lambda_{bsn})\right\}$$

$$= \left\{\lambda_{asn} - j\frac{1}{\sqrt{3}}(\lambda_{csn} - \lambda_{bsn})\right\} = \lambda_{qsn} - j\lambda_{dsn} \tag{5.268}$$

The vector rotation at the reference-frame speed converts these stator reference-frame variables into the arbitrary reference-frame variables. Similarly, to recover the phase variables in stator reference frames from the arbitrary reference-frame variables, the procedure is reversed. First, the variables are separated into the real and imaginary values, which are then identified as the q and d axes variables in the arbitrary reference frames. Using the qdo-to-abc axes transformation T_{abc}^{-1} (developed in section 5.7.3), the variables in the stator reference frames are recovered. Alternately, it could also be derived compactly as

$$\begin{aligned}
\lambda_{asn} &= \text{Re}\left\{e^{j\omega_{cn}t}\lambda_{sn}^c\right\} \\
\lambda_{bsn} &= \text{Re}\left\{e^{j\omega_{cn}t}e^{j\frac{4\pi}{3}}\lambda_{sn}^c\right\} \\
\lambda_{csn} &= \text{Re}\left\{e^{j\omega_{cn}t}e^{j\frac{2\pi}{3}}\lambda_{sn}^c\right\}
\end{aligned} \quad (5.269)$$

Applying the space-phasor definition given in the above to the expressions given in the matrix equation (5.265), the following system of equations is derived in arbitrary reference frames:

$$\begin{aligned}
\frac{d\lambda_{sn}^c}{d\tau} + \left(\frac{1}{\tau_s'} + j\omega_{cn}\right)\lambda_{sn}^c &= \frac{k_r}{\tau_s'}\lambda_{rn}^c + v_{sn}^c \\
\frac{d\lambda_{rn}^c}{d\tau} + \left[\frac{1}{\tau_r'} + j(\omega_{cn} - \omega_{rn})\right]\lambda_{rn}^c &= \frac{k_s}{\tau_r'}\lambda_{sn}^c
\end{aligned} \quad (5.270)$$

where the voltage and current phasors are similarly defined. The signal-flow graph for these equations is shown in Figure 5.23. It is very compact and lends itself elegantly to a similarity with the dc machine, with which the engineers are much more familiar than they are with the multiphase ac machines. It consists of two time constants, corresponding to the armature and field of the dc machine, but the windings are coupled in the induction machine, unlike in the dc

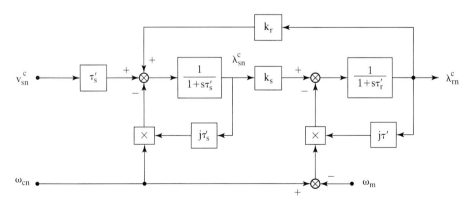

Figure 5.23 Signal-flow graph of the space-phasor-modeled induction machine

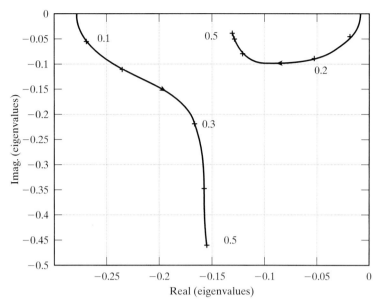

Figure 5.24 Loci of the eigenvalues in stator reference frames, with the rotor speed varied from 0 to 0.5 p.u. in a space-phasor model

machine. In the dc machine, a coupling exists only between the field and armature but not vice versa. The space-phasor modeling is uniquely suitable only for induction machines, because of its symmetry between the stator and the rotor. Such symmetry does not exist in the permanent-magnet synchronous machines, where the direct and quadrature axis inductances are not equal, because of the saliency of the rotor, and therefore the space-phasor model is not attempted in that case.

5.11.5 Root Loci of the Space-Phasor Induction-Machine Model

The root loci of the space-phasor-model-based induction machine for the same constants used in the previous figures are shown in the stator reference frames in Figure 5.24. Note that there exist only two eigenvalues in this system, and they correspond to one stator and rotor eigenvalue from the qd model with their complex conjugates removed. However, the elegance of this model comes into play in the compactness of its expression and the insight gained through it in comparison to the qd model. The eigenvalues for any arbitrary reference frames are obtained by adding the reference-frames speed to the eigenfrequencies of the roots of the stator reference frames, much like the qd model-based eigenvalues.

5.11.6 Expression for Electromagnetic Torque

The electromagnetic torque in S.I. units is found to be

$$T_e = \frac{3}{2} \frac{P}{2} \frac{L_m}{\Delta_1} \{\lambda_{qs}^c \lambda_{dr}^c - \lambda_{ds}^c \lambda_{qr}^c\} \tag{5.271}$$

and in normalized units is derived as

$$T_{en} = \frac{T_e}{T_b} = \frac{T_e}{\left(\dfrac{3}{2} \dfrac{P}{2} \dfrac{V_b I_b}{\omega_b}\right)} \tag{5.272}$$

Note that

$$V_b = \omega_b \lambda_b$$
$$\lambda_b = L_b I_b \tag{5.273}$$

where L_b is the base inductance. Substituting the above relationships into the torque expression results in

$$T_{en} = \frac{T_e}{\left(\dfrac{3}{2} \dfrac{P}{2} \dfrac{\lambda_b^2}{L_b}\right)} \tag{5.274}$$

which, on substitution of T_e, leads to the normalized torque as

$$T_{en} = \frac{L_{mn}}{(L_{sn} L_{rn} - L_{mn}^2)} \{\lambda_{qsn}^c \lambda_{drn}^c - \lambda_{dsn}^c \lambda_{qrn}^c\} \tag{5.275}$$

The normalized torque can be written compactly also in terms of the stator and rotor flux phasors as

$$T_{en} = \frac{L_{mn}}{(L_{sn} L_{rn} = L_{mn}^2)} \{\mathrm{Im}(\lambda_{sn}^c \overline{\lambda_{rn}^c})\} = \frac{L_{mn}}{\sigma L_{sn} L_{rn}} \{\mathrm{Im}(\lambda_{sn}^c \overline{\lambda_{rn}^c})\} \tag{5.276}$$

where $\overline{\lambda_{rn}^c}$ is the complex conjugate of the rotor flux linkage phasor and Im implies the imaginary component of the variable inside the parentheses. This lends itself to the interpretation that the stator and rotor flux linkage phasors are interacting to provide an electromagnetic torque, very much as the armature and field flux linkages in a dc machine interact to produce the electromagnetic torque. Similarly, the normalized torque could also be written as

$$T_{en} = (i_{qsn}^c \lambda_{dsn}^c - i_{dsn}^c \lambda_{qsn}^c) = \mathrm{Im}(i_{sn}^c \overline{\lambda_{sn}^c}) \tag{5.277}$$

This expression portrays the torque as the interaction of the stator current and stator flux linkage phasors; similarly, the following expression gives it in terms of the interaction between the stator current and rotor flux linkage phasor:

$$T_{en} = \frac{L_m}{L_r} \{\mathrm{Im}(i_{sn}^c \overline{\lambda_{rn}^c})\} \tag{5.278}$$

250 Chapter 5 Polyphase Induction Machines

Various torque expressions contribute to a variety of control schemes for induction motors, as demonstrated in Chapter 8 on vector-controlled induction-motor drives.

Example 5.8

The parameters of an induction motor are

2000 hp, 2300 V, 4 pole, 60 Hz, star-connected stator

$R_s = 0.02\ \Omega$; $R_r = 0.12\ \Omega$; $L_m = 0.1326$ H; $L_s = L_r = 0.1335$ H

$J = 10$ kg-m^2; $B = 0$; Load torque is zero.

Plot (i) the free-acceleration characteristics of the machine and (ii) the dynamic performance when the machine is started with linearly varying input voltages from 0.03 p.u. to 1 p.u. in one second. The space-phasor model of the machine in stator reference frames is recommended for solving this problem.

Solution (i) The stator reference-frames model of the induction machine in SI units derived in section 5.7.7.1 is considered here for deriving the space-phasor model. The four dq equations in (5.128) are reduced to two space-phasor equations by the following steps:

$$v_s = v_{qs} - jv_{ds} = (R_s + L_s p)(i_{qs} - ji_{ds}) + L_m p(i_{qr} - ji_{dr}) = (R_s + L_s p)i_s + L_m p i_r$$

Similarly, the rotor equations are combined as

$$v_r = v_{qr} - jv_{dr} = (R_r + L_r p - j\omega_r L_r)i_r + L_m(p - j\omega_r)i_s$$

Then these two equations are cast in state-space form as

$$\begin{bmatrix} pi_s \\ pi_r \end{bmatrix} = \frac{1}{(L_s L_r - L_m^2)} \left(\begin{bmatrix} (-L_r R_s + jL_m^2 \omega_r) & L_m(R_r - j\omega_r L_r) \\ L_m(R_s + j\omega_r L_s) & L_s(-R_r + j\omega_r L_r) \end{bmatrix} \begin{bmatrix} pi_s \\ pi_r \end{bmatrix} + \begin{bmatrix} L_r \\ -L_m \end{bmatrix} v_s \right)$$

The air gap torque is derived as

$$T_e = \frac{3}{2}\frac{P}{2} L_m(i_{qs}i_{dr} - i_{ds}i_{qr}) = \frac{3}{2}\frac{P}{2} L_m [\text{Imag}(i_s \bar{i}_r)]$$

The mechanical equation is given by

$$J\frac{d\omega_r}{dt} = \frac{P}{2} T_e$$

because the friction and load torques are zero for the given problem. The three differential equations are solved by numerical integration, and the phasor magnitudes of stator voltage, current and flux linkages and rotor flux linkages, air gap torque, and rotor speed are plotted in normalized units, as shown in Figure 5.25. The stator and rotor flux linkages phasors are derived from the stator and rotor currents as

$$\lambda_s = \lambda_{qs} - j\lambda_{ds} = (L_s i_{qs} + L_m i_{qr}) - j(L_s i_{ds} + L_m i_{dr}) = L_s i_s + L_m i_r$$

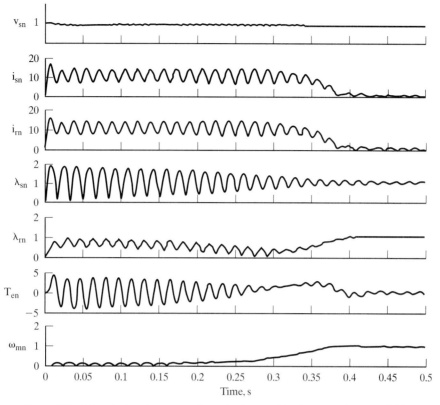

Figure 5.25 Line-start performance of an induction machine in space-phasor variables

Similarly, the rotor flux-linkages phasor is derived as

$$\lambda_r = \lambda_{qr} - j\lambda_{dr} = (L_r i_{qr} + L_m i_{qs}) - j(L_r i_{dr} + L_m i_{ds}) = L_r i_r + L_m i_s$$

Note that only phasor magnitudes are shown in the simulation results. The space-phasor simulation has reduced the total number of differential equations from 5 to 3, but the reduction comes at the expense of their being in complex variables. Even though there is no distinct advantage in simulating the system in complex variables, significant insight can be gained by viewing the phasors rather than dq components of the key variables.

(ii) Linearly varying the input voltages has limited the stator currents to 7 p.u., as shown in Figure 5.26. It has eliminated oscillations in the currents and flux linkages and hence has eliminated entirely the torque pulsations. The start-up time has increased by 75% as compared to the line start with full supply voltages. Even in this method, multiple times the rated resistive losses occur, and hence the starts per hour are restricted for large machines such as the one simulated here.

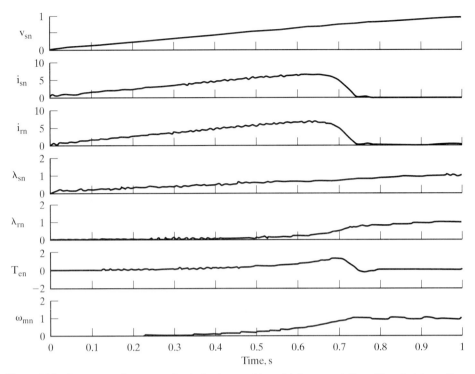

Figure 5.26 Start-up performance of an induction machine with linear variation of input-stator voltage phasor

5.11.7 Analytical Solution of Machine Dynamics

The space-vector approach not only provides compact and conceptual representation of the induction machine but also assists in the direct closed-form solution of the dynamic processes in the machine, neglecting mechanical dynamics. The solution of the stator and rotor flux-linkage phasors for a step input in the stator voltage phasor can be analytically found; it involves the solution of only two first-order differential equations with complex coefficients. This, in detail, is given below.

The roots of the system, p_1 and p_2, are found from the space-vector model derived from equation (5.272) as

$$\dot{x} = Ax + Bu$$

where

$$A = \begin{bmatrix} a_{11} & a_{12} \\ a_{21} & a_{22} \end{bmatrix}; B = \begin{bmatrix} 1 & 0 \\ 0 & 0 \end{bmatrix}; x = [\lambda_{sn}^c \quad \lambda_{rn}^c]; u = \begin{bmatrix} v_{sn}^c \\ 0 \end{bmatrix}$$

$$a_{11} = -\left(\frac{1}{\tau_s'} + j\omega_{cn}\right); a_{12} = \frac{k_r}{\tau_s'}; a_{21} = \frac{k_s}{\tau_r'}; a_{22} = -\left\{\frac{1}{\tau_r'} + j(\omega_{cn} - \omega_{rn})\right\}$$

(5.279)

The roots of the system are found from the characteristic equation, i.e., det $|A - pI| = 0$. The roots, p_1 and p_2, are evaluated as

$$p_1, p_2 = \frac{1}{2}\{a \pm \sqrt{a^2 - 4b}\}$$

where

$$a = a_{11} + a_{22}; \quad b = a_{11}a_{22} - a_{12}a_{21} \tag{5.280}$$

In taking the square root of the discriminant, which is a complex number, use has to be made of DeMoivre's theorem converted into exponential form. After having solved for the roots, and keeping the stator frequency and rotor angular speed constant, you can derive the stator and rotor flux linkage phasors for a unit step input of the stator voltage phasor of $-j1$ p.u. as

$$\lambda_{sn}^c(\tau) = \frac{1}{(p_2 - p_1)}\left[\left\{-(p_1 + a_{22})\left(\lambda_{sn}^c(0) + \frac{j}{p_1}\right)\right.\right.$$

$$\left.+ a_{12}\lambda_{rn}^c(0)\right\}e^{-p_1\tau} + \left\{(p_2 + a_{22})\left(\lambda_{sn}^c(0) + \frac{j}{p_2}\right) - a_{12}\lambda_{rn}^c(0)\right\}e^{-p_2\tau}\right] + j\frac{a_{22}}{p_1 p_2}$$

$$\lambda_{rn}^c(\tau) = \frac{1}{(p_2 - p_1)}\left[\left\{a_{21}\left(\frac{j}{p_1} + \lambda_{sn}^c(0)\right) - (p_1 + a_{11})\lambda_{rn}^c(0)\right\}e^{-p_1\tau} \tag{5.281}\right.$$

$$\left.+ \left\{-a_{21}\left(\frac{j}{p_2} + \lambda_{sn}^c(0)\right) + (p_2 + a_{11})\lambda_{rn}^c(0)\right\}e^{-p_2\tau}\right] - j\frac{a_{21}}{p_1 p_2}$$

where $\lambda_{sn}^c(0)$ and $\lambda_{rn}^c(0)$ are the initial values of the stator and rotor flux linkage phasors, respectively. Note that the normalized time is used here. The rotor and stator flux-linkage phasors are spatial distributions of the flux linkages and are available as a function of motor parameters and initial conditions. These analytical solutions are easily obtained compared to the *dq* axes model case. They help in understanding the magnetic-energy propagation within the machine during transients [14, 17].

5.11.8 Signal-Flow Graph of the Space-Phasor-Modeled Induction Motor

Based on the system of equations derived in the above, the signal-flow graph of the induction machine is drawn as shown in Figure 5.23. The symbol *s* denotes the Laplace operator in this figure. Addition of the mechanical load to the machine and load inertia, friction constant, and load torque will complete the block-diagram representation of the induction motor. In this case, the system involves the solution of a third-order system in complex variables. From this point onward, the space-phasor modeling does not provide an advantage over the *dq* model: numerical solution is resorted to in the solution of linearized systems of equations of third order in the case of space-phasor approach, as against the fifth order in the *dq* model, and the elegance of the meaningful analytical solution is lost in both cases.

Example 5.9

Consider a set of three-phase sinusoidal input voltages to an induction motor with a frequency of ω_s. Express the stator-voltage phasor in arbitrary reference frames, and, hence, find from it the stator-voltage phasors in the rotor and synchronous reference frames.

Solution The three-phase sinusoidal input voltages are defined to be

$$v_{as} = V_m \sin \theta_s; \quad v_{bs} = V_m \sin\left(\theta_s - \frac{2\pi}{3}\right); \quad v_{cs} = V_m \sin\left(\theta_s + \frac{2\pi}{3}\right)$$

The stator-voltage phasor in stator reference frames is given by

$$v_s = \frac{2}{3}(v_{as} + e^{j\frac{2\pi}{3}} v_{bs} + e^{j\frac{4\pi}{3}} v_{cs})$$

Substituting for the phase voltages from the definitions and expanding them with the exponential terms simplifies the stator-voltage phasor to

$$v_s = V_m (\sin \theta_s - j \cos \theta_s)$$

The stator-voltage phasor in the arbitrary reference frame is given by

$$v_s^c = v_s e^{-j\omega_c t} = v_s e^{-j\theta_c}$$

By substituting for the voltage phasor in stator reference frames into the above equation, the stator-voltage phasor in arbitrary reference frames is derived as

$$v_s^c = V_m (\sin \theta_s - \cos \theta_s) e^{-j\theta_c} = V_m \{-je^{j\theta_s}\} e^{-j\theta_c} = -jV_m e^{j(\theta_s - \theta_c)}$$

from which the stator-voltage phasors in rotor reference frames and synchronous frames are found by substituting $\theta_c = \theta_r$ and $\theta_c = \theta_s$, respectively, in the stator-voltage phasor expression in the arbitrary reference frames:

$$v_s^r = V_m \{\sin(\theta_s - \theta_r) - j \cos(\theta_s - \theta_r)\}$$
$$v_s^e = V_m \{0 - j1\}$$

5.12 CONTROL PRINCIPLE OF THE INDUCTION MOTOR

The control principle of the induction machine is derived in this section. A physical interpretation is given that indicates its close resemblance to the separately-excited dc machine. A transient-free operation of the induction machine is achieved only if the stator (or for that matter rotor) flux-linkages phasor is maintained constant in its magnitude and its phase is stationary with respect to the current phasor (or the current phasor is varied in such a way that it counterbalances the rate of change of the flux-linkages phasor). This fundamental theorem for the control of ac machines is proved in the following for induction machines. Such an understanding is crucial to developing appropriate control techniques and their

Section 5.12 Control Principle of the Induction Motor

implementations, including vector-controlled drives of various kinds, with and without position and speed sensors.

Consider the induction-machine model in the synchronous reference frame. In this frame, the sinusoidal variables become dc quantities, i.e., constants, and hence their derivatives, such as those of the currents and flux linkages in steady state, are all zero. This step helps in visualizing the steady state much more simply than the models in other reference frames. The phasor is defined to be the resultant of the respective q and d axes variables such as currents, voltages, and flux linkages. The stator current phasor and stator flux-linkages phasor are defined as

$$i_s = i^e_{qs} - ji^e_{ds}$$
$$\lambda_s = \lambda^e_{qs} - j\lambda^e_{ds} \qquad (5.282)$$

The input power for balanced supply voltages is given by

$$p_i = \frac{3}{2}(v^e_{qs}i^e_{qs} + v^e_{ds}i^e_{ds}) \qquad (5.283)$$

Substituting for the voltages in terms of the machine parameters, currents, and stator angular frequency, the input power is obtained as

$$p_i = \frac{3}{2}(R_s[(i^e_{qs})^2 + (i^e_{ds})^2] + \omega_s L_m[i^e_{qs}i^e_{dr} - i^e_{ds}i^e_{qr}] + i^e_{qs}[L_s p i^e_{qs} + L_m p i^e_{qr}] + i^e_{ds}[L_s p i^e_{ds} + L_m p i^e_{dr}]) \quad (5.284)$$

where p is the differential operator and ω_s is the stator frequency at which the reference frames are rotating. Recognizing the electromagnetic torque as

$$T_e = \frac{3}{2}\frac{P}{2}L_m(i^e_{qs}i^e_{dr} - i^e_{ds}i^e_{qr}) = \frac{3}{2}\frac{P}{2}(\lambda^e_{ds}i^e_{qs} - \lambda^e_{qs}i^e_{ds}) \qquad (5.285)$$

where the stator d and q axes flux linkages are given by

$$\lambda^e_{qs} = L_s i^e_{qs} + L_m i^e_{qr}$$
$$\lambda^e_{ds} = L_s i^e_{ds} + L_m i^e_{dr} \qquad (5.286)$$

gives the power input in terms of the torque and flux linkages and currents, by substituting equations (5.285) and (5.286) into equation (5.284), as

$$p_i = \frac{3}{2}R_s[(i^e_{qs})^2 + (i^e_{ds})^2] + \frac{2}{P}\omega_s T_e + \frac{3}{2}[i^e_{qs}p\lambda^e_{qs} + i^e_{ds}p\lambda^e_{ds}] \qquad (5.287)$$

By substituting the phasor in the place of the individual q and d components of the currents and stator flux linkages, and by substituting for the electromagnetic torque in terms of the stator current and stator flux-linkages phasors, the input power is compactly written as

$$p_i = \frac{3}{2}R_s i_s^2 + \frac{2}{P}\omega_s \operatorname{Im}[i_s \overline{\lambda^e_s}] + \frac{3}{2}\operatorname{Re}[i_s p\overline{\lambda^e_s}] \qquad (5.288)$$

where Re and Im are the real and imaginary part of the functions, respectively, and $\bar{\lambda}_s$ is the complex conjugate of the stator flux-linkages phasor. There are three distinct components of the input power, as is seen from equation (5.288). The first term corresponds to the stator resistance losses; the remaining two terms constitute the power crossing the stator and entering into the air gap. The second term corresponds to the sum of the rotor slip power and mechanical power, because the stator frequency is the sum of the slip and rotational frequencies. The latter part of the second term supplies shaft power and friction and windage losses in the machine. The slip power corresponds to the rotor resistance losses. Note that the electromagnetic torque is the product of the stator current and flux-linkages phasors, as in the case of the dc machine, where it is equal to the product of the field flux linkages and armature current. This expression makes the direct equivalence between the dc and induction machines complete. The understanding that came out of this is that the induction machines could also have equivalent control, i.e., decoupling control, like that of the dc machines. The only difference in the induction machine is that the variables are phasors, unlike in the dc machines, where they are scalars. In the case of the induction machine, the flux linkages have to be controlled through the stator-current phasor only, which also controls the electromagnetic torque. Thereby, only one variable, the current phasor alone, controls both torque and flux-producing channel currents in the machine, resulting in an inherent coupling of these channels. By knowing the instantaneous position of the stator flux linkages and current phasors, they can be controlled to occupy any position, with the desired magnitudes resulting in a class of control schemes known as vector control.

The third term in the input-power expression denotes the rate of change of magnetic energy. In steady state, this term is zero: the stator flux-linkages phasor is a constant, hence its derivative is zero; but this term need not be zero during transients. Only if it is made zero at all times does the power drawn by the machine go only to provide for the stator resistance losses and synchronous power to meet the load demands and rotor copper losses. This results in minimum exchange of power between the power supply and the induction machine, enabling the transient free operation of the induction machine even during dynamic load or speed changes. Such an operation is desirable when the induction machine is supplied from a finite source, such as an inverter, whose capacity is very limited. Then the key to the control of the induction machine is that this term has to be identically zero at all times of its operation. This term can be made zero only under one of the following conditions:

(i) The rate of change of the stator flux-linkages phasor is zero. This is possible only when the stator flux-linkages phasor remains constant in magnitude when seen from the synchronous reference frames. Also, for the derivative to be zero, the phase of the stator flux linkages is constant in synchronous reference frames. In stator reference frames, the phase of the stator flux-linkages phasor has to be synchronous with the current phasor, i.e., the differential velocity between them has to be zero. Therefore, the phase angle between the stator current phasor and stator flux-linkages phasor, known as torque angle, has to be a constant when viewed from both the stator and synchronous reference frames. Note that keeping the derivative of the stator-flux linkages phasor zero is not always possible; sometimes, in the case of flux weakening to increase the range of speed, it may be intentionally varied from its rated value. Then the change of the stator flux-linkages phasor is not zero, leading to a transient operation in the machine.

(ii) The real part of the product of the stator current phasor and derivative of the stator flux-linkages phasor is zero. This condition is required to be satisfied for transient-free operation during intentional flux-linkages variation in the machine. This is achieved by adjusting the current phasor in such a way that the resulting flux-linkages variation is obtained while the real part of the product of the stator current and derivative of the flux-linkages phasors becomes identically zero.

(iii) The stator current phasor is zero; this is a trivial condition and hence is neglected in further consideration.

Note that, to achieve condition (i) or (ii), the stator flux-linkages phasor has to be known in both its magnitude and its position, along with the current phasor. From this information, the relative position between them can be used to control both the electromagnetic torque and stator flux linkages, resulting in a transient-free operation of the induction machine considered in detail in Chapter 8.

5.13 REFERENCES

1. J. F. Lindsay and MH. Rashid, Electromechanics and Electrical Machinery, Prentice-Hall, Inc., Englewood Cliffs, NJ, 1986.
2. M.G. Say, The Performance and Design of Alternating Current Machines, CBS Publishers and Distributors, Delhi, India, Third Edition, 1983.
3. A. F. Puchstein, T.C. Lloyd, and A.G. Conrad, Alternating-Current Machines, John Wiley & Sons, Inc., NY, May 1960.
4. M. Liwschitz-Garik and C. C. Whipple, Electric Machinery: Volume II, A-C Machines, D. Van Nostrand Co., Inc., NY, Second Printing, 1947.
5. G. R. Slemon and A. Straughen, *Electric Machines*, Addison-Wesley Publishing Co., 1980.
6. Mulukutla S. Sarma, Electric Machines, West Publishing Co., St. Paul, MN, Second Edition, 1994.
7. N.N. Hancock, Matrix Analysis of Electrical Machinery, Pergamon Press, The Macmillan Company, 1964.
8. Paul C. Krause, *Analysis of Electrical Machinery*, McGraw-Hill Book Company, 1986.
9. T. A. Lipo and A. B. Plunkett, "A new approach to induction motor transfer functions," Conf. Record, IEEE–IAS Annual Meeting, pp. 1410–1418, Oct. 1973.
10. R. Krishnan, J. F. Lindsay, and V. R. Stefanovic, "Control characteristics of inverter-fed induction motor," IEEE Trans. on Industry Applications, vol. IA-19, no. 1, pp. 94–104, Jan./Feb. 1983.
11. Myung J. Youn and Richard G. Hoft, "Variable frequency induction motor bode diagrams," IEEE-IAS, ISPC Conf. Record, pp. 124–136, 1977.

12. R. H. Nelson, T. A. Lipo, and P.C. Krause, "Stability analysis of a symmetrical induction machine," IEEE Trans. on Power Apparatus Systems, vol. PAS-88, no. 11, pp. 1710–1717, Nov. 1969.
13. V. R. Stefanovic and T. H. Barton, "The speed torque transfer function of electric drives," IEEE Trans. on Industry Applications, vol. IA-13, no. 5, pp. 428–436, Sept./Oct. 1977.
14. J. Holtz, "On the spatial propagation of transient magnetic fields in ac machines," *IEEE Trans. on Industry Applications,* vol. 32, no. 4, pp. 927–37, July/Aug. 1996.
15. P. K. Kovacs and E. Racz, Transient Phenomena in Electrical Machines, Elsevier Science Publishers, Amsterdam, 1984.
16. D. W. Novotny and J. H. Wouterse, "Induction machine transfer functions and dynamic response by means of complex time variables," IEEE Trans. on Power Apparatus and Systems, vol. PAS-95, no. 4, pp. 1325–1333, July/August 1976.
17. J. Holtz, "The representation of ac machine dynamics by complex signal flow graphs," IEEE Trans. on Industrial Electronics, vol. 42, no. 3, pp. 263–271, June 1995.
18. Les Manz, "Motor insulation system quality for IGBT drives," IEEE Industry Applications Magazine, vol. 3, no. 1, pp. 51–55, Jan./Feb. 1997.
19. W. R. Finley and R. R. Burke, "Proper specification and installation of induction motors," Proceedings of Industry Applications Society 42nd Annual Petroleum and Chemical Industry Conference, pp. 285–97, Denver, CO, Sept. 1995.
20. NEMA MG1-1993, Motors and Generators.

5.14 DISCUSSION QUESTIONS

1. Is it possible to operate the induction machine as a generator? What is the polarity of slip in that mode of operation?
2. Cite instances in which the machine may be used to regenerate.
3. The steady-state model of the induction machine is generally assumed to have constant parameters. It is known that all those parameters vary with operating conditions. Discuss the method by which parameter variations can be considered in the computation of steady-state performance.
4. Besides core and copper losses, what other losses need to be considered in the performance evaluation of the induction motors?
5. Prove that the relative speed between stator and rotor mmfs is zero. What is the primary consequence if their speeds are unequal?
6. Discuss the shortcomings of the induction-motor dynamic model from the viewpoint of losses.
7. In this computer age, why not use the actual motor model with time-varying inductances instead of resorting to d and q axes models?
8. Discuss the suitability of stationary reference frames for the induction motor in power-system studies.
9. Can the d and q models be used for supply-voltage-unbalance studies?
10. If the input voltages/currents are rectangular or trapezoidal in wave shapes, explain how the d and q models are used to predict the steady-state behavior. (Hint: Any periodic waveform can be resolved into a fundamental and a number of harmonics with sinusoidal wave shapes.)

11. Will the type (Y or Δ) of the stator connection change the dynamic models that have been derived in this chapter?
12. Will the parameter sensitivity affect the stability of the induction motor? (Hint: Reference 11.)
13. Which is the most important transfer function for the induction motor? Why? (Hint: Reference 13.)
14. The induction motor responds differently to various inputs during dynamic situations. Discuss the merits of one input over another and how to quantify their effects. (Hint: Reference 10.)
15. The machine parameters could have wide variations due to continued operation, with resultant increase in temperature due to losses, changes in input voltages that vary the magnetic saturation, and frequency variation. Discuss the significance and magnitude of variation in parameters due to these changes.
16. The modeling does not account for stray losses. These losses are difficult to predict and usually vary from 0.5 to 1% of the input power. Discuss ways of including the stray losses in performance calculations and their significance in prediction.
17. The friction and windage losses are not included in the steady-state equivalent circuit. Could they be included in the form of resistive losses similar to the core losses?
18. How could magnetic saturation be included in the steady-state and transient modeling of induction motors?
19. Is it possible to operate the induction motor in the statically unstable region?
20. During free acceleration, the electromagnetic torque has large oscillations. What are the consequences of such oscillations?
21. Is it possible to eliminate or mitigate these oscillations with a constant-frequency ac supply?
22. With the availability of variable-voltage and variable-frequency sources, could they be used for the damping of the oscillations?

5.15 EXERCISE PROBLEMS

1. The slip speed at breakdown can be approximately represented in terms of the rotor and stator parameters. Derive this expression.
2. An induction motor has the following ratings and parameters:

    40 hp, 460 V, 4 pole, 3 phase, 60 Hz, Y connected

    $R_s = 0.22\ \Omega,\ R_r = 0.209\ \Omega,\ L_m = 0.04$ H

    $L_s = 0.0425$ H, $L_r = 0.043$ H, B = 0

    Load torque = $T_l = 0$, J = 0.124 kg − m^2, a = 2

    The statically stable slip region is required to be doubled. That can be achieved by connecting external resistors in the rotor phases. Calculate approximately the value of external rotor resistance/phase to be added.
3. Neglecting motor copper losses and core losses, prove that the efficiency of the induction motor is equal to $(1 - s)$. Discuss the implications of this derivation over the variable-speed operation of the induction motor when the applied motor voltages are varied. (More will be given in Chapter 6 on this.)

4. Capacitors are connected to the stator windings of the induction motor. Derive an expression to quantify the improvement in power factor due to the capacitors.

5. The starting torque can be increased by increasing rotor resistances with the insertion of external resistances in slip-ring induction rotor motors. Find the external resistor per phase to be introduced, in terms of the stator and rotor resistances and total leakage inductances, to double the starting torque. Consider an equivalent circuit, neglecting the magnetizing branch.

6. Consider a current source at 60 Hz supplying the induction motor given in problem 2 with rated stator currents. Draw its torque-vs.-slip characteristics and compare them with the characteristics obtained by using a voltage source at rated voltages and 60 Hz. Comment on the salient features. (More will be given in Chapter 7 on this.)

7. A NEMA D motor with the following parameters is chosen to drive a pump. The load–torque characteristic of the pump is modeled as simply proportional to the square of the rotor speed and at rated speed requires rated torque for its operation. Its speed has to be varied from rated value to the lowest value, 0.7 p.u. Compute the efficiencies at 0.7 p.u. speed when the speed is varied by using rotor resistance control and alternatively by varying the stator input voltages. The parameters and ratings of the motors are

$$100 \text{ hp}, 460 \text{ V}, 4 \text{ poles}, 3 \text{ ph}, 60 \text{ Hz, star connected}, 1050 \text{ rpm}, R_s = 0.05 \text{ }\Omega,$$
$$R_r = 0.2885 \text{ }\Omega, L_m = 0.01856 \text{ H}, L_r = L_s = 0.01936 \text{ H}.$$

8. Consider the motor given in Example 5.6, running in steady state with rated torque and supplied from a 200 V, 60 Hz, 3-phase ac main. It is desired to stop the motor quickly by cutting off supply to the stator windings and by shorting them. This is known as plugging. Analyze this operation, using dynamic simulation in the stator reference frames.

9. An induction motor is operating at rated values in steady state. The ac power supply is shut off for 0.1 second and reinstated at the end of 0.1 second. Analyze this situation from dynamic simulation results. The parameters and ratings of the motors are

$$1.5 \text{ MW}, 1800 \text{ V(line to line)}, 566 \text{ A}, 1480 \text{ rpm}, 50 \text{ Hz, star connected}, R_s = 0.01934 \text{ }\Omega,$$
$$R_r = 0.01343 \text{ }\Omega, L_m = 0.0305 \text{ H}, L_r = L_s = 0.0315 \text{ H}, B = 0.02 \text{ N·m/rad/sec}, J = 30 \text{ kg-m}^2.$$

(Hint: This is known as reclosing transients in the induction motor. It is instructive to look at the rotor fluxes during the off-time and the induced emfs in the stator windings to appreciate the phase shift between the incoming supply voltages and induced emfs.)

10. Assume an induction motor is running in steady state with rated input voltages and full-load torque. Analyze the situation when two of the three phases are interchanged by

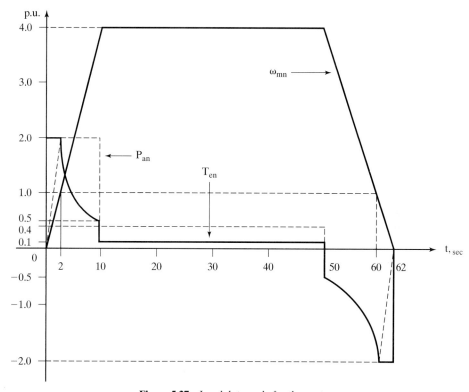

Figure 5.27 Load duty cycle for the motor

using dynamic simulation. This results in change of phase sequence to the machine. Comment on the results. The induction motor given in Example 5.6 can be used for the simulation.

11. An induction motor is desired for an application with the load duty cycle requiring the speed and air gap power shown in Figure 5.27. The absolute air gap power is maintained at 2 p.u. whenever the speed is above 1 p.u., and the load torque is only 0.1 p.u. at 4 p.u. speed. Calculate the nominal torque of the induction motor to be chosen for this application. (Hint: The rms value of the electromagnetic torque is the nominal torque.)

CHAPTER 6

Phase-Controlled Induction-Motor Drives

6.1 INTRODUCTION

The speed of the induction motor in terms of slip, frequency, and pole numbers is written as

$$\omega_m = \frac{\omega_r}{P/2} = \frac{2}{P} \cdot \omega_s (1-s) = \frac{4\pi}{P} \cdot f_s (1-s) \qquad (6.1)$$

The rotor speed of the induction motor can be varied by changing the number of poles, the slip, or the supply frequency. Pole-amplitude-modulated motors operate by changing the pole numbers, which in turn, is achieved with a change of winding connections. They require relays and circuit breakers to change winding connections. They provide a very limited but stepped (2 or 3) form of speed control, because the supply frequency is constant. They are not prevalent in practice nowadays.

The *slip* speed control is effected through the variation of applied voltage or insertion of external resistors in rotor or stator. They take different forms in implementation. As the maximum efficiency is $(1-s)$, efficiency is poor at low speeds with slip-controlled drive systems. In this chapter, slip control with stator voltage control and slip energy-recovery control are considered for study. These drives are studied with special emphasis on the converters used, control strategy, steady-state performance analysis, harmonics and their impact on the performance and rating of the motors and converters for various applications, starting provisions, feedback control, and modern developments. Speed control with frequency variation is dealt with in Chapters 7 and 8.

6.2 STATOR VOLTAGE CONTROL

6.2.1 Power Circuit and Gating

By having series-connected power switches in the induction motor, the instant of voltage application can be delayed. This is accomplished by controlling the gating / base drive signals to the power switches. A power-circuit configuration of stator-phase control methods is shown in Figure 6.1. The power switches can be SCRs, triacs, power transistors, or GTOs. The power switches are numbered to reflect the sequence of their gating control. This is very similar to the control of a three-phase controlled-bridge converter discussed in Chapter 3.

The gating signals are synchronized to the phase voltages, and they can be delayed up to a maximum of 180 degrees. The angle of delay, α, is termed the triggering angle. The gating signals are spaced at 60 degrees interval from each other. For sustained conduction, the gating signals are more than 60 degrees wide so that two power devices in two different phases conduct at a given time for current flow. The load is inductive, and the gating signals need to be maintained for the entire conduction time to ensure continued conduction (or until the current is higher than the latching current of the SCR).

6.2.2 Reversible Controller

For the phase controller shown in Figure 6.1, the power can flow only *from* the 3-phase supply *to* the machine, and it can run in only one direction for only one possible phase sequence of the output voltage. Thus, it has only one-quadrant torque–speed

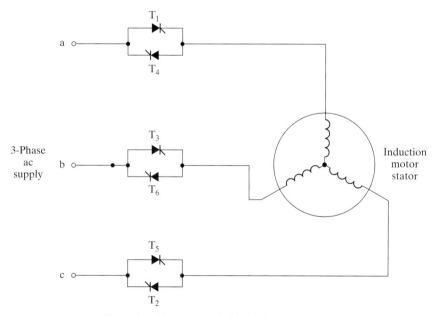

Figure 6.1 Phase-controlled induction-motor drive

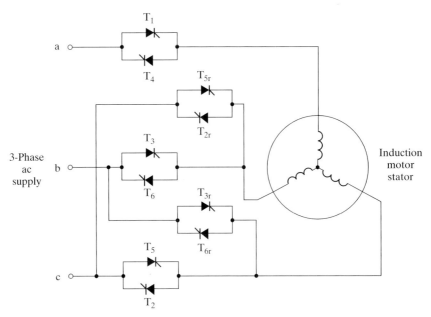

Figure 6.2 Reversible phase-controlled induction-motor drive

performance. To make this motor run in the other direction, the phase sequence of the input supply voltage is reversed. Two terminals of the stator windings need to be interchanged for phase-sequence reversal. It is accomplished with the addition of two pairs of antiparallel power switches, as shown in Figure 6.2. These power devices are denoted by an additional subscript, r. The sequence of gating for one direction of rotation (say, clockwise) is $T_1T_2T_3T_4T_5T_6$; the sequence of gating for counterclockwise rotation is $T_1T_{2r}T_{3r}T_4T_{5r}T_{6r}$. During the counterclockwise rotation, note that the devices T_2, T_3, T_5, and T_6 are not gated and hence remain open. By gating their counterparts, the phase sequence has been changed, to change the rotational direction of the motor.

The reversible phase-controlled induction-motor drive operates in the first and third quadrants of the torque–speed characteristics. It cannot operate in the II and IV quadrants and avoids regeneration. In avoiding regeneration, this motor drive becomes slow in going from one speed in one direction to another speed in the opposite direction. This is explained by using Figure 6.3.

Considering the present operating point, P_1 (ω_{m1}, T_{e1}), and the required point to be reached, $P_2(-\omega_{m2}, -T_{e2})$, the motor has to be slowed down to zero speed if stator currents are not allowed to exceed the rated values. The triggering angle is delayed so as to produce zero torque, and then the load slows down the rotor. This is shown in the diagram from P_1 to A and then to 0, respectively. At zero speed, the phase sequence is changed and the triggering angle is retarded until it can produce currents to generate the required torque trajectory, $-T_{e2}$, shown from 0 to B. Maintaining the torque at $-T_{e2}$ accelerates the rotor and load to P_2, shown as a line from B to P_2. It is assumed that T_{e1} and $-T_{e2}$ are the maximum allowable torques in their respective directions of rotation.

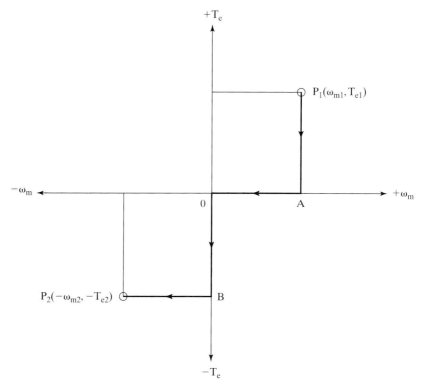

Figure 6.3 Quadrant change in the reversible phase-controlled induction-motor drive

If faster speed reversal of the motor is required, then the phase sequence of the supply is changed at P_1 instead of at 0. Then the induction motor is running opposite to the rotating magnetic field. This results in a slip greater than one, keeping the motor operation in the braking region. This would slow down the motor much faster than the load alone, but it is necessary to maintain the stator currents within safe levels, which is made possible by advancing the triggering angle. Even then, there will be excessive currents at the beginning of the phase sequence change, and they will last for a very short time. The current loop will force the triggering angle to advance to an appropriate value to keep the current below the safe level.

6.2.3 Steady-State Analysis

The torque–speed characteristics of the phase-controlled induction motor are required to evaluate its suitability for a load. The only control input is the triggering angle, α. Consider simplified but conceptual voltage and current waveforms for a phase shown in Figure 6.4. When there is no current in the stator, for instance between $(\alpha + \beta)$ and $(\pi + \alpha)$, the induced emf in the winding is reflected onto the phase terminal voltage. This induced emf is due to the air gap flux linkage, which is the resultant of stator and

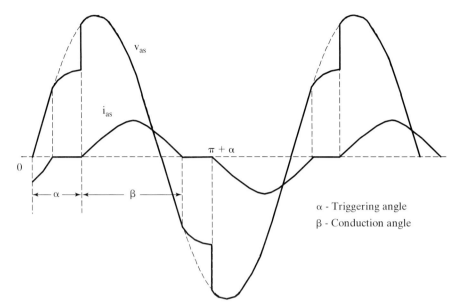

Figure 6.4 Phase voltage and current waveforms

rotor flux linkages. Due to the inductance in the rotor winding, even when stator current goes to zero, the rotor current continues to flow but decays exponentially. That rotor current contributes to the rotor and hence air gap flux linkages, resulting in the induced emf. The induced emf due to the air gap field will be sinusoidal, because the stator winding is sinusoidally distributed. Note that this reflected induced emf will be lower than the applied voltage. For a specified triggering angle, the determination of the conduction angle β is cumbersome, as it is an implicit function of the above parameters. A number of methods have been developed to compute β (and hence the steady-state currents and torque) by using state-variable techniques. They require the simulation of the motor in d and q frames and use of a computer, thus excluding a simple solution [1, 2, 3, 4]. Some simple analytical methods have come to the fore to give approximate results [2, 5, 6]. For most of the design and application studies, the accuracy of these methods is adequate. The analytical methods use the single-phase steady-state equivalent circuit of the induction motor. The performance of the induction motor with phase control is considered by one of the following methods:

(i) The applied voltage is resolved into Fourier components and their individual current responses are added to compute the resultant current [5].
(ii) The conduction angle is calculated by using the machine-equivalent circuit, and then the current is calculated [2].
(iii) A closed-form solution approach [5] is used.

Method (ii) is considered here because of its simplicity; even though method (iii) is elegant, it requires significant computational effort. Method (i) requires consideration of a large number of harmonic equivalent circuits, with higher computational burden.

6.2.4 Approximate Analysis

6.2.4.1 Motor model and conduction angle. The induction motor is modeled as an equivalent resistance, R_{im}, in series with an equivalent reactance, X_{im}, by using the equivalent circuit. These equivalent parameters are written in terms of the motor parameters and slip as

$$R_{im} = R_s + \frac{X_m^2}{\left(\frac{R_r}{s}\right)^2 + (X_m + X_{lr})^2} \cdot \frac{R_r}{s} \quad (6.2)$$

$$X_{im} = X_{1s} + \frac{\left(\frac{R_r}{s}\right)^2 + X_{lr}(X_m + X_{lr})}{\left(\frac{R_r}{s}\right)^2 + (X_m + X_{lr})^2} \cdot X_m \quad (6.3)$$

The voltage applied to this equivalent circuit consisting of R_{im} and X_{im} is

$$v(t) = V_m \sin(\omega_s t + \alpha) \quad (6.4)$$

as shown in Figure 6.4, where α is the triggering delay angle.

The stator current is then

$$i_{as}(t) = \frac{V_m}{Z_{im}}[\sin(\omega_s t + \alpha - \phi) - \sin(\alpha - \phi)e^{-\frac{\omega_s t}{\tan\phi}}], \text{ for } 0 \le \omega_s t \le \beta \quad (6.5)$$

where the induction-machine equivalent impedance and power-factor angle are

$$Z_{im} = \sqrt{R_{im}^2 + X_{im}^2} \quad (6.6)$$

$$\phi = \tan^{-1}\left(\frac{X_{im}}{R_{im}}\right) \quad (6.7)$$

The conduction angle, β, is obtained from equation (6.5) when the current becomes zero. Such a condition is denoted by

$$i_{as}\left(\frac{\beta}{\omega_s}\right) = 0 \quad (6.8)$$

and hence,

$$\sin(\beta + \alpha - \phi) - \sin(\alpha - \phi)e^{-\left(\frac{\beta}{\tan\phi}\right)} = 0 \quad (6.9)$$

The flowchart is given in Figure 6.5 to solve the equation (6.9), using the Newton–Raphson technique. When the triggering angle is less than the power-factor angle, the current will conduct for the positive half-cycle, say from α to $(\pi + \phi)$. When the negative half-cycle is applied at $(\pi + \alpha)$, and as $\alpha < \phi$, still the positive current is flowing. Note that the voltage applied to the machine is negative. As the current goes to zero, the negative half-cycle has already been activated by triggering at $(\pi + \alpha)$. Therefore, the machine receives full source voltage, and hence the current-

268 Chapter 6 Phase-Controlled Induction-Motor Drives

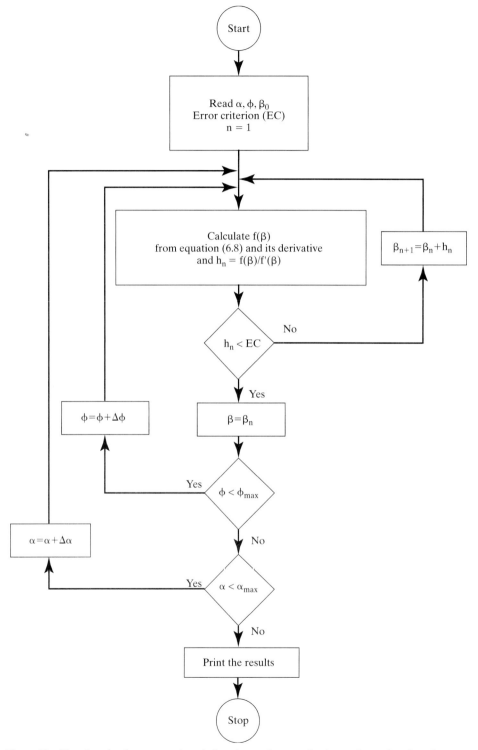

Figure 6.5 Flowchart for the computation of triggering angle vs. conduction angle as a function of power-factor angle

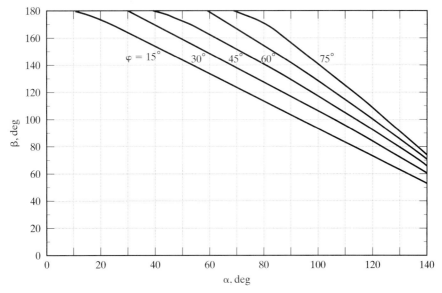

Figure 6.6 Conduction angle vs. triggering angle for various power-factor angles

conduction angle is 180 degrees. For the case when $\alpha < \phi$, the condition that $\beta = 180$ degrees has to be included in the solution of equation (6.9).

The solution of the transcendental equation (6.9) gives the value of β, and the instantaneous current $i_{as}(t)$ is evaluated from the equation (6.5). The relationship between the conduction angle and triggering angle for power-factor angles of 15°, 30°, 45°, 60°, and 75° is shown in Figure 6.6. Table 6.1 contains the relationship between ϕ and β for various α, for easy reference.

The stator voltage has to be resolved into its Fourier components, from which the fundamental current and torque developed in the motor are computed. The resolution of the stator current is derived in normalized form so that the results can be generalized.

Let the normalizing or base variables be

$$V_b = \frac{V_m}{\sqrt{2}} = \text{Rated rms phase voltage of the motor} \qquad (6.10)$$

where V_m is the peak-phase voltage

$$I_b = \text{Rated rms current of the machine} \qquad (6.11)$$

$$Z_b = \frac{V_b}{I_b} \qquad (6.12)$$

6.2.4.2 Fourier resolution of voltage. The fundamental of the input-phase voltage is derived to be

$$v_{a1} = \sqrt{(a_1^2 + b_1^2)} \sin(\omega_s t + \theta_1) \qquad (6.13)$$

TABLE 6.1 φ vs. β for various triggering angles

α, deg	10	20	30	40	50	60	70	80	90	100	110	120	130	140
φ, deg							β, deg							
9	179	169.00	159.00	149.00	139.00	129.00	119.00	109.00	99.00	89.00	78.99	68.97	58.93	48.80
12	180	172.00	162.00	152.00	142.00	132.00	122.00	111.99	101.99	91.97	81.93	71.85	61.68	51.33
15	180	175.00	165.00	155.00	145.00	134.99	124.99	114.97	104.94	94.88	84.77	74.57	64.21	53.57
18	180	178.00	168.00	158.00	147.99	137.98	127.95	117.91	107.83	97.70	87.48	77.11	66.51	55.54
21	180	180.00	171.00	160.99	150.97	140.94	130.89	120.80	110.65	100.41	90.05	79.47	68.61	57.29
24	180	180.00	173.99	163.97	153.94	143.88	133.78	123.63	113.39	103.02	92.48	81.68	70.53	58.87
27	180	180.00	176.99	166.96	156.90	146.80	136.64	126.40	116.04	105.52	94.79	83.75	72.30	60.30
30	180	180.00	180.00	169.94	159.84	149.69	139.46	129.11	118.63	107.94	96.99	85.70	73.96	61.61
33	180	180.00	180.00	172.93	162.79	152.57	142.25	131.79	121.15	110.28	99.11	87.55	75.50	62.82
36	180	180.00	180.00	175.94	165.74	155.44	145.02	134.42	123.62	112.55	101.14	89.32	76.97	63.95
39	180	180.00	180.00	178.98	168.71	158.32	147.78	137.04	126.06	114.77	103.12	91.02	78.36	65.01
42	180	180.00	180.00	180.00	171.71	161.22	150.55	139.64	128.47	116.96	105.05	92.66	79.69	66.01
45	180	180.00	180.00	180.00	174.76	164.16	153.33	142.25	130.87	119.11	106.94	94.26	80.97	66.97
48	180	180.00	180.00	180.00	177.88	167.14	156.15	144.88	133.27	121.27	108.81	95.82	82.21	67.89
51	180	180.00	180.00	180.00	180.00	170.19	159.03	147.55	135.69	123.42	110.67	97.36	83.43	68.77
54	180	180.00	180.00	180.00	180.00	173.34	161.97	150.26	138.15	125.59	112.52	98.89	84.62	69.64
57	180	180.00	180.00	180.00	180.00	176.59	165.01	153.05	140.65	127.78	114.38	100.41	85.80	70.48
60	180	180.00	180.00	180.00	180.00	180.00	168.17	155.92	143.22	130.01	116.27	101.94	86.97	71.31
63	180	180.00	180.00	180.00	180.00	180.00	171.48	158.92	145.87	132.31	118.19	103.48	88.14	72.14
66	180	180.00	180.00	180.00	180.00	180.00	174.97	162.06	148.64	134.68	120.15	105.04	89.32	72.95
69	180	180.00	180.00	180.00	180.00	180.00	178.70	165.39	151.54	137.14	122.18	106.64	90.52	73.78
72	180	180.00	180.00	180.00	180.00	180.00	180.00	168.94	154.61	139.73	124.29	108.29	91.73	74.60
75	180	180.00	180.00	180.00	180.00	180.00	180.00	172.77	157.90	142.46	126.49	109.99	92.97	75.44
78	180	180.00	180.00	180.00	180.00	180.00	180.00	176.96	161.44	145.38	128.82	111.77	94.26	76.30
81	180	180.00	180.00	180.00	180.00	180.00	180.00	180.00	165.32	148.53	131.29	113.63	95.59	77.17
84	180	180.00	180.00	180.00	180.00	180.00	180.00	180.00	169.61	151.96	133.95	115.61	96.98	78.08
87	180	180.00	180.00	180.00	180.00	180.00	180.00	180.00	174.44	155.75	136.83	117.72	98.44	79.02
90	180	180.00	180.00	180.00	180.00	180.00	180.00	180.00	180.00	160.00	140.00	120.00	100.00	90.00

where

$$a_1 = [\cos 2\alpha - \cos(2\alpha + 2\beta)]\frac{V_m}{2\pi} \quad (6.14)$$

$$b_1 = [2\beta + \sin 2\alpha - \sin(2\alpha + 2\beta)]\frac{V_m}{2\pi} \quad (6.15)$$

$$\theta_1 = \tan^{-1}\left(\frac{a_1}{b_1}\right) \quad (6.16)$$

In terms of normalized variables, the rms phase voltage is given by

$$V_{an} = \frac{V_{a1}}{V_b} = \frac{1}{\pi\sqrt{2}}[1 + 2\beta^2 + 2\beta\{\sin 2\alpha - \sin(2\alpha + 2\beta)\} - \cos 2\beta]^{1/2} \text{ p.u.} \quad (6.17)$$

A set of normalized curves between V_{an} and β for various triggering angles is shown in Figure 6.7.

6.2.4.3 Normalized currents.
The normalized fundamental of the phase current is obtained from

$$I_{an} = \frac{V_{an}}{Z_{an}} = \frac{V_{an}}{(Z_{im}/Z_b)}, \text{ p.u.} \quad (6.18)$$

The normalized rotor current is derived as

$$I_m = \frac{X_m}{\sqrt{\left(\frac{R_r}{s}\right)^2 + (X_m + X_{1r})^2}} \cdot I_{an}, \text{ p.u.} \quad (6.19)$$

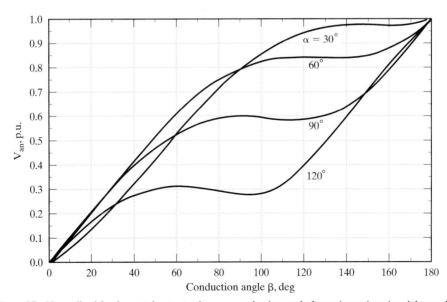

Figure 6.7 Normalized fundamental output voltage vs. conduction angle for various triggering-delay angles

6.2.4.4 Steady-state performance computation. The steady-state performance in terms of torque–speed characteristics is evaluated as in the following:

Step 1: Assume a slip and, using the equivalent circuit parameters, compute Z_{an} and ϕ.

Step 2: Given α and, from the computed power-factor angle, ϕ, read the conduction angle from the normalized graph shown in Figure 6.6 or Table 6.1.

Step 3: From β, the fundamental applied voltage is evaluated by using equation (6.17) or from Figure 6.7.

Step 4: The stator current is calculated from V_{an} and Z_{an}, and then the rotor current can be calculated from stator current and hence also the normalized torque.

Step 5: For various values of slip and α, I_{an} and T_{en} are evaluated and torque–speed characteristics are drawn.

Step 4 needs the expression to calculate the normalized electromagnetic torque, which is derived as

$$T_e = 3\frac{P}{2} \cdot \frac{I_r^2 R_r}{s\omega_s} = 3\frac{P}{2}\left(\frac{I_r}{I_b}\right)^2 \cdot \frac{I_b R_r}{V_b s\omega_s} \cdot I_b \cdot V_b = \left(3\frac{P}{2\omega_s} \cdot V_b I_b\right) I_m^2 \cdot \frac{R_m}{s} \quad (6.20)$$

$$= \frac{3V_b I_b}{\left\{\dfrac{\omega_s}{(P/2)}\right\}} \cdot I_m^2 \cdot \frac{R_m}{s}$$

Noting that

$$\text{Base torque} = T_b = \frac{3V_b I_b}{\left\{\dfrac{\omega_s}{(P/2)}\right\}} \quad (6.21)$$

$$\text{Base power} = P_b = 3V_b I_b \quad (6.22)$$

and dividing equation (6.20) by (6.21) yields the normalized torque as

$$\frac{T_e}{T_b} = T_{en} = I_m^2 \cdot \frac{R_{rn}}{s}, \text{ p.u.} \quad (6.23)$$

where the normalized rotor resistance is

$$R_{rn} = \frac{I_b R_r}{V_b} \quad (6.24)$$

I_{rn} and R_{rn} are the normalized rotor current and resistance, respectively. The mechanical power output is

$$P_{on} = I_{rn}^2 \frac{R_{rn}}{s}(1-s) \quad (6.25)$$

6.2.4.5 Limitations.
The fundamental current contributes to the useful torque of the machine, while the harmonics produce pulsating torques whose averages are zero and hence generate losses only. There will not be any triplen harmonics in the current in a star- (wye-) connected stator of the induction motor, but for the calculation of losses, thermal rating of the motor, and rating of the devices, the rms value of the current is required. It is computed approximately by using equation (6.5). Note that increasing triggering angles produce high losses in the machine that are due to the harmonics, thus affecting the safe thermal operation.

Calculation of the safe operating region of the machine requires a precise estimation of the rms current. In this regard, the approximate analysis is invalid for triggering angles greater than 135°. In such cases, it is necessary to resort to a rigorous analysis, using the dynamic equations of the motor instead of the steady-state equivalent circuit [2, 3, 4].

6.2.5 Torque–Speed Characteristics with Phase Control

Torque–speed characteristics as a function of triggering angle α, for a typical NEMA D, 75-hp, 4-pole, 460-V, 3-phase, 60-Hz machine are shown in Figure 6.8. The characteristics are necessary but not sufficient to determine the speed-control region available for a particular load. The thermal characteristics of the motor need to be incorporated to determine the feasible operating region. The thermal characteristics are dependent on the motor losses, which in turn are dependent on the stator currents. Adhering to the safe operating-current region enforces safe thermal operation. The stator current as a function of speed is derived later for use in application study. The parameters of the machine whose characteristics shown in Figure 6.8 are

$$R_s = 0.0862 \; \Omega, R_r = 0.4 \; \Omega, X_m = 11.3 \; \Omega, X_{ls} = 0.295 \; \Omega, X_{lr} = 0.49 \; \Omega,$$
$$\text{full load slip} = 0.151, \text{ and } I_b = 82.57 \text{ A}.$$

6.2.6 Interaction of the Load

The intersection of the load characteristics and torque–speed characteristic of the induction motor gives the operating point. A fan-load characteristic superimposed on the torque–speed characteristics of the induction motor whose parameters are given in Example 5.1 for various triggering angles with its upper value at 120° and lower value at 30° is shown in Figure 6.9. The feasible open-loop speed-control region is from 0.974 to 0.91 p.u., i.e., only 6.4% of the speed. To increase the speed-control region, either the operation has to be in the statically unstable region, or the load characteristics have to be modified. The latter is generally not possible. By using feedback control, the operating point can be anywhere, including the statically unstable region. Note that the speed is varied from 0.97 to 0.31 p.u. in the present case. The next step is to calculate the steady-state currents for a given load.

274　Chapter 6　Phase-Controlled Induction-Motor Drives

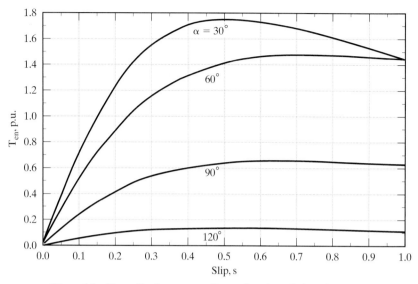

Figure 6.8　Normalized torque vs. slip as a function of triggering angle

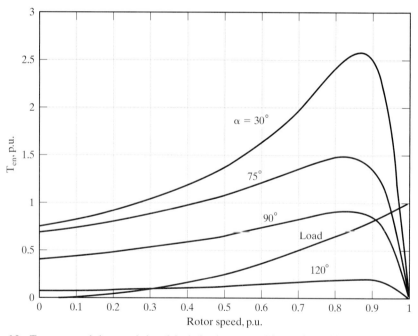

Figure 6.9　Torque–speed characteristics of the induction motor drive, and an arbitrary load to illustrate the speed-control region

6.2.6.1 Steady-state computation of the load interaction.
Neglecting mechanical losses, the load torque equals the electromagnetic torque at steady state. The load torque, in general, is modeled as

$$T_l = B_l \omega_m^k \tag{6.26}$$

where the value of k is determined by the type of load, and B_l is the load constant. For instance,

$$\begin{aligned} k &= 0, \text{ for constant load torque} \\ &= 1, \text{ for friction load} \\ &= 2, \text{ for pump, fan loads} \end{aligned} \tag{6.27}$$

Equating the load torque to the electromagnetic torque and writing it in terms of the rotor current, speed, and motor constants gives

$$3 \frac{P I_r^2 R_r}{2 s \omega_s} = B_l \omega_m^k \tag{6.28}$$

The rotor current in terms of the slip is derived from (6.28) as

$$I_r = \sqrt{\frac{B_l \omega_m^k s \omega_s}{3(P/2) R_r}} = K_r \sqrt{s(1-s)^k} \tag{6.29}$$

where

$$K_r = \sqrt{\omega_s^{k+1} \cdot \frac{B_l}{3(P/2)^{k+1} R_r}} \tag{6.30}$$

For a given value of k, the rotor current is maximum at a certain value of slip. It is found by differentiating equation (6.29) with respect to the slip and equating it to zero. The maximum currents and slip values for frictional and pump loads are

$$k = 1, \quad I_{r(max)} = 0.50 \, K_r, \text{ at } s = \frac{1}{2}$$

$$= 2, \quad I_{r(max)} = 0.385 \, K_r, \text{ at } s = \frac{1}{3} \tag{6.31}$$

The stator current is computed from the rotor-current magnitude obtained from equation (6.29) by using the induction-machine equivalent circuit. Note that the air gap voltage is the product of the rotor current and rotor impedance. The magnetizing current is found from the air gap voltage and magnetizing impedance. The stator current is realized as the phasor sum of the magnetizing and rotor currents, leading to the following expression:

$$I_{as} = I_r \left[\frac{R_r}{s} + jX_{lr} \right] \left[\frac{1}{\frac{R_r}{s} + jX_{lr}} + \frac{1}{jX_m} \right] \tag{6.32}$$

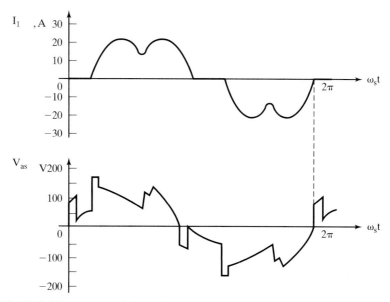

Figure 6.10 Typical line-current and phase-voltage waveforms of a phase-controlled induction motor drive

Typical steady-state phase-voltage and line-current waveforms are shown in Figure 6.10. These waveforms portray the actual condition of the motor drive, whereas the preceding analysis considered only the fundamental component.

Example 6.1

A speed-control range from 0.45 to 0.8 times full-load speed is desired for a pump drive system. A NEMA D induction motor is chosen with the following parameters:

$$150 \text{ hp}, 460 \text{ V}, 3 \text{ ph}, 60 \text{ Hz}, 4 \text{ poles, star connected}, R_s = 0.03 \ \Omega, R_r = 0.22 \ \Omega,$$
$$X_m = 10.0 \ \Omega, X_{1s} = 0.1 \ \Omega, X_{1r} = 0.12 \ \Omega, \text{full load slip} = 0.1477, \text{friction and}$$
$$\text{windage losses} = (0.01\omega_m + 0.0005\omega_m^2) \ (\text{N·m})$$

The pump load constant, B_l, is 0.027 N·m/(rad/sec)^2. Find the range of triggering angle required to achieve the desired speed variation with stator phase control.

Solution This problem is the inverse of the procedure described in the text, which outlines a method to find the torque–speed characteristics for various triggering angles and to determine the speed range available by superposing the load torque–speed characteristics on the motor characteristics. In this problem, from the load torque–speed characteristics, the determination of triggering angle range is attempted as follows.

The upper speed is 0.8 p.u. which should correspond to a triggering angle of α_1; the lower speed is 0.45 p.u., corresponding to a triggering angle of α_2. The rated speed is given by

$$\omega_m = \omega_s(1-s) \times \frac{2}{P} = (2\pi \times 60)(1 - 0.1477) \times \frac{2}{4} = 160.6 \text{ rad/s}$$

Torque corresponding to 0.8 p.u. speed ($\omega_{m1} = 128.5$ rad/sec) is given by

$$T_{l1} = B_l\omega_{m1}^2 + 0.01\omega_{m1} + 0.0005\omega_{m1}^2 = 455.5 \text{ N·m}$$

Similarly, torque corresponding to 0.45 p.u. speed ($\omega_{m2} = 72.3$ rad/sec) is

$$T_{l2} = B_l\omega_{m2}^2 + 0.01\omega_{m2} + 0.0005\omega_{m2}^2 = 144.35 \text{ N} \cdot \text{m}.$$

From the electromagnetic torques, the rotor and stator currents, stator phase angles, and phase voltages are calculated:

T_e, N·m	I_r, A	I_{as}, A	ϕ, deg	V_{as}, V
$T_{11} = 455.5$	201.3	204.2	20.6	153.2
$T_{12} = 144.35$	157.6	159.6	31.4	70.7

The triggering angles of α_1 and α_2 with respective conduction angles β_1 and β_2 yield the two stator voltages for chosen operating conditions. There are four unknowns ($\alpha_1, \alpha_2, \beta_1, \beta_2$) and two known values of v_{as}, so it is necessary to resort to Table 6.1 to find various choices of α and β for the given phase angles. They then are substituted in the voltage equation to verify that they yield the desired voltages. By that procedure,

$$\alpha_1 \cong 102° \text{ with } \beta_1 \cong 98°$$
$$\alpha_2 \cong 135° \text{ with } \beta_2 \cong 68°$$

are obtained, thus giving the range of triggering angle variation as 102° to 135°.

Example 6.2

Determine the current-vs.-slip characteristics for the phase-controlled induction motor drive whose details are given in Example 6.1. Assume fan and frictional loads, and evaluate the load constant based on the fact that 0.2 p.u. torque is developed at 0.7 p.u. speed.

Solution

$$\text{Base voltage, } V_b = \frac{V_{ll}}{\sqrt{3}} = \frac{460}{\sqrt{3}} = 265.58 \text{ V}$$

$$\text{Base current, } I_b = \frac{P_b}{3V_b} = \frac{150 \times 745.6}{3 \times 265.58} = 140.37 \text{ A}$$

$$\text{Base speed, } \omega_b = \frac{2\pi \times 60}{P/2} = \frac{2\pi \times 60}{4/2} = 188.49 \text{ rad/sec}$$

$$\text{Base torque, } T_b = \frac{P_b}{\omega_b} = \frac{150 \times 745.6}{188.49} = 593.33 \text{ N·m}.$$

Case (i): Friction load

$$\text{Load constant, } B_l = \frac{0.2 T_b}{0.7 \omega_b} = \frac{0.2 \times 593.33}{0.7 \times 188.49} = 0.8994 \text{ N·m/(rad/sec)}$$

$$K_r = \sqrt{\omega_s^2 \frac{B_l}{3\left(\frac{P}{2}\right)^2 \cdot R_r}} = 220.03$$

$$I_{rn} = \frac{K_r \sqrt{s(1-s)}}{I_b} = 1.5675\sqrt{s(1-s)} \text{ p.u.}$$

The normalized stator current is

$$I_{asn} = I_{rn}\sqrt{\left(\frac{R_r}{s}\right)^2 + X_{1r}^2}\left[\frac{1}{\frac{R_r}{s}+jX_{1r}} + \frac{1}{jX_m}\right] \text{ (p.u.)}$$

Case (ii): Fan load

Load constant, $B_l = \dfrac{0.2T_b}{(0.7\omega_b)^2} = 0.0068 \text{ Nm/(rad/sec)}^2$

$$K_r = \sqrt{\omega_s^3 \frac{B_l}{3\left(\dfrac{P}{2}\right)^3 \cdot R_r}} = 262.99$$

$$I_{rn} = \frac{K_r\sqrt{s(1-s)^2}}{I_b} = 1.8735\sqrt{s(1-s)} \text{ p.u.}$$

The stator current is computed as in the friction-load case. The normalized current-vs.-slip characteristics are shown in Figure 6.11. Note that, at synchronous speed, the current drawn is the magnetizing current: the rotor current is zero. It is to be noted that the speed-control range is very limited in phase-controlled drives for currents lower than the rated value.

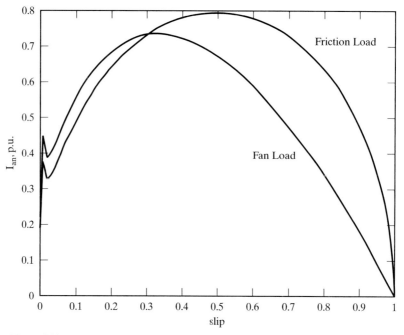

Figure 6.11 Normalized fundamental stator current vs. slip for friction and fan loads

6.2.7 Closed-Loop Operation

Feedback control of speed and current, shown in Figure 6.12, is employed to regulate the speed and to maintain the current within safe limits. The inner current-feedback loop is for the purpose of current limiting. The outer speed loop enforces the desired speed in the motor drive. The speed command is processed through a soft start/stop controller to limit the acceleration and deceleration of the drive system. The speed error is processed, usually through a PI-type controller, and the resulting torque command is limited and transformed into a stator-current command. Note that for zero torque command, the current command is not zero but equals the magnetizing current. To reduce the no-load running losses, it is advisable to run at low speed and at reduced flux level. This results in lower running costs compared to rated speed and flux operation at no load.

The current command is compared with the actual current, and its error is processed through a limiter. This limiter ensures that the control signal v_c to the phase controller is constrained to a safe level. The rise and fall of the control-signal voltage are made gradual, so as to protect the motor and the phase controller from transients. In case there is no feedback control of speed, the control voltage is increased at a preset rate. The current limit might or might not be incorporated in such open-loop drives.

6.2.8 Efficiency

Regardless of the open-loop or closed-loop operation of the phase-controlled induction motor drives, the efficiency of the motor drive is proportional to speed. The efficiency is derived from the steady-state equivalent circuit of the induction

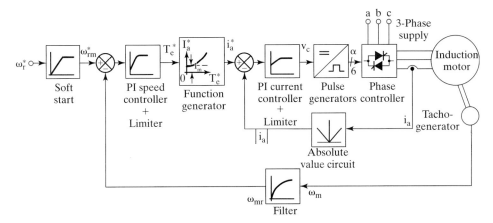

Figure 6.12 Closed-loop schematic of the phase-controlled induction motor drive

280　Chapter 6　Phase-Controlled Induction-Motor Drives

motor. Neglecting stator copper losses, stray losses, friction, and windage losses, the maximum efficiency is given as

$$\text{Efficiency, } \eta = \frac{P_s}{P_a} = \frac{P_m - P_{fw}}{P_a} = \frac{I_{rn}^2 \cdot \frac{(1-s)}{s} \cdot R_{rn}}{\frac{I_{rn}^2}{s} \cdot R_{rn}} = (1-s) = \omega_{rn} \quad (6.33)$$

Here, P_a and P_m denote the air gap and the mechanical output power of the induction motor, respectively, and P_{fw} denotes the friction and windage losses. Slip variation means speed variation, and it is seen from (6.33) that the efficiency decreases as the speed decreases. The rotor copper losses are equal to slip times the input air gap power. This imposes severe strain on the thermal capability of the motor at low operating speeds. Operation over a wide speed range will be restricted by this consideration alone more than by any other factor in this type of motor drive.

The efficiency can be calculated more precisely than is given by equation (6.33) in the following manner:

$$\eta = \frac{P_m - P_{fw}}{P_m + P_{rc} + P_{sc} + P_{co} + P_{st}} \quad (6.34)$$

where

$$\left. \begin{array}{l} P_m = \text{Power output} = \dfrac{3I_r^2 R_r (1-s)}{s} \\[4pt] P_{rc} = \text{Rotor copper losses} = 3I_r^2 R_r \\[4pt] P_{sc} = \text{Stator copper losses} = 3I_{as}^2 R_s \\[4pt] P_{co} = \text{Core losses} = 3I_c^2 R_c \end{array} \right\} \quad (6.35)$$

P_{st} is the stray losses, and the shaft output power is the difference between the mechanical power and the friction and windage losses.

This calculation yields a realistic estimate of efficiency. Energy savings are realized by reducing the motor input compared to the conventional rotor-resistance or stator-resistance control. Consider the case where speed is controlled by adding an adjustable external resistor in the stator phases. That, in turn, reduces the applied voltage to the motor windings. Considerable copper losses occur in the external resistor, R_{ex}. In the phase-controlled induction motor, the input voltage is varied without the accompanying losses, as in the case of the external resistor based speed control system. The equivalent circuit of the resistance-controlled induction motor is shown in Figure 6.13.

The equivalent normalized resistance and reactance of the induction motor are obtained from equations (6.2) and (6.3). The applied voltage is then written, from the equivalent circuit shown in Figure 6.13, as

$$V_{as} = I_{as}(R_{ex} + R_{im} + jX_{im}) \quad (6.36)$$

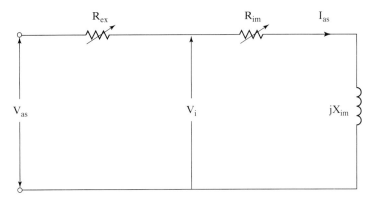

Figure 6.13 Equivalent circuit of the stator-resistance-controlled induction motor

The external resistor value for each slip is found by substituting for rated applied voltage and the stator current required to meet the load characteristics. The latter is obtained from the rotor current, calculated from load constants and slip given by (6.28) and (6.29). Then the stator current is computed from equation (6.32) for each value of slip. Summarizing these steps gives the following:

$$I_r(s) = K_r \sqrt{s(1-s)^k} \tag{6.37}$$

$$I_{as}(s) = \frac{\sqrt{\left(\frac{R_r}{s}\right)^2 + X_r^2}}{X_m} \cdot I_r(s) \tag{6.38}$$

The external resistor for each operating slip is

$$R_{ex} = \sqrt{\left(\frac{V_{as}}{I_{as}(s)}\right)^2 - X_{im}^2} - R_{im} \tag{6.39}$$

The energy savings by using phase control of the stator voltages are approximately

$$P_{ex} = 3I_{as}^2 R_{ex} \tag{6.40}$$

where I_{as} is obtained from equation (6.38) and R_{ex} is obtained from equation (6.40).

For Example 6.1, consider a fan load requiring 0.5 p.u. torque at rated speed. The speed variation possible with external stator-resistor control is 0 to 0.35 p.u. For this range of speed variation, the external resistor required and normalized power input, P_{in}, and power savings, P_{exn}, are shown in Figure 6.14(i). The efficiency of the drive with stator-phase and resistor control is shown in Figure 6.14(ii). The efficiency of the stator-phase control is computed by subtracting the external resistor losses from the input and treating that as the input in the stator-phase control case. The fact that stator-phase control is very much superior to the stator-resistor control is clearly demonstrated from Figure 6.14(ii).

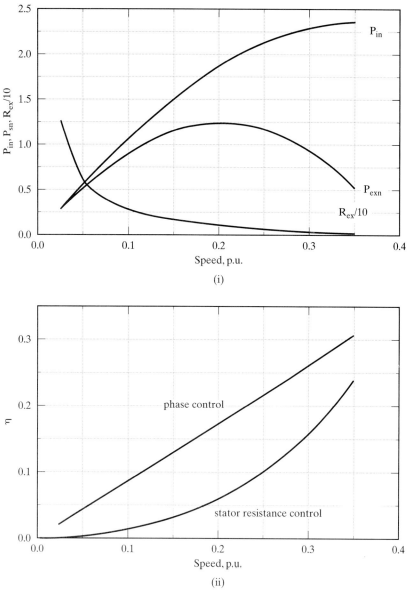

Figure 6.14 (i) External resistor, power input, and power savings vs. slip characteristics at rated voltage input (ii) Efficiency with external resistor and phase control vs. slip

6.2.9 Applications

Current-limited phase-controlled induction motor drives find applications in pumps, fans, compressors, extruders, and presses. The closed-loop drives are used in ski-lifters, conveyors, and wire-drawing machines. The phase controllers are used as solid-state contactors and also as starters in induction motors. Phase controllers are extensively used in resistance heaters, light dimmers, and solid-state relaying.

For obtaining maximum speed-control range in open-loop operation, NEMA D induction motors are used. These motors have a high rotor resistance and hence provide a high starting torque as well as a very large statically stable operating torque–speed region, but these motors have higher losses than other motors and hence have poorer efficiency.

6.3 SLIP-ENERGY RECOVERY SCHEME

6.3.1 Principle of Operation

The phase-controlled induction motor drive has a low efficiency; its approximate maximum efficiency equals the p.u. speed. As speed decreases, the rotor copper losses increase, thus reducing the output and efficiency. The rotor copper losses are

$$P_{rc} = 3I_r^2 R_r = sP_a \qquad (6.41)$$

where P_a is the air gap power. The increase in this slip power, sP_a, results in a large rotor current. This slip power can be recovered by introducing a variable emf source in the rotor of the induction motor and absorbing the slip power into it. By linking the emf source to ac supply lines through a suitable power converter, the slip energy is sent back to the ac supply. This is illustrated in Figure 6.15. By varying the magnitude of the emf source in the rotor, the rotor current, torque, and slip are controlled. The rotor current is controlled and hence the rotor copper losses, and a significant portion of the power that would have been dissipated in the rotor is absorbed by the emf source, thereby improving the efficiency of the motor drive.

6.3.2 Slip-Energy Recovery Scheme

A schematic of the slip-energy recovery in the induction motor is shown in Figure 6.16. The rotor has slip rings, and a diode-bridge rectifier is connected to the slip-ring

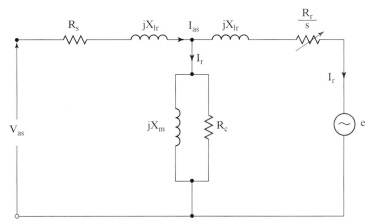

Figure 6.15 Steady-state equivalent circuit of the induction motor with an external induced-emf source in the rotor

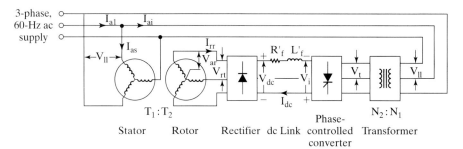

Figure 6.16 Slip-power recovery scheme

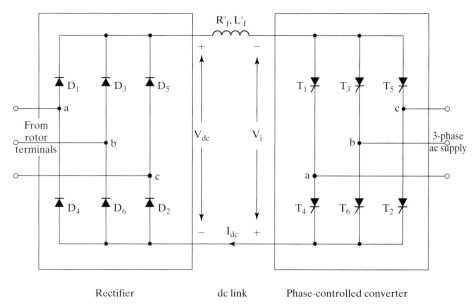

Figure 6.17 Rectifier–inverter power stage for slip-power recovery scheme

terminals. The rotor voltages are rectified and, through an inductor, fed to the phase-controlled converter. The detailed diagram of the bridge-diode rectifier and phase controlled converter are shown in Figure 6.17. The operation of the phase-controlled converter has been described in Chapter 3.

The phase-controlled converter is operated in the inversion mode by having a triggering angle greater than 90°. Then the inverter voltage V_i is in opposition to the rectified rotor voltage V_{dc} in the dc link. A current I_{dc} is established in the dc link by adjusting the triggering angle of the phase-controlled converter. The diode-bridge rectifier on the rotor side of the induction motor forces power to flow only away from the rotor windings. A transformer is generally introduced between the phase-controlled converter and the ac supply to compensate for the low turns ratio in the induction motor and to improve the overall power factor of the system. The latter assertion will be proved in the subsequent section.

6.3.3 Steady-State Analysis

The equivalent circuit is used to predict the steady-state motor drive performance. The steady-state performance is computed by proceeding logically from the rotor part of the induction motor to the stator part through the converter subsystem. The dc link current is assumed to be ripple-free.

The rotor phase voltage, V_{ar}, and the dc link current, I_{dc}, are shown in Figure 6.18(a). The current is a rectangular pulse of 120-degree duration. The magnitude of the rotor current is equal to the dc link current, I_{dc}. The phase and line voltages and phase current in the phase-controlled converter are shown in Figure 6.18(b). There is a phase displacement of α degrees between the phase voltage and current, which is the triggering angle in the converter. The magnitude of the converter line current is equal to the dc link current but has 120-degree duration for each half-cycle. The rms rotor line voltage in terms of stator line voltage is

$$V_{rt} = \left(\frac{k_{w2}T_2}{k_{w1}T_1}\right) s V_{ll} \qquad (6.42)$$

where k_{w1} are k_{w2} are the stator and rotor winding factors and T_1 and T_2 are the stator and rotor turns per phase, respectively.

Denoting the effective turns ratio by a, we write

$$\frac{1}{a} = \left(\frac{k_{w2}T_2}{k_{w1}T_1}\right) \qquad (6.43)$$

The rotor line voltage is then derived from equations (6.42) and (6.43) as

$$V_{rt} = \frac{sV_{ll}}{a} \qquad (6.44)$$

The average rectified rotor voltage is

$$V_{dc} = 1.35 V_{rt} = \frac{1.35 s V_{ll}}{a} \qquad (6.45)$$

This has to be equal to the sum of the inverter voltage and the resistive drop in the dc link inductor.

Neglecting the resistive voltage drop, we get

$$V_{dc} = -V_i \qquad (6.46)$$

where the phase-controlled-converter output voltage is given by

$$V_i = 1.35 V_t = \cos \alpha \qquad (6.47)$$

where

$$V_t = \left(\frac{N_2}{N_1}\right) V_{ll} = n_t V_{ll} \qquad (6.48)$$

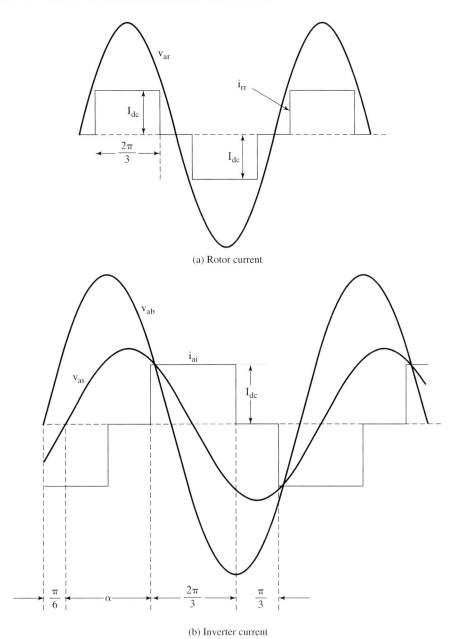

(a) Rotor current

(b) Inverter current

Figure 6.18 Rotor current and inverter current

and n_t is the turns ratio of the transformer. From equations (6.45) to (6.48), the slip is derived as

$$s = -(an_t) \cos \alpha \tag{6.49}$$

6.3.3.1 Range of slip. The triggering angle can be varied in the inversion mode from 90 to 180 degrees. The finite turn-off time of the power switches in the phase-controlled converter usually limits the triggering angle to 155 degrees. Hence, the practical range of the triggering angle is

$$90° \leq \alpha \leq 155° \tag{6.50}$$

For this range of α, the slip range is given by

$$0 \leq s \leq 0.906(an_t) \tag{6.51}$$

The turns ratio of the stator to rotor of the induction motor is usually less than unity. The dc link voltage becomes small when the induction motor is operating at around synchronous speed. This necessitates a very small range of triggering angle in the converter operation. A step-down transformer with a low turns ratio ensures a wide range of triggering angle under this circumstance.

The value of α in terms of the slip from equation (6.49) is

$$\alpha = \cos^{-1}\left\{-\frac{s}{an_t}\right\} \tag{6.52}$$

6.3.3.2 Equivalent circuit. The equivalent circuit for the slip-energy recovery scheme is derived by converting the dc link filter and phase-controlled inverter into their three-phase equivalents in steady state. The dc filter inductor is not required to be incorporated, as its effect in steady state is zero. The dc link filter resistance R_f' is converted to an equivalent three-phase resistance R_{ff} by equating their losses as follows:

$$I_{dc}^2 R_f' = 3(I_{rr})^2 R_{ff} \tag{6.53}$$

where I_{rr} is the rms value of the induction-motor rotor-phase current, which, in terms of the dc link current, is expressed as

$$I_{rr} = \sqrt{\frac{1}{\pi}\int_0^{2\pi/3} I_{dc}^2 d\theta} = I_{dc}\sqrt{\frac{2}{3}} = 0.816 I_{dc} \tag{6.54}$$

Substituting (6.54) into (6.53) gives R_{ff} as

$$R_{ff} = 0.5 R_f'$$

The rms value of the fundamental component of the rectangular rotor current is given by

$$I_{rr1} = \frac{4}{\pi\sqrt{2}} I_{dc} \sin 60° = 0.779 I_{dc} \tag{6.55}$$

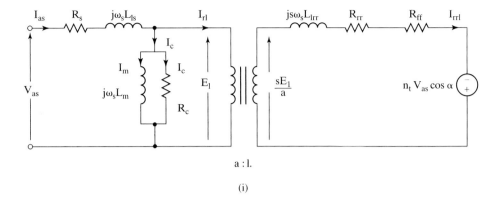

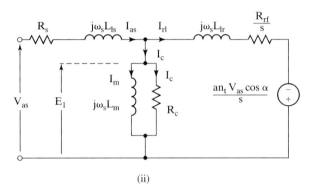

Figure 6.19 Per-phase equivalent circuit of the slip-energy-recovery-controlled induction motor drive

The power transferred through the phase-controlled inverter is

$$P_{fb} = V_i I_{dc} = (1.35 n_t V_s \cos\alpha) I_{dc} = [1.35 n_t \sqrt{3} V_{as} \cos\alpha]\left[\frac{I_{rr1}}{0.779}\right] = 3[n_t V_{as} \cos\alpha] I_{rr1} \quad (6.56)$$

This could be viewed as power transfer to a three-phase ac battery with a phase voltage of $[n_t V_{as} \cos\alpha]$ absorbing a charging current of I_{rr1} from the rotor phases of the induction motor. By integrating R_{ff} and the emf of the variable ac battery in the three-phase rotor of the induction motor, the per-phase equivalent circuit shown in Figure 6.19(i) is obtained.

The reactive power demand of the variable emf source is met not from the stator and rotor of the induction motor but through the ac supply lines. Hence, that need not appear in the equivalent circuit of the slip-energy-recovery-controlled induction motor drive. Further, note that the rotor current is in phase with the variable emf source given by $[n_t V_{as} \cos\alpha]$. The equivalent circuit is further simplified and referred to the stator in a similar manner, as described in Chapter 5, with the following steps, as shown in Figure 6.19 (ii):

$$[I_{rr1}(R_{rr} + R_{ff}) - n_t V_{as} \cos\alpha] + j s \omega_s L_{1rr} I_{rr1} = \frac{sE_1}{a} \quad (6.57)$$

The emf term is combined with the resistive voltage drop because it is in phase with I_{rr1}: the voltage component of the battery has only a real part in it, and the reactive part has been removed. This is an important point in the solution of rotor current from the derived equivalent circuit.

Multiplying (6.57) by $\frac{a}{s}$, we get

$$\left[I_{rr1}\left\{\frac{a(R_{rr} + R_{ff})}{s}\right\} - \frac{an_t V_{as}\cos\alpha}{s}\right] + j\omega_s a L_{1rr} I_{rr1} = E_1 \quad (6.58)$$

The rotor current, referred to the stator, is given by

$$I_{r1} = \frac{I_{rr1}}{a} \quad (6.59)$$

The stator-referred rotor and filter resistances and rotor leakage inductance are given by

$$R_r = a^2 R_{rr}$$
$$R_f = a^2 R_{ff} \quad (6.60)$$
$$L_{1r} = a^2 L_{1rr}$$

and let

$$R_{rf} = R_r + R_f \quad (6.61)$$

Combining (6.59), (6.60), and (6.61) with (6.58) yields

$$\frac{I_{r1}}{s}\{R_{rf} - an_t V_{as}\cos\alpha\} + j\omega_s L_{1r} I_{r1} = E_1 \quad (6.62)$$

6.3.3.3 Performance characteristics. The performance characteristics can be evaluated from the equivalent circuit derived earlier. The electromagnetic torque is evaluated from the output power and rotor speed as

$$T_e = \frac{P_m}{\omega_m} = \frac{P_a(1-s)}{\left\{\frac{\omega_s(1-s)}{P/2}\right\}} = \frac{P}{2} \cdot \frac{P_a}{\omega_s} = \frac{P}{2} \cdot \frac{1}{\omega_s}\left\{3\left[I_{r1}^2 \cdot \frac{R_{rf}}{s} - \frac{aI_{r1}n_t V_{as}\cos\alpha}{s}\right]\right\}$$

$$= \frac{3P}{2} \cdot \frac{1}{s\omega_s}[I_{r1}^2 R_{rf} - aI_{r1}n_t V_{as}\cos\alpha], \text{N·m} \quad (6.63)$$

where P_a is the air gap power, P_m is the mechanical output power, ω_m is the mechanical rotor speed, ω_s is the synchronous speed, and P is the number of poles. By neglecting the leakage reactance of the rotor, the electromagnetic torque can be written as

$$T_e = \frac{3P}{2} \cdot \frac{1}{\omega_s} \cdot I_{r1} \cdot \left[\frac{I_{r1}R_{rf}}{s} - \frac{an_t V_{as}\cos\alpha}{s}\right] \cong \frac{3P}{2} \cdot \frac{1}{\omega_s} \cdot I_{r1} \cdot E_1 \quad (6.64)$$

If stator impedance is neglected, then the applied voltage V_{as} is equal to the induced emf E_1; hence,

$$T_e \cong \frac{3P}{2} \cdot \frac{V_{as}}{\omega_s} \cdot I_{r1} \qquad (6.65)$$

and

$$I_{r1} = \frac{I_{rr1}}{a} = \frac{0.779 I_{dc}}{a} \qquad (6.66)$$

which, on substitution, gives the electromagnetic torque as

$$T_e \cong \left(\frac{1.17P}{a}\right)\left(\frac{V_{as}}{\omega_s}\right) I_{dc} = K_t I_{dc} \qquad (6.67)$$

where K_t is a torque constant given by

$$K_t = \left(\frac{1.17P}{a}\right)\left(\frac{V_{as}}{\omega_s}\right) \; (\text{N·m/A}) \qquad (6.68)$$

This shows that, for a given induction motor, the electromagnetic torque is proportional to the dc link current. An implication of this result is that torque control in this drive scheme is very similar to the torque control in a separately-excited dc motor drive by control of its armature current. Also, the term E_1/ω_s is proved to be mutual flux linkages as follows:

$$\frac{V_{as}}{\omega_s} \cong \frac{E_1}{\omega_s} = \frac{I_m X_m}{\omega_s} = \frac{I_m \omega_s L_m}{\omega_s} = I_m L_m = \lambda_m \qquad (6.69)$$

where λ_m is the mutual flux linkages, much like the field-flux linkages produced by the current in the field coils of a dc machine. Note that this is fairly constant in the induction motor over a wide range of stator current. Equation (6.65) has given a powerful insight into the torque control of this motor drive, and it will be used in the closed-loop control described in the later section. That the induction motor control does not involve the control of all the three-phase currents, as in the case of a phase-controlled induction motor, is a significant point to be noted. Sensing only one current, e.g., dc link current, is sufficient for feedback control; this simplifies the control compared to any other ac drive scheme, as will become evident when other schemes are examined in later chapters.

Efficiency of the motor drive is calculated as follows:

$$\eta = \frac{\text{Shaft power output}}{\text{Power input}} = \frac{P_m - (\text{windage + friction losses})}{\text{Power input}} = \frac{P_m - P_{fw}}{P_i} \qquad (6.70)$$

where P_{fw} is the windage and friction losses and P_i is the input power to the machine. It is calculated as the sum of the output power, rotor and stator resistive losses, core losses, and stray losses and is given by

$$P_i = P_m + 3[I_r^2(R_r + R_f) + I_c^2 R_c + I_s^2 R_s] + P_{st} \; (\text{W}) \qquad (6.71)$$

where

$$P_m = P_a(1-s) = 3[I_{r1}R_{rf} - an_tV_{as}\cos\alpha]\cdot\frac{1-s}{s}\cdot I_{r1} \qquad (6.72)$$

The stator current is

$$I_{as} = I_{r1} + I_m + I_c = |I_{as}| \angle \phi_s \qquad (6.73)$$

Angle ϕ_s is the stator phase angle. The system input-phase current is the sum of the stator input current and the real and reactive components of the current supplied to the phase-controlled inverter from the lines through the step-down transformer. This is written as

$$\begin{aligned}I_{a11} &= I_{as} - j(n_tI_{rr1}\sin\alpha) + n_tI_{rr1}\cos\alpha \\ &= I_{as}(\cos\phi_s + j\sin\phi_s) - j(an_tI_{r1}\sin\alpha) + an_tI_{r1}\cos\alpha = |I_{a11}| \angle \phi 1\end{aligned} \qquad (6.74)$$

where ϕ_1 is the input-line power-factor angle of the utility mains supply.

In all these derivations, it must be noted that the harmonic current losses in the rotor have been neglected. This could be accounted for with the harmonic equivalent circuit. Equation (6.74) clearly shows that a lower transformer-turns ratio can improve the line power factor by keeping ϕ_1 low.

The performance characteristics of the slip-energy-recovery-controlled induction motor drive are obtained by the use of the developed equivalent circuit and equations outlined above. For the purpose of solving the equivalent circuit, it could further be simplified by moving its magnetizing branch ahead of the stator impedance. This is shown in Figure 6.20.

A flowchart for the computation of performance characteristics is shown in Figure 6.21. A set of performance characteristics is shown in Figure 6.22. The motor and system details are as follows:

75 hp, 460 V, 3 ph, 60 Hz, 4 pole, Class D, $R_s = 0.0862\,\Omega$, $R_r = 0.9\,\Omega$, $L_{1s} = 0.78$ mH, $L_{1r} = 1.3$ mH, $L_m = 0.03$ H, $R_c = 150\,\Omega$, $R'_f = 0.04\,\Omega$, $a = 2.083$, $n_t = 0.6$.

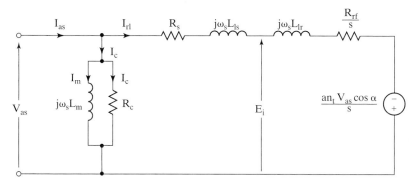

Figure 6.20 Simplified equivalent circuit of the slip-energy-recovery-controlled induction motor drive

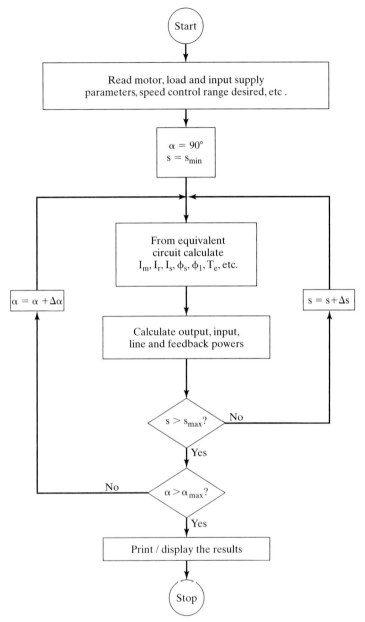

Figure 6.21 Flowchart for the computation of the slip-energy-recovery-controlled induction motor drive's steady-state performance

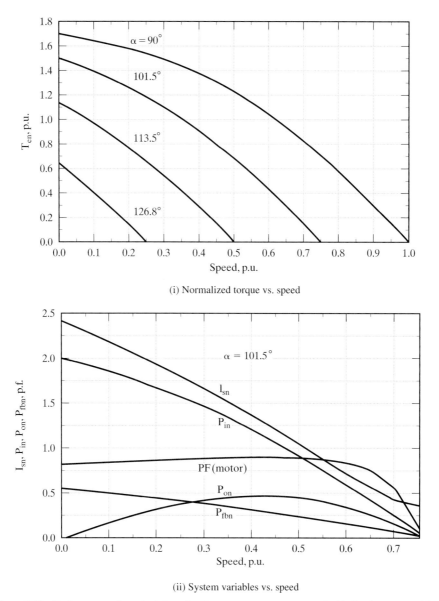

Figure 6.22 Performance characteristics of a slip-energy-recovery-controlled induction motor drive

The following procedure is used to compute I_{r1}. Write the voltage-loop equation involving the rotor emf source from Figure 6.19 (ii) as

$$\left\{ \left(R_s + \frac{R_{rf}}{s} \right) I_{r1} - \frac{a n_t V_{as} \cos \alpha}{s} \right\} + j\omega_s(L_{1s} + L_{1r}) I_{r1} = V_{as} \quad (6.75)$$

It must be noted that the variable emf source in the rotor is in phase with rotor current; hence, they are kept together as a unit in the expression. Considering only the magnitude part of the equation, the equation (6.75) is squared on both sides, and rearrangement yields a quadratic expression in I_{r1} as follows:

$$h_1 I_{r1}^2 - h_2 I_{r1} + h_3 = 0 \quad (6.76)$$

where

$$h_1 = \left(R_s + \frac{R_{rf}}{s} \right)^2 + X_{eq}^2 \quad (6.77)$$

$$X_{eq} = \omega_s(L_{1s} + L_{1r}) \quad (6.78)$$

$$h_2 = \left\{ \frac{2 a n_t V_{as} \cos \alpha}{s} \right\} \left\{ R_s + \frac{R_{rf}}{s} \right\} \quad (6.79)$$

$$h_3 = \left(\frac{a n_t V_{as} \cos \alpha}{s} \right)^2 - V_{as}^2 \quad (6.80)$$

The stator-referred fundamental rotor-current magnitude is obtained from equation (6.76) as

$$I_{r1} = \frac{h_2 + \sqrt{h_2^2 - 4 h_1 h_3}}{2 h_1} \quad (6.81)$$

The positive sign only is considered in the solution of I_{r1} in equation (6.81). Since h_2 is negative, only the addition of the discriminant yields a positive value for I_{r1}. Further, the following condition has to be imposed for the solution of I_{r1}:

$$\text{Real}\{I_{r1}\} > 0 \text{ or } I_{r1} = 0 \quad (6.82)$$

For various triggering angles, the normalized electromagnetic torque-vs.-speed characteristics are drawn by using the procedure developed above and shown in Figure 6.22 (i). For a triggering angle of 101.5 degrees, the motor power factor, stator current of the motor, output power, P_{ou}, input power, P_{in}, and feedback power from the rotor (which is the slip recovery power, P_{fbn}) are shown in Figure 6.22 (ii). Note that the motor power factor remains high for a wide speed range. As the speed decreases, the slip-energy recovery increases as well as the input power.

Example 6.3

Compare the efficiency of the slip-recovery-controlled induction motor drive with the maximum theoretical efficiency of the phase-controlled induction motor drive. The load torque is proportional to square of the speed. It is assumed that, at base speed, base torque is delivered. Consider operation at 0.6 p.u. speed. The details of the induction machine and its drive are given in the previous illustration of the torque-vs.-slip characteristics derivation.

Section 6.3 Slip-Energy Recovery Scheme

Solution The following calculations from step 1 to step 5 are for the slip-recovery-controlled induction motor drive.

Step 1: Motor filter parameter for use in equivalent circuit is calculated.

$$R_{ff} = 0.5 * R'_f = 0.5 * 0.04 = 0.02 \, \Omega$$
$$R_f = a^2 R_{ff} = 2.083^2 \times 0.02 = 0.0868 \, \Omega$$
$$R_{rf} = R_r + R_f = 0.9 + 0.0868 = 0.9868 \, \Omega$$

Step 2: Compute base values and load constant.

$$\text{Base speed, } \omega_b = \frac{2\pi f_s}{P/2} = \frac{2\pi 60}{4/2} = 188.5 \text{ rad/s}$$

$$\text{Base torque, } T_b = \frac{P_b}{\omega_b} = \frac{75 * 745.6}{188.5} = 296.66 \text{ N·m}$$

$$\text{Load constant, } B_l = \frac{T_b}{\omega_b^2} = \frac{296.66}{188.5^2} = 0.0083 \text{ N·m/(rad/s)}^2$$

Step 3: Find the triggering angle and load torque.

Rotor speed = 0.6 p.u.

Slip = 1 − speed = 1 − 0.6 = 0.4

$$\alpha = \cos^{-1}\left(-\frac{s}{an_t}\right) = \cos^{-1}\left(-\frac{0.4}{2.083 * 0.6}\right) = 1.89 \text{ rad}$$

Load torque, $T_l = B_l \omega_m^2 = 0.0083 * (0.6 * \omega_b)^2 = 106.8$ N·m.

Step 4: Solve for rotor current from the air gap-torque expression. It must be noticed that electromagnetic torque is equal to the sum of the load torque and friction and windage torque. As the latter part is not specified in the problem, they are considered to be zero; hence, the electromagnetic torque is equal to the load torque in present case.

$$T_e = 3\frac{P}{2}\frac{1}{s\omega_s}[I_{r1}^2 R_{rf} - I_{r1} an_t V_{as} \cos \alpha]$$

The only unknown in this expression is rotor current; it is solved for, yielding two values. Considering only the positive value, it is found as

$$I_{r1} = 21.12 \text{ A}$$

Step 5: From rotor current, the air gap emf is computed, from which all the other currents are evaluated. In order to do that, the reference phasor is assumed to be the rotor current. From the other currents, the stator and core losses are evaluated, and, from the rotor current, the rotor losses are calculated.

$$E_1 = \left[\frac{R_{rf}}{s}I_{r1} - \frac{an_t V_{as} \cos \alpha}{s}\right] + j\omega_s L_{1r} I_{r1} = 158.3 + j10.35 = 158.67 \angle 3.74°$$

The various currents in the circuit are

Core loss current, $I_c = \dfrac{E_1}{R_c} = 1.05 + j0.069 = 1.06\angle 3.74°$

Magnetizing current, $I_m = \dfrac{E_1}{j\omega_s L_m} = 0.91 - j14.0 = 14.03\angle -86.26°$

No load current, $I_o = I_c + I_m = 1.97 - j13.93 = 14.07\angle -81.95°$

Stator phase current, $I_s = I_{r1} + I_o = 23.09 - j13.93 = 26.97\angle -31.1°$

Stator resistive losses, $P_{sc} = 3R_s I_s^2 = 3 * 0.0862 * 26.97^2 = 188.1$ W

Rotor and dc link filter resistive losses, $P_{rc} = 3R_{rf} I_{r1}^2 = 3 * 0.9868 * 21.12^2 = 1321$ W

Core losses, $P_{co} = 3R_c I_c^2 = 3 * 150 * 1.06^2 = 503.6$ W

Power output, $P_m = T_e \omega_m = 106.6 * (0.6 * 188.5) = 12{,}079$ W

Power input, $P_i = P_m + P_{sc} + P_{rc} + P_{co} = 14{,}091$ W

Efficiency, $\eta = \dfrac{P_m}{P_i} = \dfrac{12{,}079}{14{,}091} = 0.8572$

Step 6: Comparison of efficiency

Maximum theoretical efficiency possible with the phase-controlled induction motor drive, ignoring all stator, and core losses, is given as,

$$\eta = 1 - s = 1 - 0.4 = 0.6$$

In comparison to the slip-recovery-controlled induction-motor-drive efficiency, the efficiency of the phase-controlled drive system is 25.7% smaller. When the stator resistive and core losses are considered for the phase-controlled induction motor drive, the difference in efficiency will be even higher. It is left as an exercise to the reader.

6.3.4 Starting

The ratings of the transformer, rectifier bridge, and phase-controlled converter have to be equal to the motor rating if the slip-energy-recovery drive is started from standstill by using the controlled converter. The ratings of these subsystems can be reduced in a slip-energy-recovery drive with a limited speed-control region. In such a case, resistances are introduced in the rotor for starting. As the speed comes to the minimum controllable speed, these resistances are cut out and the rotor handles only a fractional power of the motor, thus considerably reducing the cost of the system. To withstand the heat generated in the resistors during starting, liquid cooling is recommended for large machines.

6.3.5 Rating of the Converters

Auxiliary starting means are assumed in the slip-energy-recovery control for the calculation of the converter rating. If an auxiliary means is not employed, the rating of the converters is equal to the induction-motor rating. As most of the applications for slip-energy-recovery drive are for fans and pumps, they have limited operational ranges of speed and hence power. The converters handle the slip power, and therefore their ratings are proportional to the maximum slip, s_{max}. The ratings of the converters are derived separately.

6.3.5.1 Bridge-rectifier ratings. The rectifier handles the slip power passing through the rotor windings. It is estimated in terms of air gap power as

$$\text{Slip power} = sP_a = s_{max}P_a = \text{Rectifier power rating}, P_{br} \tag{6.83}$$

where P_a is the air gap power and is given approximately, by neglecting the stator impedance and rotor leakage reactance, as

$$P_a \cong 3V_{as}I_{r1} \tag{6.84}$$

Hence, the ratings of the bridge rectifier are obtained by substituting (6.84) into (6.83):

$$P_{br} = s_{max}P_a = s_{max}\{3V_{as} \cdot I_{r1}\} \tag{6.85}$$

The peak voltage and rms current ratings of the diodes are obtained from the peak rotor line voltage and rms current in the rotor phase as

$$V_d = \sqrt{2}V_s = \sqrt{2}\sqrt{3}V_{as} = 2.45V_{as} \tag{6.86}$$

$$I_d = 0.816 I_{dc} = 0.816 \times \frac{I_{rr1}}{0.779} = 1.05 I_{rr1} = 1.05 a I_{r1} \tag{6.87}$$

6.3.5.2 Phase-controlled converter. The peak voltage and rms current ratings of the power switches in the converter are

$$V_{ts} = \sqrt{2}\sqrt{3}n_t V_{as} = 2.45 n_t V_{as} \tag{6.88}$$

$$I_{ts} = 0.816 I_{dc} = 1.05 a I_{r1} \tag{6.89}$$

6.3.5.3 Filter choke. The purpose of the dc-link filter choke is to limit the ripple in the dc link current. The ripple current is caused by the ripples in the rectified rotor voltages and phase-controlled inverter voltage reflected into the dc bus. Considering only the sixth harmonic gives the equivalent circuit of the dc link shown in Figure 6.23. The sixth-harmonic component of the rectified rotor voltages is written as

$$V_{dc6} = \sqrt{a_{dc6}^2 + b_{dc6}^2} \tag{6.90}$$

where

$$a_{dc6} = -\frac{6sV_s}{a\pi\sqrt{2}}\left[\frac{1}{7}\cos 7\alpha - \frac{1}{5}\cos 5\alpha\right] \tag{6.91}$$

$$b_{dc6} = \frac{6sV_s}{a\pi\sqrt{2}}\left[\frac{1}{7}\sin 7\alpha - \frac{1}{5}\sin 5\alpha\right] \tag{6.92}$$

By noting that the triggering angle for the diode bridge is zero, the sixth-harmonic voltage is evaluated as

$$V_{dc6} = \frac{0.077 sV_s}{a} \tag{6.93}$$

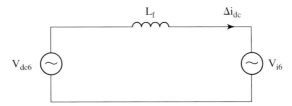

Figure 6.23 Sixth-harmonic equivalent circuit of the dc link

Similarly, the sixth-harmonic component of the inverter voltage is

$$V_{i6} = \frac{6n_t V_s}{\pi \sqrt{2}} \sqrt{\frac{1}{49} + \frac{1}{25} - \frac{2}{35} \cdot \cos 2\alpha} \qquad (6.94)$$

and its maximum occurs at

$$\alpha = 90° \qquad (6.95)$$

giving the maximum of V_{i6} as

$$V_{i6m} = 0.463 n_t V_s \qquad (6.96)$$

These two harmonics are of different frequencies (i.e., v_{dc6} at six times slip frequency and V_{i6} at six times the line frequency) and have a phase shift between them, so it is safe to consider the sum of these voltages that has to be supported by the choke for the worst-case design. If the ripple current is denoted by ΔI_{dc},

$$L_f \frac{\Delta I_{dc}}{\Delta t} = V_{dc6} + V_{i6m} \qquad (6.97)$$

where

$$\Delta t = \frac{2\pi}{6\omega_{s1}} \qquad (6.98)$$

and ω_{s1} is the angular slip frequency. This is found to be lower than the supply frequency. The slip frequency preferably has to correspond to the maximum speed of the drive and hence the minimum slip speed.

By substituting for V_{dc6} and V_{i6m} from previous equations, the filter inductor is given as

$$L_f = \frac{1}{\Delta I_{dc}} \left[\frac{\pi}{6} \cdot \frac{V_s}{\omega_{s1}} \left\{ 0.463 n_t + \frac{0.077 s}{a} \right\} \right] \qquad (6.99)$$

6.3.6 Closed-Loop Control

A closed-loop control scheme for torque and speed regulation is shown in Figure 6.24. The inner current loop is also the torque loop: the torque in a slip-energy-controlled drive is directly proportional to the dc link current. This torque loop is very similar to the phase-controlled dc motor drive's torque loop. The outer speed loop enforces the speed command, ω_r^*. In case of discrepancy between the speed and its

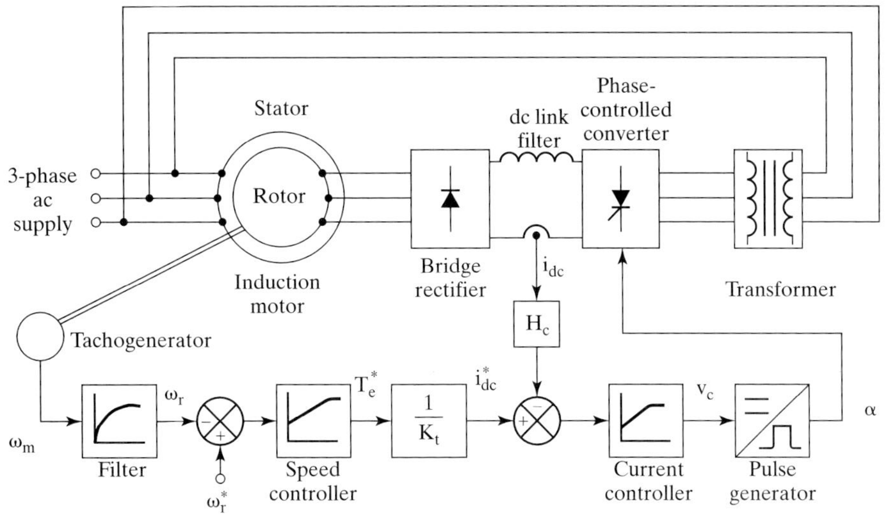

Section 6.3 Slip-Energy Recovery Scheme

Figure 6.24 Closed-loop control of the slip-energy-recovery-controlled induction motor drive

command, the speed error is amplified and processed through a proportional-plus-integral controller. Its output forms the torque command, from which the current command is obtained by using the torque constant. The current error is usually processed by a proportional-plus-integral controller to produce a control voltage, v_c. This control voltage is converted into gate pulses with a delay of α degrees. The control block of pulse generators is explained in detail in Chapter 3. The only difference between the phase-controlled dc drive and this motor drive is that the triggering angle for motoring mode is usually less than $90°$ in the dc drive and greater than $90°$ in the slip-recovery drive. Accordingly, the control signal needs to be modified from the strategy used in the dc drive. This modification of the control circuit is very minor. The control signal v_c is limited so as to limit the triggering angle α to its upper limit of α_{max}, say, 155 degrees. The similarities between the dc drive and the slip-energy-controlled drive are exploited in the design of the current and speed controllers. The fact that there is only one control variable makes simple the design of controllers in the closed-loop system. As this is of one-quadrant operation, the PI controller outputs are zero for negative speed and current errors.

6.3.7 Sixth-Harmonic Pulsating Torques

The rotor currents in the induction motor are rich in harmonics. The effect of the dominant harmonics in the form of pulsating torque is examined in this section.

Writing the rectangular rotor current referred to the stator as a sum of Fourier components yields

$$I_r = \frac{2\sqrt{3}}{a\pi} I_{dc} \left[\cos \omega_s t - \frac{1}{5} \cos 5\omega_s t + \frac{1}{7} \cos 7\omega_s t - \frac{1}{11} \cos 11\omega_s t + \ldots \right] \quad (6.100)$$

The harmonic rms currents are expressed in terms of the fundamental rms current from equation (6.100) as

$$I_{rn} = \frac{1}{n} I_{r1}, \quad n = 5, 7, 11, 13, 17, 19, \ldots \tag{6.101}$$

where n is the harmonic number, and the fundamental stator-referred rotor current in terms of the dc link current is

$$I_{r1} = \frac{2\sqrt{3}}{a\pi\sqrt{2}} I_{dc} = \frac{0.779}{a} I_{dc} \tag{6.102}$$

Pulsating torques are produced when harmonic currents interact with the fundamental air gap flux and also when the harmonic air gap fluxes interact with the fundamental rotor current. The fundamental of the rotor current and air gap flux interact to produce a steady-state dc torque. The pulsating torques produce losses, heat, vibration, and noise, all of which are undesirable in the motor drive.

The predominant harmonics in the current are the fifth and the seventh, whose magnitudes are

$$I_{r5} = \frac{1}{5} I_{r1} \tag{6.103}$$

$$I_{r7} = \frac{1}{7} I_{r1} \tag{6.104}$$

Note that the fifth harmonic rotates at 5 times synchronous speed in the backward direction with reference to the fundamental. The seventh harmonic rotates at 7 times synchronous speed in the forward direction, as seen from equation (6.100). Hence, the relative speeds of the fifth and seventh harmonics with respect to the fundamental air gap flux are six times the synchronous speed. Similarly, the fifth- and seventh-harmonic air gap fluxes have a relative speed of six times the synchronous speed with respect to the fundamental of the rotor current. The harmonic air gap flux linkages are produced by the corresponding rotor harmonic currents. For example, the fifth-harmonic air gap flux linkages are

$$\lambda_{m5} = L_m I_{m5} = L_m \cdot \frac{E_5}{5X_m} = \frac{I_{r5}}{5\omega_s} \left\{ \frac{R_r}{s_5} + j5X_{1r} \right\} \tag{6.105}$$

where the harmonic slip is given by

$$s_n = \frac{n\omega_s - \omega_r}{n\omega_s} = \frac{n\omega_s - \omega_s(1-s)}{n\omega_s} = \frac{n - (1-s)}{n} \tag{6.106}$$

For small fundamental slip s, the harmonic slip is given as

$$s_n \cong \frac{n \pm 1}{n}, \quad \begin{array}{l} + \text{ for } n = 5, 11, \ldots \\ - \text{ for } n = 7, 13, \ldots \end{array} \tag{6.107}$$

Hence, the fifth- and seventh-harmonic slips are obtained from the above as

Section 6.3 Slip-Energy Recovery Scheme

$$s_5 = \frac{6}{5} = 1.2 \tag{6.108}$$

$$s_7 = \frac{6}{7} \tag{6.109}$$

Substituting for s_5 in equation (6.105), the fifth-harmonic air gap flux linkages are

$$\lambda_{m5} = \frac{I_{r1}}{25\omega_s}\left\{\frac{R_r}{1.2} + j5X_{1r}\right\} = \frac{I_{r1}}{30\omega_s}\{R_r + j6X_{1r}\} \tag{6.110}$$

Similarly, the seventh-harmonic air gap flux linkages are derived as

$$\lambda_{m7} = \frac{I_{r1}}{42\omega_s}\{R_r + j6X_{1r}\} \tag{6.111}$$

Because

$$6X_{1r} \gg R_r \tag{6.112}$$

the fifth- and seventh-harmonic air gap flux linkages can be approximated as

$$\lambda_{m5} \cong \frac{I_{r1}}{5}L_{1r} \tag{6.113}$$

$$\lambda_{m7} \cong \frac{I_{r1}}{7}L_{1r} \tag{6.114}$$

Note that these flux linkages are leading the fundamental rotor current by 90 degrees. Representing the harmonic and fundamental variables and neglecting the leakage inductance effects, in Figure 6.25, leads to a calculation of the sixth-harmonic pulsating torques.

The fundamental of the electromagnetic torque is

$$T_{e1} = \frac{P_a}{\omega_s} \cdot \frac{P}{2} = \frac{3E_1I_{r1}\cos(\angle I_{r1} - \angle E_1)}{\omega_s} \cdot \frac{P}{2} = \frac{3 \cdot \frac{P}{2}L_m I_{m1}\omega_s I_{r1}\sin\phi_{mr}}{\omega_s} = 3 \cdot \frac{P}{2}\lambda_{m1}I_{r1}\sin\phi_{mr} \tag{6.115}$$

Because

$$\phi_{mr} = 90° \tag{6.116}$$

when the rotor leakage inductance is neglected, the fundamental torque becomes

$$T_{e1} = 3 \cdot \frac{P}{2}\lambda_{m1}I_{r1} \tag{6.117}$$

The sixth-harmonic pulsating torque is

$$T_{e6} = 3 \cdot \frac{P}{2}\{\lambda_{m1}(I_{r7} - I_{r5})\cos 6\omega_s t + \lambda_{m5}I_{r1}\sin(90° - 6\omega_s t) + \lambda_{m7}I_{r1}\sin(90° + 6\omega_s t)\} \tag{6.118}$$

$$= 3 \cdot \frac{P}{2}\{\lambda_{m1}(I_{r7} - I_{r5}) + I_{r1}(\lambda_{m5} + \lambda_{m7})\}\cos 6\omega_s t$$

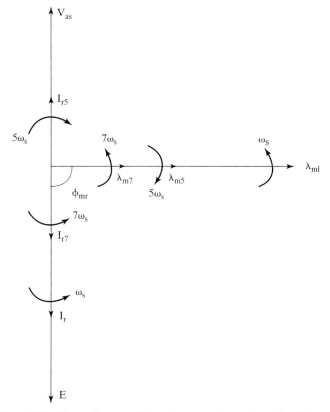

Figure 6.25 Phasor diagram of the rotor currents and air gap flux linkages

The maximum magnitude of the sixth-harmonic torque in terms of the fundamental torque is

$$\frac{|T_{e6}|}{T_{e1}} = \frac{(I_{r7} - I_{r5})}{I_{r1}} + \frac{I_{r1}L_{1r}\left(\frac{1}{5} + \frac{1}{7}\right)}{\lambda_{m1}} = -\frac{2}{35} + \frac{12}{35} \cdot \frac{I_{r1}L_{1r}}{\lambda_{m1}} \quad (6.119)$$

and the fundamental air gap flux linkage is approximated as

$$\lambda_{m1} \cong \frac{V_s}{\omega_s} \quad (6.120)$$

By substituting equation (6.120) into (6.119),

$$\frac{|T_{e6}|}{T_{e1}} = -\frac{2}{35} + \frac{12}{35} \cdot \frac{I_{r1}X_{1r}}{V_s} \quad (6.121)$$

For rated operating condition, the following assumptions are valid:

$$I_{r1} \cong 1 \text{ p.u.} \tag{6.122}$$

$$V_s = 1 \text{ p.u.} \tag{6.123}$$

and, hence,

$$\frac{|T_{e6}|}{T_{e1}} = -\frac{2}{35} + \frac{12}{35} \cdot X_{1r} \tag{6.124}$$

where X_{1r} is the p.u. leakage reactance. X_{1r} is normally less than 2% for large machines and less than 5% for small machines. From these values, it could be realized that the sixth-harmonic pulsating torque is very small and is inconsequential in fan and pump applications.

6.3.8 Harmonic Torques

As the fifth and seventh harmonics are of significant strength, it is necessary to consider and evaluate their own torques. Consider the magnitude of fifth-harmonic torque:

$$T_{e5} = 3 \cdot \frac{P}{2} \cdot \frac{I_{r5}^2 R_r (1 - s_5)}{s_5 \omega_s (1 - s)} \tag{6.125}$$

where

$$I_{r5} = \frac{I_{r1}}{5} \tag{6.126}$$

$$s_5 = \frac{-5\omega_s - \omega_s(1 - s)}{-5\omega_s} = \frac{6 - s}{5} \tag{6.127}$$

Note that the rotor time harmonics are referred to the stator for ease of computation. Substituting the equations (6.126) and (6.127) into equation (6.125) yields

$$T_{e5} = 3 \cdot \frac{P}{2} \cdot \frac{I_{r1}^2 R_r}{25 \omega_s (s - 6)} \tag{6.128}$$

Expressing T_{e5} in terms of T_{e1} gives

$$\frac{T_{e5}}{T_{e1}} = \frac{s}{25(s - 6)} \tag{6.129}$$

The maximum magnitude of this occurs at a slip of one. Its value is

$$\max\left\{\frac{T_{e5}}{T_{e1}}\right\} = -0.008 \tag{6.130}$$

Note that the seventh-harmonic torque is much less and is of opposite sign. Hence, these harmonic torques can be ignored: they are of negligible magnitude.

6.3.9 Static Scherbius Drive

The slip-energy-recovery scheme described so far allows power to flow out of the induction-motor rotor and hence restricts the speed control to subsynchronous mode, i.e., below the synchronous speed. Regeneration is also not possible with this scheme. If the converter system is bidirectional, then both regeneration and supersynchronous modes of operation are feasible. This is explained as follows.

The air gap power is assumed to be a constant and is divided into the slip and mechanical power as given by the equation

$$P_a = P_{sl} + P_m \qquad (6.131)$$

where

$$P_{sl} = sP_a \qquad (6.132)$$
$$P_m = (1 - s)P_a \qquad (6.133)$$

For a positive slip, the mechanical power is less than the air gap power, and the motor speed is subsynchronous. During this operation, the slip power is positive, and it is extracted from the rotor for feeding back to the supply. The drive system is in the subsynchronous motoring mode.

With the same assumption of constant air gap power, during regeneration the air gap power flows to the ac source. The torque is the input; hence, the mechanical power is considered to be negative. To maintain constant air gap power, the slip power becomes negative as seen from equation for negative air gap power. This results in input slip power, even though the slip remains positive. This mode of operation is subsynchronous regeneration.

For driving the rotor above the synchronous speed, the phase sequence of the rotor currents is reversed from that of the stator supply. This creates a field in the rotor opposite in direction to the stator field. The synchronous speed must be equal to the sum of the slip speed and rotor speed. Because the slip speed has become negative with respect to synchronous speed with the reversal of rotor phase sequence, the rotor speed has to increase beyond the synchronous speed by the amount of slip speed. The motor, thus, is in supersynchronous mode of operation, delivering mechanical power. In this mode, the mechanical power is greater than the air gap power. Note also that because of this the slip power is the input, as is seen from the power-balance equation of (6.131), and the slip is negative for supersynchronous operation.

To regenerate during supersynchronous operation, the slip power is extracted from the rotor. This becomes necessary, because the mechanical power is greater than the air gap power and hence the remainder constituting the slip power is available for recovery. Note that the slip remains negative and the rotor phase sequence is opposite to that of the stator. The present operation constitutes the supersynchronous regeneration mode. The various modes of operation are summarized in Table 6.2. A drive capable of all these modes is known as a Scherbius drive.

The converter system has to be bidirectional for the static Scherbius drive. Such a converter can be a cycloconverter, which is implemented with three sets of antiparallel three-phase controlled-bridge converters, discussed in Chapter 3. A

TABLE 6.2 Operational modes of static Scherbius drive

	Mode	Slip	Slip power	P_m
Subsynchronous	Motoring	+	+ (output)	$< P_a$
	Regeneration	+	− (input)	$< P_a$
Supersynchronous	Motoring	−	−	$> P_a$
	Regeneration	−	+	$> P_a$

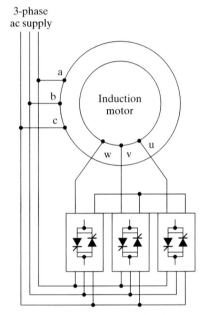

Figure 6.26 Static Scherbius drive

simple implementation of it in a block-diagram schematic is shown in Figure 6.26. This converter has a number of advantages:

(i) simple line commutation;
(ii) shaping of current, and hence power-factor improvement;
(iii) high efficiency;
(iv) suitable and compact for MW applications;
(v) capable of delivering a dc current (and hence the induction motor can be operated as a synchronous motor).

6.3.10 Applications

Slip-energy-recovery motor drives are used in large pumps and compressors in MW power ratings. It is usual to resort to an auxiliary starting means to minimize the

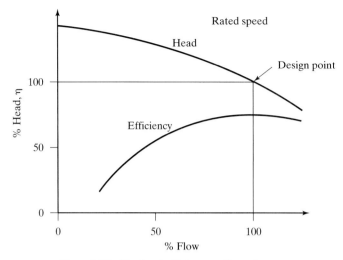

Figure 6.27 Head and efficiency vs. flow of pumps

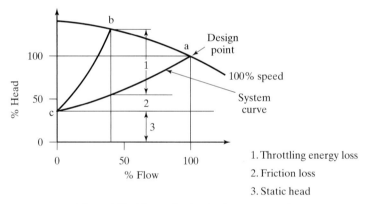

1. Throttling energy loss
2. Friction loss
3. Static head

Figure 6.28 Flow reduction by damper control

converter rating. Wind-energy alternative power systems also use a slip-recovery scheme, but the converter rating cannot be minimized, because of the large variation in operational speed. An illustrative example of such an application can be found in reference [15]. The application is briefly described here.

The pump characteristics at rated speed are shown in Figure 6.27. The nominal operating point is at 1 p.u. flow, where the efficiency is maximum. If the flow has to be reduced, then the characteristic of the pump is changed by adding a resistance to the flow in the form of damper control, shown in Figure 6.28. The new operating point, *b,* is obtained by changing the system curve to move from *ca* to *cb* by adding system resistance from a damper control. At this operating point, the input energy to the pump will not change in proportion to the flow change, resulting in poor effi-

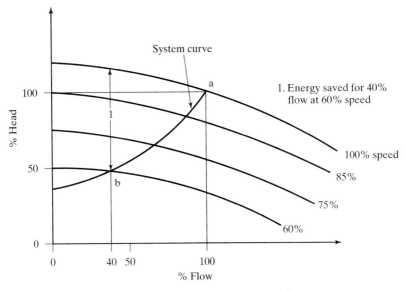

Figure 6.29 Flow control by speed variation

ciency because the input has to supply the throttling energy loss and friction energy loss. The flow can also be reduced if the pump characteristics are changed by reducing the pump speed with a variable-speed-controlled induction motor drive, and then the operating point moves as shown in Figure 6.29. The operating point can be always attempted on the highest-efficiency point by varying the speed. As the speed is reduced, the power input to the pump reduces and hence the input power to the driving motor decreases, resulting in higher energy savings compared to the damper control. Additionally, the combined efficiency of the motor and pump also improves by this method, resulting in considerable energy savings. Because the normal variation in the flow is between 60 and 100% of the rated value, the range of the variable speed control required usually is between 60 and 100% of the rated speed. Restricted-speed control requirement makes the slip-energy-recovery-controlled induction motor drive an ideal choice for this kind of application in the MW power range. It is not unusual to encounter, for example, 9.5-MW drive systems with a speed range of from 1135 to 1745 rpm and 11.5-MW drive systems with a speed range of from 757 to 1165 rpm, many of these in parallel or in series in many pumping stations. Such high-power drives get supplied from a 13.8 kV bus and are equipped with harmonic suppression circuits in the front end of the inverter and large dc reactors for smoothing in the dc link. They will have starting circuits with liquid-cooled resistors to reduce the power capability of the converters, and the inverter will be dual three-phase SCR-bridge converters in series, to keep the displacement factor above 0.88. Note that the displacement factor, which is the power factor considering only fundamentals, will be poor with a single three-phase bridge converter in the system. The converter units are force cooled by air or water in many installations.

6.4 REFERENCES

1. R.E. Bedford and V.D. Nene, "Voltage control of the three phase induction motor by thyristor switching: A time-domain analysis using the α-β-o Transformation," *IEEE Trans. on Industry and General Applications*, vol. IGA-6, pp. 553–562, Nov./Dec. 1970.
2. T.A. Lipo, "The analysis of induction motors with voltage control by symmetrical triggered thyristors," *IEEE Trans. on Power Apparatus and Systems*, vol. PAS-90, pp. 515–525, March/April 1971.
3. William McMurray, "A comparative study of symmetrical three-phase circuits for phase-controlled ac motor drives," *IEEE Trans. on Industry Applications*, vol. IA-10, pp. 403–411, May/June 1974.
4. B. J. Chalmers, S. A. Hamed, and P. Schaffel, "Analysis and application of voltage-controlled induction motors," *Conf. Record, IEEE—Second International Conference on Machines—Design and Applications*, pp. 190–194, Sept. 1985.
5. W. Shepherd, "On the analysis of the three-phase induction motor with voltage control by thyristor switching," *IEEE Trans. on Industry and General Applications*, vol. IGA-4, no. 3, pp. 304–311, May/June 1968.
6. Derek A. Paice, "Induction motor speed control by stator voltage control," *IEEE Trans. on Power Apparatus and Systems*, vol. PAS-87, pp. 585–590, Feb. 1968.
7. John Mungensat, "Design and application of solid state ac motor starter," *Conf. Record, IEEE–IAS Annual Meeting*, pp. 861–866, Oct. 1974.
8. R. Locke, "Design and application of industrial solid state contactor," *Conf. Record, IEEE–IAS Annual Meeting*, pp. 517–523, Oct. 1973.
9. M. S. Erlicki, "Inverter rotor drive of an induction motor," *IEEE Trans. on Power Apparatus and Systems*, vol. PAS-84, pp. 1011–1016, Nov. 1965.
10. A. Lavi and R. J. Polge, "Induction motor speed control with static inverter in the rotor," *IEEE Trans. Power Apparatus and Systems*, vol. PAS-85, pp. 76–84, Jan. 1966.
11. W. Shepherd and J. Stanway, "Slip power recovery in an induction motor by the use of thyristor inverter," *IEEE Trans. on Industry and General applications*, vol. IGA-5, pp. 74–82, Jan./Feb. 1969.
12. T. Wakabayashi et al., "Commutatorless Kramer control system for large capacity induction motors for driving water service pumps," *Conf. Record, IEEE–IAS Annual Meeting*, pp. 822–828, 1976.
13. A. Smith, "Static Scherbius systems of induction motor speed control," *Proc. Institute of Electrical Engineers*, vol. 124, pp. 557–565, 1977.
14. H. W. Weiss, "Adjustable speed ac drive systems for pump and compressor applications," *IEEE Trans. on Industry Applications*, vol. IA-10, pp. 162–167, Jan/Feb 1975.
15. P. C. Sen and K. H. J. Ma, "Constant torque operation of induction motors using chopper in rotor circuit," *IEEE Trans. on Industry Applications*, vol. IA-14, no. 5, pp. 408–414, Sept./Oct. 1978.
16. M. Ramamoorthy and N. S. Wani, "Dynamic model for a chopper controlled slip ring induction motor," *IEEE Trans. on Electronics and Control Instrumentation*, vol. IECI-25, no. 3, pp. 260–266, Aug. 1978.

6.5 DISCUSSION QUESTIONS

1. Is there a difference in gating/triggering control between the induction-motor phase controller and the phase-controlled rectifier used in the dc motor drives?
2. In the phase-controlled induction motor drive, there could be 3 or 2, 2 or 1, 1 or 0 phases conducting, depending on the load. Envision the triggering angles and conduction patterns in these modes of operation.
3. The power switches can be connected in a number of ways to the star- and delta-connected stator of the induction motors. Enumerate the variations.
4. Is there an optimal configuration of the phase controller to be used in a delta-connected stator?
5. A single-phase capacitor-start induction motor is to have a phase controller. Discuss the best arrangement of its connection to the motor winding.
6. Discuss the configuration of the phase controller for a capacitor-start and a capacitor-run single-phase induction motor.
7. Discuss the merits and demerits of using a phase controller as a step-down transformer.
8. The relationship between the triggering angle and the fundamental of the output rms voltage is nonlinear in a phase-controlled induction motor drive. How can it be made linear? Is there any advantage to making it linear?
9. The phase controller is placed in series with the rotor windings as shown in Figure 6.30. Discuss the following:
   (i) the advantages of this scheme over the stator-phase-controlled induction motor drive;
   (ii) the current and voltage ratings of the power switches and their comparison to the stator-phase controller;

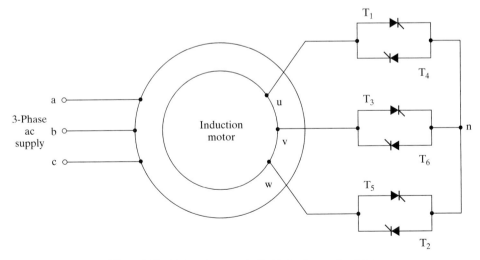

Figure 6.30 Rotor-phase-controlled induction motor drive

310 Chapter 6 Phase-Controlled Induction-Motor Drives

   (iii) the improvement in the input power factor;
   (iv) the improvement in line-current waveform and the reduction of harmonics;
   (v) the disadvantages of this scheme;
   (vi) the reversibility of the motor drive;
   (vii) the control-signal generation for this scheme;
   (viii) any soft-start possibility;
   (ix) closed-loop control and its requirements.

10. Similarly to a phase-controlled converter (used in dc motor drives), whose α is limited in the inversion mode, will there be a need to limit the maximum value of α in the phase-controlled induction motor drive? Explain.

11. Compare the performance of the rotor-phase- and chopper-controlled induction motor drives.

12. Discuss the performance, analysis, design, merits, and demerits of the rotor chopper-controlled induction motor drive shown in Figure 6.31. (Hint: References 15 and 16.)

13. Compare the stator-phase and slip-energy-recovery-controlled induction motor drives on the basis of
    (i) input power factor
    (ii) efficiency
    (iii) cost
    (iv) control complexity
    (v) pulsating torque (6th harmonic)
    (vi) feedback control
    (vii) range of speed control

14. What is the effect of the inductor in the dc link?

15. Suppose that the inductor in the dc link is replaced by an electrolytic capacitor. How would this affect the performance of the phase-controlled converter and consequently the slip-energy recovery?

16. The slip-energy-controlled drive, with a current loop controlling the dc link current, behaves like a separately-excited dc motor. Will a change in the load affect the flux and hence the torque constant?

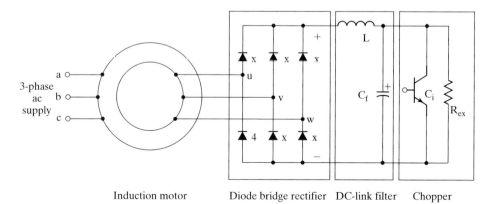

Induction motor   Diode bridge rectifier   DC-link filter   Chopper

Figure 6.31 Rotor chopper-controlled induction motor drive

17. Discuss the feasibility of replacing the thyristor bridge converter with a GTO- or transistor-based converter to control the harmonics fed to the utility.
18. The diode-bridge rectifier in the dc link can be replaced by a GTO converter. By suitable control of this controlled converter, is it possible to control the current to eliminate the 6th harmonic in the dc link and reduce the size of the dc inductor?
19. At the time of starting (assuming that there is no auxiliary starting means), what is the value of the triggering angle in the phase-controlled converter?
20. The following applications need a variable-speed motor drive:
    (i) a 20,000-hp pump;
    (ii) a 50-hp fan.
    The options available are stator-phase- and slip-energy-recovery-controlled induction motor drives. Which one is to be recommended for the above applications? Explain the reasons for your choice.
21. For open-loop speed control, NEMA class D induction motors are preferred in phase-controlled motor drives. Is the same choice valid for open-loop slip-energy-recovery-controlled-induction motor drives?
22. What are the modifications required in the controller of the phase-controlled converter used in a dc motor drive for adaptation to the slip-energy-recovery-controlled induction motor drive?
23. Can a slip-energy-controlled induction motor be reversed in speed? How is it done?

6.6 EXERCISE PROBLEMS

1. Determine the range of speed control obtained with a phase-controlled induction motor drive. The details are given below.
   1 hp, 2 pole, 230 V, 3 phase, star connected, 60 Hz, $R_s = 1\,\Omega$, $R_r = 6.4\,\Omega$, $X_m = 75\,\Omega$, $X_{lr} = 3.4\,\Omega$, $X_{ls} = 3\,\Omega$, and full-load slip is 0.1495. Load is a fan requiring 0.5 p.u. torque at 3200 rpm. α is limited to a maximum of 135 degrees.
2. Calculate and draw the efficiency of the drive given in Example 1 as a function of rotor speed.
3. A 1000-hp induction motor is to be started with a phase controller. The current is limited to 1 p.u. Find the starting torque and the range of triggering-angle variation to run the motor from standstill to 0.9 p.u. speed at a constant load torque of 0.05 p.u. The motor details are as follows:
   Base kVA = 850 kVA, Base voltage = 2700 V, Rated power = 1000 hp, Rated frequency = 50 Hz, P = 4, $R_s = 0.041\,\Omega$, $R_r = 0.1663\,\Omega$, $X_m = 80\,\Omega$, $X_{lr} = 0.2\,\Omega$, $X_{ls} = 0.2\,\Omega$, full-load slip = 0.0801 (base speed = full load speed).
4. Calculate the impact of rotor resistance increase by 80% on the speed-control range and efficiency of the drive given in problem 3.
5. The problem given in Example 1 is driving a fan load. At 1500 rpm, the load torque is 100%. Determine the open-loop stable operating-speed range for the triggering angles of $45° < \alpha < 135°$.
6. Determine approximately the magnitude of the two predominant harmonics present in the phase-controlled induction motor drive. Are these harmonics load-dependent?
7. A movie theater has been reconverted to a house one-fourth of its original seating. The ventilation fans have to be accordingly derated to reduce the operational costs. Phase

controllers are introduced to reduce the speed and control the induction motors driving the fans depending upon the audience in the theater. There are four fans in the theater. The new speed of the motors is to be at three-fourths of rated speed. Before the conversion, the motor supplied rated power at rated speed to the fan. (From this, you may find the load constant.)

(i) Calculate the efficiency at this speed for the following motor details:

18.6 kW, 4 pole, 3 phase, Y-connected, 60 Hz, 230 V, $R_s = 0.08\ \Omega$, $R_r = 0.4\ \Omega$, $X_m = 5.54\ \Omega$, $X_s = 5.7\ \Omega$, $X_r = 5.76\ \Omega$, Full load slip = 0.344, friction and windage losses = 1% of its rating.

(ii) Find the triggering angle required to operate the motor at its new maximum speed.

(iii) Assume that the induction motor has slip-ring rotor and that external resistors could be introduced in the rotor phases to control the speed. Find the per-phase external resistor required (referred to the stator) to operate it at its new maximum speed. Compare its operational efficiency with that of the phase controller.

(iv) Which solution would you prefer between the phase controller and the external resistors for this particular problem? Discuss it briefly.

8. A small ski-lift conveyor is driven by a phase-controlled induction motor. The motor-load constants follow:

NEMA D, 5 hp, 2 pole 208 V, 60 Hz, star connected, $R_s = 0.6\ \Omega$, $R_r = 1.4\ \Omega$, $M = 61$ mH, $L_{lr} = 1.9$ mH, $L_{ls} = 1.9$ mH, full-load slip = 0.1892.

The load torque varies from 5 to 10 N·m, depending on the number of passengers. Calculate the drive efficiency as a function of speed for load torques of 5 and 10 N·m and compare the savings achieved over the rotor-resistance-controlled induction motor drive.

9. It is desired to have a continuous speed control from 0.6 to nearly 1.0 p.u. speed for a pump application. The motor parameters are given in problem 3. The pump has a requirement of 0.8 p.u. torque at 0.95 p.u. speed. Calculate the efficiency vs. speed and compare it with the solution of problem 3. The rotor-to-stator turns ratio is assumed to be 0.65 and a slip-energy-recovery scheme is used. The transformer turns ratio is 0.5.

10. Calculate the ratings of the converter subsystem given in problem 9.

11. Draw the torque–speed characteristics as a function of triggering angles for problem 9.

CHAPTER 7

Frequency-Controlled Induction Motor Drives

7.1 INTRODUCTION

The speed of the induction motor is very near to its synchronous speed, and changing the synchronous speed results in speed variation. Changing the supply frequency varies the synchronous speed. The relationship between the synchronous speed and the frequency is given by

$$n_s = \frac{120 f_s}{P} \tag{7.1}$$

where n_s is the synchronous speed in rpm, f_s is the supply frequency in Hz, and P is the number of poles. The utility supply is of constant frequency, so the speed control of the induction motor requires a frequency changer to change the speed of the rotor. The frequency changers, circuit configurations, operation and input sources are discussed briefly in this chapter. The interaction of the frequency changers and the induction motor is studied in detail. They pertain to the steady-state performance, harmonics and their impact on the performance, control of harmonics, and drive-control strategies and their impact on drive performance. The enhancement of motor-drive efficiency is an integral aspect of the study and is considered with illustrative examples. The dynamic models of the various drive systems are derived and their performance simulated and analyzed as they pertain to design aspects.

7.2 STATIC FREQUENCY CHANGERS

There are basically two types of static frequency changers: direct and indirect. The direct frequency changers are known as cycloconverters; they convert an ac supply of utility frequency to a variable frequency. The cycloconverter for a single phase

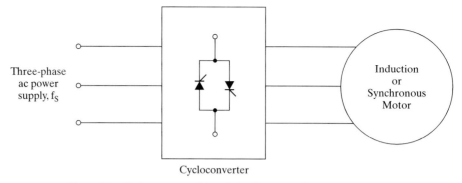

Figure 7.1 Cycloconverter-driven induction or synchronous motor drive

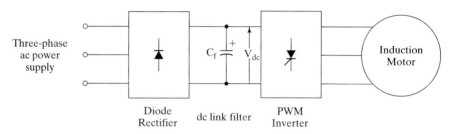

Figure 7.2 PWM inverter fed induction motor drive

consists of an antiparallel dual-phase-controlled converter. It can be a single- or three-phase-controlled converter; the latter version is very common in practice. Symbolically, a cycloconverter-driven ac motor drive is shown in Figure 7.1. The output frequency has a range of from 0 to $0.5f_s$. For better waveform control of the output voltage, the frequency is limited to $0.33f_s$. The smaller range of frequency variation is suitable for low-speed, large-power applications, such as ball mills and cement kilns. For a majority of applications, a wide range of frequency variation is desirable. In that case, indirect frequency conversion methods are appropriate. The indirect frequency changer consists of a rectification (ac to dc) and an inversion (dc to ac) power conversion stage. They are broadly classified depending on the source feeding them: voltage or current sources. In both these sources, the magnitude should be adjustable. The output frequency becomes independent of the input supply frequency, by means of the dc link.

If rectification is uncontrolled, the voltage and its frequency are controlled in the inverter; a pulse-width-modulated inverter-fed induction motor drive is shown in Figure 7.2. The dc link filter consists of a capacitor to keep the input voltage to the inverter constant and to smooth the ripples in the rectified output voltage. The dc link voltage cannot reverse; it is a constant, so this is a voltage-source drive. The advantage of using the diode-bridge rectifier in the front end is that the input line power factor is nearly unity, but there is also a disadvantage, in that the power cannot be recovered from the dc link for feeding back into the input supply. The regen-

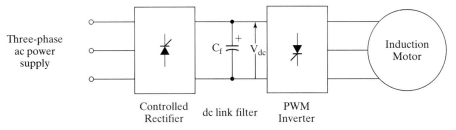

Figure 7.3 Variable-voltage, variable-frequency (VVVF) induction motor drive

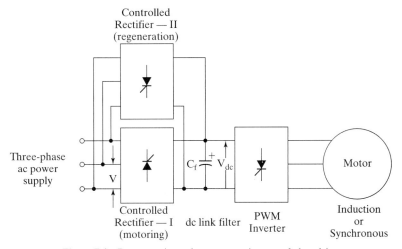

Figure 7.4 Regenerative voltage-source inverter-fed ac drive

erative power has to be handled by some other controller arrangement and is discussed later.

Separating the magnitude and frequency control functions in the controlled rectifier and inverter, respectively, gives a configuration shown in Figure 7.3. This configuration is known as a variable-voltage, variable-frequency induction motor drive. It has a disadvantage: the power factor is low at low voltages, from phase control. Refer to Chapter 3 on the operation of the phase-controlled rectifier. To recover the regenerative energy in the dc link, the direction of the dc link current to the phase-controlled rectifier has to be reversed; the dc link voltage cannot reverse through the antiparallel diodes across the inverter bridge. Hence, an antiparallel-controlled rectifier is required to handle the regenerative energy, as shown in Figure 7.4. Similarly to the voltage-source induction motor drives, the current-source drives have a PWM control and variable-current variable-frequency control.

In the PWM current source, shown in Figure 7.5, instantaneous current control on the ac side is enforced by a set of fast-acting current-control loops. Inverter switching is based on the current error signals to force the actual currents to track their

316 Chapter 7 Frequency-Controlled Induction Motor Drives

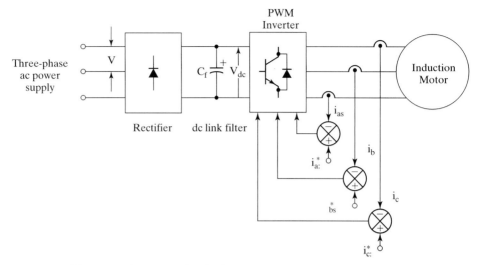

Figure 7.5 Current-regulated voltage-source-driven induction motor drive

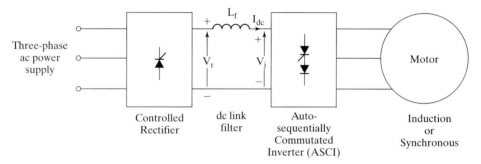

Figure 7.6 Current-source inverter-driven induction or synchronous motor drive

respective commanded values. The input source is voltage; because of this fact, this arrangement is sometimes referred to as a *current-regulated induction motor drive*.

In the variable-current, variable-frequency (VCVF) systems, the current magnitude and frequency control are exercised independently by the controlled rectifier with an inner current loop and autosequentially commutated inverter (ASCI), respectively. This arrangement is capable of four-quadrant operation and is discussed in detail in later subsections. To maintain a current source, the dc link filter has an inductor and a current loop enforcing the dc link current command, very much like the inner current loop in the phase-controlled dc motor drive. The current source drive is shown in Figure 7.6. The disadvantages of this drive scheme are the need for a large dc link inductor and a set of commutation capacitors. A general schematic of classification of the static frequency changers is shown in Figure 7.7.

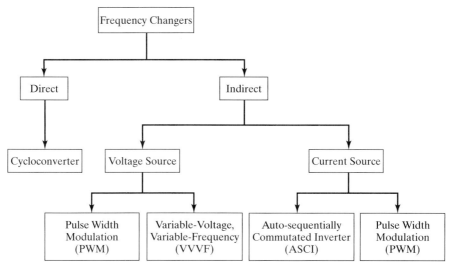

Figure 7.7 Classification of frequency changers

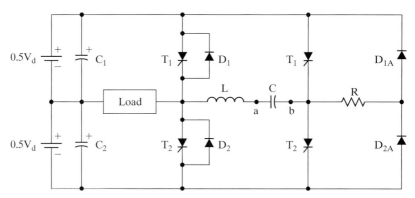

Figure 7.8 Half-bridge voltage-source inverter

7.3 VOLTAGE-SOURCE INVERTER

A commonly used SCR inverter is the modified McMurray inverter in industrial motor drives. The configuration and principle of operation of this inverter are explained in this section. For ease of development, a half-bridge inverter is considered. Then it is extended to include a full-bridge inverter. The transistorized inverter is discussed, and a comparison between SCR and transistor inverters is given that is based on the number of switches.

7.3.1 Modified McMurray Inverter

A half-bridge modified McMurray inverter is shown in Figure 7.8. The inductor L and capacitor C are known as the commutating inductor and capacitor, respectively.

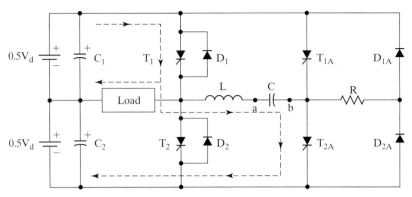

Figure 7.9a (a) Half-bridge voltage-source inverter charging

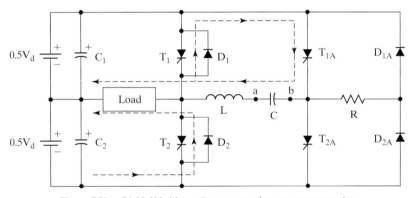

Figure 7.9b (b) Half-bridge voltage-source inverter commutation

This inverter accepts a dc voltage (source) and provides an alternating voltage of variable frequency across its load. Its operation is described as follows.

The circuit arrangement assumes that the snubbers are placed across the devices to limit dv/dt and its effects. At the time of starting, let the voltage across the commutating capacitor C be zero.

Step 1: T_1 and T_{2A} are gated on. The sequence of operation is shown in Figure 7.9(a). A current is established from the dc source which partly goes through switch T_1 and the load; the rest goes through T_1, L, C, and T2A. The capacitor C is charged with *a* positive and *b* negative. The capacitor will charge to a voltage greater than twice the source voltage (V_d) in an LC circuit. The polarity of *b* and the anode of T2A become very much negative to the cathode of T2A, thus turning off the device. Note that the current through the load is maintained. The excess charge in the capacitor due to its overvoltage will forward bias D_1 if load current is zero or flows through the load and D_{2A}, circulating a current through the bottom source voltage $0.5V_d$, L, C, and R. When the capacitor voltage is equal to the source voltage, this current ceases.

Step 2: To turn T_1 off, T_{1A} is turned on as shown in Figure 7.9(b), thus applying a positive voltage from capacitor C to the cathode of T_1, forcing it to cease conduc-

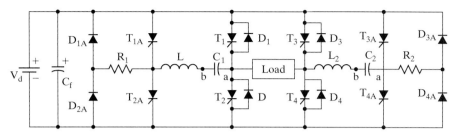

Figure 7.10 Full-bridge voltage-source inverter

tion. An oscillation through L, C, T_1, and T_{1A} occurs. When the capacitor current exceeds load current, the current in T_1 completely ceases. The current in excess of load current flows through D_1 now. This maintains a negative voltage across T_1, to let it recover to withstand forward voltage. The voltage across C is reversed during commutation. Assuming that the load is reactive, the load current will be in the same direction, thus forcing D_2 to be forward biased. Because of the action of diode D_2, the load voltage will become negative. Similarly, the negative-half-cycle output-voltage operation can be realized.

The frequency of the load voltage is determined by the rate at which T_1 and T_2 are enabled. This half-bridge inverter does not fully utilize the source voltage. A half-bridge inverter is placed on each side of the load to make a full-bridge inverter, shown in Figure 7.10.

7.3.2 Full-Bridge-Inverter Operation

The main thyristors are commutated by the auxiliary thyristors, which are denoted by an additional A in their subscripts. The operation of this circuit is explained here.

Step 1: T_1, T_{2A}, T_4, and T_{3A} are gated on. C_1 and C_2 are charged with *a* positive and *b* negative. Load voltage and load current are positive. The excess charge in C_1 is drained through $V_d, D_{2A}, R_1, L_1, C_1$, and D_1 if there is no load current and through the load if there is a load current. The excess charge in C_2 is drained through V_d, D_4, L_2, C_2, R_2, and D_{3A}.

Step 2: To provide a zero voltage across the load for part of the positive half cycle, turn T_{1A} on and thus turn T_1 off. The load is shorted via D_2, load, and T_4, thus driving the voltage to zero. The zeroing of load voltage is used to change the effective volt–sec across the load.

Step 3: T_1 and T_4 are turned off by turning on T_{1A} and T_{4A}, respectively, to prepare for the negative half-cycle across the load. The priming of commutating capacitors and gating on of T_2 and T_3 take place in a manner similar to the positive-half-cycle operation described in steps 1 and 2.

Note that this single-phase full-bridge inverter is identical to the four-quadrant chopper described in Chapter 4, except that the present one is an SCR version. A three-phase inverter is made of three such single-phase inverters.

A chopper with self-commutating power switches, such as a transistor, has been studied in Chapter 4. It is recalled that a four-quadrant chopper is also a single-phase

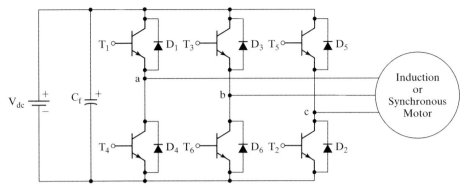

Figure 7.11 Voltage-source inverter with transistors

TABLE 7.1 Comparison of three-phase SCR and self-commutating inverters

Comparison Aspect	SCR Inverter	Self-Commutating Inverter
SCRs/Transistors	24	6
Diodes	24	6
Inductors/Capacitors	6/6	–
Fundamental RMS output phase voltage	$0.78 V_d$	$0.45 V_d$

inverter. A combination of three of the two-quadrant choppers, i.e., with two power switches, results in a three-phase inverter. The voltage available to one phase is always less than the full source or dc link voltage in such a case. A three-phase self-commutating inverter with transistor switches is shown in Figure 7.11. Note the simplicity of the configuration and the minimum use of power devices: the transistors are self-commutating. A comparison of power switches and other factors for the SCR and self-commutating inverters is given in Table 7.1. Note that the auxiliary thyristors are not rated to be equal to the main thyristors: they operate only for a fraction of time compared to the main SCRs. The same argument applies equally to the rating of the auxiliary diodes.

7.4 VOLTAGE-SOURCE INVERTER-DRIVEN INDUCTION MOTOR

7.4.1 Voltage Waveforms

A generic self-commutating three-phase inverter is shown in Figure 7.12 in schematic, with its output connected to the three stator phases of a wye-connected induction motor. The gating signals and the resulting line voltages are shown in Figure 7.13. The power devices are assumed to be ideal: when they are conducting, the voltage across them is zero; they present an open circuit in their blocking mode. The phase voltages are derived from the line voltages in the following manner by assuming a balanced three-phase system.

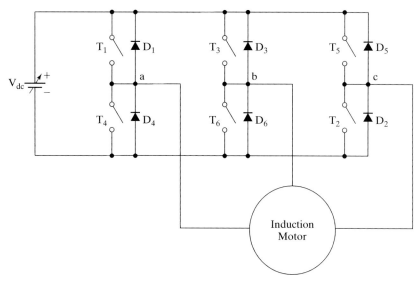

Figure 7.12 A schematic of the generic inverter-fed induction motor drive

The line voltages in terms of the phase voltages in a three-phase system with phase sequence *abc* are

$$V_{ab} = V_{as} - V_{bs} \tag{7.2}$$

$$V_{bc} = V_{bs} - V_{cs} \tag{7.3}$$

$$V_{ca} = V_{cs} - V_{as} \tag{7.4}$$

where V_{ab}, V_{bc}, and V_{ca} are the various line voltages and V_{as}, V_{bs}, and V_{cs} are the phase voltages. Subtracting equation (7.4) from equation (7.2) gives

$$V_{ab} - V_{ca} = 2V_{as} - (V_{bs} + V_{cs}) \tag{7.5}$$

In a balanced three-phase system, the sum of the three phase voltages is zero:

$$V_{as} + V_{bs} + V_{cs} = 0 \tag{7.6}$$

Using equation (7.6) in (7.5) shows that the difference between line voltages V_{ab} and V_{ca} is

$$V_{ab} - V_{ca} = 3V_{as} \tag{7.7}$$

from which the phase *a* voltage is given by

$$V_{as} = \frac{V_{ab} - V_{ca}}{3} \tag{7.8}$$

Similarly, the *b* and *c* phase voltages are

$$V_{bs} = \frac{V_{bc} - V_{ab}}{3} \tag{7.9}$$

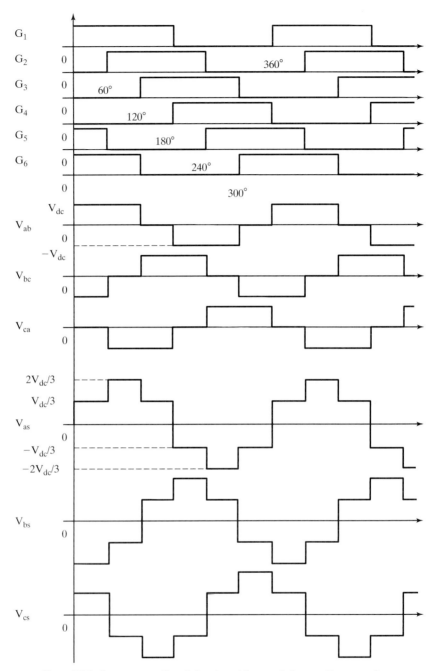

Figure 7.13 Inverter gate (base) signals and line- and phase-voltage waveforms

$$V_{cs} = \frac{V_{ca} - V_{bc}}{3} \qquad (7.10)$$

The phase voltages derived from line voltages are shown in Figure 7.13. Although the line-to-line voltages are 120 electrical degrees in duration, the phase voltages are six-stepped and of quasi-sine waveforms. These periodic voltage waveforms, when resolved into Fourier components, have the following form:

$$v_{ab}(t) = \frac{2\sqrt{3}}{\pi} V_{dc} \left(\sin \omega_s t - \frac{1}{5} \sin 5\omega_s t + \frac{1}{7} \sin 7\omega_s t - \ldots \right) \qquad (7.11)$$

$$v_{bc}(t) = \frac{2\sqrt{3}}{\pi} V_{dc} \left\{ \sin(\omega_s t - 120°) - \frac{1}{5} \sin(5\omega_s t - 120°) + \frac{1}{7} \sin(7\omega_s t - 120°) - \ldots \right\} \qquad (7.12)$$

$$v_{ca}(t) = \frac{2\sqrt{3}}{\pi} V_{dc} \left\{ \sin(\omega_s t + 120°) - \frac{1}{5} \sin(5\omega_s t + 120°) + \frac{1}{7} \sin(7\omega_s t + 120°) - \ldots \right\} \qquad (7.13)$$

The phase voltages are shifted from the line voltages by 30 degrees, and their magnitudes are $\frac{2}{\pi} V_{dc}$. Only the fundamental produces useful torque, and hence only it needs to be considered for the steady-state performance evaluation of inverter-fed ac motor drives. In this regard, the fundamental rms phase voltage for the six-stepped waveform is

$$V_{ph} = \frac{V_{as}}{\sqrt{2}} = \frac{2}{\pi} \cdot \frac{V_{dc}}{\sqrt{2}} = 0.45 V_{dc} \qquad (7.14)$$

7.4.2 Real Power

The dc link transfers real power to the inverter and induction motor. Assuming that the harmonic powers are negligible, and considering only the fundamental input power to the induction motor,

$$P_i = V_{dc} I_{dc} = 3 V_{ph} I_{ph} \cos \phi_1 \qquad (7.15)$$

where I_{dc} is the average steady dc link current, I_{ph} is the phase current, and ϕ_1 is the fundamental power-factor angle in the induction motor. Substituting for V_{ph} from equation (7.14) into (7.15), we get

$$I_{dc} = 1.35 I_{ph} \cos \phi_1 \qquad (7.16)$$

7.4.3 Reactive Power

The induction motor requires reactive power for its operation. Since the dc link does not supply the reactive power, this reactive power is supplied by the inverter. It is done by the switching in the inverter, and hence the inverter can be considered a reactive power generator. The facility to turn on and off the phase currents allows the inverter to vary the phase angle between the current and voltages. This ability has been endowed by solid-state power switching and has enormous impact on the control of ac machines. This subject is treated in subsequent sections and chapters. The fundamental input reactive power demand of the induction motor is

$$Q_i = 3V_{ph}I_{ph} \sin \phi_1 \qquad (7.17)$$

The inverter is therefore rated for $3V_{ph}I_{ph}$ and in volt-amp, which is the apparent power. If the load power factor is low, then the inverter rating goes up.

7.4.4 Speed Control

Speed control is achieved in the inverter-driven induction motor by means of variable frequency. Apart from frequency, the applied voltage needs to be varied, to keep the air gap flux constant and not let it saturate. This is explained as follows.

The air gap induced emf in an ac machine is given by

$$E_1 = 4.44 k_{\omega 1} \phi_m f_s T_1 \qquad (7.18)$$

where $k_{\omega 1}$ is the stator winding factor, ϕ_m is the peak air gap flux, f_s is the supply frequency, and T_1 is the number of turns per phase in the stator. Neglecting the stator impedance, $R_s + jX_{ls}$, the induced emf approximately equals the supply-phase voltage. Hence,

$$V_{ph} \cong E_1 \qquad (7.19)$$

The flux is then written as

$$\phi_m \cong \frac{V_{ph}}{K_b f_s} \qquad (7.20)$$

where

$$K_b = 4.44 k_{\omega 1} T_1 \qquad (7.21)$$

If K_b is constant, flux is approximately proportional to the ratio between the supply voltage and frequency. This is represented as

$$\phi_m \propto \frac{V_{ph}}{f_s} \propto K_{vf} \qquad (7.22)$$

where K_{vf} is the ratio between V_{ph} and f_s.

From equation (7.22), it is seen that, to maintain the flux constant, K_{vf} has to be maintained constant. Therefore, whenever stator frequency is changed to obtain speed control, the stator input voltages have to be changed accordingly to maintain the air gap flux constant. This particular requirement compounds the control problem and sets it apart from the dc motor control, which requires only the voltage control. The implications of such a requirement during the transients made the dc drives preferable over the ac drives until the 1970s.

Section 7.4 Voltage-Source Inverter-Driven Induction Motor

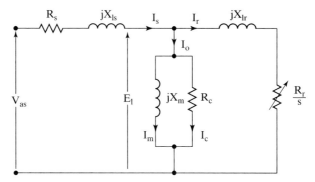

Figure 7.14 Equivalent circuit of the induction motor

A number of control strategies have been formulated, depending on how the voltage-to-frequency ratio is implemented:

 (i) Constant volts / Hz control
 (ii) Constant slip-speed control
 (iii) Constant air gap flux control
 (iv) Vector control

These control strategies are considered individually, the first three in the following sections; the fourth is reserved for the next chapter.

7.4.5 Constant Volts/Hz Control

7.4.5.1 Relationship between voltage and frequency. The applied phase voltage from the equivalent circuit shown in Figure 7.14 is

$$V_{as1} = E_1 + I_{sl}(R_s + jX_{ls}) \tag{7.23}$$

where I_{sl} is the fundamental stator phase current.

The dependence of the phase voltage on the stator impedance drop in p.u. is derived as follows.

$$\frac{V_{as1}}{V_b} = \frac{E_1}{V_b} + \frac{I_{sl}}{V_b}(R_s + jX_{ls}) \tag{7.24}$$

or

$$V_{asn} = E_{1n} + I_{sln}(R_{sn} + jX_{lsn}) \tag{7.25}$$

where

$$V_{asn} = \frac{V_{as1}}{V_b} \text{ (p.u.)}$$

$$E_{1n} = \frac{E_1}{V_b} = \frac{j(L_m I_m)\omega_s}{\lambda_b \omega_b} = j\left(\frac{\lambda_m}{\lambda_b}\right)\left(\frac{\omega_s}{\omega_b}\right) = j\lambda_{mn}\omega_{sn} \tag{7.26}$$

$$I_{sn} = \frac{I_{sl}}{I_b}, \text{p.u.} \tag{7.27}$$

$$R_{sn} = \frac{I_b R_s}{V_b}, \text{p.u.} \tag{7.28}$$

$$X_{1n} = \frac{I_b L_{1s} \omega_s}{V_b} = L_{1sn} \omega_{sn}, \text{p.u.} \tag{7.29}$$

$$\lambda_{mn} = \frac{\lambda_m}{\lambda_b}, \text{p.u.} \tag{7.30}$$

The p.u. fundamental input phase voltage is written as

$$V_{asn} = I_{sn} R_{sn} + j\omega_{sn}(\lambda_{mn} + L_{1sn} I_{sn}) \text{ (p.u.)} \tag{7.31}$$

where L_{lsn} is the p.u. stator leakage inductance and ω_{sn} is the p.u. stator frequency. Substituting equation (7.31) into (7.30) gives the normalized input-phase stator voltage:

$$V_{asn} = \sqrt{(I_{sn} R_{sn})^2 + \omega_{sn}^2 (\lambda_{mn} + L_{1sn} I_{sn})^2} \text{ (p.u.)} \tag{7.32}$$

where R_{sn} is the p.u. stator resistance and λ_{mn} is the p.u. airgap flux linkages.

For constant air gap flux linkages of 1 p.u., the p.u. applied voltage vs. p.u. stator frequency is shown in Figure 7.15. The stator resistance and leakage inductance are 0.026 and 0.051 p.u., respectively, for this graph. The graphs are shown for 0.25-, 0.5-, 1-, and 2-p.u. stator currents. Even though, for constant flux, the relationship between the applied voltage and the stator frequency appears linear, note that it

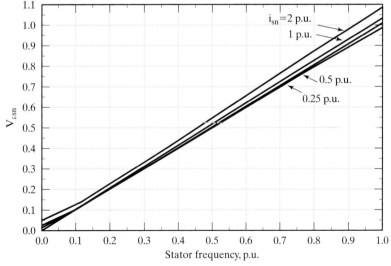

Figure 7.15 Normalized stator phase voltage vs. stator frequency for various stator currents

does not go through the origin; we need a small voltage to overcome the stator resistance at zero frequency.

Figure 7.15 demonstrates that the volt/Hz ratio needs to be adjusted in dependence on the frequency, the air gap flux magnitude, the stator impedance, and the magnitude of the stator current. Such a complex implementation is not desirable for low-performance applications, such as fans and pumps; there it is usual to have a preprogrammed volts-to-frequency relationship, as shown in Figure 7.16. This allows for an offset voltage at zero stator frequency to overcome the stator resistance drop, which is kept as an adjustable parameter in the inverters. The relationship between the applied phase voltage and frequency in general is written as

$$V_{as} = V_o + K_{vf}f_s \tag{7.33}$$

where

$$V_o = I_{sl}R_s \tag{7.34}$$

V_o is the offset voltage to overcome the stator resistive drop. This can be converted into the dc link voltage by converting equation (7.14) into normalized form and combining it with equation (7.33). It is carried out as follows.

$$V_{asn} = \frac{V_{ph}}{V_b} = 0.45 \times \frac{V_{dc}}{V_b} = 0.45 V_{dcn} \tag{7.35}$$

$$V_{on} = \frac{V_o}{V_b} \text{ (p.u.)}$$

$$E_{1n} = \frac{E_1}{V_b} = \frac{K_{vf}f_s}{K_{vf}f_b} = f_{sn}$$

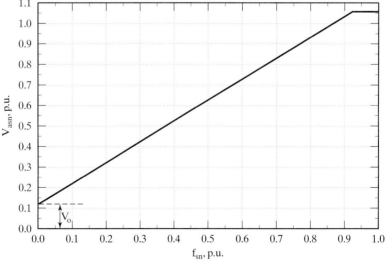

Figure 7.16 General implementation of the voltage-to-frequency profile in inverter-fed induction motor drives

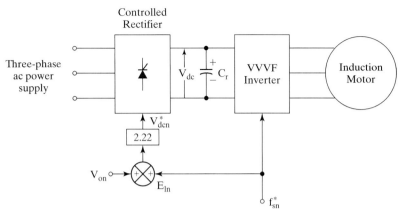

Figure 7.17 Implementation of volts/Hz strategy in inverter-fed induction motor drives

Hence,

$$0.45 V_{dcn} = V_{on} + f_{sn} = V_{on} + E_{ln} \qquad (7.36)$$

or

$$V_{dcn} = 2.22\{V_{on} + f_{sn}\} \qquad (7.37)$$

where V_{dcn} is the dc link voltage in p.u. and E_{ln} is the normalized induced emf.

7.4.5.2 Implementation of volts/Hz strategy. An implementation of the constant volts/Hz control strategy for the inverter-fed induction motor in open loop is shown in Figure 7.17. This type of variable-speed drive is used in low-performance applications where precise speed control is not necessary. The frequency command f_{sn}^* is enforced in the inverter and the corresponding dc link voltage is controlled through the front-end converter. The offset voltage, V_{on}, is added to the voltage proportional to the frequency, and they are multiplied by 2.22 to obtain the dc link voltage.

Some problems encountered in the operation of this open-loop drive are the following:

1. The speed of the motor cannot be controlled precisely, because the rotor speed will be less than the synchronous speed. Note that stator frequency, and hence the synchronous speed, is the only variable controlled in this drive.
2. The slip speed, being the difference between the synchronous and electrical rotor speed, cannot be maintained: the rotor speed is not measured in this drive scheme. This can lead to operation in the unstable region of the torque–speed characteristics.
3. The effect discussed in 2 can make the stator currents exceed rated current by many times, thus endangering the inverter–converter combination.

Section 7.4 Voltage-Source Inverter-Driven Induction Motor

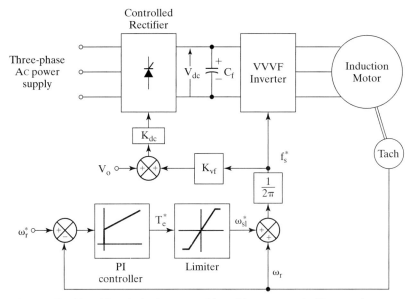

Figure 7.18 Closed-loop induction motor drive with constant volts/Hz control strategy

These problems are, to an extent, overcome by having an outer speed loop in the induction motor drive, shown in Figure 7.18. The actual rotor speed is compared with its commanded value, ω_r^*, and the error is processed through a controller, usually a PI, and a limiter to obtain the slip-speed command, ω_{sl}^*. The limiter ensures that the slip-speed command is within the maximum allowable slip speed of the induction motor. The slip-speed command is added to electrical rotor speed to obtain the stator frequency command. Thereafter, the stator frequency command is processed as in an open-loop drive. K_{dc} is the constant of proportionality between the dc load voltage and the stator frequency.

In the closed-loop induction motor drive, the limits on the slip speed, offset voltage, and reference speed are externally adjustable variables. This external adjustment allows the tuning and matching of the induction motor to the converter and inverter and the tailoring of its characteristics to match the load requirements.

Example 7.1

Find the relationship between the dc link voltage and the stator frequency for the closed-loop implementation of a volts/Hz inverter-fed induction motor drive. The motor parameters are as follows:

$$5 \text{ hp}, 200 \text{ V}, 60 \text{ Hz}, 3 \text{ phase, star connected, 4 pole, 0.86 pf and 0.82 efficiency.}$$

$$R_s = 0.277 \ \Omega, R_r = 0.183 \ \Omega, X_m = 20.30 \ \Omega, X_{ls} = 0.554 \ \Omega, X_{lr} = 0.841 \ \Omega$$

Solution The constants required for the implementation of the closed-loop drive are: $K_{vf}, V_o,$ and the maximum slip speed.

330 Chapter 7 Frequency-Controlled Induction Motor Drives

(i) Maximum slip speed $\cong \dfrac{R_r}{(L_{lr} + L_{ls})} \cong \dfrac{0.183}{(0.84 + 0.554)/377.0} = 49.5$ rad/sec

(ii) $V_o = I_{sl}R_s$

where

I_{sl} = Rated stator phase current = $\dfrac{hp \times 745.6}{3V_{ph} \times pf \times \text{efficiency}} = \dfrac{5 \times 745.6}{3 \times 115.5 \times 0.86 \times 0.82} = 15.26$ A

Hence

$$V_o = 15.26 \times 0.277 = 4.23 \text{ V}$$

(iii) $K_{vf} = \dfrac{(V_{ph} - V_o)}{f_s} = \dfrac{\left(\dfrac{200}{\sqrt{3}} - 4.23\right)}{60} = 1.854$ volts/Hz

(iv) The dc link voltage in terms of stator frequency is given by

$$V_{dc} = 2.22 V_{as} = 2.22\{V_o + K_{vf}f_s\} = 2.22\{4.23 + 1.854 f_s\} = (9.4 + 4.12 f_s), \text{V}$$

7.4.5.3 Steady-state performance. The steady-state performance of the constant-volts/Hz-controlled induction motor drive is computed by using the fundamental applied phase voltage given in the expression (7.33). In the equivalent circuit, the following steps are taken to compute the steady-state performance:

Step 1: Start with minimum stator frequency and zero slip.
Step 2: Calculate magnetizing, core-loss, rotor, and stator phase currents.
Step 3: Calculate the electromagnetic torque, power output, copper, and core losses.
Step 4: Calculate input power factor and efficiency.
Step 5: Increment the slip; then go to step 2, unless $s = s_{max}$.
Step 6: Increment the stator frequency; then go to step 1, unless $f_s = f_{s\,max}$.

A set of drive-torque-vs.-speed characteristics is given in Figure 7.19 for the motor constants given in Example 7.1. The dissimilarity between the torque–speed characteristics for various stator frequencies is quite notable. The peak torque in the motoring region decreases as the stator frequency decreases, contrary to the generator action. Note that the offset voltage is set at zero for this figure. The effects of offset-voltage variations are shown in Figure 7.20 for the same motor at a stator frequency of 15 Hz. The offset voltage increases the induced emf and rotor current, resulting in enhanced torque.

7.4.5.4 Dynamic simulation. The dynamic simulation of the constant-volts/Hz-controlled induction motor drive is developed in this section. The models of the motor, inverter, dc link, controlled rectifier, and drive controllers are developed, and the equations are combined to obtain the transient response. As a

Section 7.4 Voltage-Source Inverter-Driven Induction Motor

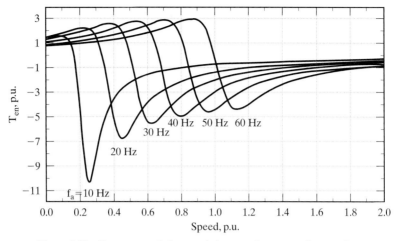

Figure 7.19 Torque–speed characteristics at various stator frequencies

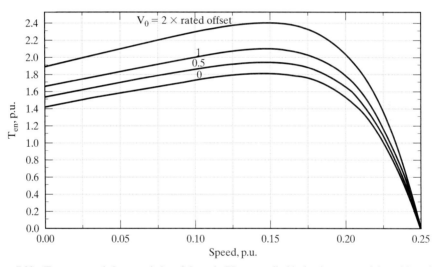

Figure 7.20 Torque-speed characteristics of the volts/Hz-controlled induction motor drive with various voltage offsets at $f_s = 15$ Hz

corollary, a direct method of evaluating steady-state performance without going through initial transients is also outlined. The simulation considers the actual voltage waveforms without ignoring harmonic contents in the applied voltages.

Motor: The induction motor model in synchronously rotating reference frames is chosen, because it gives many advantages. The same model after perturbations can be used for small-signal response, since the input and output variables become dc quantities (considering the fundamental alone).

The model is given in Chapter 5. The equations of the induction motor in synchronous reference frame are as follows:

$$\begin{bmatrix} v_{qs}^e \\ v_{ds}^e \\ 0 \\ 0 \end{bmatrix} = \begin{bmatrix} R_s + L_s p & \omega_s L_s & L_m p & \omega_s L_m \\ -\omega_s L_s & R_s + L_s p & -\omega_s L_m & L_m p \\ L_m p & (\omega_s - \omega_r) L_m & R_r + L_r p & (\omega_s - \omega_r) L_r \\ -(\omega_s - \omega_r) L_m & L_m p & -(\omega_s - \omega_r) L_r & R_r + L_r p \end{bmatrix} \begin{bmatrix} i_{qs}^e \\ i_{ds}^e \\ i_{qr}^e \\ i_{dr}^e \end{bmatrix} \quad (7.38)$$

and the electromechanical system equation is

$$J \frac{d\omega_r}{dt} = \frac{P}{2}(T_e - T_l) - B\omega_r$$

and electromagnetic torque is

$$T_e = \frac{3}{2} \frac{P}{2} L_m (i_{qs}^e i_{dr}^e - i_{ds}^e i_{qr}^e) \quad (7.39)$$

The synchronous reference frames voltage vector is

$$v_{qdo}^e = [T_{abc}^e] v_{abc} \quad (7.40)$$

where the transformation from *abc* to *qdo* variables is given by

$$[T_{abc}^e] = \frac{2}{3} \begin{bmatrix} \cos\theta_s & \cos(\theta_s - 2\pi/3) & \cos(\theta_s + 2\pi/3) \\ \sin\theta_s & \sin(\theta_s - 2\pi/3) & \sin(\theta_s + 2\pi/3) \\ \frac{1}{2} & \frac{1}{2} & \frac{1}{2} \end{bmatrix} \quad (7.41)$$

$$v_{qdo}^e = [v_{qs}^e \quad v_{ds}^e \quad v_o]^t \quad (7.42)$$

$$v_{abc} = [v_{as} \quad v_{bs} \quad v_{cs}]^t \quad (7.43)$$

where the symbols are explained in Chapter 5.

Input voltages: The input phase voltages are the six stepped waveforms shown in Figure 7.13. From the transformation given in (7.40) to (7.43), the *d* and *q* stator voltages in synchronously rotating reference frames are as follows.

For the interval $0 \leq \theta_s \leq \frac{\pi}{3}$, the *d* and *q* axes voltages are

$$v_{qsI}^e(\theta_s) = \frac{2}{3} V_{dc} \cos\left(\theta_s + \frac{\pi}{3}\right) \quad (7.44)$$

$$v_{dsI}^e(\theta_s) = \frac{2}{3} V_{dc} \sin\left(\theta_s + \frac{\pi}{3}\right) \quad (7.45)$$

where

$$\theta_s = \omega_s t \quad (7.46)$$

and for the interval $\frac{\pi}{3} \leq \theta_s \leq \frac{2\pi}{3}$, the voltages are

$$v_{qsII}^e(\theta_s) = \frac{2}{3} V_{dc} \cos(\theta_s) \tag{7.47}$$

$$v_{dsII}^e(\theta_s) = \frac{2}{3} V_{dc} \sin(\theta_s) \tag{7.48}$$

but the equations (7.47) and (7.48) can be written as

$$v_{qsII}^e(\theta_s) = \frac{2}{3} V_{dc} \cos\left(\theta_s + \frac{\pi}{3} - \frac{\pi}{3}\right) \tag{7.49}$$

$$v_{dsII}^e(\theta_s) = \frac{2}{3} V_{dc} \sin\left(\theta_s + \frac{\pi}{3} - \frac{\pi}{3}\right) \tag{7.50}$$

Equations (7.44), (7.45), (7.49), and (7.50) signify that q and d axis voltages are similar for each 60° and that they are periodic. They are shown in Figure 7.21. The symmetry of these voltages is exploited to find the steady state directly. Note that this voltage representation considers the actual waveforms of the inverter; i.e., the fundamental and higher-order harmonics are included.

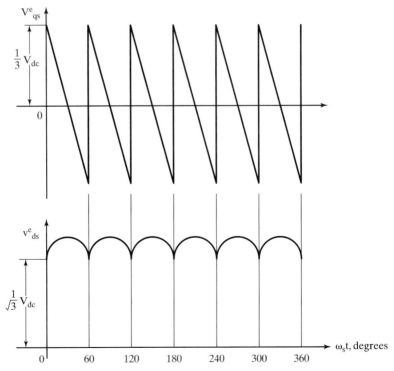

Figure 7.21 The q and d axis voltages in the synchronous reference frames

334 Chapter 7 Frequency-Controlled Induction Motor Drives

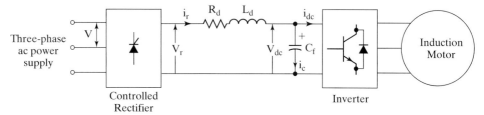

Figure 7.22 DC link; details of the motor drive

DC link: For a realistic dc link, as shown in Figure 7.22, the relationships are

$$v_{dc} = \frac{1}{C_f p}(i_r - i_{dc}) \tag{7.51}$$

$$v_{dc} > 0, i_r \geq 0 \tag{7.52}$$

with the constraint that the dc link voltage never become less than zero. The controlled rectifier's output voltage is given by

$$v_r = 1.35 V_s \cos \alpha \tag{7.53}$$

which, in terms of the link parameters, is written as

$$v_r = v_{dc} + (R_d + pL_d)i_r \tag{7.54}$$

By neglecting the inverter and cable losses, the dc link power can be equated to the input power of the induction motor; it is given by

$$v_{dc} i_{dc} = \frac{3}{2}(v_{qs}^e v_{qs}^e + v_{ds}^e i_{ds}^e) \tag{7.55}$$

and the zero-sequence power, $v_o i_o$, is zero for a balanced three-phase system. By substituting for v_{qs}^e and v_{ds}^e from (7.44) and (7.45) into (7.55), the dc link current is defined as

$$i_{dc} = i_{qs}^e \cos\left(\theta_s + \frac{\pi}{3}\right) + i_{ds}^e \sin\left(\theta_s + \frac{\pi}{3}\right) \tag{7.56}$$

Stator-referred dc link: The dc link and stator q and d axes voltages are at different magnitudes. For uniformity of treatment, the dc link variables need to be referred to the stator of the induction motor. It is done by defining a fictitious dc link whose relationship to the actual dc link variables is as follows:

$$v'_{dc} = \frac{2}{3} v_{dc} \tag{7.57}$$

$$i'_{dc} = i_{dc} \tag{7.58}$$

and hence the impedance transformation ratio is

$$\frac{v'_{dc}}{i'_{dc}} = \frac{2}{3} \frac{v_{dc}}{i_{dc}}$$

i.e., expressed in terms of the impedances as

$$\frac{Z'_{dc}}{Z_{dc}} = \frac{2}{3} \tag{7.59}$$

where the fictitious and actual dc link impedances are defined, respectively, as follows:

$$Z'_{dc} = \frac{V'_{dc}}{i'_{dc}} \tag{7.60}$$

$$Z_{dc} = \frac{V_{dc}}{i_{dc}} \tag{7.61}$$

From the impedance-transformation ratio, the dc link filter constants can be written as

$$R'_d = \frac{2}{3} R_d \tag{7.62}$$

$$L'_d = \frac{2}{3} L_d \tag{7.63}$$

$$C'_f = \frac{3}{2} C_f \tag{7.64}$$

Similarly,

$$v'_r = \frac{2}{3} v_r \tag{7.65}$$

and the fictitious rectified dc voltage is

$$v'_r = v'_{dc} + (R'_d + pL'_d)i'_r \tag{7.66}$$

The power balance is expressed as,

$$v'_r i'_r = (v'_{dc} + (R'_d + pL'_d)i'_r)i'_r = v'_{dc}i'_r + R'_d(i'_r)^2 \tag{7.67}$$

and the fictitious input power to the inverter is

$$v'_{dc}i'_{dc} = \frac{2}{3} v_{dc}i_{dc} = \frac{2}{3}\left[\frac{3}{2}\{v^e_{qs}i^e_{qs} + v^e_{ds}i^e_{ds}\}\right] = v^e_{qs}i^e_{qs} + v^e_{ds}i^e_{ds} \tag{7.68}$$

This expression makes the dc link compatible to the motor model, in the sense that the sum of the q and d axis powers is equal to the dc link power. The d and q axis voltages then can be written in terms of the fictitious variables as

$$v^e_{qs} = \frac{2}{3} v_{dc} \cos\left(\theta_s + \frac{\pi}{3}\right) = \frac{2}{3}\frac{3}{2} v'_{dc} \cos\left(\theta_s + \frac{\pi}{3}\right) = \{v'_r - (R'_d + L'_d p)i'_r\}\cos\left(\theta_s + \frac{\pi}{3}\right) \tag{7.69}$$

Similarly, the d axis voltage is derived as

$$v^e_{ds} = \frac{2}{3} v'_r \sin\left(\theta_s + \frac{\pi}{3}\right) - (R'_d + L'_d p)i'_r \sin\left(\theta_s + \frac{\pi}{3}\right) \tag{7.70}$$

336 Chapter 7 Frequency-Controlled Induction Motor Drives

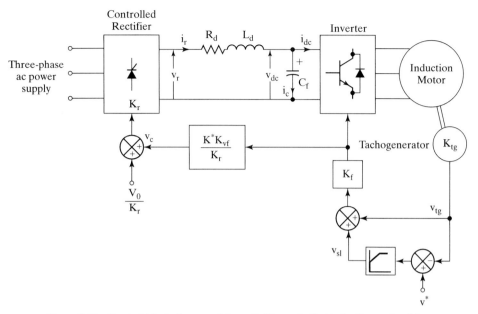

Figure 7.23 Control-block diagram of the volts/Hz-controlled induction motor drive

Substituting equations (7.69) and (7.70) into equation (7.38) yields a set of equations for simulating the dynamics of the volts/Hz-controlled induction motor drive.

Controller: The drive diagram, with its blocks, is shown in Figure 7.23. The external input v* is

$$-V_{cm} < v^* < +V_{cm}, V \qquad (7.71)$$

where V_{cm} is the maximum control voltage and is either ± 10 or ± 5 V, v* is proportional to the commanded speed of the motor, and the proportionality constant K* is defined as

$$K^* = \frac{v^*}{\omega_r^*} = \frac{V_{cm}}{\max\{\omega_r^*\}}, \text{volt}/(\text{rad/sec}) \qquad (7.72)$$

where ω_r^* is the commanded electrical rotor speed in rad/sec. The tachogenerator in the associated gain block is adjusted to have its maximum output corresponding to $\pm V_{cm}$ for control compatibility. Hence the gain of the tachogenerator and the filter is

$$K_{tg} = \frac{V_{cm}}{\omega_m} = \frac{V_{cm}}{\omega_r/(P/2)} = \frac{P}{2} \cdot K^* \qquad (7.73)$$

The maximum slip speed corresponds to the maximum torque of the induction motor; to avoid statically unstable and low-efficiency operation, the motor is not allowed to exceed the maximum slip speed, whose value, therefore, is

$$\max\{v_{sl}\} = K^* \omega_{sl(\max)} (V) \qquad (7.74)$$

The sum of the slip-speed signal and the rotor electrical-speed signal corresponds to the synchronous speed; hence, the gain of the frequency transfer block is

$$K_f = \frac{1}{2\pi K^*} \text{ (Hz/volt)} \quad (7.75)$$

The stator frequency and dc link voltage are written as

$$f_s = K_f\left(K^*\omega_{sl} + \frac{P}{2}K^*\omega_m\right) = K_f K^*(\omega_r + \omega_{sl}), \text{ Hz} \quad (7.76)$$

The control voltage of the output rectifier, obtained from a previous derivation, is

$$v_c = \frac{2.22}{K_r}(V_o + K_{vf}f_s)$$

Hence, the output of the rectifier is

$$v_r = 2.22K_r\{V_o + K_f K^* K_{vf}(\omega_r + \omega_{sl})\} \quad (7.77)$$

where K_r is the gain of the controlled rectifier, and the offset voltage and slip speed are generalized as follows:

$$V_o = I_{sr}R_s \quad (7.78)$$

and

$$\omega_{sl} = f_n\{v^* - v_{tg}\} = f_n\{v^* - \omega_m K_{tg}\} = f_n\{K^*\omega_r^* - K^*\omega_r\} \quad (7.79)$$

where I_{sr} is the rated stator current and f_n is the speed controller function.

The controller usually is a PI controller in tandem with a phase lead–lag network.

The synchronous speed, then, is given in terms of the controller constants and variables as

$$\omega_s = 2\pi f_s = 2\pi K_f K^*(\omega_r + \omega_{sl}) \quad (7.80)$$

By using equations (7.77) to (7.80), (7.69), and (7.70) in the equation (7.38), the modeling of the inverter-fed induction motor drive is obtained. Solving these equations by using fourth-order Runge–Kutta gives the time responses of the rectifier, inverter, and motor variables. They can be effectively employed to study the dynamic interaction of the drive system and to design a suitable controller and compensators.

Simulation Results: The drive given in Example 7.1 is simulated for a one-p.u.-step speed command from standstill. All the important variables of the induction motor drive are shown in Figure 7.24 in parts (i) and (ii). Notable is the torque behavior in this drive. Torque is unable to follow its command transiently, because the rotor flux linkages oscillate, resulting in large stator current and torque transients. In an actual drive system this could be avoided by limiting the stator currents. The controlled rectifier currents reach peaks of 14 p.u. This could be controlled by increasing the filter inductance or by reducing the rate of change of

338 Chapter 7 Frequency-Controlled Induction Motor Drives

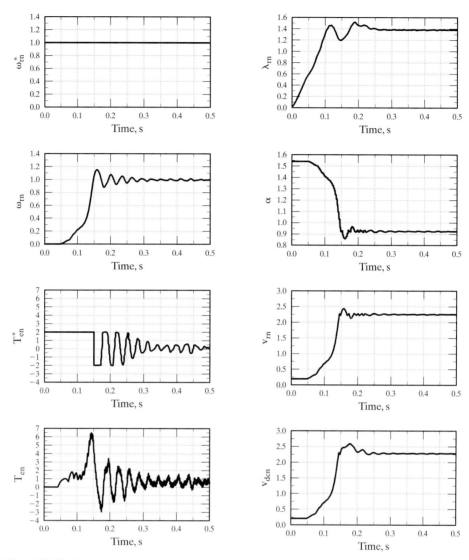

Figure 7.24(i) Dynamic performance of the volts/Hz-controlled induction motor drive system—Part (i)

the dc link voltage. The ac input line current reaches a maximum of 5.5 p.u., and the current in the capacitor has a maximum of 9 p.u.. On the inverter side, the stator current has a maximum value of 5 p.u.. Although this is smaller than the controlled rectifier currents, it is not acceptable; it increases the inverter rating to multiple times that of the induction motor. In practice, current limiting is set in the controller to keep this under control.

7.4.5.5 Small-signal responses. The small-signal responses are important in evaluating the stability and bandwidth of the motor drive. They are assessed by

Section 7.4   Voltage-Source Inverter-Driven Induction Motor   339

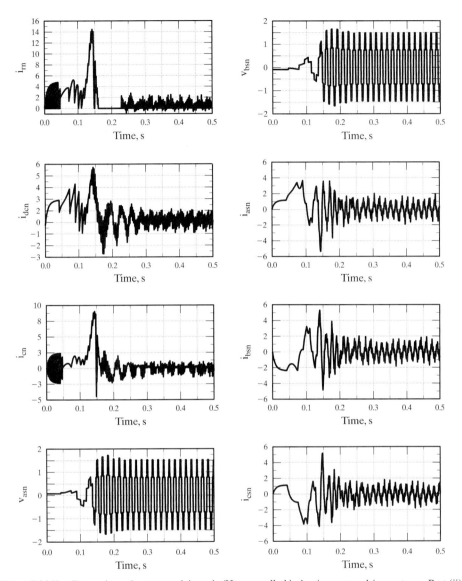

Figure 7.24(ii)   Dynamic performance of the volts/Hz-controlled induction motor drive system—Part (ii)

considering only the fundamental of the input voltages, contrary to the dynamic simulation. Consideration of fundamentals only reduces the inputs and outputs to dc quantities; hence, a perturbation around the steady-state operating point results in small-signal equations.

The steady-state operating point is determined by neglecting the harmonics in the input line voltages given in equations (7.11) to (7.13). The d and q voltages are calculated, by using the transformation given in equation (7.41), from the phase voltages, which, in turn, are derived from line voltages. It is worth mentioning at this juncture that choosing the phase voltages in the following form will result in simple d and q voltages.

$$v_{as} = \frac{2}{\pi} V_{dc} \cos \omega_s t \tag{7.81}$$

$$v_{bs} = \frac{2}{\pi} V_{dc} \cos\left(\omega_s t - \frac{2\pi}{3}\right) \tag{7.82}$$

$$v_{cs} = \frac{2}{\pi} V_{dc} \cos\left(\omega_s t + \frac{2\pi}{3}\right) \tag{7.83}$$

resulting in

$$v_{qs}^e = \frac{2}{\pi} V_{dc} \tag{7.84}$$

$$v_{ds}^e = 0$$

Denoting the steady-state values by an additional subscript '0', the steady-state currents are evaluated from equation (7.71) by making $p = 0$ and by maintaining ω_r constant. The equations then become algebraic; hence, the currents and subsequently the torque can be evaluated. The steady-state currents, neglecting the dc-link filter dynamics, are

$$\begin{bmatrix} i_{qso}^e \\ i_{dso}^e \\ i_{qro}^e \\ i_{dro}^e \end{bmatrix} = \begin{bmatrix} R_s & \omega_s L_s & 0 & \omega_s L_m \\ -\omega_s L_s & R_s & -\omega_s L_m & 0 \\ 0 & (\omega_s - \omega_r)L_m & R_r & (\omega_s - \omega_r)L_r \\ -(\omega_s - \omega_r)L_m & 0 & -(\omega_s - \omega_r)L_r & R_r \end{bmatrix}^{-1} \begin{bmatrix} v_{qso}^e \\ v_{dso}^e \\ 0 \\ 0 \end{bmatrix} \tag{7.85}$$

$$T_e = \frac{3}{2}\frac{P}{2} L_m(i_{qso}^e i_{dro}^e - i_{dso}^e i_{qro}^e)$$

Neglecting the dc-link filter dynamics, we perturb the motor equations around a steady-state operating point, and small-signal equations are derived similar to those in Chapter 5, section 5.9. Likewise, the loop equations describing the relationship between the speed command and voltage command and stator frequency are perturbed to obtain the small-signal equations. Combining the motor and feedback-loop small-signal equations gives the system equations. They can be used to study various transfer functions, stability, and small-signal transient responses and can be used for purposes of speed-controller design.

7.4.5.6 Direct steady-state evaluation. The steady state has been calculated so far by using the fundamental of the input voltages only. The steady-state performance for the actual input voltages, including harmonics, is necessary to select the rating of the converter–inverter switches and for computation of losses and derating of the induction motor. The steady state then is calculated either by using steady-state-harmonic equivalent circuits and summing the responses, or directly, by matching the boundary conditions. The harmonic equivalent-circuit approach has the conceptual advantage of simplicity but carries the disadvantage that its accuracy is limited by the number of harmonics considered in the input voltages. The direct steady-state evaluation overcomes this disadvantage but used to be limited by the requirement of a computer for solution. This method is derived and discussed in the section that follows.

The direct method exploits the symmetry of input voltages and currents in the steady state. They are symmetric over a given interval, in this case 60°, so their boundaries are matched to extract an elegant solution. Considering the input voltages, it has been proven in section 7.4.5.4 that they are periodic for every 60 electrical degrees. Hence, for a linear system (the case whenever speed is constant in the induction motor), the response must also be periodic. That is, the stator and rotor currents are periodic. The derivations are given below.

The induction motor equations in synchronously rotating reference frame are written in state-variable form as

$$\dot{X}_1 = A_1 X_1 + B_1 U_1 \quad (7.86)$$

where

$$X_1 = [i_{qs}^e \ i_{ds}^e \ i_{qr}^e \ i_{dr}^e]^t \quad (7.87)$$

$$A_1 = Q^{-1} P_1 \quad (7.88)$$

$$B_1 = Q^{-1} \quad (7.89)$$

$$u_1 = [v_{qs}^e \ v_{ds}^e \ 0 \ 0]^t \quad (7.90)$$

$$Q = \begin{bmatrix} L_s & 0 & L_m & 0 \\ 0 & L_s & 0 & L_m \\ L_m & 0 & L_r & 0 \\ 0 & L_m & 0 & L_r \end{bmatrix} \quad (7.91)$$

$$P_1 = \begin{bmatrix} -R_s & -\omega_s L_s & 0 & -\omega_s L_m \\ \omega_s L_s & -R_s & \omega_s L_m & 0 \\ 0 & -(\omega_s - \omega_r)L_m & -R_r & -(\omega_s - \omega_r)L_r \\ (\omega_s - \omega_r)L_m & 0 & (\omega_s - \omega_r)L_r & -R_r \end{bmatrix} \quad (7.92)$$

From equations (7.44) and (7.45), the voltages are written in state-space form as

$$\begin{bmatrix} pv_{qs}^e \\ pv_{ds}^e \end{bmatrix} = \begin{bmatrix} 0 & -\omega_s \\ \omega_s & 0 \end{bmatrix} \begin{bmatrix} v_{qs}^e \\ v_{ds}^e \end{bmatrix} \quad (7.93)$$

Equations (7.86) and (7.87) are combined to give

$$\begin{bmatrix} \dot{X}_1 \\ \dot{X}_2 \end{bmatrix} = \begin{bmatrix} A_1 & B_1 \\ 0 & S \end{bmatrix} \begin{bmatrix} X_1 \\ X_2 \end{bmatrix} \quad (7.94)$$

where

$$X_2 = [v_{qs}^e \ v_{ds}^e]^t \quad (7.95)$$

$$S = \begin{bmatrix} 0 & -\omega_s \\ \omega_s & 0 \end{bmatrix} \quad (7.96)$$

and matrices A_1 and B_1 are of compatible dimensions. The equation (7.94), in a compact form, is expressed as

$$\dot{X} = AX \tag{7.97}$$

where

$$X = [X_1 \ X_2]^t \tag{7.98}$$

$$A = \begin{bmatrix} A_1 & B_1 \\ 0 & S \end{bmatrix} \tag{7.99}$$

The solution of equation (7.97) is written as,

$$X(t) = e^{At}X(0) \tag{7.100}$$

where the initial steady-state vector $X(0)$ is to be evaluated to compute $X(t)$ and the electromagnetic torque. It is found by the fact that the state vector has periodic symmetry; hence,

$$X\left(\frac{\pi}{3\omega_s}\right) = S_1 X(0) \tag{7.101}$$

where S_1 is evaluated later.

The boundary condition for the currents is

$$X_1\left(\frac{\pi}{3\omega_s}\right) = X_1(0) \tag{7.102}$$

The boundary-matching condition for the voltage vector is obtained by expanding (7.49) and (7.50) and substituting (7.44) and (7.45) into them. The direct axis voltage is

$$v_{qsII}^e(\theta_s) = \frac{2}{3}V_{dc}\cos\left(\theta_s + \frac{\pi}{3} - \frac{\pi}{3}\right) = \frac{2}{3}V_{dc}\left\{\cos\left(\theta_s + \frac{\pi}{3}\right)\cos\frac{\pi}{3} + \sin\left(\theta_s + \frac{\pi}{3}\right)\sin\frac{\pi}{3}\right\}$$

$$= \frac{1}{2}v_{qsI}^e(\theta_s) + \frac{\sqrt{3}}{2}v_{dsI}^e(\theta_s) \tag{7.103}$$

Similarly,

$$v_{dsII}^e = \frac{1}{2}v_{dsI}^e - \frac{\sqrt{3}}{2}v_{qsI}^e \tag{7.104}$$

Hence,

$$X_2\left(\frac{\pi}{3\omega_s}\right) = \begin{bmatrix} \frac{1}{2} & \frac{\sqrt{3}}{2} \\ -\frac{\sqrt{3}}{2} & \frac{1}{2} \end{bmatrix} X_2(0) = S_2 X_2(0) \tag{7.105}$$

Section 7.4 Voltage-Source Inverter-Driven Induction Motor

where

$$S_2 = \begin{bmatrix} \dfrac{1}{2} & \dfrac{\sqrt{3}}{2} \\ -\dfrac{\sqrt{3}}{2} & \dfrac{1}{2} \end{bmatrix} \quad (7.106)$$

S_1 is obtained from the equations (7.102) and (7.105) as

$$S_1 = \begin{bmatrix} I & 0 \\ 0 & S_2 \end{bmatrix} \quad (7.107)$$

where I is a 4 × 4 identity matrix.

Substituting equation (7.101) into (7.100), we get

$$X\left(\frac{\pi}{3\omega_s}\right) = S_1 X(0) = e^{A\left(\frac{\pi}{3\omega_s}\right)} X(0) \quad (7.108)$$

Hence,

$$[S_1 - e^{A\left(\frac{\pi}{3\omega_s}\right)}] X(0) = 0 \quad (7.109)$$

or

$$W X(0) = 0 \quad (7.110)$$

where

$$W = [S_1 - e^{A\left(\frac{\pi}{3\omega_s}\right)}] = \begin{bmatrix} W_1 & W_2 \\ W_3 & W_4 \end{bmatrix} \quad (7.111)$$

where W_1 is 4 × 4, W_2 is 2 × 4, W_3 is 2 × 4 and W_4 is 2 × 2. It can be proven that W_3 is a null matrix. Expanding only the upper row in equation (7.110) gives the following relationship:

$$W_1 X_1(0) + W_2 X_2(0) = 0 \quad (7.112)$$

from which the steady-state current vector $X_1(0)$ is obtained as

$$X_1(0) = -W_1^{-1} W_2 X_2(0) \quad (7.113)$$

Having evaluated the initial current vector, we could use it in equation (7.100) to evaluate currents for one full cycle and the electromagnetic torque. A sample set of steady-state waveforms is shown in Figures 7.25(i) and (ii). The peaking of the current at no-load and then its becoming quasi-sinusoidal for full load is significant to determine the peak rating of the devices in the inverter. Note that this approach is suitable for steady-state calculation of six-stepped voltage inputs, irrespective of the control strategy used. The electromagnetic torque has a sixth-harmonic ripple in the six-stepped-voltage-fed induction motor drive that deserves scrutiny. Its cause and effects are considered in later sections.

344 Chapter 7 Frequency-Controlled Induction Motor Drives

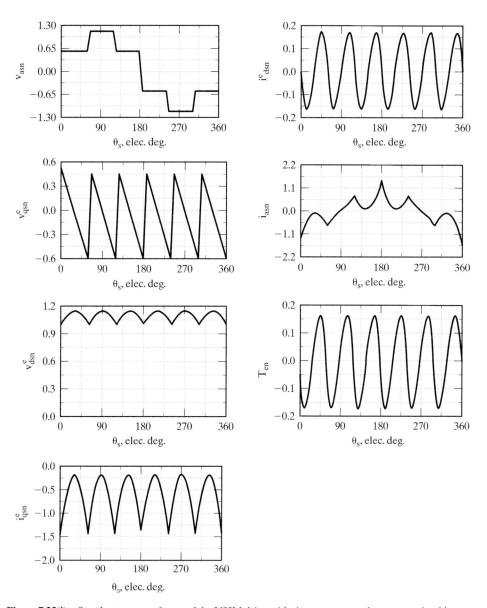

Figure 7.25(i) Steady-state waveforms of the VSIM drive with six-step stator voltages at no-load in p.u.

Section 7.4   Voltage-Source Inverter-Driven Induction Motor   345

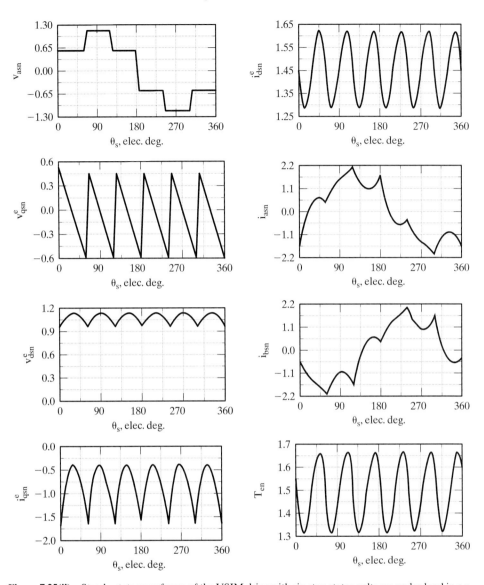

Figure 7.25(ii)   Steady-state waveforms of the VSIM drive with six-step stator voltages under load in p.u.

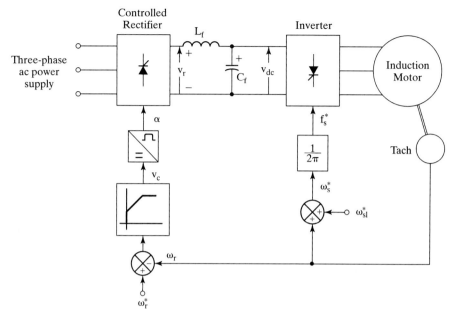

Figure 7.26 Constant-slip-speed drive strategy

7.4.6 Constant Slip-Speed Control

7.4.6.1 Drive strategy. The slip speed of the induction motor is maintained constant; hence, for various rotor speeds, the slip will be varying, as is seen from the following expressions:

$$\omega_s = \omega_r + \omega_{sl} \quad (7.114)$$

$$\omega_{sl} = s\omega_s = \text{constant} \quad (7.115)$$

from which the slip is obtained as

$$s = \frac{\omega_{sl}}{\omega_s} = \frac{\omega_{sl}}{\omega_r + \omega_{sl}} \quad (7.116)$$

The varying slip control places the drive operation on the static torque–speed characteristics. To maintain the slip speed constant, it is necessary to know the rotor speed, so this scheme involves rotor-speed estimation or measurement for feedback control. The stator frequency is obtained by summing the slip speed and the electrical rotor speed. The required input voltages to the induction motor are made to be a function of the speed-error signal, as is shown in Figure 7.26. In place of a proportional controller, a PI controller eliminates the steady-state error in the rotor speed. Note that a negative speed error clamps the bus voltage at zero, and triggering angles of greater than 90° are not allowed. Given the configuration of the drive, it cannot regenerate: the stator electrical speed is always maintained greater than the

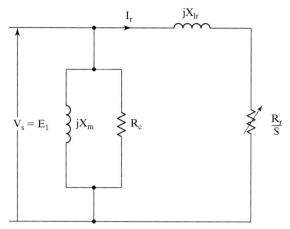

Figure 7.27 Simplified equivalent circuit considered for the steady-state analysis of the slip-controlled induction motor drive

rotor speed. Hence, this drive is restricted to one-quadrant operation only. Alternate implementation of dc link voltage control is effected by implementing a volts/Hz controller, using f_s^*. Note that this alternate scheme will not give a closed-loop speed control.

7.4.6.2 Steady-state analysis. The characteristics of the constant-slip-speed induction motor are derived in this section from the simplified equivalent circuit of the induction motor. Only the fundamental of the applied voltages is considered at this stage. Considering the rotor and magnetic circuit shows that the equivalent circuit amounts to the one shown in Figure 7.27. The slip speed is a constant; hence, in terms of the slip speed, the rotor current is derived as

$$I_r = \frac{E_1}{\left(\dfrac{R_r}{s} + jX_{lr}\right)} = \frac{E_1/\omega_s}{\left(\dfrac{R_r}{\omega_{sl}} + jL_{lr}\right)} \quad (7.117)$$

and the electromagnetic torque is

$$T_e = \frac{P}{2} \cdot \frac{P_a}{\omega_s} = 3 \cdot \frac{P}{2} \cdot \frac{I_r^2 R_r}{s\omega_s} = 3 \cdot \frac{P}{2} \cdot \frac{I_r^2 R_r}{\omega_{sl}} \quad (7.118)$$

Substituting for I_r from equation (7.117) into equation (7.118) yields

$$T_e = 3\frac{P}{2} \cdot \frac{E_1^2}{\omega_s^2} \cdot \frac{\left(\dfrac{R_r}{\omega_{sl}}\right)}{\left(\dfrac{R_r}{\omega_{sl}}\right)^2 + (L_{lr})^2} \quad (7.119)$$

By rearranging all the constants into one term, we get

$$T_e = K_{tv}\left(\frac{E_1^2}{\omega_s^2}\right) \quad (7.120)$$

where the torque constant for this strategy is defined as

$$K_{tv} = \frac{3\dfrac{P}{2}\left(\dfrac{R_r}{\omega_{sl}}\right)}{\left(\dfrac{R_r}{\omega_{sl}}\right)^2 + (L_{1r})^2} \quad (7.121)$$

Neglecting stator impedance amounts to making the air gap emf equal to the applied stator voltage; hence, the torque is given by

$$T_e \cong K_{tv}\left(\frac{V_s}{\omega_s}\right)^2 \quad (7.122)$$

where V_s is the stator voltage per phase.

Note that the torque is independent of rotor speed, implying its capability to produce a torque even at zero speed. This feature is essential in many applications where a starting or holding torque needs to be produced, such as in robotics.

A diagram of torque vs. stator phase voltage is shown in Figure 7.28 for three slip speeds. Here, the stator impedance has not been neglected, and its consideration has reduced the torque, notably at low speeds. The motor data has been taken from Example 7.1.

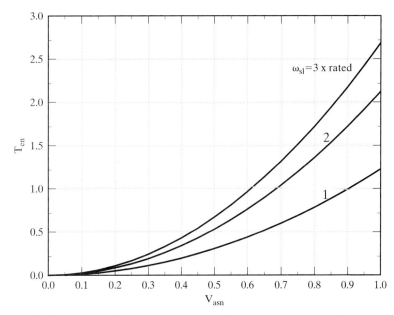

Figure 7.28 Torque vs. applied voltage for various slip speeds at rated stator frequency in p.u.

Example 7.2

Draw the performance characteristics of the constant-slip-speed-controlled induction motor drive given in Example 7.1. The slip speed is maintained at 9 rad/sec. The converter combination is a controlled rectifier and a six-step voltage source inverter. The drive uses a volts/Hz strategy for controlling the dc link voltage.

Solution The steps involved are as follows:

$$\omega_r = \omega_m \times \frac{P}{2}$$

$$\omega_s = \omega_r + \omega_{sl}$$

$$K_{vf} = \frac{V_{s(rated)}}{f_s} = \frac{V_{s(rated)}}{(\omega_s/2\pi)}$$

where

$$V_s = 0.45 V_{dc} = K_{vf} f_s$$

By using the fundamental equivalent circuit, the stator, magnetizing, and rotor currents are computed. From the currents, the motor power factor, torque, output, and efficiency are evaluated. The performance characteristics are shown in Figure 7.29.

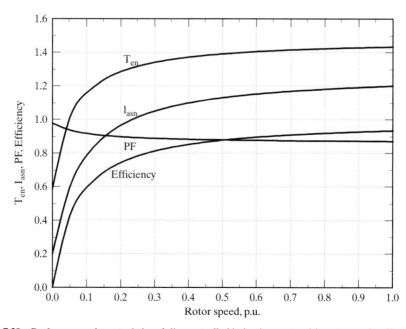

Figure 7.29 Performance characteristics of slip-controlled induction motor drive at ω_{sl} = 9 rad/sec in p.u.

7.4.7 Constant Air Gap Flux Control

7.4.7.1 Principle of operation. Constant air gap flux control resolves the induction motor into an equivalent separately-excited dc motor in terms of its speed of response but not in terms of decoupling of the flux and torque channels. Having constant air gap flux linkages amounts to

$$\lambda_m = L_m i_m = \frac{E_1}{\omega_s} \quad (7.123)$$

which, substituted into equation (7.119), yields the electromagnetic torque as

$$T_e = 3 \frac{P}{2} \cdot \lambda_m^2 \cdot \frac{\left(\frac{R_r}{\omega_{sl}}\right)}{\left(\frac{R_r}{\omega_{sl}}\right)^2 + (L_{1r})^2} \quad (7.124)$$

Assuming the air gap flux linkage is maintained constant, the torque is

$$T_e = K_{tm} \cdot \frac{\left(\frac{R_r}{\omega_{sl}}\right)}{\left(\frac{R_r}{\omega_{sl}}\right)^2 + (L_{1r})^2} \quad (7.125)$$

where the torque constant in the control strategy is written as

$$K_{tm} = 3 \frac{P}{2} \cdot \lambda_m^2 \quad (7.126)$$

Now the electromagnetic torque is dependent only on the slip speed, as is seen from equation (7.125). Such a feature signifies a very important phenomenon, in that the slip speed can be varied instantly, making the torque response instantaneous. A fast torque response paves the way for a high-performance motor drive, suitable for demanding applications, thus replacing the separately-excited dc motor drives. The above facts are true only if the air gap flux is maintained constant. That task is compounded by, for example, the saturation of the machine or the need for a sensor to measure the air gap flux. Even if air gap flux is regulated accurately, the torque is not a linear function of the slip speed. Hence, for a torque drive, the slip speed has to be programmed to generate linear characteristics between torque and its commanded value. Note that this programming of the slip speed has to account for the sensitivity of the rotor resistance and leakage inductance. In particular, the rotor resistance will usually have a wide variation (from 0.8 to 2 times its nominal value at ambient temperature) and this complicates the control task.

7.4.7.2 Drive strategy. The fact that the constant air gap flux requires control of the magnetizing current necessitates stator current control of the induction motor, apart from its stator frequency. A voltage-source inverter drive with inner current loops would transform it into a variable-current, variable-frequency source.

Section 7.4 Voltage-Source Inverter-Driven Induction Motor 351

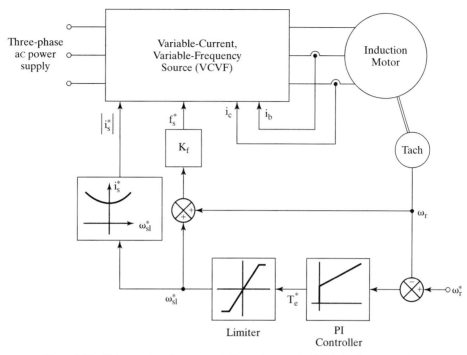

Figure 7.30 Drive strategy for constant-air gap-flux-controlled induction motor drive

More of this is to come in later sections. Assuming that such a current-regulated variable-frequency source is available, the drive strategy is shown in Figure 7.30.

The slip-speed command is generated from the speed-control error, which in turn is added to the rotor speed signal to provide the frequency-command signal. The same slip-speed command determines the stator current magnitude, maintaining the air gap flux constant. This aspect is shown as a function generator block. For reasons of saturation and sensitivity of rotor constants, it is preferable to have on-line computation of the stator current from the slip-speed command. The current command is appropriately translated into three-phase stator-current commands. Stator-current feedback control ensures that the stator-current commands are enforced both in magnitude and in phase.

Example 7.3

Draw the steady-state performance characteristics of a constant-air gap-flux-controlled induction motor drive. The motor data are given in Example 7.1.

Solution The various steps involved in the steady-state performance computation are as follows:

Step 1: Calculate the rated value of air gap flux from the equivalent circuit including stator impedence when the motor has a rated electromagnetic torque. Calculate magnetizing current and hence $\frac{E_1}{\omega_s}$.

Step 2: For a given stator frequency, calculate the air gap emf.

Step 3: Starting with a slip, and from the air gap emf, calculate the rotor current, and then find the stator current, power factor, torque, applied stator voltage, and efficiency. The equations are

$$I_r = \frac{E_1}{\left(\dfrac{R_r}{s} + jX_{lr}\right)} \quad (7.127)$$

$$I_s = I_r + I_m \quad (7.128)$$

$$V_{as} = E_1 + I_s(R_s + j\omega_s L_{1s}) \quad (7.129)$$

$$T_e = 3I_r^2 \cdot \frac{R_r}{s\omega_s} \cdot \frac{P}{2} \quad (7.130)$$

$$\text{Power factor} = \cos\phi = \frac{\text{Real}(V_{as}I_s^*)}{|V_{as}I_{as}|} \quad (7.131)$$

$$\text{Efficiency} = \frac{\text{Mechanical power output}}{\text{Electrical power input}}, \; s > 0$$

$$= \frac{\text{Electrical power output}}{\text{Mechanical power input}}, \; s < 0 \quad (7.132)$$

$$= 0 \quad\quad\quad\quad\quad\quad\quad\quad\quad\quad\quad\quad , \; s = 0$$

Step 4: Increment the slip, and go to step 3.
Step 5: Increment the stator frequency, and go to step 2.
Step 6: Draw torque vs. speed for various stator frequencies and rotor and stator currents and voltage and torque vs. speed for rated stator frequency.

The normalized electromagnetic torque vs. speed for various stator frequencies is shown in Figure 7.31. The torque characteristic is uniquely symmetric for both the motoring and generating mode and for every stator frequency. The asymmetry in the torque characteristics for the motoring and generating regions in the volts/Hz-controlled drive is due to the lack of control of air gap flux linkages. The stator current and voltage vs. speed are shown in Figure 7.32 for rated stator frequency.

Constant air gap flux linkages result in higher demand for stator phase voltages, as is seen in Figure 7.32. This is due to the increasing rotor and hence stator currents and the consequent increase in voltage drops across the stator impedances. For torques less than 2 p.u., the increase in the stator voltage, although very small, could exceed the dc bus supply. This necessitates proper selection of the induction-motor voltage rating for an available dc supply. High-transient-torque demands of four to five p.u. in emerging applications such as electric vehicle propulsion need to consider the stator voltage demand at higher speeds.

Example 7.4

Derive stator current magnitude in terms of the motor parameters, slip speed, and magnetizing current to implement a constant-air gap-flux-linkages drive system.

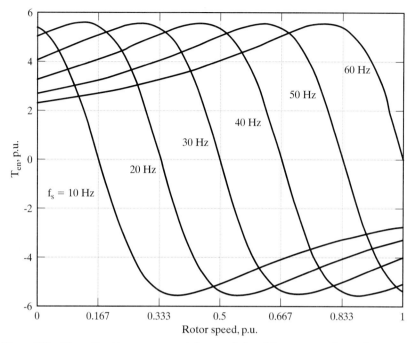

Figure 7.31 Normalized torque vs. speed characteristics of the constant-air gap-flux-controlled induction motor drive for different stator frequencies

Solution The equivalent circuit shown in Figure 7.27 is used to derive the stator current magnitude, as follows:

The rotor current is

$$I_r = \frac{E_1}{\left(\dfrac{R_r}{s} + jX_{lr}\right)} = \frac{jX_m I_m}{\left(\dfrac{R_r}{s} + jX_{lr}\right)} = \frac{jL_m I_m s\omega_s}{R_r + js\omega_s L_{1r}} = \frac{jL_m I_m \omega_{sl}}{R_r + j\omega_{sl}L_{1r}}$$

$$I_s = I_r + I_m = I_m\left[1 + \frac{jL_m\omega_{sl}}{R_r + j\omega_{sl}L_{1r}}\right] = I_m\left[\frac{R_r + j\omega_{sl}(L_{1r} + L_m)}{R_r + j\omega_{sl}L_{1r}}\right]$$

$$|I_s| = I_m\sqrt{\frac{R_r^2 + (\omega_{sl}L_r)^2}{R_r^2 + (\omega_{sl}L_{1r})^2}}$$

The stator current is dependent upon the rotor self-inductance and leakage inductance and on the resistance of the rotor apart from the slip-speed and magnetizing current. Because the flux linkages are controlled, the variation of the rotor self-inductance and leakage inductance will not vary significantly, but the rotor resistance due to the temperature and slip frequency variations will significantly vary, and, in that case, the stator current has to be made a function of the rotor resistance to maintain the flux constant. It is an important consideration in the design of the controller in this drive scheme.

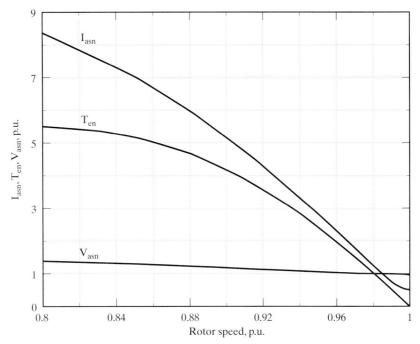

Figure 7.32 Torque, voltage, and current vs. speed at rated stator frequency

7.4.8 Torque Pulsations

7.4.8.1 General. Six-stepped-voltage waveforms are rich in harmonics. These time harmonics produce respective rotor current harmonics, which in turn interact with the fundamental air gap flux, generating harmonic torque pulsations. The torque pulsations are undesirable: they generate audible noise, speed pulsations, and losses, thus decreasing the thermal capability of the motor and eventually derating the motor. Even though the magnitude of the torque pulsations can be evaluated from the steady-state computation by using boundary-matching conditions, it is simple to calculate the magnitude of each harmonic torque pulsation by using the harmonic equivalent circuit of the induction motor. This method has the advantage of singling out the dominant torque pulsation; when its source is identified, methods can be devised to control it.

7.4.8.2 Calculation of torque pulsations. By using the Fourier series of the line voltages given in equations (7.11) to (7.13), the fundamental, fifth, and seventh harmonics of the phase voltages are derived as

$$V_{as1} = \frac{2}{\pi} V_{dc} \sin(\omega_s t - 30°) \tag{7.133}$$

$$V_{as5} = \frac{2}{5\pi} V_{dc} \sin(-5\omega_s t - 30°) \tag{7.134}$$

Section 7.4 Voltage-Source Inverter-Driven Induction Motor

$$V_{as7} = \frac{2}{7\pi} V_{dc} \sin(7\omega_s t - 150°) \quad (7.135)$$

The fifth harmonic is rotating opposite to the fundamental, whereas the seventh is in the same direction as the fundamental. Therefore, the air gap flux linkages due to the fifth- and seventh-harmonic currents field, are revolving at six times the synchronous speed relative to the fundamental air gap flux.

The sixth-harmonic torque pulsation is created by

(i) the fundamental air gap flux linkages, interacting with the fifth- and seventh-harmonic rotor currents;

(ii) the fundamental rotor current, interacting with the fifth- and seventh-harmonic air gap flux linkages.

The various variables are computed, as follows, from the two equivalent-circuit diagrams shown in Figure 7.33.

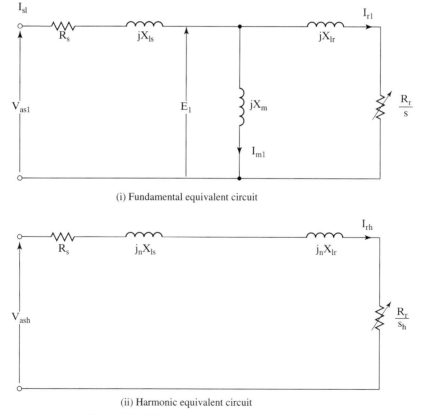

Figure 7.33 Equivalent circuit of the induction motor

The harmonic slip for a harmonic of order h, derived in Chapter 6, is given as

$$s_h \cong \frac{h \pm 1}{h}, \begin{cases} + \text{ for h odd} \\ - \text{ for h even} \end{cases} \quad (7.136)$$

Hence,

$$s_5 \cong \frac{6}{5}$$
$$s_7 \cong \frac{6}{7} \quad (7.137)$$

The fundamental, fifth-, and seventh-harmonic mutual flux linkages are

$$\lambda_{m1} = L_m I_{m1} \quad (7.138)$$

$$\lambda_{m5} = L_m I_{m5} = \frac{I_{r5}}{5\omega_s}\left(\frac{R_r}{s_5} + j5X_{lr}\right) \quad (7.139)$$

$$\lambda_{m7} = \frac{I_{r7}}{7\omega_s}\left(\frac{R_r}{s_7} + j7X_{lr}\right) \quad (7.140)$$

and the harmonic rotor currents are given by

$$I_{r5} = \frac{V_{as5}}{\left(R_s + \dfrac{R_r}{s_5}\right) + j5(X_{ls} + X_{lr})} \quad (7.141)$$

$$I_{r7} = \frac{V_{as7}}{\left(R_s + \dfrac{R_r}{s_7}\right) + 7(jX_{ls} + X_{lr})} \quad (7.142)$$

but at frequencies above 0.3 p.u., the rotor peak currents can be approximated as

$$I_{r5} \cong \frac{V_{as5}}{5(X_{ls} + X_{lr})} \cong \frac{2V_{dc}}{5\pi}\left(\frac{1}{5X_{eq}}\right) \cong \left(\frac{2V_{dc}}{\pi}\right)\left(\frac{1}{25X_{eq}}\right) \quad (7.143)$$

where the equivalent leakage reactance is given as

$$X_{eq} = (X_{ls} + X_{lr}) \quad (7.144)$$

and

$$I_{r7} \cong \left(\frac{2V_{dc}}{\pi}\right)\left(\frac{1}{49X_{eq}}\right) \quad (7.145)$$

By substituting these approximated values of rotor currents into equations (7.139) and (7.140) and neglecting the resistances in comparison to the harmonic leakage reactances, the harmonic peak flux linkages are found to be

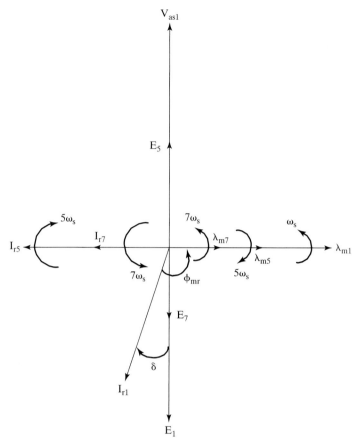

Figure 7.34 Phasor diagram of the induction motor, with harmonic components

$$\lambda_{m5} \cong I_{r5}L_{1r} \cong \frac{2V_{dc}}{25\pi\omega_s}\left(\frac{L_{1r}}{L_{eq}}\right) \tag{7.146}$$

$$\lambda_{m7} \cong \frac{2V_{dc}}{49\pi\omega_s}\left(\frac{L_{1r}}{L_{eq}}\right) \tag{7.147}$$

where

$$L_{eq} = \frac{X_{eq}}{\omega_s} \tag{7.148}$$

These variables are represented in a phasor diagram, as shown in Figure 7.34. The fundamental torque is then computed as

$$T_{e1} = 3\frac{P}{2}\lambda_{m1}I_{r1}\sin\phi_{mr} \tag{7.149}$$

and the peak of the fundamental mutual flux linkages is

$$\lambda_{m1} \cong \frac{V_{as1}}{\omega_s} = \frac{2V_{dc}}{\pi \omega_s} \tag{7.150}$$

Substituting equation (7.150) into equations (7.146) and (7.147) yields the harmonic flux linkages in terms of fundamental flux linkages:

$$\lambda_{m5} = \frac{\lambda_{m1}}{25}\left(\frac{L_{lr}}{L_{eq}}\right) \tag{7.151}$$

$$\lambda_{m7} = \frac{\lambda_{m1}}{49}\left(\frac{L_{lr}}{L_{eq}}\right) \tag{7.152}$$

The sixth-harmonic torque in the anticlockwise direction is

$$T_{e6} = \frac{3}{2}\frac{P}{2}[\lambda_{m1}(I_{r7} - I_{r5})\sin 6\omega_s t + I_{r1}\{\lambda_{m7}\sin(6\omega_s t + 90° + \delta) + \lambda_{m5}\sin(-6\omega_s t + 90° + \delta)\}] \tag{7.153}$$

where

$$90° + \delta = \phi_{mr} \tag{7.154}$$

δ is very small in practice; assuming it is zero, the sixth-harmonic torque is approximated as

$$T_{e6} \cong \frac{3}{2}\frac{P}{2}[\lambda_{m1}(I_{r7} - I_{r5})\sin 6\omega_s t + I_{r1}(\lambda_{m7} + \lambda_{m5})\cos 6\omega_s t] \tag{7.155}$$

The torque is divided by an additional factor of 2 because the flux linkage and rotor currents are peak values. Normalizing the sixth-harmonic torque in terms of its fundamental torque yields

$$T_{e6n} = \frac{T_{e6}}{T_{e1}} = \frac{(I_{r7} - I_{r5})}{I_{r1}\sin\phi_{mr}}\sin 6\omega_s t + \left(\frac{L_{lr}}{L_{eq}}\right)\frac{\left(\frac{1}{25} + \frac{1}{49}\right)}{\sin\phi_{mr}}\cos 6\omega_s t \tag{7.156}$$

$$\cong \frac{(I_{r7} - I_{r5})}{I_{r1}}\sin 6\omega_s t + (0.0604)\left(\frac{L_{lr}}{L_{eq}}\right)\cos 6\omega_s t$$

Similarly, twelfth-harmonic pulsating torque can be evaluated from the eleventh- and thirteenth-harmonic input voltages. In most of the applications, the dominant-pulsating-torque calculation is sufficient.

These pulsating torques are smoothed by the rotor and load inertia at high speeds, but, at low speeds, they might induce speed ripples, which are undesirable. They could excite and resonate the critical frequencies of the motor drive. In such cases, corrective measures have to be taken. Elimination of undesirable pulsating torques is discussed subsequently.

Section 7.4 Voltage-Source Inverter-Driven Induction Motor 359

Example 7.5

Calculate the sixth-harmonic pulsating torque for the motor drive given in Example 7.1, neglecting stator impedance. The operating points are at 60 Hz with 0 and 1 p.u. slip speeds.

Solution

Case (i) $\omega_{sl} = 0, s = 0 \therefore I_{r1} = 0$

There is slip speed in regard to the time-harmonic input voltages, which in turn generates rotor harmonic current and pulsating torque. The rms values are calculated; accordingly, the torque expression has the term 3 and not, as in the previous derivation, 3/2. The rotor harmonic currents and air gap flux linkages are

$$\lambda_{m1} = \frac{E_1}{\omega_s} \cong \frac{V_{as1}}{\omega_s} \cong \frac{200/\sqrt{3}}{2\pi \times 60} \cong 0.306 \text{ Wb-turn}$$

$$\lambda_{m5} \cong I_{r5} L_{1r}$$
$$\lambda_{m7} \cong I_{r7} L_{1r}$$

where

$$I_{r5} \cong \frac{V_{as5}}{5X_{eq}} = \frac{\left(\frac{V_{as1}}{5}\right)}{5X_{eq}} = \frac{200/\sqrt{3}}{25 \times 2\pi \times 60 \times 0.0037} = 3.31 \text{ A}$$

$$I_{r7} = \frac{V_{as1}}{49X_{eq}} = \frac{200/\sqrt{3}}{49 \times 1.395} = 1.69 \text{ A}$$

Hence,

$$\lambda_{m5} = 3.31 \times 0.0022 = 0.0073 \text{ Wb-turn}$$
$$\lambda_{m7} = 1.69 \times 0.0022 = 0.0038 \text{ Wb-turn}$$

$$T_{e6} = 3\frac{P}{2}[\lambda_{m1}(I_{r7} - I_{r5})\sin 6\omega_s t + I_{r1}(\lambda_{m7} + \lambda_{m5})\cos 6\omega_s t]$$

$$= 3 \times \frac{4}{2}\{(0.306)(1.69 - 3.31)\sin 6\omega_s t\} = -2.97 \sin 6\omega_s t$$

Peak absolute value of $T_{e6} = 2.97 \text{ N·m} = \frac{2.97}{20.0} = 0.149 \text{ p.u.}$

Case (ii) $\omega_{sl} = 6.882 \text{ rad/sec}$

$$\lambda_{m1} = 0.306 \text{ Wb-turn}$$

and

$$s = \frac{\omega_{sl}}{\omega_s} = \frac{6.882}{377} = 0.0183$$

$$\therefore I_{r1} \cong \frac{V_{as1}}{\frac{R_r}{s} + jX_{lr}} = \frac{115.5}{\frac{0.183}{0.0183} + j0.837} = \frac{115.5}{10 + j0.837} = 11.51 \text{ A}$$

The rotor harmonic currents and harmonic air gap flux linkages are the same as in the case where slip is equal to zero. Hence, the sixth-harmonic pulsating torque is computed as

$$T_{e6} = 3 \times \frac{4}{2} \{(0.306)(1.69 - 3.31) \sin 6\omega_s t + 11.51(0.0073 + 0.0038) \cos 6\omega_s t\}$$

$$= -2.97 \sin 6\omega_s t + 0.766 \cos 6\omega_s t$$

$$\text{Peak value of } T_{e6} = \sqrt{2.97^2 + 0.766^2} = 3.06 \text{ N} \cdot \text{m} = \frac{3.06}{20} = 0.153 \text{ p.u.}$$

Note that these calculations neglect the stator impedance drop, because, at rated frequency, the effect of the stator impedance drop is small compared to that of the induced emf in the air gap. At low operating frequencies, the stator impedance drop will dominate, and hence reduce the induced emf and hence the torque pulsations too. The torque pulsations will fatigue the shaft and eventually lead to failure. This type of motor drive is unsuitable for position applications, because of high torque ripples.

7.4.8.3 Effects of time harmonics. The harmonic voltages produce harmonic currents that, in turn, generate not only torque pulsation but also increased losses in the form of copper and core losses. The net effect of the torque pulsation on the resultant average torque is zero, but the harmonic losses add to the heating of the machine. Since the induction machines are usually designed for sinusoidal inputs, the additional losses due to the harmonics tax the thermal capability of the motor.

For a given cooling arrangement of the motor, the allowable temperature rise for safe operation is prescribed. To maintain the motor within its class of operation, the motor then needs to be derated. It is quite unavoidable with converter-controlled motors, but the magnitude of derating can be kept to a minimum by controlling the harmonic contents of the voltage and current inputs to the induction motor.

The induction motor has the following losses:

(i) stator copper loss;
(ii) rotor copper loss;
(iii) core loss;
(iv) friction and windage loss;
(v) stray load loss.

The presence of harmonics causes the stator and rotor copper losses to increase. As for the core loss, its increase is due to higher flux density caused by the harmonic components. Note that the fundamental and harmonic magnetizing currents increase the air gap mmf, and hence there is an increase in the flux density over the fundamental flux density. Compared to the harmonic copper losses, the increase in core loss is usually negligible for frequencies up to 100 Hz. For very high-speed machines operating at hundreds of Hz, the increase in core loss is considerable.

The friction and windage losses are independent of the harmonics. The stray losses are dependent on harmonics as some of stray-loss components are directly influenced by them. Its components are

(i) rotor zigzag loss,
(ii) stator end loss,
(iii) rotor end loss, and
(iv) other undefined losses.

Even though these losses constitute only a minor portion of the total losses, they are considerably increased by the harmonics. Their dependence on the harmonics is quantified and made available in references.

For standard machines, the total losses of inverter-fed machines can increase nearly as much as 50% compared to the losses resulting from a sinusoidal source. While this should be taken as a guideline, it is necessary to evaluate the losses for individual cases to appraise their suitability for specific applications.

Example 7.6

Consider the motor drive given in Example 7.1. Calculate the increase in stator and rotor copper losses and compare them with sinusoidal-source-produced losses at rated slip speed. Neglect any other loss and harmonics greater than 19.

Solution

$$I_{m1} = \frac{\lambda_{m1}}{L_m} = \frac{0.306}{0.05383} = 5.68 \text{ A}$$

$$I_{r1} = \frac{V_{as1}}{\frac{R_r}{s} + jX_{lr}} \cong \frac{115.5}{10} = 11.55 \text{ A}$$

$$I_{s1} \cong 11.55 + j5.68 = 12.87 \text{ A}$$

The stator and rotor resistive losses due to the fundamentals alone are

$$3I_{s1}^2 R_s = 3 \times (12.87)^2 \times 0.277 = 137.7 \text{ W}$$
$$3I_{r1}^2 R_s = 3 \times (11.55)^2 \times 0.183 = 73.2 \text{ W}$$

$$I_{rh} \cong \frac{V_{as1}}{h^2 X_{eq}}$$

$$I_{r5} = 3.31 \text{ A}$$

$$I_{r7} = 1.69 \text{ A}$$

$$I_{r11} = \frac{V_{as11}}{11 \times X_{eq}} = \frac{115.5/11}{11 \times 1.395} = 0.68 \text{ A}$$

$$I_{r13} = \frac{115.5/13}{13 \times 1.395} = 0.49 \text{ A}$$

$$I_{r17} = \frac{115.5/17}{17 \times 1.395} = 0.29 \text{ A}$$

$$I_{r19} = \frac{115.5/19}{19 \times 1.395} = 0.23 \text{ A}$$

$$I_{har} = \sqrt{I_{r5}^2 + I_{r7}^2 + I_{r11}^2 + I_{r13}^2 + I_{r17}^2 + I_{r19}^2}$$

$$= \sqrt{3.31^2 + 1.69^2 + .68^2 + 0.49^2 + 0.29^2 + 0.23^2} = 3.83 \text{ A}$$

$$\text{Stator current, } I_s = \sqrt{I_{s1}^2 + I_{har}^2} = \sqrt{12.87^2 + 3.83^2} = 13.43 \text{ A}$$

$$3I_s^2 R_s = 3 \times 13.43^2 \times 0.277 = 149.82 \text{ W}$$

$$I_r \cong \sqrt{I_{r1}^2 + I_{har}^2} = \sqrt{11.55^2 + 3.83^2} = 12.16 \text{ A}$$

The stator and rotor resistive losses, including harmonics from 5th to 19th, are

$$3I_r^2 R_r = 3 \times 12.16^2 \times 0.183 = 81.29 \text{ W}$$

Increase in stator and rotor copper losses = $(149.82 - 137.9) + (81.29 - 73.2) = 20.2$ W

$$\% \text{ increase over sinusoidal losses} = \frac{20.2}{137.7 + 73.2} \times 100 = 9.57\%$$

7.4.9 Control of Harmonics

7.4.9.1 General. Undesirable pulsating torques and additional losses in the induction motor are generated by harmonic input voltages. Some of the dominant harmonics can be eliminated selectively by waveshaping the inverter output voltages. It can be done either by phase-shifting two or more inverters and summing their outputs or by pulse-width-modulating the output voltages. Both approaches are described in this section.

7.4.9.2 Phase-shifting control. If the output voltages of two or more inverters fed from a common dc source are phase-shifted and summed through a set of output transformers, a multistep voltage can be generated. The amount of phase shift and magnitude of individual inverter voltages determine the harmonic contents of the resultant voltages. For example, consider the inverters and their waveforms shown in Figure 7.35. The three inverter voltages are equally spaced from each other and summed with different gains through a transformer. The turns ratios of the transformers are chosen to be

$$K_1 = 1$$
$$K_2 = 2 \cos \theta \quad (7.157)$$

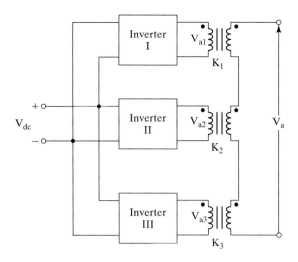

(i) Inverter arrangement

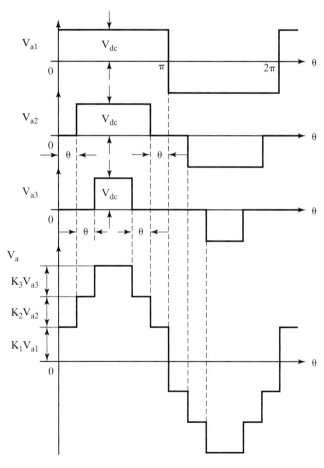

(ii) Summation of voltages

Figure 7.35 Stepped-waveform generation

$$K_3 = 2\cos 2\theta$$

It is generalized for x inverters as

$$K_x = 2\cos(x-1)\theta, \quad x \geq 2 \qquad (7.158)$$

where θ is the phase shift of the inverter voltages, defined as

$$\theta = \frac{2\pi}{(\text{steps in the waveforms})} = \frac{2\pi}{n} = \frac{2\pi}{4x} = \frac{\pi}{2x} \qquad (7.159)$$

where n is the number of steps in the voltage. Each inverter introduces 4 steps; therefore, x inverters introduce $4x$ steps into the output voltage. The output of the fundamental voltage for three steps is obtained as

$$V_{a1} = 4\frac{V_{dc}}{\pi}[K_1 + K_2\cos\theta + K_3\cos 2\theta]\sin\omega_s t \qquad (7.160)$$

Substituting for the turns ratio from equation (7.157) into (7.160) gives the fundamental voltage as

$$V_{a1} = 12\frac{V_{dc}}{\pi}\sin\omega_s t \qquad (7.161)$$

Similarly, the fifth and seventh harmonic voltages are

$$V_{a5} = \frac{4V_{dc}}{5\pi}[K_1 + K_2\cos 5\theta + K_3\cos 10\theta]\sin 5\omega_s t = 0 \qquad (7.162)$$

$$V_{a7} = \frac{4V_{dc}}{7\pi}[K_1 + K_2\cos 7\theta + K_3\cos 14\theta]\sin 7\omega_s t = 0 \qquad (7.163)$$

The eleventh and thirteenth harmonic voltages are,

$$V_{a11} = \frac{V_{a1}}{11} \qquad (7.164)$$

$$V_{a13} = \frac{V_{a1}}{13} \qquad (7.165)$$

Note that there is a complete cancellation of harmonics in the case of the fifth and seventh. Hence, phase-shifting neutralizes harmonics below the sideband frequencies of the n step voltages. In this case, the minimum sideband frequencies are $n-1$ and $n+1$, i.e., 11 and 13.

The voltage magnitude of the multistepped waveforms is varied by controlling the magnitude of dc source voltage. This has the advantage of keeping the harmonic contents to a minimum. Also, the voltage magnitude can be controlled by phase-shifting the various inverter voltages instead of keeping it a constant, but the total harmonic distortion will increase with phase-shifting.

This technique of phase-shifting and neutralizing the undesirable harmonics involves the use of transformers at the output stage that increases the cost and space requirements of the inverter drive, but it has the advantage of paralleling many inverters of small capacities to obtain a larger inverter capability. This technique avoids also the paralleling of devices to obtain higher capacity.

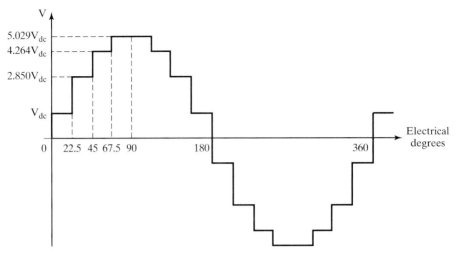

Figure 7.36 Sixteen-stepped waveform to neutralize harmonics up to the fifteenth

---

Example 7.7

Using the phase-shifting principle, find the number of inverters, their phase shifts, and the respective turns ratio to suppress harmonics lower than the fifteenth.

Solution To suppress harmonics lower than the fifteenth amounts to an acceptable minimum sideband frequency of $n-1$, where n is the number of steps in the voltage waveform.

(i) Therefore, $n = 15 + 1 = 16$ steps

(ii) Number of inverters: $x = n/4 = 16/4 = 4$

(iii) Phase shift: $\theta = \dfrac{360°}{n} = \dfrac{360°}{16} = 22.5°$

(iv) $K_1 = 1$

$$K_2 = 2\cos\theta = 2\cos 22.5° = 1.847$$
$$K_3 = 2\cos 2\theta = 2\cos 45° = 1.414$$
$$K_4 = 2\cos 3\theta = 2\cos 67.5° = 0.765$$

The output-voltage waveform is shown in Figure 7.36.

7.4.9.3 Pulse-width modulation (PWM). The control of harmonics and variation of the fundamental component can be achieved by chopping the input voltage. A number of pulse-width-modulation schemes have been in use in the inverter-fed induction motor drives. All of these PWM schemes aim to maximize the fundamental and selectively eliminate a few lower harmonics. Some are discussed in this section.

Figure 7.37 shows the schematic of an inverter and waveforms for a phase mid-pole voltage. The intersections of the carrier signal v_c (usually a bidirectional triangular

366 Chapter 7 Frequency-Controlled Induction Motor Drives

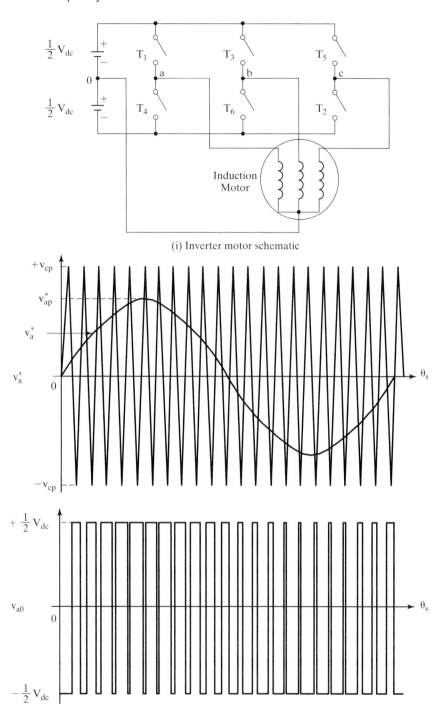

Figure 7.37 Sinusoidal pulse-width modulation

waveform) and the commanded fundamental v_a^* (usually a sinusoidal waveform) provide the switching signals to the base drive of the inverter switches. The switching logic for one phase is summarized as

$$v_{a0} = \frac{1}{2}V_{dc}, \quad v_c < v_a^* \quad (7.166)$$

$$= -\frac{1}{2}V_{dc}, \quad v_c > v_a^*$$

The fundamental of this midpoint voltage is

$$v_{a01} = \frac{V_{dc}}{2} \cdot \frac{v_{ap}^*}{v_{cp}} \quad (7.167)$$

where v_{ap}^* is the peak value of the *a* phase reference or command signal and v_c is the peak value of the triangular carrier signal.

The resulting output has the frequency and phase of the reference signal. The modulation index or ratio is defined by

$$m = \frac{v_{ap}^*}{v_{cp}} \quad (7.168)$$

which, substituted into equation (7.167), gives

$$v_{a01} = m\frac{V_{dc}}{2} \quad (7.169)$$

Varying modulation index changes the fundamental amplitude, and varying the frequency of reference v_a^* changes the output frequency. The ratio between the carrier and reference frequencies, f_c/f_s, changes the harmonics. To eliminate a large number of lower harmonics, f_c has to be very high. Note that this will entail high switching losses and a considerable derating of the supply voltage, because having many turn-on and turn-off intervals sizably reduces the voltage available for output. It is usual to have a fixed value for f_c as say from 9 up to the base frequency of the induction motor, and then reduced values of f_c as higher-harmonic torque pulsations do not significantly affect the drive performance at high speed. The carrier and reference signals have to be synchronized to eliminate the beat-frequency voltage appearing at the output. If the f_c ratio is very high, then synchronization is not very critical. A typical relationship between the carrier and reference frequencies is shown in Figure 7.38. At low frequencies, less than 40 Hz, the carrier frequency could be fixed or a variable to have the synchronization feature.

An alternative sinusoidal pulse-width-modulation strategy is shown in Figure 7.39. The reference contains the fundamental and third harmonic. As the third harmonic is canceled in a star-connected system, the fundamental is enriched in this strategy.

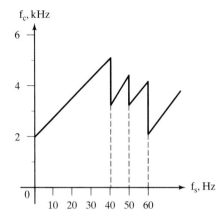

Figure 7.38 Relationship between carrier and fundamental stator frequency

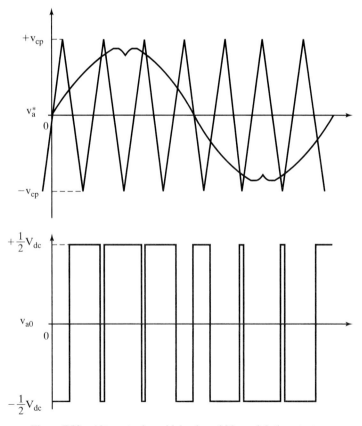

Figure 7.39 Alternate sinusoidal pulse width modulation strategy

7.4.10 Steady-State Evaluation with PWM Voltages

Irrespective of the control strategies employed in the induction motor drive, the input voltages are periodic in steady state. Hence, direct steady-state performance evaluation is possible by matching boundary conditions. In this section, PWM voltage inputs are considered for steady-state performance evaluation of the induction motor drive system. The PWM can be generated in any number of ways: sine-triangle, trapezoidal-triangle, space vector, sampled asymmetric method modulation strategies, etc. Because of their symmetry for either half-wave or full-wave, the boundary-matching technique is ideal for evaluating the steady-state current vector directly, without going through the dynamic simulation from start-up. The algorithm for this follows and is illustrated with three-phase sampled asymmetric modulated voltage inputs to the induction motor whose parameters are given in Example 7.1.

7.4.10.1 PWM voltage generation. The PWM voltages are generated from the sequences of pulses whose turn-on times are given by the expression

$$t(i) = \frac{1 \pm m \sin[a(i)]}{2f_c}, \quad \begin{array}{l} + \text{ for even value of n} \\ - \text{ for odd value of n} \end{array} \quad (7.170)$$

where n is the ratio between the carrier and modulation frequencies, $t(i)$ is the ith pulse width, m is the modulation ratio, f_c is the carrier frequency, and

$$a(i) = \frac{2\pi i}{n}, \text{ rad}; \quad i = 1, 2, ..., n \quad (7.171)$$

The pulse widths are spread equally on either side of the pulse centers given by the expression,

$$p_c(i) = \frac{2i - 1}{2} \frac{2\pi}{n} \quad (7.172)$$

The pulse widths $t(i)$ could be expressed in electrical radians as

$$p_w(i) = t(i)f_s, \text{ rad} \quad (7.173)$$

where f_s is the modulation frequency, i.e., the fundamental frequency desired for motor input voltages. The positions of the pulses are found from $p_c(i)$ and $p_w(i)$. During on-time, the midpole voltage is half of the dc link voltage, $0.5V_{dc}$; during off-time, the midpole voltage is negative half of the dc link voltage, $-0.5V_{dc}$. From the midpole voltages, the line voltages and, in turn from them, the phase voltages are derived by using equations (7.8) to (7.10). In a voltage-source inverter, note that complementary switching in each phase leg is used to provide for a definite predetermined voltage across the load irrespective of the load current.

Further, in digital implementations, the pulse widths are approximated, depending on the number of bits involved in the digital controller. For illustration, the following parameters are chosen:

$$m = 0.5, f_s = 60 \text{ Hz}, n = 9, f_c = nf_s = 540 \text{ Hz}, V_b = 163.3 \text{ V}$$

The pulse widths and their approximations for implementation are given in the following table

i	$p_w(i)$, deg.	approximated $p_w(i)$, deg.
1	20	20
2	26.43	26
3	29.85	30
4	28.66	29
5	23.41	23
6	16.57	17
7	11.33	11
8	10.16	10
9	13.59	14

For the direct steady-state evaluation of the current vector and hence of the performance of the induction motor drive, the PWM phase and d and q axes voltages in stator reference frames are obtained in p.u., as is shown in Figure 7.40. Note that the d and q axes voltages are obtained from the following relationships:

$$v_{qs} = \frac{2}{3}[v_{as} - 0.5(v_{bs} + v_{cs})] \quad (7.174)$$

$$v_{ds} = \frac{1}{\sqrt{3}}[v_{cs} - v_{bs}] \quad (7.175)$$

and the zero-sequence component is zero because the set of voltages is balanced. The next step, then, is to use these voltages in the machine model to obtain the steady-state current vector.

7.4.10.2 Machine model. The induction-machine model in the stator reference frames can be written in the following form:

$$V = (R + Lp)i + G\omega_r i \quad (7.176)$$

where

$$V = [v_{as} \quad v_{ds} \quad 0 \quad 0]^t \quad (7.177)$$

$$i = [i_{qs} \quad i_{ds} \quad i_{qr} \quad i_{dr}]^t \quad (7.178)$$

$$R = \begin{bmatrix} R_s & 0 & 0 & 0 \\ 0 & R_s & 0 & 0 \\ 0 & 0 & R_r & 0 \\ 0 & 0 & 0 & R_r \end{bmatrix} \quad (7.179)$$

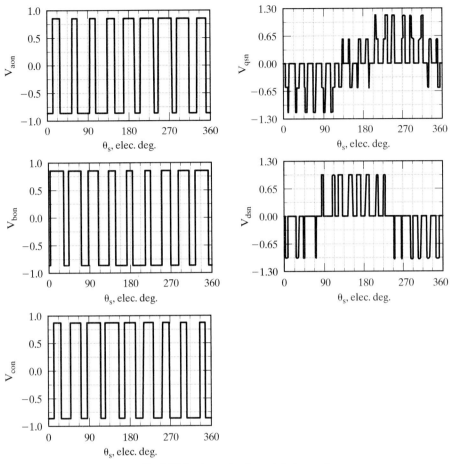

Figure 7.40 PWM voltages in *abc* and *dq* frames in p.u.

$$L = \begin{bmatrix} L_s & 0 & L_m & 0 \\ 0 & L_s & 0 & L_m \\ L_m & 0 & L_r & 0 \\ 0 & L_m & 0 & L_r \end{bmatrix} \quad (7.180)$$

$$G = \begin{bmatrix} 0 & 0 & 0 & 0 \\ 0 & 0 & 0 & 0 \\ 0 & -L_m & 0 & -L_r \\ L_m & 0 & L_r & 0 \end{bmatrix} \quad (7.181)$$

which is cast in state-space form as

$$\dot{X} = AX + Bu \quad (7.182)$$

where

$$A = -L^{-1}[R + \omega_r G] \quad (7.183)$$
$$B = L^{-1} \quad (7.184)$$
$$X = i \quad (7.185)$$
$$u = V \quad (7.186)$$

The next step is to use this set of state-space equations to obtain the steady-state current vector directly.

7.4.10.3 Direct evaluation of steady-state current vector by boundary-matching technique. The state-space equations can be discretized as follows. The solution for the current vector is

$$X(t) = e^{At}X(0) + \int_0^t e^{A(t-\tau)}Bu(\tau)d\tau \quad (7.187)$$

In one sampling interval, T_s, the input and state variables are constant, by discretization. Hence,

$$X(T_s) = e^{AT_s}X(0) + e^{AT_s}\int_0^{T_s} e^{-A\tau}d\tau Bu(0) \quad (7.188)$$

which gives rise to a solution of the form

$$X(T_s) = e^{AT_s}X(0) + (e^{AT_s} - I)A^{-1}Bu(0) \quad (7.189)$$

and it could be generalized for the kth sampling interval (and omitting the T_s in the parentheses for simplicity):

$$X(k+1) = \Phi X(k) + Fu(k) \quad (7.190)$$

where

$$\Phi = e^{AT_s} \text{ and } F = (\Phi - I)A^{-1}B \quad (7.191)$$

Note that $X(k)$ can be calculated in terms of $X(k-1)$ and $u(k-1)$ and so on, as in the following:

$$X(1) = \Phi X(0) + Fu(0) \quad (7.192)$$
$$X(2) = \Phi X(1) + Fu(1) \quad (7.193)$$

and, substituting for $X(1)$ from the previous relationship and expanding the equation,

$$X(2) = \Phi^2 X(0) + \Phi Fu(0) + Fu(1) \quad (7.194)$$

and similarly, for the $(k+1)$th sampling interval, the current vector is

$$X(k+1) = \Phi^{k+1}X(0) + \Phi^k Fu(0) + \Phi^{k-1}Fu(1) + \ldots + Fu(k) \quad (7.195)$$

The last expression, by symmetry of the wave forms, must be equal to the initial vector itself if the $(k+1)$th sampling interval corresponds to 360 electrical degrees. Equating these, we get

Section 7.4 Voltage-Source Inverter-Driven Induction Motor

$$X(k + 1) = X(0) \quad (7.196)$$

but X(k+1) contains the term X(0); rearranging these, the steady-state initial vector X(0) is obtained as

$$X(0) = [I - \Phi^{(k+1)}]^{-1}\{\Phi^k Fu(0) + \Phi^{k-1} Fu(1) + \ldots + Fu(k)\}$$

$$= [I - \Phi^{(k+1)}]^{-1}\left\{\sum_{j=0..k} \Phi^j Fu(k - j)\right\} \quad (7.197)$$

where I is the 4 × 4 identity matrix. For the example under consideration, the sampling time corresponds to 2 electrical degrees, given as

$$T_s = \frac{2}{360 f_s}, s \quad (7.198)$$

The exponential of AT_s can be evaluated from the series with 8 to 15 terms. The value of k is given by

$$k + 1 = 360/2 = 180 \quad (7.199)$$

which gives k = 179. For wave forms with half-wave symmetry, the final and initial values are related by other than identity, as is shown in the section on the six-step inverter-fed induction motor drive.

7.4.10.4 Computation of steady-state performance. The current vector is evaluated for the entire cycle from the discretized state-space equation discussed and derived above. The electromagnetic torque is computed as

$$T_e(k) = \frac{3}{2}\frac{P}{2} L_m \{i_{qs}(k) i_{dr}(k) - i_{ds}(k) i_{qr}(k)\} \quad (7.200)$$

where the currents are the components of the state vector X(k). By inverse transformation, the phase currents are computed as

$$i_{as}(k) = i_{qs}(k) \quad (7.201)$$
$$i_{bs}(k) = -0.5 i_{qs}(k) - 0.866 i_{ds}(k) \quad (7.202)$$

The stator q and d axes currents, a and b phase currents, and electromagnetic torque are shown in Figure 7.41 for the induction machine whose parameters are given in Example 7.1, but running with a slip of 0.05 for the present illustration. A significant difference between these current wave forms and those of the six-step inverter-fed induction motor drives is to be noted: the currents have become more sinusoidal, with the superposed switching ripples. Further, the rate of change of currents and sharpness of the peaks have become smaller by comparison, contributing to lower overall rms current, resulting in lower stator copper losses and better thermal performance of the machine. Even though the switching harmonic torque has higher magnitude, note that its frequency has increased in proportion to the PWM frequency, indicating that it is easier to filter them with the mechanical inertia of the machine and its load than that of the sixth-harmonic ripple present in the six-step inverter-fed machines.

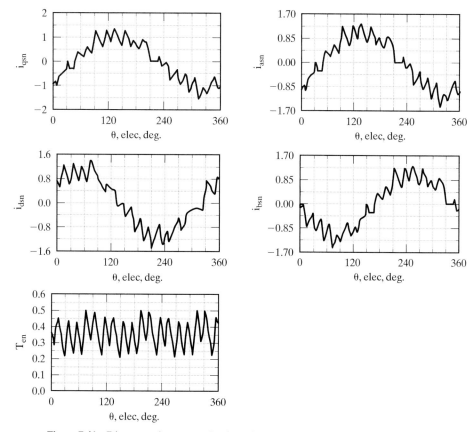

Figure 7.41 Direct steady-state evaluation of PWM-fed induction motor drive in p.u.

Example 7.8

Compute the transient response of the PWM-based induction motor drive with volts/Hz control strategy. The motor and PWM details are as follows:

100 hp, 460 V, 6 pole, 60 Hz, 3 phase, star connected induction motor.

Efficiency = 0.82, Power factor = 0.86, $R_s = 0.0551\ \Omega$, $R_r = 0.311\ \Omega$, $L_m = 0.02066$ H, $L_{ls} = 0.0007798$ H, $L_{lr} = 0.00007798$ H, $J = 0.8$ kg-m², $B_1 = 0.15$ N·m/(rad/sec).

$$\text{PWM: } \frac{f_c}{f_s} = 9.$$

Solution The steady state of the PWM-fed induction motor was calculated with the preset PWM voltages given in equations (7.170) to (7.173). Alternately, the PWM voltages can be generated from the intersections of the carrier and modulation signals. The transient response of such a PWM-based induction motor drive with volts/Hz control strategy and on open-loop speed control is computed by using the dynamic model of the induction motor. The offset voltage, V_o, and the volts-to-frequency constant, K_{vf}, are calculated by using stator full-

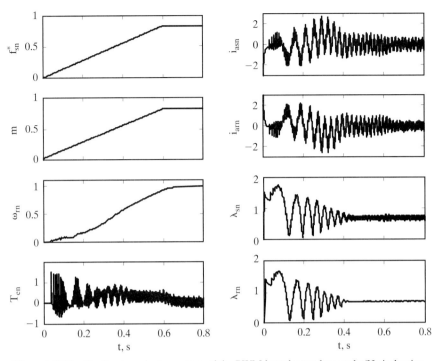

Figure 7.42(i) Starting transient response of the PWM-based open-loop volts/Hz induction motor drive in normalized units.

load current. A ramp frequency command is given, which serves as a soft start, and the modulation ratio is calculated as

$$m = \frac{2}{V_{dc}} (V_o + K_{vf} f_s)$$

The phase command voltage is also given as

$$v_{as}^* = m v_c \sin(\omega_s t)$$

where v_c is the absolute peak carrier voltage and in this case is taken to be 10 V. The intersection of the carrier and the command voltages is obtained by sampling every microsecond. Then the midpole voltages, i.e., the midpoints of dc link voltage and the midpoint of inverter phases are obtained from the intersections. The midpole voltages are $\pm \dfrac{V_{dc}}{2}$, where V_{dc} is the dc link voltage.

In this example, it is assumed that $V_{dc} = \sqrt{2} \times 460$ V.

$$V_b = \sqrt{2} \times \frac{460}{\sqrt{3}} V, \quad T_b = \frac{P}{2} \frac{P_b}{\omega_b}, \quad P_b = 100 \text{ hp}, \quad f_b = 60 \text{ Hz}, \quad I_b = \frac{2}{3} \frac{P_b}{V_b}, \quad \omega_b = 2\pi f_b.$$

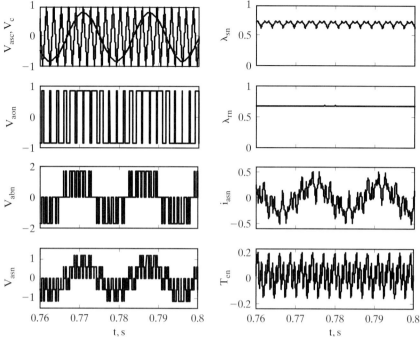

Figure 7.42(ii) Steady-state response of the PWM-based open-loop volts/Hz induction motor drive

From the midpole voltages, the line voltages are calculated as

$$V_{ab} = V_{ao} - V_{bo}$$
$$V_{bc} = V_{bo} - V_{co}$$
$$V_{ca} = V_{co} - V_{ao}$$

from which the phase voltages are derived as

$$V_{as} = \frac{V_{ab} - V_{ca}}{3}$$
$$V_{bs} = \frac{V_{bc} - V_{ab}}{3}$$
$$V_{cs} = \frac{V_{ca} - V_{bc}}{3}$$

The *dq* voltages are derived from the phase voltages in synchronous reference frames and the motor equations are integrated to obtain the currents and torque. The *abc* currents are obtained from the synchronous-reference-frame *dq* currents by the inverse transformation. The transient response for the start-up is shown in Figure 7.42(i). The modulation ratio follows the command frequency with the adjusted magnitude. The stator and rotor current magnitudes are smaller than those of the six-step inverter-fed volts/Hz-controlled drive. The soft start helps this reduc-

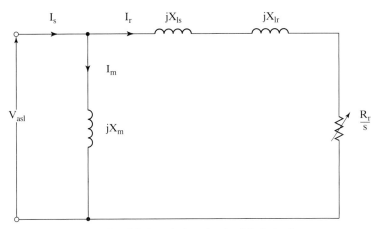

Figure 7.43 Simplified equivalent circuit of the induction motor

tion in current magnitude. The stator and rotor flux linkages reach steady state at around 0.5 s. The rotor torque has large oscillations, caused mainly by the stator and rotor flux-linkage oscillations. The rotor speed has a slow rise until the flux linkages are settled, and then the speed rises smoothly.

When the motor has reached steady state, the waveforms for a few cycles are shown in Figure 7.42(ii). These waveforms show the generation of PWM signals, the midpole voltages, a line-to-line and a phase voltage, a stator phase current, stator and rotor flux linkages, and electromagnetic torque. The motor drive is operating at no-load but overcomes the friction and hence requires only a small average torque as shown in the figure. All the variables in Figures 7.42(i) and (ii) are in normalized units.

7.4.11 Flux-Weakening Operation

7.4.11.1 Flux weakening. The air gap flux of the induction motor is maintained constant by keeping the ratio between the induced air gap emf and stator frequency a constant. The voltage input to the inverter is limited, and, at its maximum value, it is designed to give rated flux operation at rated frequency of the induction motor. For operation at higher than rated frequency, the dc voltage input to the inverter is clamped and thereafter the induction motor is fed with a constant voltage. This leads to the flux weakening. Consider the induction-motor equivalent circuit given in Figure 7.43. Stator resistance is neglected in this equivalent circuit. The air gap flux linkage is

$$\lambda_m = \frac{V_{as1}}{\omega_s} \quad (7.203)$$

With increasing stator frequency, the air gap flux linkage is decreasing. This has an interesting consequence on the performance of the motor drive. Very similar to the dc motor drives, the ac motors are also flux-weakened for operation above rated speed. At high speed, the output of the motor will be exceeded if the torque is not programmed to vary inversely proportionally to the speed. The torque is programmed as

$$T_e \propto \frac{1}{\omega_m} \qquad (7.204)$$

or

$$T_e = \frac{\text{rated output}}{\omega_m} = \frac{P_m}{\omega_m} \qquad (7.205)$$

where P_m is the output power.

Equation (7.205) is further written as

$$T_e = \frac{P_m}{\omega_s(1-s)} \cdot \frac{P}{2} \qquad (7.206)$$

By writing electromagnetic torque in terms of air gap flux linkage, voltage, and motor parameters, and by using a simple equivalent circuit with X_m being very large, we get

$$T_e = 3\frac{P}{2} \cdot \frac{V_{as1}^2}{\left(\frac{R_r}{s}\right)^2 + X_{eq}^2} \cdot \left(\frac{R_r}{s\omega_s}\right) = 3\frac{P}{2} \cdot \lambda_m \cdot \omega_s \cdot \frac{V_{as1}}{\left(\frac{R_r}{s}\right)^2 + X_{eq}^2} \cdot \left(\frac{R_r}{s\omega_s}\right)$$

$$= 3\frac{P}{2} \cdot \lambda_m \cdot \frac{V_{as1}}{\left(\frac{R_r}{s}\right)^2 + X_{eq}^2} \cdot \left(\frac{R_r}{s}\right) \qquad (7.207)$$

where

$$X_{eq} = X_{lr} + X_{ls} \qquad (7.208)$$

Equating (7.206) and (7.207), we get

$$P_m = 3\lambda_m \cdot \omega_s \cdot \frac{V_{as1}}{\left(\frac{R_r}{s}\right)^2 + X_{eq}^2} \cdot \left(\frac{R_r}{s}\right)(1-s) \qquad (7.209)$$

From equation (7.209), it is inferred that, to maintain the air gap power constant, the air gap flux linkage has to be varied inversely proportionally to the stator frequency. The stator voltage input is a constant during flux weakening, so, if slip s is small, then the air gap power becomes a constant, but maintaining air gap power constant does not guarantee that the shaft power output is constant. To do so, the slip also has to be regulated. The flux-weakening and hence torque-programming is inherent in the induction motor by simple variation of stator frequency, and this makes the induction motor drive attractive in high-speed applications. This is true only in an approximate sense. An accurate control and maintenance of constant shaft power output during flux-weakening is an involved process, taken up in detail in Chapter 8.

7.4.11.2 Calculation of slip. Because the input phase voltage is a constant, the induction motor is controlled by varying its slip speed. For a given air gap power, the slip is obtained as follows.

The electromagnetic torque is given by

$$T_e = 3 \frac{V_{as1}^2}{\left(\dfrac{R_r}{s}\right)^2 + L_{eq}^2 \omega_s^2} \cdot \left(\frac{R_r}{s\omega_s}\right) \quad (7.210)$$

The rated value or torque is

$$T_{er} = 3 \frac{V_{as1}^2}{\left(\dfrac{R_r}{s_r}\right)^2 + L_{eq}^2 \omega_{sr}^2} \cdot \left(\frac{R_r}{s\omega_{sr}}\right) \quad (7.211)$$

where the subscript r denotes the rated value of the corresponding variables.

The normalized torque is

$$T_{en} = \frac{T_e}{T_{er}} = \left(\frac{s}{s_r}\right)\left(\frac{1}{\omega_{sn}}\right)\frac{(R_r^2 + s_r^2 X_{eq}^2)}{(R_r^2 + s^2 \omega_{sn}^2 X_{eq}^2)} \quad (7.212)$$

where

$$\omega_{sn} = \frac{\omega_s}{\omega_{sr}} \quad (7.213)$$

and ω_{sn} is the normalized stator frequency. (Its maximum value is usually a fixed value at the design stage of the inverter.)

Also, the normalized torque in the field-weakening mode is given by

$$T_{en} = \frac{P_{mn}}{\omega_{mn}} = \frac{P_{mn}}{(\omega_m/\omega_{mr})} = P_{mn}\frac{(1 - s_r)}{(1 - s)}\cdot\frac{\omega_{sr}}{\omega_s} = P_{mn}\frac{(1 - s_r)}{(1 - s)}\cdot\frac{1}{\omega_{sn}}, \text{p.u.} \quad (7.214)$$

where P_{mn} is the normalized output power in p.u.

Equating (7.212) and (7.214), we get

$$P_{mn}\frac{(1 - s_r)s_r}{R_r^2 + s_r^2 X_{eq}^2} = \frac{(1 - s)s}{R_r^2 + s^2 \omega_{sn}^2 X_{eq}^2} \quad (7.215)$$

By noting that the left-hand side of equation (7.214) is a constant and by defining

$$a = \frac{R_r}{X_{eqr}} \quad (7.216)$$

$$b = \frac{P_{mn}(1 - s_r)s_r}{a^2 + s_r^2} \quad (7.217)$$

equation (7.217) is written as

$$\frac{(1 - s)s}{a^2 + s^2 \omega_{sn}^2} = b \quad (7.218)$$

from which the slip is solved for as

$$s = \frac{1}{2(1 + b\omega_{sn}^2)} \pm \frac{1}{2}\sqrt{\frac{1}{(1 + b\omega_{sn}^2)^2} - \frac{4ba^2}{(1 + b\omega_{sn}^2)}} \quad (7.219)$$

Note that in open-loop operation, this slip must be less than the slip corresponding to the breakdown torque or peak torque, denoted as s_m. The constraint for open-loop operation, then, is

$$s < s_m \quad (7.220)$$

Usually, this is accompanied by another constraint, that is, the stator current should not exceed a certain value. In most cases, this is the rated value of the stator current, denoted by I_{sr}. Then the constraint is expressed as

$$I_s < I_{sr} \quad (7.221)$$

where

$$I_s = \frac{V_{as1}}{\left(\frac{R_r}{s}\right) + jL_{eq}\omega_s} + \frac{V_{as1}}{j\omega_s L_m} \quad (7.222)$$

Equation (7.219) for slip gives two values, the smaller value being the statically stable region of the torque–speed characteristics and the larger value corresponding to the operating point in the unstable region of the torque–speed characteristics. For regeneration, the rated slip s_r has to be negative, and hence the slip calculated from equation (7.219) will yield negative values; P_{mn} will be negative for this condition.

7.4.11.3 Maximum stator frequency.

The maximum operating stator frequency for the motor drive is evaluated from the fact that the slip cannot be a complex value; hence, from equation (7.219),

$$\frac{1}{(1 + b\omega_{sn}^2)^2} \geq \frac{4a^2 b}{(1 + b\omega_{sn}^2)} \quad (7.223)$$

The maximum stator frequency is

$$\omega_{sn(max)} = \frac{\sqrt{1 - 4a^2 b}}{2ab}, \text{ p.u.} \quad (7.224)$$

from which the maximum rotor speed is obtained as

$$\omega_{mn(max)} = \omega_{sn(max)} \cdot \frac{(1 - s)}{(1 - s_r)}, \text{ p.u.} \quad (7.225)$$

Example 7.9

Calculate the stator current magnitude and slip at maximum stator frequency when the induction motor drive is in the flux-weakening mode of operation and delivering rated power. The motor and drive details are given in Example 7.1.

Solution Air gap power is assumed to be equal to rated value.

$s_r = 0.0183$

$$a = \frac{R_r}{X_{eq}} = \frac{0.183}{(0.554 + 0.841)} = 0.131$$

$$b = P_{mn} \frac{(1 - s_r)s_r}{a^2 + s_r^2} = \frac{1(1 - 0.0183)(0.0183)}{(0.131)^2 + (0.0183)^2} = 1.024$$

$$\omega_{sn(max)} = \frac{\sqrt{1 - 4a^2b}}{2ab} = \frac{\sqrt{1 - 4 \times 0.131^2 \times 1.024}}{2 \times 0.131 \times 1.024} = 3.59 \text{ p.u.}$$

$$s = \frac{1}{2(1 + b\omega_{sn}^2)} + \frac{1}{2}\sqrt{\frac{1}{(1 + b\omega_{sn}^2)^2} - \frac{4a^2b}{(1 + b\omega_{sn}^2)}} = \frac{1}{2(1 + 1.024 \times 3.59^2)} = 0.0359 \text{ p.u.}$$

$\omega_s = \omega_{sn} \times \omega_{sr} = 3.59 \times 377 = 1353.4 \text{ rad/sec}$

$V_{as1} = 115.5 \text{ V}$

$$I_s = V_{as1}\left\{\frac{1}{\left(\frac{R_r}{s}\right) + jL_{eq}\omega_s} + \frac{1}{j\omega_s L_m}\right\} = 115.5\left\{\frac{1}{\frac{0.183}{0.0183} + j5.0} + \frac{1}{j1353.4 \times 0.0538}\right\}$$

$= 9.24 - j6.2 \text{ A}$

$|I_s| = 11.12 \text{ A}$

$$\omega_{mn(max)} = \omega_{sn(max)} \cdot \frac{(1 - s)}{(1 - s_r)} = 3.59 \times \frac{(1 - 0.0359)}{(1 - 0.0183)} = 3.525 \text{ p.u.}$$

7.5 CURRENT-SOURCE INDUCTION MOTOR DRIVES

7.5.1 General

In a current-source drive, the input currents are six-stepped waveforms. Amplitude and frequency are variables, as in voltage-source drives. Current-source drives have a distinct advantage over voltage-source drives: the electrical apparatus is current-sensitive, and torque is directly related to the current rather than the voltage. Hence, control of current ensures the direct and precise control of the electromagnetic torque and drive dynamics. Current-source variable-frequency supplies are realized either with an autosequentially commutated inverter (hereafter referred to as ASCI) or with current-regulated inverter drives. Their operation is explained in the following sections. The closed-loop four-quadrant current-source inverter-fed induction motor and its performance are discussed. The method of computing steady-state and dynamic per-

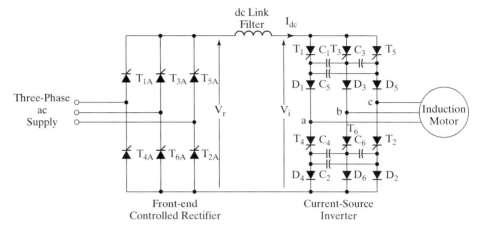

Figure 7.44 Current-source induction motor drive

formance is given. The effect of lower harmonics on the performance of the motor drive is evaluated, and control of harmonics is explained in this section.

7.5.2 ASCI

The ASCI-fed induction motor drive is shown in Figure 7.44. The converter system has a controlled rectifier for providing the ac-to-dc conversion and an inverter for dc-to-ac power conversion. The dc output voltage is fed to the autosequentially commutated current-source inverter (ASCI) through a filter inductor. This inductor is provided to maintain the dc link current at a steady value. The operation of the controlled rectifier and current source is given in Chapter 3. The principle of operation, to facilitate the understanding of the current-source drive, is given below.

7.5.2.1 Commutation. The sequence of firing the ASCI is $T_1, T_2, T_3, T_4, T_5, T_6$ for the phase sequence *abc* in the induction motor. At any time, two SCRs are conducting, and they are turned on at an interval of 60 electrical degrees. Let T_1 and T_2 conduct, and let the dc link current be a constant. The capacitor C_1 is charged positive at the plate connecting it to the cathode of T_1 and negative at the plate connecting it to the cathode of T_3. The current is following the path T_1, D_1, phase *a* winding, phase *c* winding, D_2, T_2, dc source, as shown in Figure 7.45(i). To commutate phase *a* current, T_3 is gated on. With that, the capacitor voltage is applied across T_1, which reverse biases it. The current will continue to flow through T_3, C_1, D_1, *a* phase, *c* phase, D_2, T_2, dc source. The SCR T_1 is turned off. The current flow reverses the charge across C_1. This is shown in Figure 7.45(ii).

The reversing charge in the capacitor C_1 forward biases D_3; hence, the dc link current is split through both *a* and *b* phases, as shown in Figure 7.45(iii). The voltage across C_1 becomes increasingly negative with respect to D_1, which reverse biases it and cuts off conduction. The dc link current goes through only *b* and *c* phases, as shown in Figure 7.45(iv).

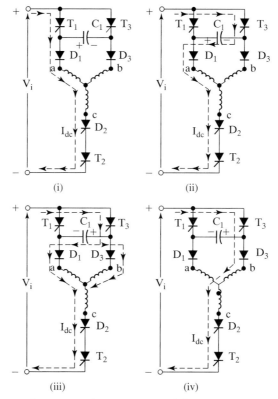

Figure 7.45 Commutation sequence in an autosequentially commutated current-source inverter

The current in phase *a* has been commutated completely. Likewise, gating one SCR will commutate the parallel SCR. The stator phase currents are shown in Figure 7.46 assuming that the stator is Y-connected. For purposes of analysis, it is usual to neglect the rise and fall times of the current and consider only ideal rectangular blocks of current with 120-electrical-degree duration.

The current commutation is slow; hence, converter-grade SCRs with a large turn-off time are sufficient for the ASCI. This reduces the cost of the devices. The transfer of current from one phase to another produces a voltage spike due to the leakage inductances in the motor. Such voltage spikes are undesirable, and they increase the SCR voltage ratings. The commutation capacitors absorb part of this commutation energy and reduce the magnitude of the voltage spikes.

7.5.2.2 Phase-sequence reversal. The phase sequence of the inverter output currents is reversed by changing the sequence of firing the SCRs. The sequence of firing for phase sequence *abc* is $T_1, T_2, \ldots, T_6$; for phase sequence *acb*, it is $T_1, T_6, T_5, T_4, T_3, T_2$. Note that the phase-sequence reversal command is determined by the speed command and actual speed of the induction motor. An abrupt phase-sequence reversal in the induction motor would result in plugging, accompanied by large stator

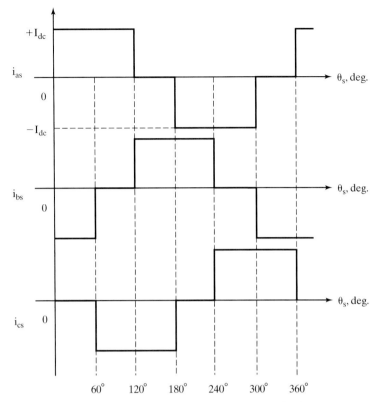

Figure 7.46 Stator currents in a star-connected induction motor fed from a current source

currents. Therefore, a great amount of caution and control is required to initiate such an action.

7.5.2.3 Regeneration. To understand regeneration, it is instructive to go through the motoring operation of the current-source induction motor drive shown in Figure 7.47. In the motoring mode, the electrical power input is positive, and the rectified dc voltage and current are positive. The reflected voltage of the induction motor at the input of the inverter, V_i, is positive, and it opposes the dc input voltage V_r in the dc link. Note that the induction motor drive is in the forward motoring mode with speed and torque in the clockwise direction. In the regeneration mode, the rotor speed remains in the same direction but the torque is reversed. This is achieved by making the slip speed negative. It results in the negative induced emfs in the stator phases of the induction motor, which appear as a negative voltage at the inverter input, as is shown in Figure 7.48. For stable operation, the rectified voltage V_r is reversed, thus enabling the absorption of regenerative energy. The input dc link current is positive, and dc link voltage is made negative, resulting in the input dc link power being negative, implying that the induction machine is sending power to the input ac mains.

The same operation results during the regeneration when the rotor speed is in the counter clockwise direction. Thus, four-quadrant operation is achieved with one

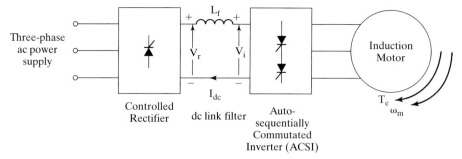

Figure 7.47 Forward motoring of the current-source induction motor drive

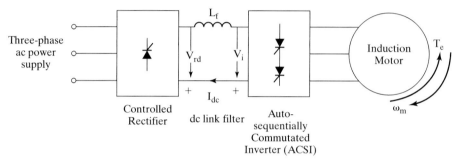

Figure 7.48 Regeneration in the current-source induction motor drive

controlled rectifier, as opposed to a dual controlled rectifier in the case of the voltage-source-inverter induction motor drive.

7.5.2.4 Comparison of converters for ac and dc motor drives. A comparison of converters for dc and ac motor drives is timely to assess the relative cost. The converters are assumed to use SCRs only. The motor drive has to be four-quadrant in operation. The induction motor drives with voltage and current source (ASCI) are compared with the dc motor drives. The comparison is given in Table 7.2. From the table, it is inferred that the ASCI-fed induction motor drive is more economical than its voltage-source counterpart but is more expensive than the dc motor drive. When the cost of the motor is included for comparison, particularly above 100 hp, the ASCI-fed induction motor is likely to be the economical motor drive.

7.5.3 Steady-State Performance

The steady-state performance of the current-source induction motor (hereafter referred to as CSIM) drive can be evaluated approximately by using the fundamental of the input currents or exactly by matching boundary conditions. The approximate solution by using the equivalent circuit of the induction motor lends insight into the influence of the motor parameters on the drive performance. The exact solution using boundary-matching conditions offers no such insight into the drive performance. Both solution methods are developed in this section.

TABLE 7.2 Comparison of SCR converters for four-quadrant dc and voltage- and current-source induction motor drives

Serial Number	Items	DC Drive	Induction Motor Drives	
			Voltage Source	Current Source
1	SCRs	12	36 (full wave)	12
2	Diodes	0	24	6
3	Commutating Reactors	0	6	—
4	Filter Capacitors	0	1	0
5	Commutating Capacitors	0	6 (small)	6 (large)
6	DC Link Filter Inductor	0	0	1 (large)

Equivalent Circuit Approach. The equivalent circuit of the induction motor with constant stator current is shown in Figure 7.49. The rotor and magnetizing currents as a function of stator current are

$$I_r = \frac{jL_m}{\frac{R_r}{s\omega_s} + jL_r} \cdot I_s \quad (7.226)$$

$$I_m = \frac{\frac{R_r}{s\omega_s} + jL_{lr}}{\frac{R_r}{s\omega_s} + jL_r} \cdot I_s \quad (7.227)$$

where the fundamental stator rms current is

$$I_s = \frac{\sqrt{2}\sqrt{3}}{\pi} \cdot I_{dc} = 0.779 \, I_{dc} \quad (7.228)$$

and I_{dc} is the dc link current.

The electromagnetic torque is

$$T_e = 3 \cdot \frac{P}{2} \cdot \frac{L_m^2}{\left(\frac{R_r}{s\omega_s}\right)^2 + L_r^2} \cdot \frac{R_r}{s\omega_s} \cdot I_s^2 \quad (7.229)$$

and the maximum torque occurs at

$$s = \frac{R_r}{\omega_s L_r} \quad (7.230)$$

Substituting equation (7.230) into (7.229) yields

$$T_{e(max)} = \frac{3}{2} \cdot \frac{P}{2} \cdot \frac{L_m^2}{\left(\frac{R_r}{s\omega_s}\right)} \cdot I_s^2 = \frac{3}{2} \cdot \frac{P}{2} \cdot \frac{L_m^2}{L_r} \cdot I_s^2 \quad (7.231)$$

Section 7.5 Current-Source Induction Motor Drives

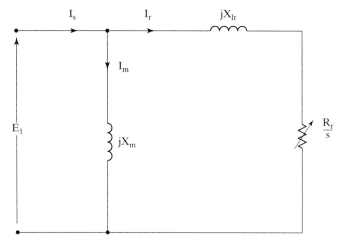

Figure 7.49 Induction-motor equivalent circuit with constant stator current

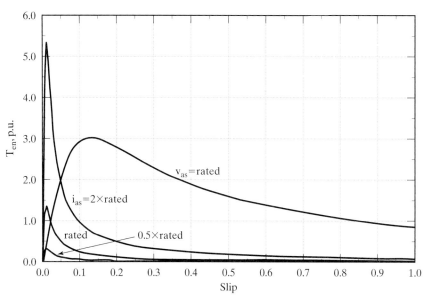

Figure 7.50 Torque–speed characteristics of the current-source induction motor drive for 0.5, 1, and 2 times the rated stator currents and the torque-vs.-slip characteristic at rated stator voltages

The maximum torque, when saturation is considered, becomes much smaller compared to the unsaturated case. From the equivalent circuit, the steady-state torque–speed characteristics are drawn for various values of stator current and shown in Figure 7.50. To contrast them, the torque-vs.-speed characteristics of the voltage-fed induction motor is also drawn in the same figure for nominal stator voltages. This characteristic reflects operation at rated air gap flux linkages if the stator impedance is neglected. The rated air gap-flux-linkages operation is available for the current-source drive only in its statically unstable portion of the characteristic, as is seen from the intersections of the voltage-source and current-source torque–speed characteristics. Points to the left of the intersections indicate operation with higher flux linkages than rated value and, hence, deep saturation. To operate the current-source motor drives at rated flux linkages, correlation between the slip speed and stator current amplitude has to be achieved. Further, these operating points are in the statically unstable region, so stabilization is required in the form of feedback control. That both of these features are unnecessary in the voltage-source inverter-fed induction motor drives is to be taken note of. It further simplifies the control and sensor requirements in the voltage-source drives, thus making them attractive for low-cost, low-performance applications. The steady-state performance without neglecting harmonics is computed by boundary-matching conditions, as given below, and various features of the steady-state performance will be highlighted from these results.

Example 7.10

An induction motor is fed from a current source with base current. The desired operating point is at rated flux that is obtained with a voltage-source feed at base voltage. Find the slip and electromagnetic torque at the desired operating point. The machine details are as follows:

$$40 \text{ hp}, 460 \text{ V}, 4 \text{ pole}, 3 \text{ phase}, 60 \text{ Hz}, Y \text{ connected}$$

$$R_r = 0.209 \, \Omega \, ; L_m = 0.04 \text{ H}; L_s = 0.0425 \text{ H}; L_r = 0.043 \text{ H}$$

Ignore the stator resistance and consider that the magnetizing branch is placed ahead of the stator impedance in the equivalent circuit in order to simplify the analytical expressions in this example.

Solution

$$\omega_s = 2\pi f_s = 2\pi * 60 = 376.99 \text{ rad/sec}$$
$$X_m = L_m \omega_s = 15.45 \, \Omega$$
$$X_s = L_s \omega_s = 16.02 \, \Omega$$
$$X_r = L_r \omega_s = 16.21 \, \Omega$$
$$X_{ls} = X_s - X_m; X_{lr} = X_r - X_m;$$
$$X_{eq} = X_{ls} + X_{lr} = 1.32 \, \Omega$$

Base Values:

$$\text{Base voltage} = 460/\sqrt{3} = 265.6 \text{ V}$$
$$\text{Base current} = P_h/(3V_b) = 40*745.6/(3*265.6) = 37.4 \text{ A}$$

For approximate solution, the induction-motor equivalent circuit shown in Figure 7.49 is chosen as per the problem formulation.

Solution Method The operating point is at the intersection of the voltage and current source torque-vs.-slip characteristics of the induction motor; it satisfies the base current operation from a current source and as well a base flux operation. To find that operating point, the air gap torque expressions for both the current and voltage source are derived and equated to each other. It will yield a solution for slip, and then substitution of the slip in either the current source or voltage source air gap torque expressions will evaluate the operating air gap torque of the machine. The following steps in the solution are given.

For base voltage operation, the air gap torque is

$$T_{ev} = 3\frac{P}{2}I_r^2\frac{R_r}{s\omega_s} = 3\frac{P}{2}\frac{V_{as}^2}{\left(R_s + \frac{R_r}{s}\right)^2 + X_{eq}^2}\frac{R_r}{s\omega_s}$$

For current-source operation with stator current of base value, the air gap torque is derived from the following:

$$I_s = I_m + I_r$$

$$jX_m I_m = I_r\left(R_s + \frac{R_r}{s} + jX_{eq}\right)$$

and then, finding the rotor current in terms of the stator current, we get

$$I_r = \frac{jX_m}{\left(R_s + \frac{R_r}{s} + jX_{eq} + jX_m\right)}I_s$$

which, when substituted into the torque expression, yields the final relationship:

$$T_{ec} = 3\frac{P}{2}I_r^2\frac{R_r}{s\omega_s} = 3\frac{P}{2}\frac{X_m^2 I_s^2}{\left(R_s + \frac{R_r}{s}\right)^2 + (X_m + X_{eq})^2}\frac{R_r}{s\omega_s}$$

By equating the torque expressions for current-source and voltage-source operation, the unknown slip is derived. Further, neglecting the stator resistance and keeping the applied voltage as base voltage and the applied current as base current gives the slip as

$$s = \pm R_r\sqrt{\frac{V_b^2 - X_m^2 I_b^2}{(X_m X_{eq} I_b)^2 - (V_b[X_m + X_{eq}])^2}} = 0.0244$$

The air gap torque is obtained by substituting this slip in T_{ec} or T_{ev}:

$$T_e = 128.26 \text{ N·m}$$

7.5.4 Direct Steady-State Evaluation of Six-Step Current-Source Inverter-Fed Induction Motor (CSIM) Drive System

It is assumed that the stator currents are alternating and rectangular, with 120 electrical degrees of conduction. The rise and fall times of the currents are negligible but would be considered for the computation of the voltage spikes generated by the leakage inductances of the induction motor. The technique for the evaluation of the direct steady-state initial current vector (and hence of the performance of the CSIM drive) is identical to that developed for the six-step voltage-fed induction motor drive described in section 7.4.5.6.

Stator Currents The stator phase currents are shown in Figure 7.46. Viewing them in the synchronous reference frames by using the transformation T^e_{abc} given by the equation (7.41) shows that the quadrature and direct axis stator currents are as follows:

(i) Interval I: $0 < \theta_s < 60°$

$$i_{as} = I_{dc}$$
$$i_{bs} = -I_{dc} \quad (7.232)$$
$$i_{cs} = 0$$

giving the quadrature axis stator current as

$$i^e_{qs} = \frac{2}{3}\left[i_{as}\cos\theta_s + i_{bs}\cos\left(\theta_s - \frac{2\pi}{3}\right) + i_{cs}\cos\left(\theta_s + \frac{2\pi}{3}\right)\right] \quad (7.233)$$

yielding for the first interval the quadrature axis current, which is indicated by an additional subscript of I:

$$i^e_{qsI} = aI_{dc}\cos\left(\theta_s + \frac{\pi}{6}\right) \quad (7.234)$$

where

$$a = \frac{2}{\sqrt{3}} \quad (7.235)$$

Similarly, for the direct axis stator current in the synchronous reference frames, we obtain

$$i^e_{dsI} = aI_{dc}\sin\left(\theta_s + \frac{\pi}{6}\right) \quad (7.236)$$

(ii) interval II: $\frac{\pi}{3} < \theta_s < \frac{2\pi}{3}$

The quadrature and direct axis stator currents by transformation then are given with an additional subscript of II:

$$i^e_{qsII} = aI_{dc}\cos\left(\theta_s - \frac{\pi}{6}\right)$$
$$i^e_{dsII} = aI_{dc}\sin\left(\theta_s - \frac{\pi}{6}\right) \quad (7.237)$$

and they could be written to resemble the first interval quadrature and direct axis stator currents as

$$i^e_{qsII} = aI_{dc}\cos\left(\left\{\theta_s + \frac{\pi}{6}\right\} - \frac{\pi}{3}\right)$$
$$i^e_{dsII} = aI_{dc}\sin\left(\left\{\theta_s + \frac{\pi}{6}\right\} - \frac{\pi}{3}\right) \quad (7.238)$$

In this form, it is evident that the q and d axes currents are periodic; hence, the currents in the second interval can be obtained from the currents in the first interval. Thus knowledge of the currents in one 60-degree interval is sufficient to reconstruct the current for one cycle. Expanding the right-hand side of the second-interval q and d axes currents results in the terms of the first interval q and d currents as

$$i_{qsII}^e = aI_{dc}\left[\cos\left(\theta_s + \frac{\pi}{6}\right)\cos\frac{\pi}{3} + \sin\left(\theta_s + \frac{\pi}{6}\right)\sin\frac{\pi}{3}\right] = \frac{1}{2}\left\{aI_{dc}\cos\left(\theta_s + \frac{\pi}{6}\right)\right\}$$

$$+ \frac{\sqrt{3}}{2}\left\{aI_{dc}\sin\left(\theta_s + \frac{\pi}{6}\right)\right\} = \frac{1}{2}i_{qsI}^e + \frac{\sqrt{3}}{2}i_{dsI}^e \qquad (7.239)$$

Similarly, the direct axis stator current in the second interval is derived in terms of the first interval q and d axes currents as,

$$i_{dsII}^e = -\frac{\sqrt{3}}{2}i_{qsI}^e + \frac{1}{2}i_{dsI}^e \qquad (7.240)$$

Combining the above two equations in matrix form and generalizing between any two consecutive intervals of 60 degrees yields

$$\begin{bmatrix} i_{qs}^e\left(\theta_s + \frac{\pi}{3}\right) \\ i_{ds}^e\left(\theta_s + \frac{\pi}{3}\right) \end{bmatrix} = \begin{bmatrix} \frac{1}{2} & \frac{\sqrt{3}}{2} \\ -\frac{\sqrt{3}}{2} & \frac{1}{2} \end{bmatrix} \begin{bmatrix} i_{qs}^e(\theta_s) \\ i_{ds}^e(\theta_s) \end{bmatrix} \qquad (7.241)$$

Now, because they are continuous functions, they are differentiable, thus providing the following relationship in state-space form among themselves:

$$X_a = S_1 X_1 \qquad (7.242)$$

where the matrix S_1 and vector X_1 are given by

$$X_1 = [i_{qs}^e \quad i_{ds}^e]^t \qquad (7.243)$$

$$S_1 = \begin{bmatrix} 0 & -\omega_s \\ \omega_s & 0 \end{bmatrix} \qquad (7.244)$$

Machine Equations With stator currents as the inputs, the performance evaluation requires only the solution of rotor currents from the rotor voltage equations. Note that the stator currents are discontinuous in real time and rotor flux linkages are continuous. The rotor equations with rotor flux linkages as variables are used in the solution of the rotor variables and hence in the computation of the electromagnetic torque. The rotor equations in rotor flux linkages are derived as follows. The rotor equations in the synchronous reference frames are

$$R_r i_{qr}^e + p(L_r i_{qr}^e + L_m i_{qs}^e) + \omega_{sl}(L_r i_{dr}^e + L_m i_{ds}^e) = 0 \qquad (7.245)$$
$$R_r i_{dr}^e + p(L_r i_{dr}^e + L_m i_{ds}^e) - \omega_{sl}(L_r i_{qr}^e + L_m i_{qs}^e) = 0 \qquad (7.246)$$

392 Chapter 7 Frequency-Controlled Induction Motor Drives

Defining the rotor flux linkages in the q and d axes as

$$\lambda_{qr}^e = L_r i_{qr}^e + L_m i_{qs}^e \qquad (7.247)$$

$$\lambda_{dr}^e = L_r i_{dr}^e + L_m i_{ds}^e \qquad (7.248)$$

and substituting these into the rotor equations yields the equations in rotor flux linkages as

$$p\lambda_{qr}^e + \omega_{sl}\lambda_{dr}^e + R_r i_{qr}^e = 0 \qquad (7.249)$$

$$p\lambda_{dr}^e - \omega_{sl}\lambda_{qr}^e + R_r i_{dr}^e = 0 \qquad (7.250)$$

and the rotor currents are obtained from equations (7.247) and (7.248) as

$$i_{qr}^e = \frac{1}{L_r}(\lambda_{qr}^e - L_m i_{qs}^e) \qquad (7.251)$$

$$i_{dr}^e = \frac{1}{L_r}(\lambda_{dr}^e - L_m i_{ds}^e) \qquad (7.252)$$

Substitution of these rotor currents into the rotor flux-linkages equations results in the following equations in rotor flux linkages:

$$p\lambda_{qr}^e + \frac{R_r}{L_r}\lambda_{qr}^e + \omega_{sl}\lambda_{dr}^e - R_r\frac{L_m}{L_r}i_{qs}^e = 0 \qquad (7.253)$$

$$-\omega_{sl}\lambda_{qr}^e + p\lambda_{dr}^e + \frac{R_r}{L_r}\lambda_{dr}^e - R_r\frac{L_m}{L_r}i_{ds}^e = 0 \qquad (7.254)$$

Combining these equations with the state equations of the stator currents casts the system equations in a compact state-space form:

$$\dot{X} = AX \qquad (7.255)$$

where the state vector X and state matrix A are given by

$$X = [\lambda_{qr}^e \ \lambda_{dr}^e \ i_{qs}^e \ i_{ds}^e]^t \qquad (7.256)$$

$$A = \begin{bmatrix} -\dfrac{R_r}{L_r} & -\omega_{sl} & \dfrac{R_r L_m}{L_r} & 0 \\ \omega_{sl} & -\dfrac{R_r}{L_r} & 0 & \dfrac{R_r L_m}{L_r} \\ 0 & 0 & 0 & -\omega_s \\ 0 & 0 & \omega_s & 0 \end{bmatrix} \qquad (7.257)$$

The solution of the state-space equation is given by

$$X(t) = e^{At}X(0) \qquad (7.258)$$

The symmetry of the rotor flux linkages and the stator currents for every 60 electrical degrees shows that the state vector at 60 degrees could be equated to the boundary-matching condition times the initial vector, as follows:

$$X\left(t = \frac{\pi}{3\omega_s}\right) = e^{A\frac{\pi}{3\omega_s}}X(0) = VX(0) \qquad (7.259)$$

where

$$V = \begin{bmatrix} I & 0 \\ 0 & S_1 \end{bmatrix} \qquad (7.260)$$

where I is an identity matrix of 2×2 dimensions and S_1 has been defined above. Rearranging the above equation as a function of $X(0)$ yields

$$\left[V - e^{A\frac{\pi}{3\omega_s}}\right]X(0) = 0 \qquad (7.261)$$

We can rewrite this equation in a compact way as

$$WX(0) = 0 \qquad (7.262)$$

where

$$W = \left[V - e^{A\frac{\pi}{3\omega_s}}\right] = \begin{bmatrix} w_1 & w_2 \\ w_3 & w_4 \end{bmatrix} \qquad (7.263)$$

The submatrices w_1, w_2, w_3, and w_4 are of 2×2 dimensions, from which the initial steady-state rotor flux-linkages vector is obtained as

$$x_1(0) = -w_1^{-1}w_2 x_2(0) \qquad (7.264)$$

where the vectors x_1 and x_2 are defined as

$$x_1 = [\lambda_{qr}^e \quad \lambda_{dr}^e]^t \qquad (7.265)$$
$$x_2 = [i_{qs}^e \quad i_{ds}^e]^t \qquad (7.266)$$

Once the initial steady-state vector is found, compute the exponential for a small interval of time corresponding to 2 or 3 electrical degrees from the exponential series, considering 7 to 10 terms (depending on the accuracy of the solution desired), and store this matrix for repetitive computation of the state variables via the following recursive relationship:

$$X(nT) = e^{AT}X(\{n - 1\}T) \text{ for } n = 0, 1, \ldots, k \qquad (7.267)$$

where T is the small sampling time of interest and k is related to it by

$$k = \frac{\pi}{3\omega_s T} \qquad (7.268)$$

Note that by storing n of the X vectors and using the boundary-matching condition matrix V, the state vectors for the rest of the five of the 60-degree intervals are generated, and from them the electromagnetic torque is found to be

$$T_e(t) = \frac{3}{2}\frac{P}{2}\frac{L_m}{L_r}(i_{qs}^e \lambda_{dr}^e - i_{ds}^e \lambda_{qr}^e) \qquad (7.269)$$

The stator phase voltages are computed from the stator equations in the synchronous reference frames by using the transformation on the synchronous d and q axes voltages; these are given by

$$v_{qs}^e = (R_s + L_s p)i_{qs}^e + \omega_s L_s i_{ds}^e + L_m p i_{qr}^e + \omega_s L_m i_{dr}^e \quad (7.270)$$

$$v_{ds}^e = -\omega_s L_s i_{qs}^e + (R_r + L_s p)i_{ds}^e - \omega_s L_m i_{qr}^e + L_m p i_{dr}^e \quad (7.271)$$

$$i_{abc} = [T_{abc}^e]^{-1} i_{qdo} \quad (7.272)$$

The q and d axes stator currents are piecewise continuous, so their derivatives can be found:

$$p i_{qs}^e = -\omega_s i_{ds}^e + \delta\left(m\frac{\pi}{3}\right), \quad m = 0, 1, 2, \ldots 6 \quad (7.273)$$

$$p i_{ds}^e = \omega_s i_{qs}^e - \delta\left(m\frac{\pi}{3}\right), \quad m = 0, 1, 2, \ldots 6 \quad (7.274)$$

where $\delta(\theta)$ is the impulse function at discontinuities.

By substituting these into the stator equations and derivatives of the rotor quadrature and direct axis currents from the rotor equations in terms of the stator current derivatives and rotor currents, the stator q and d axes voltages are obtained as

$$v_{qs}^e = R_s i_{qs}^e + \left(L_s - \frac{L_m^2}{L_r}\right)\delta(0) + \frac{L_m^2}{L_r}\omega_r i_{ds}^e - \frac{R_r L_m}{L_r} i_{qr}^e + \omega_r L_m i_{dr}^e \quad (7.275)$$

$$v_{ds}^e = -\frac{L_m^2}{L_r}\omega_r i_{qs}^e + R_s i_{ds}^e - \left(L_s - \frac{L_m^2}{L_r}\right)\delta(0) - L_m \omega_r i_{qr}^e - \frac{R_r L_m}{L_r} i_{dr}^e \quad (7.276)$$

where the impulse functions are of same magnitude as at the initial value.

From these voltages, the phase voltages are computed, and all of the variables are shown in Figure 7.51. Note that the impulse function is multiplied by the stator transient leakage inductance, which is in terms of the stator and rotor self-inductances and the mutual inductance. The magnitude of these spikes is limited by the commutating capacitors in the inverter; hence, their values in steady state are constant. Note that these spikes occur predominantly at 6 times in the stator phase voltages corresponding to the discontinuities in the phase currents. As it is not possible to have currents with zero rise and fall times, the voltage spikes are more of a function of the circuit and induction motor parameters. These spikes have to be taken into account for rating the inverter devices and commutating capacitors and for specifying the insulation of the induction motor.

The rotor flux linkages are constants except at the 60th-degree instants, where they seem to have very small changes. The rotor currents have a phase displacement of 180 degrees from their respective stator counterparts in synchronous reference

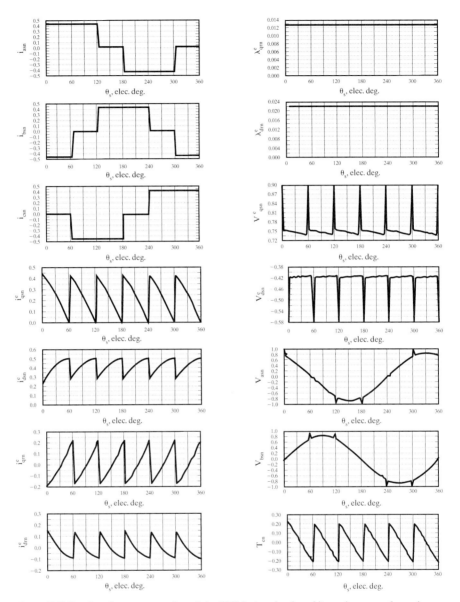

Figure 7.51(i) Steady-state operation of the CSIM at no-load, and its various waveforms in p.u.

frames but maintain the same periodicity. The q and d axes voltages have the spikes at the commutating current intervals but of opposite polarity, and the phase voltages are almost sinusoidal with superposed voltage spikes due to the current discontinuities. The electromagnetic torque has a large sixth-harmonic pulsation, as is seen from the figure, and this is one of the significant disadvantages in the six-step CSIM drive system.

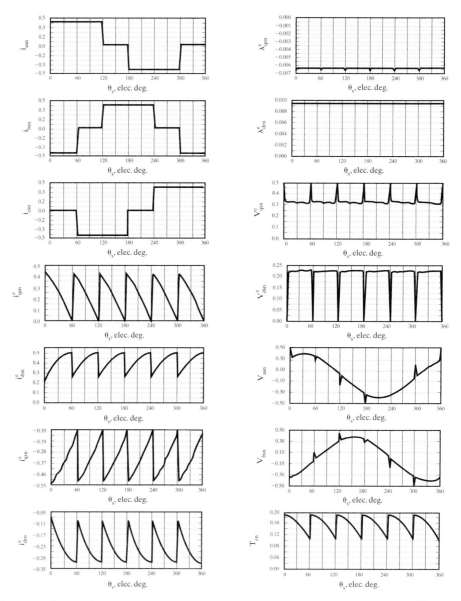

Figure 7.51(ii) Steady-state waveforms of the CSIM drive with load (ω_{sl} = 6.65 rad/sec, I_{dc} = 6.5 A, T_{eo} = 0.159 p.u.)

7.5.5 Closed-Loop CSIM Drive System

The schematic of the CSIM drive system is shown in Figure 7.52. The induction motor is supplied by an autosequentially commutated inverter (ASCI) with step currents. The control inputs to the inverter are current and frequency commands. The inverter control circuit enforces the stator frequency instantaneously by switching of the inverter power devices. The inverter input current is regulated by the

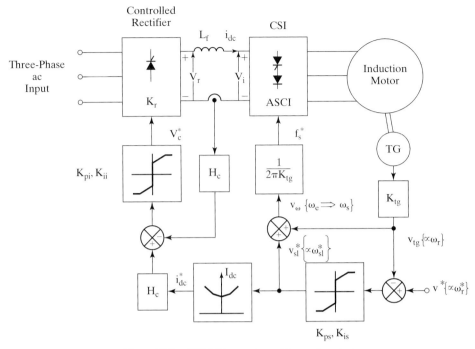

Figure 7.52 CSIM drive system with speed control

dc link input voltage, v_r, obtained from the phase-controlled rectification of the three-phase ac supply voltages. The control of v_r is exercised by feedback control of the inverter input current, with the feedback loop consisting of the controlled rectifier, dc link inductor, inverter input port, inverter input current, gain of the current transducer, inverter input current command, and control voltage. This inner current loop provides the variable current source for feeding into the inverter. The time lag involved in the switching of the controlled rectifier forces the dc link to be provided with a large inductor to maintain a constant current and to provide protection during inverter and motor short-circuits.

The frequency command is synthesized from the summation of the rotor speed and the slip-speed command, given by the proportional voltages v_{tg} and v_{sl}, respectively. The slip-speed command is generated from the error between the reference speed and the rotor speed, given proportionally by the signals V* and v_{tg}, respectively. The slip speed is limited, to limit the dc link current within the safe operational bounds and to regulate the operation of the induction motor in the high-efficiency region. The slip-speed command provides the inverter input-current command through a function generator whose design is considered subsequently. Note that the inverter input-current command has to be positive, irrespective of the slip-speed command signal. This is due to the facts that the controlled rectifier allows current in one direction only and that the regeneration is handled by the reversal of the inverter input voltage (and, hence, by the controlled-rectifier output voltage). The reversal of the input inverter voltage occurs because of the regeneration induced by the negative slip speed in the induction machine. Note that the inverter input voltage reflects the machine phase voltages. The reversal of the voltage at the inverter input results in instantaneous increase in dc link current, which results in a negative current error. This negative current error produces a negative control voltage, making the controlled rectifier produce a negative voltage across its output so as to oppose the inverter input voltage and thereby maintain the dc link current at its commanded value. The reversal of the converter voltage is made possible by increasing the triggering-angle delay to the converter, resulting in the operation of the converter in its inverter mode. During this time, note that the converter output voltage is negative but its output current is positive, thus producing a negative power, implying that power from the machine is returned to the ac mains via the dc link.

7.5.6 Dynamic Simulation of the Closed-loop CSIM Drive System

The dynamic simulation of the CSIM drive system is considered under the following assumptions:

(i) The stator currents have negligible rise and fall times.
(ii) The voltage spikes due to the step currents are constant.
(iii) The time delay in the inverter is neglected, even though it could have a substantial effect at high speeds.

The voltage spikes are a function of the machine leakage inductances and commutating capacitors and hence are determined by the machine and inverter designers. The spikes do not, in general, affect the electromagnetic torque. The assumption that they are of fixed magnitude will not affect the dynamic simulation results, except in the case of the selection and specification of the inverter and controlled-rectifier switch ratings.

Various subsystem models are derived subsequently, and then they are integrated, to obtain the system equations with the limiter constraints in the speed and current controllers.

Stator Currents: The reference frames considered are synchronous and the stator currents are generalized as in the steady-state modeling section:

$$i_{qs}^e = g_c i_{dc} \tag{7.277}$$

$$i_{ds}^e = g_s i_{dc} \tag{7.278}$$

where

$$g_c = \frac{2}{\sqrt{3}} \cos\left(\theta_s + \frac{\pi}{6} - \{k-1\}\frac{\pi}{3}\right) \tag{7.279}$$

$$g_s = \frac{2}{\sqrt{3}} \sin\left(\theta_s + \frac{\pi}{6} - \{k-1\}\frac{\pi}{3}\right) \tag{7.280}$$

where

$$k = 1, 2, \ldots, 6, 1, 2, \ldots \tag{7.281}$$

defines the corresponding 60-degree intervals of stator phase angle and i_{dc} is the inverter input current. The inverter input current is the dc link current, unlike in the case of the VSIM drive systems.

The derivatives of the stator currents are obtained here for subsequent use in the simulation. They are the same as those derived in the direct steady-state evaluation section, except that they have additional terms involving the derivative of the inverter input current:

$$pi_{qs}^e = g_c p i_{dc} + i_{dc} p g_c + \delta\left(\{k-1\}\frac{\pi}{3}\right) = g_c p i_{dc} + i_{dc}\{-\omega_s g_s\} + \delta\left(\{k-1\}\frac{\pi}{3}\right) \quad (7.282)$$

Similarly, the derivative of the d axis current is

$$pi_{ds}^e = g_s p i_{dc} + \omega_s g_c i_{dc} - \delta\left(\{k-1\}\frac{\pi}{3}\right) \quad (7.283)$$

where δ is the impulse contributing to the voltage spikes at the instants of current commutation occurring at stator phase angles every 60 electrical degrees.

DC Link and Power Equivalence The input inverter power is equal to the power input to the machine (neglecting losses in the inverter), and it is represented as

$$v_i i_{dc} = \frac{3}{2}[v_{qs}^e i_{qs}^e + v_{ds}^e i_{ds}^e] + v_o i_o \quad (7.284)$$

The zero-sequence current is zero, because the stator currents are balanced three-phase inputs and hence the zero-sequence power becomes zero. Substituting for the d and q axes currents in terms of the inverter input current yields

$$v_i = \frac{3}{2}[g_c v_{qs}^e + g_s v_{ds}^e] \quad (7.285)$$

This shows the dependence of the inverter input voltage on the d and q axes machine voltages; it compacts the two stator equations into one equation in terms of the inverter input voltage. The next step in the modeling is to find the stator d and q axes voltages, which involve stator and rotor currents and their derivatives. Rotor flux linkages are employed, as in the direct steady-state evaluation, because they are continuous even in the face of the discontinuous stator currents. From these variables, the stator voltages are derived.

Rotor Equations The rotor equations in terms of the rotor flux linkages are

$$p\lambda_{qr}^e = -\frac{R_r}{L_r}\lambda_{qr}^e - \omega_{sl}\lambda_{dr}^e + \frac{R_r L_m}{L_r}i_{qs}^e \quad (7.286)$$

$$p\lambda_{dr}^e = -\frac{R_r}{L_r}\lambda_{dr}^e + \omega_{sl}\lambda_{qr}^e + \frac{R_r L_m}{L_r}i_{ds}^e \quad (7.287)$$

where λ_{qr}^e and λ_{dr}^e are as defined in equations (7.247) and (7.248).

Stator Voltages The stator q and d axes voltages are derived in terms of the rotor flux linkages and stator currents as

$$v_{qs}^e = (R_s + L_s p)i_{qs}^e + \omega_s L_s i_{ds}^e + L_m p i_{qr}^e + \omega_s L_m i_{dr}^e \quad (7.288)$$

Substituting for the rotor currents in terms of the rotor flux linkages and stator currents from the equations (7.251) and (7.252) results in the q axis stator voltage:

$$v_{qs}^e = R_s i_{qs}^e + a_1 p i_{qs}^e + a_1 \omega_s i_{ds}^e + \frac{L_m}{L_r} p \lambda_{qr}^e + \frac{L_m}{L_r} \omega_s \lambda_{dr}^e \quad (7.289)$$

where

$$a_1 = L_s - \frac{L_m^2}{L_r} \quad (7.290)$$

The stator currents are substituted for in terms of the inverter input current and their derivatives, yielding

$$v_{qs}^e = R_s g_c i_{dc} + a_1 g_c p i_{dc} + \frac{L_m}{L_r} p \lambda_{qr}^e + \frac{L_m}{L_r} \omega_s \lambda_{dr}^e + a_1 \delta\left(\{k-1\}\frac{\pi}{3}\right) \quad (7.291)$$

Similarly, the d axis stator voltage is

$$v_{ds}^e = R_s g_s i_{dc} + a_1 g_s p i_{dc} + \frac{L_m}{L_r} p \lambda_{dr}^e - \frac{L_m}{L_r} \omega_s \lambda_{qr}^e - a_1 \delta\left(\{k-1\}\frac{\pi}{3}\right) \quad (7.292)$$

These two equations are substituted into the inverter input voltage equation for subsequent use in the dc link equation, given by

$$L_f p i_{dc} = v_r - v_i \quad (7.293)$$

where v_r is the product of the controlled-rectifier gain and v_c^*. Its real-time modeling is described in Chapter 3. Noting that v_i has as derivative the inverter input current; rearranging this equation in terms of the system variables results in the following state equation of the inverter input current:

$$pi_{dc} = \frac{-1}{2a_1 + L_f}\left[2R_s i_{dc} + 1.5\frac{L_m}{L_r}\{g_c p \lambda_{qr}^e + g_s p \lambda_{dr}^e\}\right.$$

$$\left. + 1.5\frac{L_m}{L_r}\omega_s\{g_c \lambda_{dr}^e - g_s \lambda_{qr}^e\} + 1.5 a_1(g_c - g_s)\delta\left(\{k-1\}\frac{\pi}{3}\right) - v_r\right] \quad (7.294)$$

Load Interaction The machine and load interaction is given by the equation with inertia and friction constant:

$$J p \omega_m + B \omega_m = T_e - T_l \quad (7.295)$$

where the electromagnetic torque is given by

$$T_e = \frac{3}{2}\frac{P}{2}L_m[i_{qs}^e i_{dr}^e - i_{ds}^e i_{qr}^e] \quad (7.296)$$

which, in terms of the rotor flux linkages, is given by

$$T_e = \frac{3}{2} \frac{P}{2} \frac{L_m}{L_r} [\lambda_{dr}^e g_c - \lambda_{qr}^e g_s] i_{dc} \qquad (7.297)$$

In terms of the rotor electrical speed, equation (7.295) becomes, by the multiplication of the pole-pairs,

$$p\omega_r = -\frac{B}{J}\omega_r + \frac{1}{J}\frac{3}{2}\frac{P^2}{2^2}\frac{L_m}{L_r}\{\lambda_{dr}^e g_c - \lambda_{qr}^e g_s\} i_{dc} - \frac{1}{J}\frac{P}{2}T \qquad (7.298)$$

Equations (7.291) to (7.297) describe the machine, inverter, and load whereas the stator frequency and rectifier output-voltage variables are obtained from the controller structure consisting of the reference speed, speed-controller and current-controller constants, and slip-speed-to-inverter current-command transfer block. These are obtained as follows.

Frequency Command The stator frequency is generated by the rotor speed loop and the slip-speed command, generated by the speed error between the reference and actual speed. The tachogenerator, with its conditioning circuit, is modeled as a gain block, but, in practice, it will contain a first-order filter. The tachogenerator output is proportional to the rotor speed, and the slip-speed command is obtained by passing the speed error through a proportional and integral controller with the gains of K_{ps} and K_{is}, respectively. The slip-speed signal is limited by the fact that the operation of the induction motor has to be constrained within the high-efficiency region of the slip torque, which corresponds to the statically stable portion of the voltage-source-fed induction-motor characteristics. The equations of relevance are the following:

$$v_{tg} = K_{tg}\omega_r \qquad (7.299)$$

$$v_{sl}^* = K_{ps}(v^* - v_{tg}) + K_{is}\int(v^* - v_{tg})dt \qquad (7.300)$$

where v^* is the external speed reference.

Then, the stator frequency command is given by

$$f_s^* = \frac{v_{tg} + v_{sl}^*}{2\pi K_{tg}} \qquad (7.301)$$

The slip-speed command is then obtained by dividing equation (7.300) by K_{tg}. The next step is to generate the command signal for the controlled rectifier; this step requires the generation of the inverter input-current command.

DC-Link Current Command From the induction motor steady-state equivalent circuit, the stator rms current is written in terms of the magnetizing current as

$$I_s = I_m + I_r = I_m + \frac{j\omega_s L_m I_m}{\frac{R_r}{s} + j\omega_s L_{lr}} = I_m \left[1 + \frac{js\omega_s L_m}{R_r + js\omega_s L_{lr}}\right] \qquad (7.302)$$

Note that $\omega_{sl} = s\omega_s$; hence,

$$I_s = I_m \left[\frac{R_r + j\omega_{sl}L_r}{R_r + j\omega_{sl}L_{lr}} \right] \qquad (7.303)$$

The relationship between the fundamental stator phase and dc link currents is

$$i_{dc} = \frac{I_s}{0.778} = 1.285 I_s \qquad (7.304)$$

Then the inverter input-current command is obtained by combining the last two expressions:

$$i_{dc}^* = 1.285 I_m \left| \frac{R_r + j\omega_{sl}L_r}{R_r + j\omega_{sl}L_{lr}} \right| = 1.285 \, I_m \frac{\sqrt{R_r^2 + (\omega_{sl}L_r)^2}}{\sqrt{R_r^2 + (\omega_{sl}L_{lr})^2}} \qquad (7.305)$$

Thus, the link-current command is a function of only the slip-speed command, assuming that the magnetizing current is kept constant. An on-line computation of the inverter input-current command would facilitate the variation of the magnetizing current as a function of slip speed, to optimize the efficiency in steady-state operation.

Controlled-Rectifier Voltage Command The inverter current command is compared to the measured value of the inverter input current to produce an error signal, which is amplified through a proportional-plus-integral current controller whose gains are K_{pi} and K_{ii}, respectively. The output of the current controller is limited, to provide the safe operation of the converter during inversion. The output of the current controller actuates the controlled rectifier to provide a proportional output voltage, v_r, to generate the inverter input current to match its command. The control-voltage signal and the rectifier output voltage are given by

$$v_c^* = K_{pi}(i_{dc}^* - i_{dc})H_c + K_{ii} \int \{i_{dc}^* - i_{dc}\} H_c \, dt \qquad (7.306)$$

where H_c is the current transducer gain. Note that the inverter input current command and slip speed were derived above, and they should be substituted into this expression to obtain the voltage-command signal.

System Integration and Equations The equations of the various subsystems can now be assembled in state-space form for simulation. The following definitions are made to get them into state-space form:

$$\begin{aligned} x_1 &= i_{dc} \\ x_2 &= \lambda_{qr}^e \\ x_3 &= \lambda_{dr}^e \\ x_4 &= \omega_r \\ x_5 &= \int (v^* - K_{tg}\omega_r) \, dt \\ x_6 &= \int H_c(i_{dc}^* - i_{dc}) \, dt \end{aligned} \qquad (7.307)$$

Section 7.5 Current-Source Induction Motor Drives

The system differential equations in terms of the states defined are

$$px_1 = \frac{-1}{2a_1 + L_f} \left[2R_s x_1 + 1.5 \frac{L_m}{L_r} \{g_c px_2 + g_s px_3\} \right.$$
$$\left. + 1.5 \frac{L_m}{L_r} \omega_s \{g_c x_3 - g_s x_2\} + 1.5 a_1 (g_c - g_s) \delta \left(\{k-1\} \frac{\pi}{3} \right) - v_r \right]$$

$$px_2 = -\frac{R_r}{L_r} x_2 - \omega_{sl} x_3 + \frac{R_r L_m}{L_r} g_c x_1$$

$$px_3 = \omega_{sl} x_2 - \frac{R_r}{L_r} x_3 + \frac{R_r L_m}{L_r} g_s x_1 \qquad (7.308)$$

$$px_4 = \frac{13}{J2} \left(\frac{P}{2}\right)^2 \frac{L_m}{L_r} \{g_c x_3 - g_s x_2\} x_1 - \frac{B}{J} x_4 - \frac{P}{2J} T_l$$

$$px_5 = v^* - K_{tg} x_4$$

$$px_6 = H_c(i_{dc}^* - x_1)$$

Note that these equations have only the inverter input-current command, slip frequency, stator frequency, and external speed reference as external inputs; they are described by the following equations in terms of the states as

$$\omega_{sl} = \frac{v_{sl}^*}{K_{tg}} = \frac{K_{is} x_5 + K_{ps}(V^* - K_{tg} x_4)}{K_{tg}} = \frac{K_{is} x_5 + K_{ps} px_5}{K_{tg}} \qquad (7.309)$$

$$\omega_s = \omega_r + \omega_{sl} = x_4 + \frac{K_{is} x_5 + K_{ps}(V^* - K_{tg} x_4)}{K_{tg}} = x_4 + \frac{K_{is} x_5 + K_{ps} px_5}{K_{tg}} \qquad (7.310)$$

The inverter input-current command is derived from equations (7.305) and (7.309) as

$$i_{dc}^* = 1.285 I_m \frac{\sqrt{(R_r K_{tg})^2 + L_r^2 \{K_{is} x_5 + K_{ps} px_5\}^2}}{\sqrt{(R_r K_{tg})^2 + L_{lr}^2 \{K_{is} x_5 + K_{ps} px_5\}^2}} \qquad (7.311)$$

The voltage-command signal is given by

$$v_c^* = K_{pi} px_6 + K_{ii} x_6 \qquad (7.312)$$

The slip-speed and voltage-command signals are limited; their limits are given as

$$-V_{cm} < v_c^* < +V_{cm} \qquad (7.313)$$
$$-V_{sm} < v_{sl}^* < +V_{sm} \qquad (7.314)$$

These limiters are applied to the relevant state variables and their derivatives during computation.

A sample CSIM drive system simulation is shown in Figure 7.53 for the induction motor whose parameters are given in Example 5.1; the system parameters are given below.

$V_{ll} = 240$ V, $L_f = 0.1$ H, $K_{ps} = 10$, $K_{is} = 1$, $K_{pi} = 10$, $K_{ii} = 0.1$, speed command = 1 p.u.

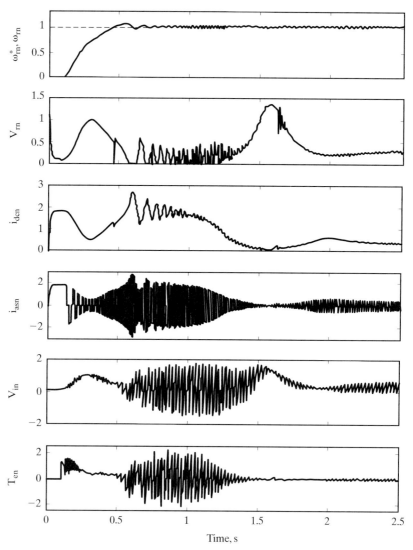

Figure 7.53(i) Normalized transient response for 1-p.u.-step speed command, from standstill

A one-p.u.-step command is given to the motor drive system. The transient responses are shown in Figure 7.53(i). The machine takes approximately 0.1 s to raise the rotor flux to rated value. Until then, the stator current is used entirely to set up the flux. The machine is accelerated to 1-p.u. speed in 0.4 s thereafter. The machine has a 0.2 p.u. load; only friction has to be overcome. To reach steady state, the drive goes through a severe oscillation. This is one of the characteristic disadvantages of the ASCI drive with the adopted control strategy. Almost an additional 0.2 s is taken to reach steady state. The steady-state waveforms are shown in Figure 7.53(ii). The cusps in the front-end converter's output are due to

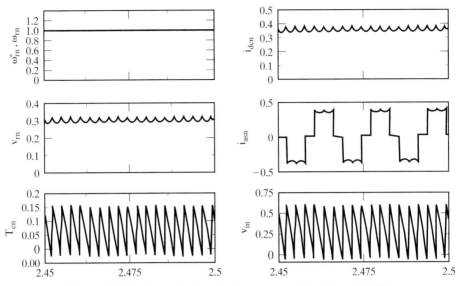

Figure 7.53(ii) Normalized steady-state responses from Figure 7.54(i).

the v_c^* variations, and the output of the converter is modeled as the product of the v_c^* and gain of the converter. These cusps in v_{rn}, and hence in i_{dcn}, are transmitted to the phase currents and are shown in the phase *a* current. The torque shows large pulsations at six times the fundamental frequency, and so does the inverter input voltage.

7.6 APPLICATIONS

Voltage-source induction motor drives are used as general-purpose drives in such applications as fans, pumps, packaging, conveyor, hand tools, and appliances. These applications do not require positioning capabilities, and precise torque control in the 0-to-5-Hz range is not critical.

Use of software-based controllers and the increasing use of modern control theory in the synthesis of the controllers has resulted recently in a far wider control range with these drives. Because of this factor, these drives have made immense strides in the marketplace by enlarging their share.

The hard-current source drives are popular in limited high-power (>100 hp) applications, such as paper mills and pulp, sugar, and rubber centrifuges, where reliability and ruggedness are major considerations. The soft current-source induction motor drives have eliminated the disadvantages of the hard current-source drives, and, with higher controllers such as vector controllers, have moved into demanding precision applications such as machine-tool servo and spindle drives.

7.7 REFERENCES

1. V. R. Stefanovic, "Static and dynamic characteristics of induction motors operating under constant airgap flux control," *Proceedings of the Annual Meeting of the IEEE Industry Applications Society*, pp. 143–147, Oct. 1974.
2. K. P. Phillips, "Current source converter for a.c. motor drives," *1971— 6th annual meeting of the IEEE Industry and General Applications Group*, pp. 385–92, Oct. 1971.
3. T. A. Lipo, "Simulation of a current source inverter drive," *IEEE Transactions on Industrial Electronics and Control Instrumentation*, vol. IECI-26, no. 2, pp. 98–103, May 1979.
4. R. Krishnan, W. A. Maslowski, and V. R. Stefanovic, "Control principles in current source induction motor drives," *IAS Annual Meeting*, 1980, pp. 892, 605–17, Oct. 1980.
5. R. Krishnan, J. F. Lindsay, and V. R. Stefanovic, "Design of angle-controlled current source inverter-fed induction motor drive," *IEEE Transactions on Industry Applications*, vol. IA-19, no. 3, pt. 1, pp. 370–8, May–June, 1983.
6. D. A. Bradley, C. D. Clarke, R. M. Davis, and D. A. Jones, "Adjustable frequency inverters and their application on variable speed drives," *Proceedings of IEE (London)*, vol. III, no. 11, pp. 1833–1846, Nov. 1964.
7. W. Charlton, "Matrix method for the steady-state analysis of inverter-fed induction motors," *Proceedings of the IEE*, vol. 120, no. 3, pp. 363–4, March 1973.
8. T. A. Lipo and F.G. Turnbull, "Analysis and comparison of two types of square wave inverter drives," *Proceedings of the Seventh Annual Meeting of the IEEE Industry Applications Society*, pp. 127–36, Oct. 1972.
9. R. Magureanu, "A state variable analysis of inverter-fed a.c. machines," *Revue Roumaine des Sciences Techniques, Serie Electrotechnique et Energetique*, vol. 18, no. 4, pp. 663–78, 1973.
10. M. Ramamoorty, "Steady state analysis of inverter driven induction motors using harmonic equivalent circuits," *8th Annual Meeting of the IEEE Industry Applications Society*, pp. 437–40, Oct. 1973.
11. W. Farrer, "Quasi-sine-wave fully regenerative inverter," *Proceedings of the Institution of Electrical Engineers*, vol. 120, no. 9, pp. 969–76, Sept. 1973.
12. E. A. Klingshirn and H. E. Jordan, "Polyphase induction motor performance and losses on nonsinusoidal voltage source," *IEEE Transactions on Power Apparatus and Systems*, vol. PAS-87, pp. 624–631, March 1968.
13. V. B. Honsinger, "Induction motors operating from inverters," *Conference Record of the Industry Applications Society IEEE-IAS 1980 Annual Meeting*, pp. 1276–85, Oct. 1980.
14. K. Venkatesan and J. F. Lindsay, "Comparative study of the losses in voltage and current source inverter fed induction motors," *Conference Record, Industry Applications Society IEEE IAS Annual Meeting*, pp. 644–9, Oct. 1981.
15. P. C. Krause and T. A. Lipo, "Analysis and simplified representations of a rectifier-inverter induction motor drive," *IEEE Transactions on Power Apparatus and Systems*, vol. PAS-88, no. 5, pp. 588–96, May 1969.
16. S. D. T. Robertson and K. M. Hebbar, "Torque pulsations in induction motors with inverter drives," *IEEE Transactions on Industry and General Applications*, vol. IGA-7, no. 2, pp. 318–23, March–April 1971.
17. Tung-Hai Chin and H. Tomita, "Analysis of torque behavior of squirrel cage induction motors driven by controlled current inverter and evaluation method for torque ripple," *Transactions of the Institute of Electrical Engineers of Japan*, vol. 97, no. 6, pp. 127–33, Nov.–Dec. 1977.

18. L. H. Walker and P. M. Espelage, "A high performance controlled current inverter drive," *Conference Record, Industry Application Society IEEE–IAS Annual Meeting*, pp. 928–36, Oct. 1979.
19. R. Palaniappan, J. Vithayathil, and S.K. Datta, "Principle of a dual current source converter for AC motor drives," *IEEE Transactions on Industry Applications*, vol. IA-15, no. 4, pp. 445–52, July–Aug. 1979.
20. G. K. Creighton, "Current source inverter fed induction motor torque pulsations," *IEE Proceedings B (Electric Power Applications)*, vol. 127, no. 4, pp. 231–40, July 1980.
21. F. Fallside and A. T. Wortley, "Steady-state oscillation and stabilisation of variable-frequency inverter-fed induction-motor drives," *Proceedings of the Institution of Electrical Engineers*, vol. 116, no. 6, pp. 991–9, June 1969.
22. B. Mokrytzki, "The controlled slip static inverter drive," *IEEE Transactions on Industry General Applications*, vol. IGA-4, no. 3, pp. 312–317, May/June 1968.
23. A. Schonung and H. Stemmler, "Static frequency changers with 'subharmonic' control in conjunction with reversible variable-speed ac drives," *The Brown Boveri Review*, vol. 51, pp. 555–577, Aug./Sept. 1964.
24. V. R. Stefanovic, "Variable frequency induction motor drive dynamics," Ph.D. Thesis, McGill University, Aug. 1975.
25. R. Krishnan, V. R. Stefanovic, and J.F. Lindsay, "Control characteristics of inverter-fed induction motor," *IEEE Transactions on Industry Applications*, vol. IA-19, no. 1, pp. 94–104, Jan.–Feb. 1983.
26. B. Ramaswami, R. Venkataraman, and J. Holtz, "Design of variable speed induction motor drive with a current-fed inverter," *Electric Machines and Electromechanics*, vol. 5, no. 6, pp. 523–42, Nov.–Dec. 1980.

7.8 DISCUSSION QUESTIONS

1. Why is it that the variable-frequency requirement is always accompanied by a variable-voltage requirement in ac machines?
2. Inverters provide variable voltage and variable frequency. Will they provide current control? If so, how is it accomplished?
3. What is the greatest shortcoming in a voltage-source inverter?
4. What is the greatest shortcoming in a current-source inverter?
5. Compare voltage- and current-source inverters in terms of their reliability, control complexity, and harmonics.
6. For induction motor control, compare the suitability of the voltage- and current-source inverters in terms of their control complexity and reliability.
7. Compare the performance features of six-step and PWM control of voltage-source inverters in terms of harmonics, control complexity, and ease of implementation.
8. Compare the features of the PWM and hysteresis control of inverters, and identify, with justification, the most desirable control in most of the applications.
9. If the pulse-width modulation ratio is increased beyond one, what happens to the output voltage? If there is a side effect to this mode of operation, identify it. (Hint: Refer to a power electronics text.)
10. What is the maximum rms voltage obtainable with a six-step operation in a voltage-source inverter? DC link voltage is V_{dc}. Justify your assumptions.

408 Chapter 7 Frequency-Controlled Induction Motor Drives

11. What is the maximum rms voltage obtainable under PWM operation by using triangular carrier frequency and sine modulation signal in a voltage-source inverter? Justify your assumptions.
12. Discuss the differences between a hard (ASCI) and a soft (PWM-based) current-source inverter.
13. If the line-to-line voltages are unbalanced in the inverter output, are the phase voltages determined by equations (7.8), (7.9), and (7.10)?
14. An inverter has to be rated in apparent power rather than in real power. Explain why.
15. Discuss the effects of unbalanced voltages on the operation of the induction motor.
16. How are unbalanced voltages detected and corrected in the inverter?
17. Discuss the merits of the volts/Hz control strategy.
18. Draw the control and power schematics of the volts/Hz drive that uses a PWM inverter.
19. Discuss the difference between the six-step and the PWM inverter with volts/Hz control for an induction motor.
20. Offset voltage is a variable dependent on the stator impedance and stator current. Discuss a method for an accurate implementation of offset-voltage control in volts/Hz-controlled induction motor drive.
21. Discuss the parameter sensitivity of the offset voltage that is due to stator resistance in volts/Hz controller.
22. Dynamic response of a volts/Hz-controlled induction motor drive is characterized by large starting stator currents. Explain why.
23. How can this starting-current magnitude be controlled?
24. Discuss the cause of starting-current transient in a volts/Hz-controlled induction motor.
25. Explain the distinct advantages of direct-steady-state-performance evaluation of the induction motor compared to the method using an equivalent circuit.
26. Discuss the important advantage of slip-speed control over the volts/Hz control strategy.
27. Can the direct-steady-state evaluation method be used for predicting the performance of the constant-slip-speed-controlled induction motor drive?
28. Compare the principle of operation of constant-air gap-flux-linkages control as against that of the volts/Hz and constant-slip-speed controls.
29. Implementation of constant air gap flux requires stator current control. Justify this statement.
30. Will constant air gap flux be maintained by using the drive schematic shown in Figure 7.30, even during dynamic conditions?
31. Discuss the merits and demerits of the multistep and PWM-inverter control.
32. How will the PWM voltages, as compared to six-step voltages, affect the torque pulsation both in magnitude and in frequency?
33. PWM voltages eliminate lower-order harmonics but increase higher-order harmonics. How does this fact affect motor losses?
34. Is PWM applicable to hard current-source inverters also?
35. Which of the following will have a larger sixth-harmonic torque pulsation: (a) six-step voltage source, (b) six-step current source? Induction motor is operating at rated current/rated voltage and at rated speed.
36. The load has no impact on peak-to-peak value of sixth-harmonic torque pulsation in a voltage-source inverter-fed induction motor. Why is that?

37. Load decreases the severity of the sixth-harmonic torque pulsation in the current-source-inverter-fed induction motor drive, as is seen from Figures 7.51(i) and 7.51(ii). Explain the reason for this.

38. The inner speed loop in the induction motor drive acts to keep the machine in a statically stable region and therefore in synchronous lock. Without a speed sensor, how can such a loop be implemented?

7.9 EXERCISE PROBLEMS

1. Consider the machine parameters described in problem 7, section 6.6. Instead of using phase control for the induction motor drive, variable-frequency control with a constant-volts/Hz strategy is used. Find the efficiency of the motor drive with this drive system for the same application. [Hint: From the operating speed, torque, and volts/Hz relationship, the slip is derived through the solution of a quadratic equation.]

2. Assume that the price of energy is $0.05/kWH (kWH — kilowatthour). The fan considered in problem 1 is running eight hours per day for 356 days in a year. The price of the phase controller is $2000 and the price of the variable-frequency drive with constant-volts/Hz control is $3500. It is planned to have one of the variable-speed drives installed for the problem studied. If the variable-frequency drive is to be chosen instead of the phase controller, what is the payback period for the difference in the initial investment involved? Consider that the capital is acquired interest-free for this problem. (In practice, there is an interest for the loan.)

3. Consider the 5-hp induction motor described in Example 5.1. Find its slip speed, stator and rotor current magnitudes, efficiency, and power factor when it is delivering 12 Nm air gap torque for two control strategies:
    (i) volts/Hz control
    (ii) slip-speed control

4. A 100-hp induction motor is to drive a fan in a warehouse. It is found that the air volume has to be reduced during the off-peak business hours. That corresponds to an operational speed of 65% of the rated speed. The motor details are given next:

    100 hp, 460 V, 3 phase, 60 Hz, 2 poles, star connected squirrel cage induction motor, $R_s = 0.05\ \Omega$, $R_r = 0.0288\ \Omega$, $X_m = 10.5\ \Omega$, $X_{ls} = 0.3\ \Omega$, $X_{lr} = 0.405\ \Omega$, Rated speed = 3555 rpm.

    Find the efficiency of the motor drive when it is operated from a six-step inverter with a volts/Hz control strategy for the following two cases:
    (i) with no offset voltage
    (ii) with an offset voltage of 4.7 V

    The fan has a rated torque requirement at rated speed.

5. Prove that, in an air gap-flux-controlled induction motor drive,
    (i) air gap torque is a function of only magnetizing current and slip speed, because these two can be independently varied in this drive system, and
    (ii) the maximum torque occurs at the slip frequency $\omega_{sl} = \dfrac{R_r}{L_{lr}}$.

6. In a volts/Hz-controlled induction motor, only applied voltage and stator frequency are control variables.

(i) Determine the slip speed at which the air gap torque is maximum for this motor drive. The magnetizing impedance is considered to be very large compared to stator and rotor impedances.

(ii) Compare the efficiency of this drive at maximum torque to the air gap-flux-controlled induction motor drive.

7. Consider the problem in Example 7.5. Include the effects of stator impedance, and recalculate the magnitude of the sixth-harmonic torque pulsation.

8. Find the range of rotor speed for which the air gap power is constant for the machine considered in Example 7.8. Use the formula developed in the text. Also, calculate the actual range of speed, with the consideration that the stator phase voltage and current are constrained to be equal to or less than 1 p.u..

9. The slip frequency is set as the product of the rated slip and stator frequency. Flux weakening is introduced beyond rated frequency with

$$\omega_s^* = \omega_r + \omega_{sl}^* = \omega_r + s_r \omega_s^*$$

resulting in $\omega_s^* = \dfrac{\omega_r}{1 - s_r}$. Find the range of speed for which the air gap power is at rated value for this flux-weakening strategy.

10. A six-step current-source inverter is feeding the induction motor whose details are given in Example 5.1. Calculate the magnitude of the sixth-harmonic torque pulsation when the machine is delivering rated torque at a stator frequency of 10 Hz.

11. Simulate a PWM-based, volts/Hz-controlled induction motor drive with a stator current limit.

12. Simulate the PWM current-source-based induction motor drive with slip-speed limit. Comment on the salient differences between this simulation and the ASCI drive results given in Figure 7.53.

13. Develop an algorithm to evaluate the losses in a machine supplied from a PWM voltage-source inverter.

14. The current-command generator is sensitive to rotor-resistance variation in current-source drives. Develop a method that is not on-line-computation-based to generate the current command when the rotor-resistance estimate is available as a signal.

CHAPTER 8

Vector-Controlled Induction Motor Drives

8.1 INTRODUCTION

The various control strategies for the control of the inverter-fed induction motor have provided good steady-state but poor dynamic response. From the traces of the dynamic responses, the cause of such poor dynamic response is found to be that the air gap flux linkages deviate from their set values. The deviation is not only in magnitude but also in phase. The variations in the flux linkages have to be controlled by the magnitude and frequency of the stator and rotor phase currents and their instantaneous phases. So far, the control strategies have utilized the stator phase current magnitude and frequency and not their phases. This resulted in the deviation of the phase and magnitudes of the air gap flux linkages from their set values.

The oscillations in the air gap flux linkages result in oscillations in electromagnetic torque and, if left unchecked, reflect as speed oscillations. This is undesirable in many high-performance applications, such as in robotic actuators, centrifuges, servos, process drives, and metal-rolling mills, where high precision, fast positioning, or speed control are required. Such requirement will not be met with the sluggishness of control due to the flux oscillations. Further, air gap flux variations result in large excursions of stator currents, requiring large peak converter and inverter ratings to meet the dynamics. An enhancement of peak inverter rating increases cost and reduces the competitive edge of ac drives in the marketplace, in spite of excellent advantages of the ac drives over dc drives.

Separately-excited dc drives are simpler in control because they independently control flux, which, when maintained constant, contributes to an independent control of torque. This is made possible with separate control of field and armature currents which, in turn, control the field flux and the torque independently. Moreover, the dc motor control requires only the control of the field or armature current magnitudes, providing a simplicity not possible with ac machine control. By contrast, ac induction motor drives require a coordinated control of stator current magnitudes, frequencies, and their phases, making it a complex control. As with the dc drives, independent control of the flux and torque is possible in ac drives. The stator current phasor can be resolved, say, along the rotor flux linkages, and the component along the rotor flux linkages is the field-producing current, but this requires the position of the rotor flux linkages at every instant; note that this is dynamic, unlike in the dc machine. If this is available, then the control of ac machines is very similar to that of separately-excited dc machines. The requirement of phase, frequency, and magnitude control of the currents and hence of the flux phasor is made possible by inverter control. The control is achieved in field coordinates (hence the name of this control strategy, *field-oriented control*); sometimes it is known as *vector control*, because it relates to the phasor control of the rotor flux linkages.

Vector control made the ac drives equivalent to dc drives in the independent control of flux and torque and superior to them in their dynamic performance. These developments positioned the ac drives for high-performance applications, hitherto reserved for separately-excited dc motor drives. This chapter describes the basic principles, classifications, derivation, modeling, analysis, and design of vector-control schemes. The parameter sensitivity of vector-control schemes is analyzed, and methods for its compensations are described. The design of the speed controller for a vector controller is systematically derived by using symmetric optimum technique.

8.2 PRINCIPLE OF VECTOR CONTROL

To explain the principle of vector control, an assumption is made that the position of the rotor flux linkages phasor, λ_r, is known. λ_r is at θ_f from a stationary reference, θ_f is referred to as field angle hereafter, and the three stator currents can be transformed into q and d axes currents in the synchronous reference frames by using the transformation

$$\begin{bmatrix} i_{qs}^e \\ i_{ds}^e \end{bmatrix} = \frac{2}{3} \begin{bmatrix} \sin\theta_f & \sin\left(\theta_f - \frac{2\pi}{3}\right) & \sin\left(\theta_f + \frac{2\pi}{3}\right) \\ \cos\theta_f & \cos\left(\theta_f - \frac{2\pi}{3}\right) & \cos\left(\theta_f + \frac{2\pi}{3}\right) \end{bmatrix} \begin{bmatrix} i_{as} \\ i_{bs} \\ i_{cs} \end{bmatrix} \qquad (8.1)$$

from which the stator current phasor, i_s, is derived as

$$i_s = \sqrt{(i_{qs}^e)^2 + (i_{ds}^e)^2} \qquad (8.2)$$

and the stator phasor angle is

$$\theta_s = \tan^{-1}\left\{\frac{i_{qs}^e}{i_{ds}^e}\right\} \qquad (8.3)$$

where i_{qs}^e and i_{ds}^e are the q and d axes currents in the synchronous reference frames that are obtained by projecting the stator current phasor on the q and d axes, respectively. That the current phasor magnitude remains the same regardless of the reference frame chosen to view it is evident from Figure 8.1. The current phasor i_s

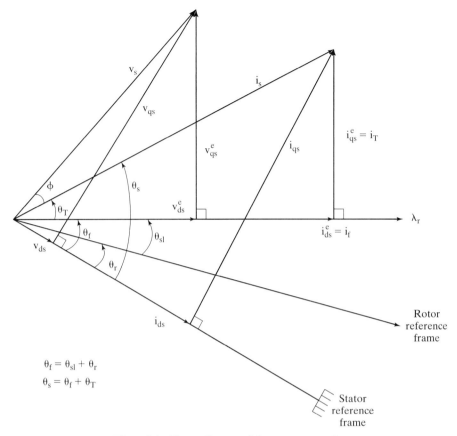

Figure 8.1 Phasor diagram of the vector controller

produces the rotor flux λ_r and the torque T_e. The component of current producing the rotor flux phasor has to be in phase with λ_r. Therefore, resolving the stator current phasor along λ_r reveals that the component i_f is the field-producing component, shown in Figure 8.1. The perpendicular component i_T is hence the torque-producing component. By writing rotor flux linkages and torque in terms of these components as

$$\lambda_r \propto i_f \tag{8.4}$$
$$T_e \propto \lambda_r i_T \propto i_f i_T \tag{8.5}$$

it can be seen that i_f and i_T have only dc components in steady state, because the relative speed with respect to that of the rotor field is zero: the rotor flux-linkages phasor has a speed equal to the sum of the rotor and slip speeds, which is equal to the synchronous speed. Orientation of λ_r amounts to considering the synchronous reference frames, and hence the flux- and torque-producing components of current are dc quantities. Because they are dc quantities, they are ideal for use as control variables; the bandwidth of the computational control circuits will have no effect on the processing of these dc control signals.

Crucial to the implementation of vector control, then, is the acquiring of the instantaneous rotor flux phasor position, θ_f. This field angle can be written as

$$\theta_f = \theta_r + \theta_{sl} \tag{8.6}$$

where θ_r is the rotor position and θ_{sl} is the slip angle. In terms of the speeds and time, the field angle is written as

$$\theta_f = \int (\omega_r + \omega_{sl}) dt = \int \omega_s \, dt \tag{8.7}$$

Vector-control schemes are classified according to how the field angle is acquired. If the field angle is calculated by using terminal voltages and currents or Hall sensors or flux-sensing windings, then it is known as *direct vector control*. The field angle can also be obtained by using rotor position measurement and partial estimation with only machine parameters but not any other variables, such as voltages or currents; using this field angle leads to a class of control schemes known as *indirect vector control*. The direct and indirect vector-controlled schemes are explained in the following sections.

Vector control is summarized by the following algorithm.

(i) Obtain the field angle.
(ii) Calculate the flux-producing component of current, i_f^*, for a required rotor flux linkage λ_r^*. By controlling only this field current, the rotor flux linkages are controlled. It is very similar to the separately-excited dc machine, in that the field current controls the field flux; the armature current has no impact on it.
(iii) From λ_r^* and the required T_e^*, calculate the torque-producing component of stator current, i_T^*. Controlling the torque-producing component current when the rotor flux linkages phasor is constant gives an independent control of electromagnetic torque. It is very similar to the case of the armature current's controlling the electromagnetic torque in a separately-excited dc machine with the field current maintained constant. Steps (ii) and (iii) enable a complete decoupling of flux- from torque-producing channels in the induction machine.
(iv) Calculate the stator-current phasor magnitude, i_s^*, from the vector sum of i_T^* and i_f^*.
(v) Calculate torque angle from the flux- and torque-producing components of the stator-current commands, $\theta_T = \tan^{-1} \dfrac{i_T^*}{i_f^*}$
(vi) Add θ_T and θ_f to obtain the stator current phasor angle, θ_s.
(vii) By using the stator-current phasor angle and its magnitude, θ_s and i_s^*, the required stator-current commands are found by going through the *qdo* transformation to *abc* variables:

$$i_{as}^* = i_s^* \sin \theta_s$$
$$i_{bs}^* = i_s^* \sin \left(\theta_s - \frac{2\pi}{3} \right) \tag{8.8}$$
$$i_{cs}^* = i_s^* \sin \left(\theta_s + \frac{2\pi}{3} \right)$$

(viii) Synthesize these currents by using an inverter; when they are supplied to the stator of the induction motor, the commanded rotor flux linkages and torque are produced.

The constants of proportionality involved in the flux and torque will be derived in later sections. The correspondence between the separately-excited dc motor and the induction motor is complete: i_f and i_T correspond to the field and armature currents of the dc machine, respectively. Even though the induction motor does not have separate field and armature windings, finding equivalent field and armature currents as components of the stator-current phasor has resulted in the decoupling of flux- from torque-producing channels in a machine that is highly coupled. Unlike the scalar control involved in dc machines, phasor or vector control is employed in induction machines. In the dc machine, the field and armature are fixed in space by the commutator, whereas, in the induction machine, no such additional component exists to separate the field (to produce the flux) from the armature (to produce the torque) channels to the optimum space angle of 90 electrical degrees between them. In the place of the commutator, the induction machine (and for that matter any ac machine) acquires the functionality of the commutator with an inverter. The inverter controls both the magnitude of the current and its phase, allowing the machine's flux and torque channels to be decoupled by controlling precisely and injecting the flux- and torque-producing currents in the induction machine to match the required rotor flux linkages and electromagnetic torque. The phasor control of current further adds to the complexity of computation involving phase and magnitude and of transformations to orient i_f and i_T with respect to rotor flux linkages. Note that the orientation need not be on rotor flux linkages; the computations can be carried out in stator or rotor or arbitrary reference frames. Synchronous reference frames are often used for the sake of freeing the signal-processing circuits from high bandwidth requirements, as is explained elsewhere.

8.3 DIRECT VECTOR CONTROL

8.3.1 Description

A block diagram of the direct vector-control scheme with a current-source inverter is shown in Figure 8.2. The electrical rotor speed, ω_r, is compared to the reference speed, ω_r^*, and the error is amplified and limited to generate the reference (command) torque, T_e^*. The rotor flux linkages reference λ_r^* is derived from the rotor speed via an absolute-value function generator. λ_r^* is kept at 1 p.u. for from 0- to 1-p.u. rotor speed; beyond 1-p.u. speed, it is varied as a function of the rotor speed. This is to ensure that the rotor speed is extended beyond the base speed with the available dc voltage to the inverter, by weakening the rotor flux linkages, thus reducing the induced emf to lower than that of the available output voltage from the inverter. By reducing the rotor flux linkages for the same

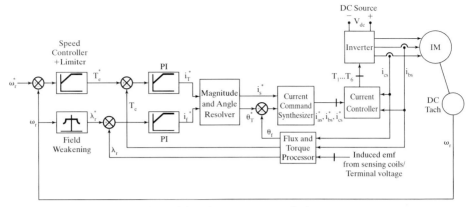

Figure 8.2 Direct vector-control scheme

torque-producing component of the stator current, the electromagnetic torque is reduced, which, in combination with the increasing rotor speed, can be controlled to produce the constant power output, say at rated value of the machine in steady state. The torque and rotor flux-linkages references are compared to the torque, T_e, and the rotor flux-linkages, λ_r, respectively. Their errors are amplified and limited to generate the reference torque- and flux-producing components of stator current, i_T^* and i_f^*, respectively. Phasor addition of i_f^* and i_T^* yields the stator-current phasor reference i_s^*, and the angle between i_T^* and i_f^* gives the torque-angle reference, θ_T^*. The sum of torque angle and field angle gives the position of the stator-current phasor, θ_s. Together with i_s^*, this generates the stator-phase current references, i_{as}^*, i_{bs}^*, and i_{cs}^*. These stator-phase-current requests are simplified by using an inverter and current feedback loops. The phase-current control loops can use one of the following switching techniques:

1. PWM
2. hysteresis
3. space-vector modulation (SVM)

Refer to Chapter 7 for a description of PWM, and hysteresis switching schemes and SVM scheme may be found in references and in some manner in this chapter. By one of these control techniques, the inverter output currents are made to correspond to the reference inputs. The feedback variables θ_f, T_e, and λ_r are obtained from the flux and torque processor block. This is the key to the direct vector-control scheme. The realization of this block will be examined in detail.

8.3.2 Flux and Torque Processor

The inputs to this block are two stator phase currents and one set of the following: either (i) terminal voltages or (ii) induced emf from the flux-sensing coils or Hall sensors. The selection of (i) or (ii) will determine the computational algorithm. Hence, for these inputs, computational steps are developed in the following.

8.3.2.1 Case (i): Terminal voltages. By using terminal voltages, the air gap torque, flux, and field angle can be computed with either rotor or stator flux linkages. The choice has significant impact on the performance of the drive system in terms of its parameter sensitivity, as is seen from the following development.

A. Rotor-Flux-Based Calculator Two line-to-line voltages can be measured, from which the phase (line to neutral) voltages can be computed (provided that the voltages are balanced). The q and d stator voltages in the stator reference frames are obtained from the phase voltages as

$$\begin{aligned} v_{qs} &= v_{as} \\ v_{ds} &= \frac{1}{\sqrt{3}}(v_{cs} - v_{bs}) \end{aligned} \quad (8.9)$$

Similarly, the currents are obtained in the same way; these equations hold true for them, too. From the stator-reference-frame equations of the induction motor, the stator equations are

$$\begin{aligned} v_{qs} &= (R_s + L_s p)i_{qs} + L_m p i_{qr} \\ v_{ds} &= (R_s + L_s p)i_{ds} + L_m p i_{dr} \end{aligned} \quad (8.10)$$

from which the rotor currents, i_{qr} and i_{dr}, can be computed as

$$\begin{aligned} i_{qr} &= \frac{1}{L_m}\left\{\int (v_{qs} - R_s i_{qs})dt - L_s i_{qs}\right\} \\ i_{dr} &= \frac{1}{L_m}\left\{\int (v_{ds} - R_s i_{ds})dt - L_s i_{ds}\right\} \end{aligned} \quad (8.11)$$

From all the stator and rotor currents, torque, the flux and field angle can be computed as follows:

$$\begin{aligned} T_e &= \frac{3}{2}\frac{P}{2}L_m(i_{qs}i_{dr} - i_{ds}i_{qr}) \\ \lambda_{qr} &= L_r i_{dr} + L_m i_{ds} \\ \lambda_{dr} &= L_r i_{qr} + L_m i_{qs} \\ \lambda_r &= \sqrt{(\lambda_{qr})^2 + (\lambda_{dr})^2} \\ \theta_f &= \tan^{-1}\left(\frac{\lambda_{qr}}{\lambda_{dr}}\right) \end{aligned} \quad (8.12)$$

Equations (8.9) to (8.12) can be implemented with computational circuits, as shown in Figure 8.3. The same diagram could be used for the code development for implementation in a single-chip processor. Note the circuit's dependence on the motor parameters R_s, L_s, L_r, and L_m. The changes in stator resistance could be tracked

418 Chapter 8 Vector-Controlled Induction Motor Drives

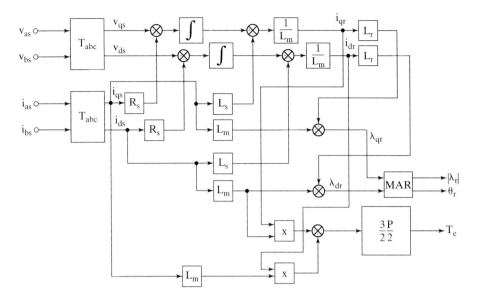

Note : MAR — Magnitude and Angle Resolver Block

Figure 8.3 Flux and torque processor implementation

indirectly with inexpensive temperature sensors. As for the inductances, they would have no significant variations in this scheme since flux control is implemented. Hence, parameter sensitivity would not greatly plague the accuracy of the measurement and calculation of flux, torque, and field angle.

B. Stator-Flux-Based Calculator Computational steps and dependence on many motor parameters could be very much reduced by using the stator flux linkages and calculating the electromagnetic torque, using only the stator flux linkages and stator currents. Then only stator resistance is employed in the computation of the stator flux linkages, thereby removing the dependence of mutual and rotor inductances of the machine on its calculation. The steps involved are summarized briefly as follows.

$$\lambda_{ds} = \int (v_{ds} - R_s i_{ds}) dt$$

$$\lambda_{qs} = \int (v_{qs} - R_s i_{qs}) dt$$

$$\lambda_s = \sqrt{(\lambda_{qs})^2 + (\lambda_{ds})^2} \angle \theta_{fs} \qquad (8.13)$$

$$\theta_{fs} = \tan^{-1}\left(\frac{\lambda_{qs}}{\lambda_{ds}}\right)$$

$$T_e = \frac{3}{2}\frac{P}{2}(i_{qs}\lambda_{ds} - i_{ds}\lambda_{qs})$$

In this case, the flux loop can be closed with the stator flux linkages instead of the rotor flux linkages. The accuracy of the computation might not be high, even though this algorithm depends only on the stator resistance rather than on many other motor parameters, as is proved by an example in the following. The sensitivity of the stator resistance variation and its impact on the accuracy of the stator flux linkages and hence on the electromagnetic torque is high when the stator voltages are small and of comparable magnitude to the resistive voltage drops. This is the case at low speeds; hence, dynamic operation at such speeds is very poor under this scheme.

Example 8.1

Find the errors in the stator flux-linkages magnitude and electromagnetic torque computations when the stator resistance has risen from its nominal value by 50%. The motor parameters are as follows:

$$R_s = 0.277\ \Omega;\ R_r = 0.183\ \Omega;\ L_m = 0.0538\ H;\ L_s = 0.0553\ H;\ L_r = 0.05606\ H;\ P = 4$$

Operating points: (i) $\omega_s = 16.932$ rad/sec; $\omega_{sl} = 6.932$ rad/sec; $V = 17.657$ (line to line) rms
(ii) $\omega_s = 306.932$ rad/sec; $\omega_{sl} = 6.932$ rad/sec; $V = 172.33$ (line to line) rms

Solution For ease of computation, the synchronous reference frame is employed.

Operating point (i): The synchronous-frame q and d axes voltages are

$$V_{qs}^e = \frac{\sqrt{2}}{\sqrt{3}} V = 14.416\ v;\ V_{ds}^e = 0$$

The currents are found as,

$$\begin{bmatrix} I_{qs}^e \\ I_{ds}^e \\ I_{qr}^e \\ I_{dr}^e \end{bmatrix} = \begin{bmatrix} R_{sl} & L_s\omega_s & 0 & L_m\omega_s \\ -L_s\omega_s & R_{sl} & -L_m\omega_s & 0 \\ 0 & L_m\omega_{sl} & R_r & L_r\omega_{sl} \\ -L_m\omega_{sl} & 0 & -L_r w_{sl} & R_r \end{bmatrix}^{-1} \begin{bmatrix} V_{qs}^e \\ V_{ds}^e \\ 0 \\ 0 \end{bmatrix} = \begin{bmatrix} 17.641 \\ 5.176 \\ -15.722 \\ 2.459 \end{bmatrix}$$

where $R_{sl} = 1.5 R_s$.

These machine currents are available through current transducers to the flux and torque processor. The stator flux linkages are then calculated, with the nominal value of the stator resistance in the calculator circuit as

$$\lambda_{qs}^e(R_s) = -\frac{V_{ds}^e - R_s I_{ds}^e}{\omega_s};\quad \lambda_{ds}^e(R_s) = \frac{V_{qs}^e - R_s I_{qs}^e}{\omega_s}$$

$$\lambda_s(R_s) = \sqrt{\{\lambda_{qs}^e(R_s)\}^2 + \{\lambda_{ds}^e(R_s)\}^2}$$

from which, upon substitution of the stator currents and voltages for stator resistances of R_s and R_{sl}, the stator flux linkages are obtained as

$$\lambda_{qs}^e(R_s) = 0.085\ Wb\text{–Turn};\ \lambda_{ds}^e(R_s) = 0.563\ Wb\text{–Turn};\ \lambda_s(R_s) = 0.569\ Wb\text{–Turn}$$

$$\lambda_{qs}^e(R_{sl}) = 0.127\ Wb\text{–Turn};\ \lambda_{ds}^e(R_{sl}) = 0.419\ Wb\text{–Turn};\ \lambda_s(R_{sl}) = 0.437\ Wb\text{–Turn}$$

$$T_e(R_s) = \frac{3}{2}\frac{P}{2}\{\lambda_{ds}^e(R_s)I_{qs}^e - \lambda_{qs}^e(R_s)I_{ds}^e\}$$

from which the torque is calculated as

$$T_e(R_s) = 28.473 \text{ N·m; and } T_e(R_{s1}) = 20.179 \text{ N·m}$$

The errors in the flux and torque are

$$\partial\lambda_s = \lambda_s(R_{s1}) - \lambda_s(R_s) = -0.132 \text{ Wb-Turn}$$
$$\partial T_e = T_e(R_{s1}) - T_e(R_s) = -8.249 \text{ N·m}$$

These errors are considerable; the flux error is approximately 25% and torque error is 41%, compared to rated values. Consider the second operating point.

Operating point (ii): The currents at R_{s1} are found, similarly to the above procedure,

$$I_{qs}^e = 15.67 \text{ A}; I_{ds}^e = 9.614 \text{ A}$$

The flux linkages are

$$\lambda_{qs}^e(R_s) = 0.009 \text{ Wb-Turn}; \lambda_{ds}^e(R_s) = 0.444 \text{ Wb-Turn}; \lambda_s(R_s) = 0.444 \text{ Wb-Turn}$$

$$\lambda_{qs}^e(R_{s1}) = 0.013 \text{ Wb-Turn}; \lambda_{ds}^e(R_{s1}) = 0.437 \text{ Wb-Turn}; \lambda_s(R_{s1}) = 0.437m \text{ Wb-Turn}$$

The torques are

$$T_e(R_s) = 20.636 \text{ N·m}, T_e(R_{s1}) = 20.178 \text{ N·m}$$

The errors in the stator flux linkages and torque are

$$\partial\lambda_s = \lambda_s(R_{s1}) - \lambda_s(R_s) = -0.007 \text{ Wb-Turn}$$
$$\partial T_e = T_e(R_{s1}) - T_e(R_s) = -0.458 \text{ N·m}$$

These errors are very small; they constitute less than 2% of rated values.

8.3.2.2 Case (ii): Induced EMF from flux-sensing coils or hall sensors.

Two sets of sensing coils can be placed in stator slots having 90 electrical degrees displacement, and one set can be placed on the mmf axis of one phase, say phase *a*. These coils can be concentric, making the layout easier. The sensing coils are isolated from the power circuit, to help in tying the outputs of those coils to the logic level-control circuits directly. Two Hall sensors can also be placed very similarly to the sensing coils; Hall sensors reflect the rate of change of the stator flux linkages. Let the induced emfs of the *q* and *d* axes be e_{qs} and e_{ds}, respectively. The stator *q* and *d* axes flux linkages, λ_{qs} and λ_{ds}, respectively, are defined as

$$\lambda_{qs} = L_s i_{qs} + L_m i_{qr}$$
$$\lambda_{ds} = L_s i_{ds} + L_m i_{dr} \qquad (8.14)$$

In terms of induced emfs, they are

$$\lambda_{qs} = \int e_{qs} \, dt$$
$$\lambda_{ds} = \int e_{ds} \, dt \qquad (8.15)$$

from which the rotor q and d axes currents are derived as,

$$i_{qr} = \frac{\lambda_{qs}}{L_m} - \frac{L_s}{L_m}i_{qs} = \frac{1}{L_m}\{\int e_{qs}dt - L_s i_{qs}\} \qquad (8.16)$$

$$i_{dr} = \frac{\lambda_{ds}}{L_m} - \frac{L_s}{L_m}i_{ds} = \frac{1}{L_m}\{\int e_{qs}dt - L_s i_{qs}\} \qquad (8.17)$$

From the rotor q and d axes currents and stator flux linkages, the rotor flux linkages and torque are obtained as

$$\begin{aligned}
\lambda_{qr} &= L_r i_{qr} + L_m i_{qs} \\
\lambda_{dr} &= L_r i_{dr} + L_m i_{ds} \\
\lambda_r &= \sqrt{(\lambda_{qr})^2 + (\lambda_{dr})^2} \angle \theta_f \\
\theta_f &= \tan^{-1}\left(\frac{\lambda_{qr}}{\lambda_{dr}}\right) \\
T_e &= \frac{3}{2}\frac{P}{2}(\lambda_{ds}i_{qs} - \lambda_{qs}i_{ds})
\end{aligned} \qquad (8.18)$$

Note that the torque equation does not involve machine parameters for its calculation. The rotor flux linkages and the field angle are dependent on machine parameters: L_m, L_s, and L_r. When there is a change (due to saturation) in these parameters, they will introduce an error in the computation of λ_r and θ_f. In particular, the error in the computation of field angle will generate significant errors in the vector control; that is the most crucial information for control. The flux and torque processor realized with terminal or sensing coils or Hall sensors have some merits and demerits. These are discussed in the following.

The merits of these forms of measuring and computing the rotor flux linkages, its position, and the electromagnetic torque are as follows.

(i) The sensing schemes use only electronic transducers and do not use any with moving or rotating parts, such as synchros or optical encoders. The absence of moving parts in the transducers makes the reliability of these schemes more robust than from employing mechanical/optical transducers.

(ii) Further, the costly process of mechanical mounting and the loss of valuable space and volume inside or outside the machine enclosure for the rotating sensor parts are avoided. The compactness afforded increases the overall power density of the motor drive system. A saving in labor and parts makes these sensing schemes very attractive at the low-cost, low-power end of industrial applications.

Some disadvantages are as follows:

(i) At zero stator frequency, there is no induced emf in all the measurement schemes. The result is that neither flux linkages nor their positions are available for vector control. Therefore, torque production at zero speed is not precisely controllable, thus making it unsuitable for positioning applications such as servos.

(ii) The same problem in a different form appears at low speeds. At these speeds, the induced-emf signals can be so small that signal-processing circuits cannot use them; they are comparable to quantization errors in digital circuits and will be affected by drift in the analog circuits. This, combined with factor (i) makes the direct vector-control drive unsuitable for precise positioning and low-speed operation.

(iii) The installation of sensing windings or of Hall-effect sensors, even though it is an inexpensive production process, adds to the number of wires coming out of the machine frame. This is not acceptable in high-volume applications, such as HVAC, because of the cost involved in the hermetic sealing. Extra wires are also not desirable in high-reliability applications, such as defense actuators and nuclear-plant pump drives.

(iv) In the case of using voltage and current transducers, the filtering required to obtain the fundamental at high frequencies will produce a large phase shift and inaccuracy in the computation of field angle, which will deteriorate the decoupling of flux and torque controls. It is relevant to observe here that voltage sensing with galvanic isolation can be easily and cost-effectively realized from the logic-level switching signals of the inverter, the phase currents, and an indication of the dc link voltage magnitude. The interested reader is referred to [25].

8.3.3 Implementation with Six-Step Current Source

In the above development, the form of current-source inverter was not discussed. It could be a six-step inverter or an inverter with sinusoidal current outputs. The six-step current-source inverter can be realized either with an ASCI or with a PWM inverter with inner current loops, discussed in Chapter 7. The sinusoidal inverter is realized from a PWM inverter with inner current loops. The switching of the inverter can be done with different modulation techniques. This section considers the six-step current-source inverter for direct vector-control implementation. Some drives use an ASCI implementation for its advantages in cost and reliability.

The simulation results of direct vector control with a six-step current-source inverter is shown in Figures 8.4(i) and (ii). Ideal current source is assumed with negligible rise and fall times. One-p.u. speed is commanded from standstill, which produces the maximum torque command. The rotor flux linkages are at 1 p.u. when the speed command is applied. The torque command is limited to ±2 p.u. In a six-step current-source inverter, a high amount of sixth-harmonic torque pulsation is present. A feedback of such a torque results in larger oscillations in torque in a direct vector-controlled induction motor. To avoid such a problem, either the feedback control of torque and flux linkages is deleted or the filtered torque signal is used for feedback control. These two cases are shown in Figures 8.4(i) and (ii), respectively. In both the control schemes, the key results are similar, except in the case of flux linkages experiencing a deviation in the open-loop case. Also, because of the torque

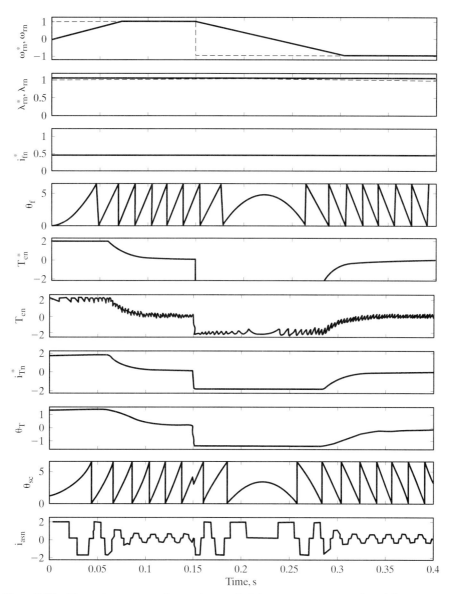

Figure 8.4(i) Dynamic responses of a speed-controlled six-step current-source-based direct vector-controlled induction motor drive (no T_e, λ_r feedback)

feedback, there is additional switching introduced in the phase currents. The common performance features are summarized in the following.

The electromagnetic torque rises instantaneously, and its average reaches the command value. A small swing in rotor flux linkages is seen, due to the fact that the stator-current phasor can assume only seven distinct states and hence is not smooth. As the speed error is decreased, the torque command and, hence, the magnitudes of the stator currents decrease, as is clearly seen in the figure.

424 Chapter 8 Vector-Controlled Induction Motor Drives

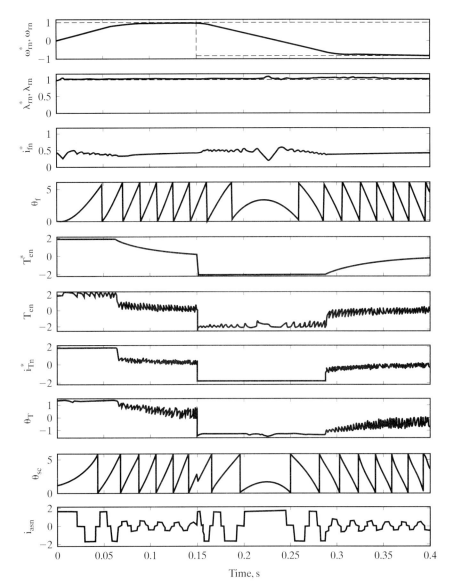

Figure 8.4(ii) Dynamic responses of a speed-controlled six-step current-source-based direct vector-controlled induction motor drive (with T_e, λ_r feedback loops included)

Figure 8.5 shows the simulation results of a sinusoidal PWM inverter-controlled induction motor torque drive. A step command of one p.u. bidirectional load torque is applied. The torque-producing current rises in a very short time to match the load torque, and the response is almost instantaneous. As the speed is allowed to vary freely, the stator frequency follows the speed as seen from the stator current. The rotor flux linkages are steady and are unperturbed by the torque changes.

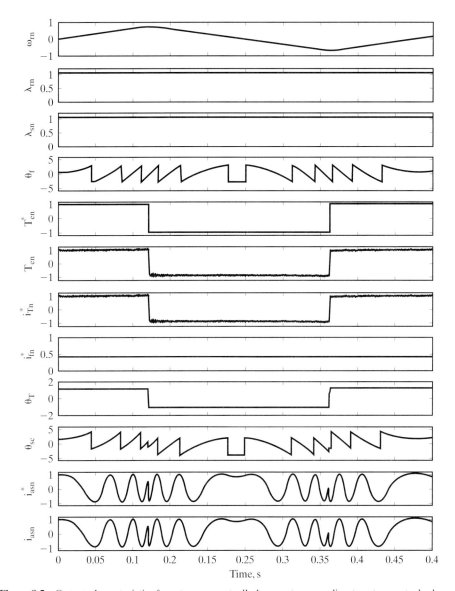

Figure 8.5 Output characteristics for a torque-controlled current-source direct vector-control scheme

8.3.4 Implementation with Voltage Source

Figure 8.6 shows the vector-control implementation with a voltage-source inverter. The current commands $i_{qs}^e{}^*$ and $i_{ds}^e{}^*$ are generated, very much as with the vector control with current-source inverter. The dq current errors set the required stator voltage commands $v_{qs}^e{}^*$ and $v_{ds}^e{}^*$. By using the transformation matrix, the two axis voltage commands are converted into three-phase voltage commands, which will

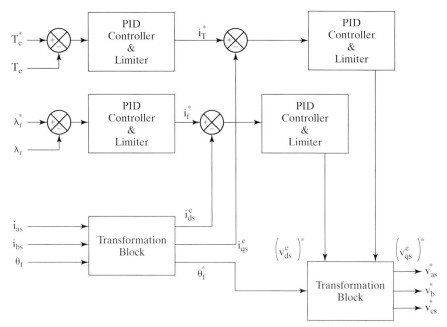

Figure 8.6 Functional block diagram of a voltage-source direct vector controller

be amplified by the inverter using any one of the switching techniques. The simulation results for a speed-controlled drive using PWM voltage source are shown in Figure 8.7. An 0.75-p.u. bidirectional speed command is given to the drive system when the machine has rated rotor flux linkages. During acceleration and deceleration, the torque is maintained at 2 p.u. and −2 p.u., respectively. Note the smooth acceleration and deceleration speed profiles. The zero-speed crossover is affected, with no oscillations in current, torque, flux, and speed, by means of the precise control of rotor flux linkages.

8.3.5 Direct Vector (Self) Control in Stator Reference Frames with Space-Vector Modulation

A novel way of implementing the direct vector-controlled induction motor drive with voltage-source inverter is presented in this section from reference [46]. This method uses feedback control of torque and stator flux, which are computed from the measured stator voltages and currents. As the method does not use a position or speed sensor to control the machine and uses its own electrical output currents and resulting terminal voltages, this is also referred as a direct self-control scheme. The method uses a stator reference model of the induction motor for its implementation, thereby avoiding the trigonometric operations in the coordinate transformations of the synchronous reference frames. This is one of the key advantages of the control scheme. The scheme uses stator flux-linkages control. That enables the flux-weakening operation of the motor to be straightforward compared to rotor flux-weakening, in that the stator flux is directly proportional to the induced emf, whereas the rotor flux does not have the same relationship. This reduces the depen-

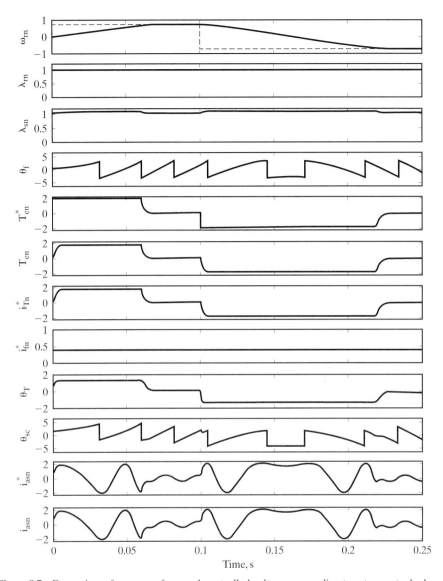

Figure 8.7 Dynamic performance of a speed-controlled voltage-source direct vector-control scheme

dence of the scheme on many motor parameters, thus making it a robust scheme in the flux-weakening region. The scheme depends only on stator resistance and on no other parameters. For its flux and torque control, this control scheme requires the position of the flux phasor, which is difficult to obtain at low and zero speeds from measured voltages and currents by using a torque processor. At such low speeds, these signals are very small, making scaling and accuracy problematic; also, the stator-resistance variations introduce significant errors in the flux-phasor computation. Invariably the low-speed operation suffers with this scheme without stator-resistance compensation.

The implementation of the scheme requires flux linkages and torque computations, plus generation of switching states through a feedback control of the torque and flux directly without inner current loops.

The stator q and d axes flux linkages are

$$\lambda_{qs} = \int (v_{qs} - R_s i_{qs})\, dt \qquad (8.19)$$

$$\lambda_{ds} = \int (v_{ds} - R_s i_{ds})\, dt \qquad (8.20)$$

where the direct and quadrature axis components are obtained from the abc variables by using the transformation,

$$i_{qs} = i_{as} \qquad (8.21)$$

$$i_{ds} = \frac{1}{\sqrt{3}} (i_{cs} - i_{bs}) \qquad (8.22)$$

This transformation is applicable for voltages and flux linkages as well. To obtain uniformly rotating stator flux, note that the motor voltages have to be varied uniformly without steps too. This imposes a requirement of continuously variable stator voltages with infinite steps, which is not usually met by the inverter because it has only finite switching states.

Switching States of the Inverter Consider the inverter shown in Figure 8.8. The terminal voltage a with respect to negative of the dc supply is considered, and V_a is determined by a set of switches, S_a, consisting of T_1 and T_4 as shown in the Table 8.1. When the switching devices T_1 and T_4 and their antiparallel diodes are off, V_a is indeterminate. Such a situation is not encountered in practice and, hence, has not been considered. The switching of S_b and S_c sets for line b and c can be similarly derived. The total number of switching states possible with S_a, S_b, and S_c is eight; they are elaborated in Table 8.2 by using the following relationships:

$$\left. \begin{array}{l} v_{ab} = v_a - v_b \\ v_{bc} = v_b - v_c \\ v_{ca} = v_c - v_a \end{array} \right\} \qquad (8.23)$$

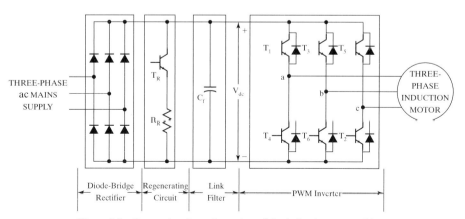

Figure 8.8 Power-circuit configuration of the induction motor drive

Section 8.3 Direct Vector Control

TABLE 8.1 Switching states of inverter phase leg a

T_1	T_4	S_a	V_a
on	off	1	V_{dc}
off	on	0	0

TABLE 8.2 Inverter switching states and machine voltages

States	S_a	S_b	S_c	V_a	V_b	V_c	V_{ab}	V_{bc}	V_{ca}	V_{as}	V_{bs}	V_{cs}	V_{qs}	V_{ds}
I	1	0	0	V_{dc}	0	0	V_{dc}	0	$-V_{dc}$	$\tfrac{2}{3}V_{dc}$	$-\tfrac{1}{3}V_{dc}$	$-\tfrac{1}{3}V_{dc}$	$\tfrac{2}{3}V_{dc}$	0
II	1	0	1	V_{dc}	0	V_{dc}	V_{dc}	$-V_{dc}$	0	$\tfrac{1}{3}V_{dc}$	$-\tfrac{2}{3}V_{dc}$	$\tfrac{1}{3}V_{dc}$	$\tfrac{1}{3}V_{dc}$	$\tfrac{V_{dc}}{\sqrt{3}}$
III	0	0	1	0	0	V_{dc}	0	$-V_{dc}$	V_{dc}	$-\tfrac{1}{3}V_{dc}$	$-\tfrac{1}{3}V_{dc}$	$\tfrac{2}{3}V_{dc}$	$-\tfrac{1}{3}V_{dc}$	$\tfrac{V_{dc}}{\sqrt{3}}$
IV	0	1	1	0	V_{dc}	V_{dc}	$-V_{dc}$	0	V_{dc}	$-\tfrac{2}{3}V_{dc}$	$\tfrac{1}{3}V_{dc}$	$\tfrac{1}{3}V_{dc}$	$-\tfrac{2}{3}V_{dc}$	0
V	0	1	0	0	V_{dc}	0	$-V_{dc}$	V_{dc}	0	$-\tfrac{1}{3}V_{dc}$	$\tfrac{2}{3}V_{dc}$	$-\tfrac{1}{3}V_{dc}$	$-\tfrac{1}{3}V_{dc}$	$-\tfrac{V_{dc}}{\sqrt{3}}$
VI	1	1	0	V_{dc}	V_{dc}	0	0	V_{dc}	$-V_{dc}$	$\tfrac{1}{3}V_{dc}$	$\tfrac{1}{3}V_{dc}$	$-\tfrac{2}{3}V_{dc}$	$\tfrac{1}{3}V_{dc}$	$-\tfrac{V_{dc}}{\sqrt{3}}$
VII	0	0	0	0	0	0	0	0	0	0	0	0	0	0
VIII	1	1	1	V_{dc}	V_{dc}	V_{dc}	0	0	0	0	0	0	0	0

and machine phase voltages for a balanced system are

$$\left.\begin{aligned} v_{as} &= \frac{(v_{ab} - v_{ca})}{3} \\ v_{bs} &= \frac{(v_{bc} - v_{ab})}{3} \\ v_{cs} &= \frac{(v_{ca} - v_{bc})}{3} \end{aligned}\right\} \quad (8.24)$$

and q and d axes voltages are given by

$$\left.\begin{aligned} v_{qs} &= v_{as} \\ v_{ds} &= \frac{1}{\sqrt{3}}(v_{cs} - v_{bs}) = \frac{1}{\sqrt{3}} v_{cb} \end{aligned}\right\} \quad (8.25)$$

The stator q and d voltages for each state are shown in Figure 8.9. The limited states of the inverter create distinct discrete movement of the stator-voltage phasor, v_s consisting of the resultant of v_{qs} and v_{ds}. An almost continuous and uniform flux phasor is feasible with these discrete voltage states, one due to their integration over time as seen from equations (8.19) and (8.20).

For control of the voltage phasor both in its magnitude and phase, the requested voltage vector's phase and magnitude are sampled, say once every

430 Chapter 8 Vector-Controlled Induction Motor Drives

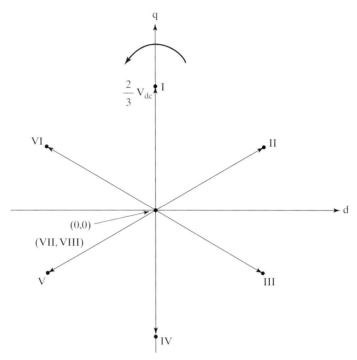

Figure 8.9 The inverter output voltages corresponding to switching states

switching period. The phase of the requested voltage vector identifies the nearest two nonzero voltage vectors. The requested voltage vector can be synthesized by using fractions of the two nearest voltage vectors, which amounts to applying these two vectors, one at a time, for a fraction of the switching period. The nearest zero voltage vector to the two voltage vectors is applied for the remaining switching period. The duty cycle for each of the voltage vectors is determined by the phasor projection of the requested voltage vector onto the two nearest voltage vectors. This method of controlling the input voltages to the machine through a synthesis of voltage phasor rather than the individual line-to-line voltages has the advantages of: (i) not using pulse-width modulation carrier-frequency signals, (ii) higher fundamental voltages compared to sine-triangle PWM, based controllers, (iii) given the switching frequency, the switching losses are minimized, and (iv) lower voltage and current ripples. This method of switching the inverter is known as space vector modulation (SVM) and many variations are available in literature.

Flux control A uniform rotating stator flux is desirable, and it occupies one of the sextants (in the phasor diagram shown in Figure 8.10) at any time. The stator-flux phasor has a magnitude of λ_s, with an instantaneous position of θ_{fs}. The corresponding d and q axes components are λ_{ds} and λ_{qs}, respectively. Assuming that a feedback of stator flux is available, its place in the sextant is identified from its position. Then the influencing voltage phasor is identified by giving a $90°$ phase shift. For example, if the stator-flux phasor is in sextant <2>, the right influencing

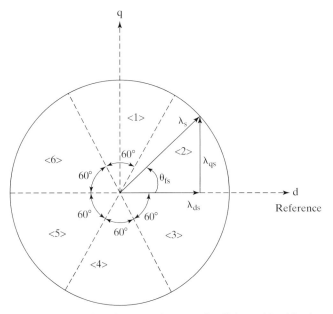

Figure 8.10 Division of sextants for stator flux-linkages identification

voltage phasor has to be either VI or I. Voltage phasor I is $90° - \theta_{fs}$ and VI is $150° - \theta_{fs}$ from the flux phasor. One of these two sets increases λ_s, the other decreases λ_s. This is found from the following explanation obtained from Figure 8.11.

Consider the effect of switching voltage phasor set I, v_I, and phasor set VI, v_{VI}. As seen from the phasor diagram, in the case of set I, the flux phasor increases in magnitude from λ_s to λ_{sI}; in the case of set VI, it decreases to λ_{sVI}. This implies that the closer voltage-phasor set increases the flux and the farther voltage-phasor set decreases the flux, but note that both of them advance the flux phasor in position. Similarly for all other sextants, the switching logic is developed. A flux error, λ_{er}, thus determines which voltage phasor has to be called, and this flux error is converted to a digital signal with a window comparator with a hysteresis of $\delta\lambda_s$ (let it be S_λ). The switching logic to realize S_λ from λ_{er} is given in the following:

Condition	S_λ
$\lambda_{er} > \lambda_s$	1
$\lambda_{er} \leq \lambda_s$	0

Over and above the flux control, the control of electromagnetic torque is required for a high-performance drive. It is achieved as follows.

Torque control Torque control is exercised by comparison of the command torque to the torque measured from the stator flux linkages and stator currents as

$$T_e = \frac{3}{2} \cdot \frac{P}{2} (\lambda_{ds} i_{qs} - \lambda_{qs} i_{ds}) \qquad (8.25)$$

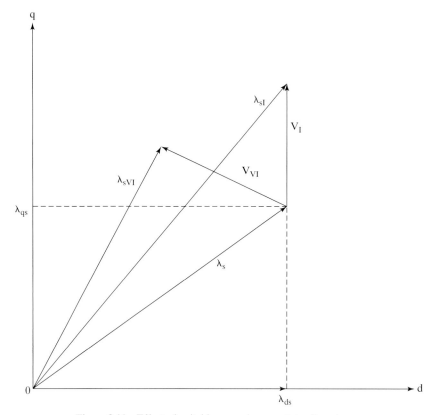

Figure 8.11 Effect of switching v_I and v_{VI} on stator-flux phasor

The error torque is processed through a window comparator to produce digital outputs, S_T, as follows:

Condition	S_T
$(T_e^* - T_e) > \delta T_e$	1
$-\delta T_e < (T_e^* - T_e) < \delta T_e$	0
$(T_e^* - T_e) < -\delta T_e$	-1

where δT_e is the torque window acceptable over the commanded torque. When the error exceeds δT_e, it is time to increase the torque, denoting it with a +1 signal. If the torque error is between positive and negative torque windows, then the voltage phasor could be at zero state. If the torque error is below $-\delta T_e$, it amounts to calling for regeneration, signified by -1 logic signal.

Interpretation of S_T is as follows: when it is 1 amounts to increasing the voltage phasor, 0 means to keep it at zero, -1 requires retarding the voltage phasor behind the flux phasor to provide regeneration. Combining the flux error output S_λ, the torque error output S_T, and the sextant of the flux phasor S_θ, a switching table can be realized to obtain the switching states of the inverter; it is given in Table 8.3. The algorithm for S_θ is shown in Table 8.4.

TABLE 8.3 Switching states for possible S_λ, S_T, and S_θ

		S_θ					
S_λ	S_T	<1>	<2>	<3>	<4>	<5>	<6>
1	1	VI (1,1,0)	I (1,0,0)	II (1,0,1)	III (0,0,1)	IV (0,1,1)	V (0,1,0)
1	0	VIII (1,1,1)	VII (0,0,0)	VIII (1,1,1)	VII (0,0,0)	VIII (1,1,1)	VII (0,0,0)
1	−1	II (1,0,1)	III (0,0,1)	IV (0,1,1)	V (0,1,0)	VI (1,1,0)	I (1,0,0)
0	1	V (0,1,0)	VI (1,1,0)	I (1,0,0)	II (1,0,1)	III (0,0,1)	IV (0,1,1)
0	0	VII (0,0,0)	VIII (1,1,1)	VII (0,0,0)	VIII (1,1,1)	VII (0,0,0)	VIII (1,1,1)
0	−1	III (0,0,1)	IV (0,1,1)	V (0,1,0)	VI (1,1,0)	I (1,0,0)	II (1,0,1)

TABLE 8.4 Flux-phasor sextant logic (S_θ)

θ_{fs}	Sextant
$0 \leq \theta_{fs} \leq \pi/3$	<2>
$-\pi/3 \leq \theta_{fs} \leq 0$	<3>
$-2\pi/3 \leq \theta_{fs} \leq -\pi/3$	<4>
$-\pi \leq \theta_{fs} \leq -2\pi/3$	<5>
$2\pi/3 \leq \theta_{fs} \leq \pi$	<6>
$\pi/3 \leq \theta_{fs} \leq 2\pi/3$	<1>

Consider the first column corresponding to S_θ = <1>. The switching states of the inverter are indicated in parentheses; they correspond to S_a, S_b, and S_c. The flux error signal indicates 1, which means the flux is less than its request value and therefore the flux phasor has to be increased. At the same time, torque error is positive, asking for an increase. Merging these two with the position of the flux phasor in <1>, the voltage phasor I and VI satisfy the requirements only if the flux is within the first 30° of the sextant <1>. In the second 30°, note that the voltage phasor I will increase the flux-phasor magnitude but will retard it in phase. This will result in a reduction of the stator frequency and reversal of the direction of torque. The control requires the advancement of the flux phasor in the same direction (i.e., counterclockwise in this discussion); that could be satisfied only by voltage phasor VI in this 30°. Voltage phasor VI is the only one satisfying the uniform requirements throughout the sextant <1>, so the voltage phasor VI is chosen for S_T and S_λ to be equal to +1 with the flux phasor in sextant <1>. When the torque error is zero, the only logical choice is to apply zero line voltages; because the previous state had two +1 states, it is easy to achieve zero line voltages by choosing the switching state VIII with all ones. If the torque error becomes negative with $S_T = -1$, the machine has to regenerate, with a simultaneous increase in flux phasor due to $S_\lambda = 1$; hence, the voltage phasor is retarded close to the flux phasor, and, hence, switching state II(1, 0, 1) is selected.

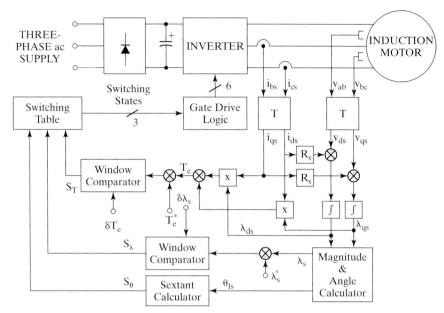

Figure 8.12 Block-diagram schematic of the direct torque (self) induction motor drive

If $S_\lambda = 0$ (i.e., the flux phasor has exceeded its request by the hysteresis window amount, $\delta\lambda_s$), then it has to be decreased to match its request value by choosing a voltage phasor away from flux phasor, i.e., V. This accelerates the flux phasor and increases the slip speed, resulting also in an increase of electromagnetic torque, thus satisfying $S_T = 1$ demand. When $S_T = 0$, reach zero line-voltage states by going to VII, because V contained two zero states in it. If $S_T = -1$, then regenerate, increasing the negative torque but decreasing the flux phasor; this is achieved by retarding the voltage phasor behind the flux phasor, but far away from it, and hence III is chosen. Note that it has two zeros in it, and, therefore, transition from VII to III requires a change of only one switch signal.

Implementation The drive scheme is realized as shown in Figure 8.12. Further, to reduce the voltage transducers, the information of line and phase voltages could be obtained from a single dc-link voltage transducer and the gate drive signals. Similarly, the phase currents can be reconstructed from a single dc-link current transducer and the gate drive signals of the inverter. In all, only two transducers are required, both of which are electrical; it does not require moving parts, thus making this control scheme robust and reliable. Further, the cost of the drive-system control is very low compared to that of position sensor based vector-controlled induction motor drives.

Performance The dynamic performance of this drive scheme is shown for both the torque- and the speed-controlled drive in Figures 8.13(a) and (b), respectively. The motor details are given in Example 5.4. The hysteresis windows for the torque and flux linkages are 0.05 p.u. and 0.001 p.u., respectively. The proportional and integral gains of the speed controller are 50 and 0.05, respectively. The base values of torque, speed, flux linkage, current, and voltage are 20.3 N·m, 377 rad/sec, 0.4213 Wb-turn,

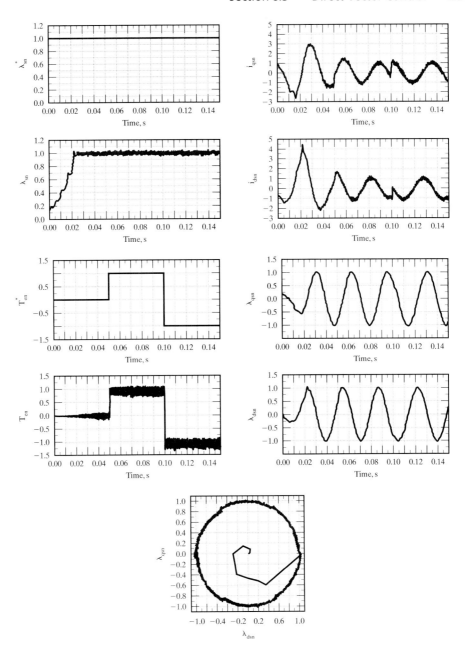

Figure 8.13(a) Dynamic performance of the torque-controlled drive system in normalized units

18.94 A, and 200 V, respectively. The dc-link voltage is set at 285 V. The rotor is free of load for both torque- and speed-controlled operation.

For the torque drive, note that the machine starts from standstill and, hence, takes a time of 25 ms to establish the stator flux linkages with high currents of 4.5 p.u. This is unusual and is not attempted in practice. The stator flux phasor

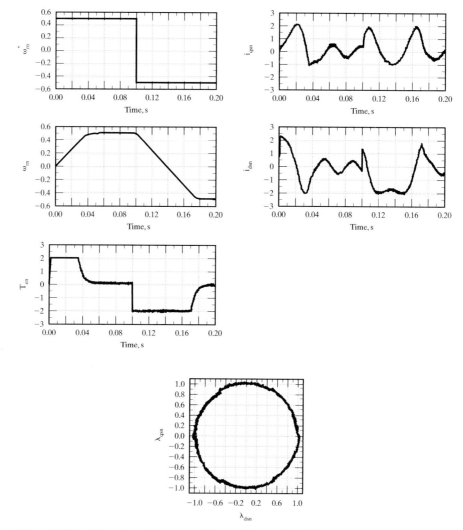

Figure 8.13(b) Dynamic performance of the speed-controlled drive system in normalized units

becomes uniform, as seen from its plot. The torque response is almost instantaneous, with considerable switching frequency ripples on it. The ripple torques have no effect on the speed, as shown from the speed-controlled drive.

The speed command is a step of 0.5 p.u. at starting and a step of 0.5 to −0.5 p.u. at 0.1 s. It is assumed that the machine has rated flux linkages at the time of application of the speed command. The electromagnetic torque is limited to ±2 p.u., which provides a faster acceleration, as is seen from the Figure 8.13(b). The rotor speed takes a smooth ramp profile with no oscillations, while the current is limited to two p.u. The locus of the stator flux-linkages phasor is almost a uniform circle, even during large speed changes and hence torque commands, thus showing the complete decoupling of the flux- from the torque-producing channels in the drive system.

Example 8.2

It is noticed that stator frequency is not a distinct input in the direct self-control scheme. How can that be extracted for possible use in speed-sensorless control application?

Solution The synchronous speed is the speed of the stator flux-linkages phasor. It may be recalled that the synchronous frequency is equal to the stator frequency. It is obtained as the derivative of the stator flux-linkages-phasor position. As the stator flux linkages and its d and q components are available, this task is made easier in the control scheme. The synchronous speed, then, is derived as

$$\omega_s = \frac{d\theta_{fs}}{dt} = \frac{d}{dt}\left\{\tan^{-1}\left(\frac{\lambda_{qs}}{\lambda_{ds}}\right)\right\}$$

$$= \frac{\dot{\lambda}_{qs}\lambda_{ds} - \dot{\lambda}_{ds}\lambda_{qs}}{\lambda_{ds}^2 + \lambda_{qs}^2}$$

$$= \frac{(v_{qs} - R_s i_{qs})\lambda_{ds} - (v_{ds} - R_s i_{ds})\lambda_{qs}}{\lambda_s^2}$$

Note that differentiation of the q and d axes stator flux linkages is avoided with measured stator voltages and computed resistive voltage drops with measured currents. The denominator is zero to start with, so caution must be exercised at start. After the initial start mode, the denominator, which is the stator flux-linkages phasor, is always nontrivial. As the stator voltages are pulsed, the stator frequency signal is fraught with noise and, therefore, filtering is required to enable its use in feedback control applications.

Parameter sensitivity The scheme is sensitive to stator resistance instrumented in the controller. Note that, in comparison to many other schemes, the parameter dependency of this scheme is restricted to only one parameter; other control schemes are dependent on many more parameters. The stator-resistance change has a wide variation (from 0.75 to 1.7 times its nominal value), due to a large extent to temperature variation and to a smaller degree to stator-frequency variation. It deteriorates the drive performance by introducing errors in the estimated flux linkage's magnitude and position and hence in the electromagnetic torque estimation, particularly at low speeds. Note that, at low speeds, the stator-resistance voltage drops constitute a significant portion of the applied voltages.

A few control schemes have been proposed to overcome this parameter sensitivity. A partial, operating-frequency-dependent hybrid flux estimator has been proposed for stator-resistance tuning; it has as problems convergence and slowness of response. Adjustment of the stator resistance in conformance to the difference between the flux current and its command has the problem of identifying the actual flux current. Finding stator resistance from the steady-state voltage equation has the shortcoming of using the direct axis flux linkage, which itself has been affected by stator-resistance variations. The use of stator-current phasor error with PI and fuzzy estimators demonstrates good performance in the tuning of the stator resistance. The fuzzy estimator seems to outperform the PI-based estimator, thereby necessitating the use of intelligent control techniques to synthesize the adaptation mechanism.

There is a possible instability problem, due to parameter sensitivity, that arises in the direct self-control scheme. The drive system becomes unstable if the instrumented stator resistance in the controller is higher than the machine resistance. Such cases can arise in practice. Two possible scenarios follow:

(i) The stator resistance can be lower than its controller-set-point nominal value in an externally housed drive system in colder climates; the operating temperature could be different from the motor temperature at starting. The controller resistance might correspond to the operating temperature that is justifiable for operation in the parameter-uncompensated system.

(ii) If the drive system has parameter adaptation and if its performance is poor, then the estimated stator resistance might at times be higher than the actual stator resistance, leading to instability.

This section presents a solution: to track the stator resistance, so that the performance degradation and a possible instability problem can be avoided. A signal proportional to stator-resistance change is developed by using the error between the reference and actual stator current phasor. A simple proportional–integral (PI) controller is used in the stator-current phasor feedback loop to adapt the stator resistance instrumented in the controller. An analytic expression to evaluate the stator current command from the torque and stator flux linkage commands [35] is used here. The performance of the controller is shown, with dynamic simulation results for a wide variety of operating conditions, including flux-weakening.

Instability Due to Parameter Mismatch A mismatch between the controller-set stator resistance and its actual value in the machine can create instability, as is shown in Figure 8.14. This figure shows the simulations for a step stator resistance change from 100% to 80% of its nominal value at 0.5 s, with a rated torque command applied at 0.1 s. The drive system becomes unstable. This could be reasoned out as follows: As the motor resistance decreases in the machine, its current increases for the same applied voltages, which increases the flux and electromagnetic torque. The controller has the opposite effect: the increased currents, which are inputs to the system, cause increased stator-resistance voltage drops in the calculator, resulting in lower flux linkages and electromagnetic torque estimation. They are compared with their command values, giving larger torque and flux-linkages errors, resulting in the commanding of larger voltages, causing larger currents, all leading to a run-off condition.

The instability result of a step change is not realistic; the stator resistance does not in practice change in a step manner. A linearly decreasing stator resistance is simulated, and the performance is shown in Figure 8.15. Even for such a gradual change of stator resistance, note that the system becomes unstable. The controller-calculated electromagnetic torque and stator flux-linkages are almost equal to reference values and contrary to the real situation in the machine. But the S_T, S_λ, and S_θ signals to command the voltage vector signify an increase in voltage magnitude and a decrease in slip speed. Hence, any scheme using reference torque and flux linkages for parameter compensation would not be effective. For example, the air gap power-feedback control for parameter compensation in the indirect vector-controlled drive system is successful, but it will not work in this drive system.

The parameter mismatch between the controller and the machine also results in a nonlinear relationship between torque and its reference, making it an imperfect

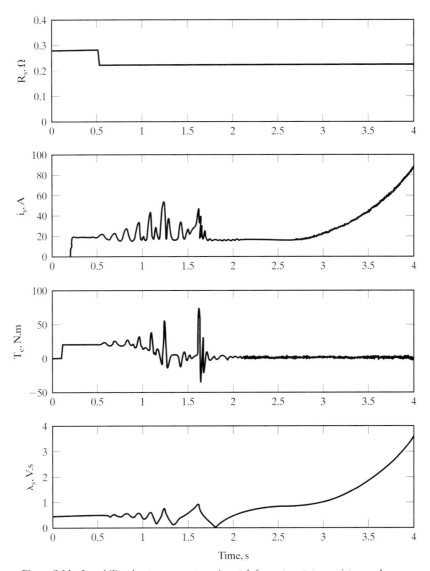

Figure 8.14 Instability due to parameter mismatch for a step stator-resistance change

torque amplifier. This will have undesirable consequences in a torque drive and to a smaller extent in the speed-controlled drive systems. The motor-resistance adaptation is essential to overcome instability and to guarantee a linear torque amplifier in the direct torque-controlled drive. A stator-resistance parameter adaptation scheme to achieve these objectives is described in the next section.

Stator-Resistance Compensation

A. Scheme A block-diagram schematic of the stator-resistance compensation scheme is shown in Figure 8.16; its incorporation in the drive schematic is given in

440 Chapter 8 Vector-Controlled Induction Motor Drives

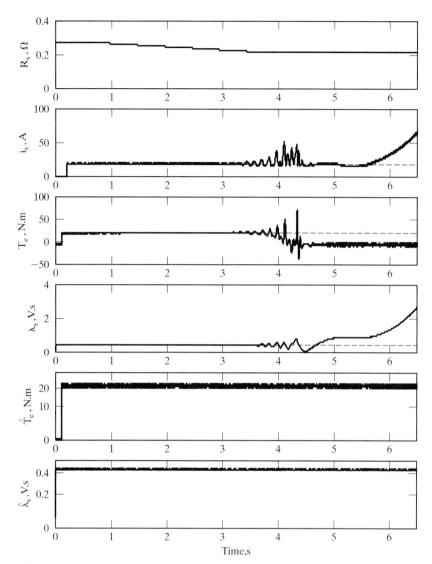

Figure 8.15 Instability due to parameter mismatch for a linearly decreasing stator-resistance change

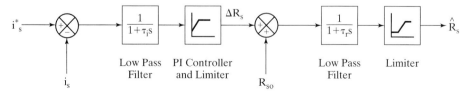

Figure 8.16 Block-diagram schematic of the adaptive stator-resistance compensator

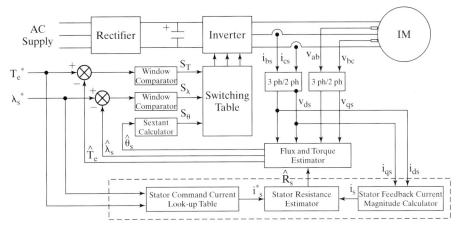

Figure 8.17 Stator-resistance adaptation in the DTC drive

Figure 8.17 and shown in dotted lines. This technique is based on the principle that the error between the measured stator-feedback current-phasor magnitude i_s and its command i_s^* is proportional to the stator-resistance variation.

The incremental stator resistance for correction is obtained through a PI controller and limiter. The current error is processed through a low-pass filter that has very low cutoff frequency, in order to remove high-frequency components contained in the stator-feedback current. The low-pass filter does not generate any adverse effect on the stator-resistance adaptation if the filter time constant is chosen to be smaller than the adaptation time constant. Incremental stator resistance, ΔR_s, is continuously added to the previously estimated stator resistance, R_{so}. The final estimated value, $\hat{R}_s$, is obtained as the output of another low-pass filter and limiter. This low-pass filter is necessary for a smooth variation of the estimated resistance value. The final signal is the updated stator resistance and can be used directly in the controller. The above algorithm requires the stator current-phasor command for implementation. An analytic procedure to compute the stator current command from the torque and stator flux-linkages commands is presented in reference [35].

B. Dynamic Simulation Results Dynamic simulations are performed to validate the performance of the torque-controlled drive system with the compensation technique. The induction motor and drive system details used in the simulation are given next:

5 HP, 200 V, 60 Hz, 3 Phase, Y Connection, 4 Poles, 1766.9 rpm, $R_s = 0.277 \, \Omega$, $R_r = 0.183 \, \Omega$, $L_s = 0.0553$ H, $L_r = 0.056$ H, $L_m = 0.0538$ H, $J = 0.01667$ kg-m², $B = 0$ N·m-s/rad, Rated $\lambda_s = 0.433$ V-s.

The proportional and integral gains of the adaptation controller are 0.01 and 0.0005, respectively. The stator-resistance sampling time is 5 ms, and the cutoff frequencies for the current and resistance low-pass filters are 2 and 0.24 Hz, respectively.

In the implementation of the drive algorithm, only six nonzero voltage vectors are used, and two zero voltage vectors are excluded. It is noted that this has no impact on the basic performance of the system. Because the stator-resistance voltage drop is a significant portion of the applied voltage at low speed, the performance of the system

442 Chapter 8 Vector-Controlled Induction Motor Drives

at low speed, from variation in the stator resistance, deteriorates much more compared with the performance at high speed. Therefore, the rotor speed is limited to 0.1 p.u. in the simulations, to illustrate the performances in the most affected region.

Figure 8.18 shows the simulation results for a step change in stator resistance in a parameter-uncompensated torque-drive system and in a compensated system. The system controller has the nominal value of stator resistance; after 0.5 s, stator resistance is changed to twice its nominal value, and the corresponding effects are studied. Rated

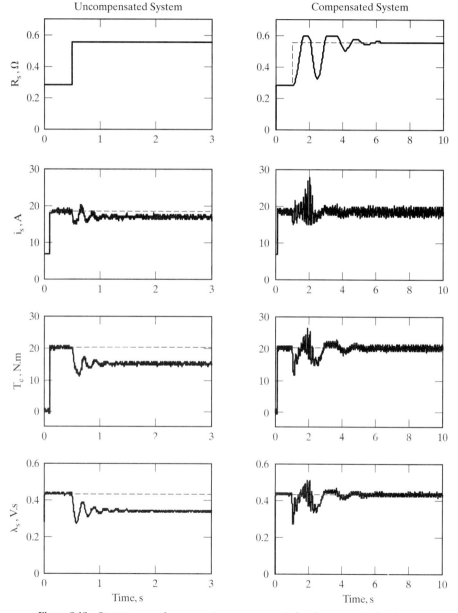

Figure 8.18 Step response for parameter-uncompensated and compensated systems

torque command is applied at 0.1 s after the stator flux linkage has reached the steady state. Parameter adaptation is initiated after 0.2 s. In the uncompensated system, right after the change of resistance, the generated electromagnetic torque decreases by nearly 25% in steady state, with much higher drop during the transient. The stator flux linkages and stator current show similar worsening. In the compensated system, it is noticed that stator-resistance estimate suffers in the initial transient state and converges gradually to its final actual value in steady state. All the other variables also have initial transient state but reach their final values in steady state. A step variation in the stator resistance is rather an extreme test and not a significant case encountered in practice.

In actual operating conditions, the rate of change of temperature is very slow and so is that of the stator resistance. Figure 8.19 covers this situation. Stator

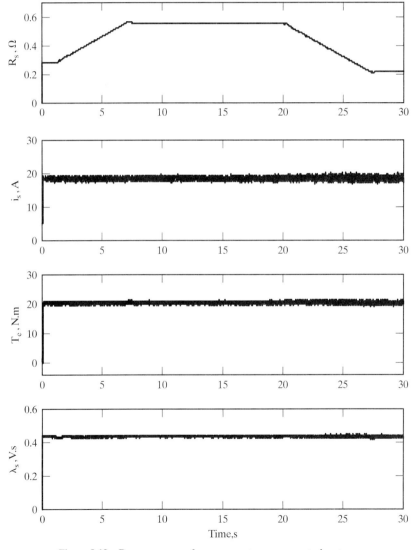

Figure 8.19 Ramp response for a parameter-compensated system

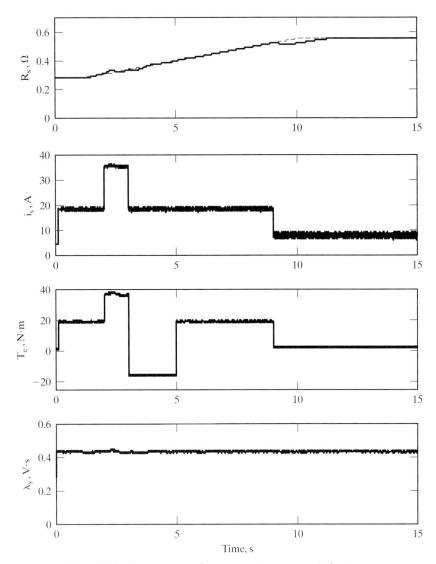

Figure 8.20 Ramp response for a parameter-compensated system

resistance is increased linearly from nominal value to twice its nominal value in 7 s, then decreased linearly from twice to 0.8 times its nominal value. The stator-resistance estimate, stator current, torque, and stator flux linkages are tracking the references very closely and experience no oscillation or transient during tracking. Note that the instability problem encountered with no parameter adaptation is solved by using this adaptive compensation scheme.

Step changes in torque command from 1 to 2 p.u. and then from 2 p.u. to −1 p.u. while the stator resistance is being ramped up have no adverse effects on the drive system, as is shown in Figure 8.20. The flux linkages hardly vary during these

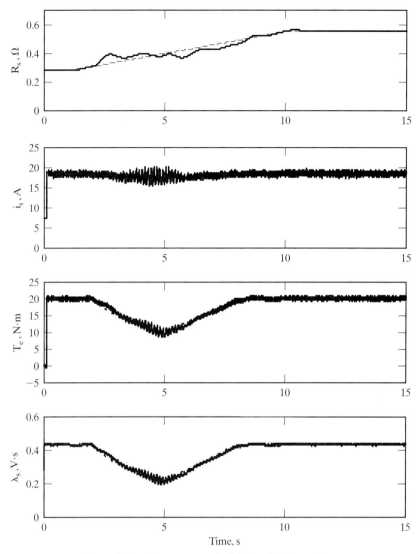

Figure 8.21 Ramp response in field-weakening mode

changes, and torque linearity is perfectly maintained between the torque and its command in this adaptation scheme.

The dynamic performance in the flux-weakening range is shown in Figure 8.21. The stator resistance is ramped from nominal value to twice its nominal value. Stator flux-linkages command and torque command are proportionally decreased from and increased linearly again to original reference values. The tracking of motor variables and stator resistance is achieved with hardly any transients, thus proving the effectiveness of the adaptive controller in the flux-weakening region also.

8.4 DERIVATION OF INDIRECT VECTOR-CONTROL SCHEME

In this section, the indirect vector controller is derived from the dynamic equations of the induction machine in the synchronously rotating reference frames. Reference could be made to the derivation of dynamic equations of the induction machine in Chapter 5. To simplify the derivation, a current-source inverter is assumed. In that case, the stator phase currents serve as inputs; hence, the stator dynamics can be neglected. In turn, that leads to omitting the stator equations from further consideration.

The rotor equations of the induction machine containing flux linkages as variables are given by

$$R_r i_{qr}^e + p\lambda_{qr}^e + \omega_{sl}\lambda_{dr}^e = 0 \tag{8.27}$$
$$R_r i_{dr}^e + p\lambda_{dr}^e - \omega_{sl}\lambda_{qr}^e = 0 \tag{8.28}$$

where

$$\omega_{sl} = \omega_s - \omega_r \tag{8.29}$$
$$\lambda_{qr}^e = L_m i_{qs}^e + L_r i_{qr}^e \tag{8.30}$$
$$\lambda_{dr}^e = L_m i_{ds}^e + L_r i_{dr}^e \tag{8.31}$$

In these equations, the various symbols denote the following: R_r, the referred rotor resistance per phase; L_m, the mutual inductance per phase; L_r, the stator-referred rotor self-inductance per phase; i_{dr}^e and i_{qr}^e, the referred direct and quadrature axes currents, respectively; and p, the differential operator d/dt. ω_{sl} is slip speed in rad/sec, ω_s is electrical stator frequency in rad/sec, ω_r is electrical rotor speed in rad/sec, and λ_{dr}^e and λ_{qr}^e are rotor direct and quadrature axes flux linkages, respectively.

The resultant rotor flux linkage, λ_r, also known as the rotor flux-linkages phasor, is assumed to be on the direct axis, to reduce the number of variables in the equations by one. Moreover, it corresponds with the reality that the rotor flux linkages are a single variable. Hence, aligning the *d* axis with rotor flux phasor yields

$$\lambda_r = \lambda_{dr}^e \tag{8.32}$$
$$\lambda_{qr}^e = 0 \tag{8.33}$$
$$p\lambda_{qr}^e = 0 \tag{8.34}$$

Substituting equations (8.32) to (8.34) in (8.27) and (8.28) causes the new rotor equations to be

$$R_r i_{qr}^e + \omega_{sl}\lambda_r = 0 \tag{8.35}$$
$$R_r i_{dr}^e + p\lambda_r = 0 \tag{8.36}$$

The rotor currents in terms of the stator currents are derived from equations (8.30) and (8.31) as

Section 8.4 Derivation of Indirect Vector-Control Scheme

$$i_{qr}^e = -\frac{L_m}{L_r} i_{qs}^e \qquad (8.37)$$

$$i_{dr}^e = \frac{\lambda_r}{L_r} - \frac{L_m}{L_r} i_{ds}^e \qquad (8.38)$$

Substituting for d and q axes rotor currents from equations (8.37) and (8.38) into equations (8.35) and (8.36), the following are obtained:

$$i_f = \frac{1}{L_m}[1 + T_r p]\lambda_r \qquad (8.39)$$

$$\omega_{sl} = K_{it}\left[\frac{L_r}{T_r}\right]\left[\frac{T_e}{\lambda_r^2}\right] = K_{it} R_r \left[\frac{T_e}{\lambda_r^2}\right] = \frac{L_m}{T_r} \frac{i_T}{\lambda_r} \qquad (8.40)$$

where

$$i_f = i_{ds}^e \qquad (8.41)$$

$$i_T = i_{qs}^e \qquad (8.42)$$

$$T_r = \frac{L_r}{R_r} \qquad (8.43)$$

$$K_{it} = \frac{2}{3}\frac{2}{P} \qquad (8.44)$$

The q and d axes currents are relabeled as torque- (i_T) and flux-producing (i_f) components of the stator-current phasor, respectively. T_r denotes the rotor time constant. The equation (8.39) resembles the field equation in a separately-excited dc machine, whose time constant is usually on the order of seconds. Likewise, that the induction-motor rotor time constant is also on the order of a second is to be noted.

Similarly, by the same substitution of the rotor currents from (8.37) and (8.38) into the torque expression, the electromagnetic torque is derived as

$$T_e = \frac{3}{2}\frac{P}{2}\frac{L_m}{L_r}(\lambda_{dr}^e i_{qs}^e - \lambda_{qr}^e i_{ds}^e) = \frac{3}{2}\frac{P}{2}\frac{L_m}{L_r}(\lambda_{dr}^e i_{qs}^e) = K_{te}\lambda_r i_{qs}^e = K_{te}\lambda_r i_T \qquad (8.45)$$

where the torque constant K_{te} is defined as

$$K_{te} = \frac{3}{2}\frac{P}{2}\frac{L_m}{L_r} \qquad (8.46)$$

Note that the torque is proportional to the product of the rotor flux linkages and the stator q axis current. This resembles the air gap torque expression of the dc motor, which is proportional to the product of the field flux linkages and the armature current. If the rotor flux linkage is maintained constant, then the torque is simply proportional to the torque-producing component of the stator current, as in the case of the separately-excited dc machine with armature current control, where the torque is proportional to the armature current when the field current is constant. Similar to the dc machine armature time constant, which is on the order of a few milliseconds, the time constant of the torque current is proved to be also on the same order in a later section and is equal to the stator-transient time constant. The rotor flux linkages and air gap torque equations given in (8.40) and (8.45), respectively, complete the transformation of the induction machine into an equivalent separately-excited dc machine from a control point of view.

The stator-current phasor is the phasor sum of the d and q axes stator currents in any frames; it is given by

$$i_s = \sqrt{(i_{qs}^e)^2 + (i_{ds}^e)^2} \tag{8.47}$$

and the dq axes to abc phase-current relationship is obtained from

$$\begin{bmatrix} i_{qs}^e \\ i_{ds}^e \end{bmatrix} = \frac{2}{3} \begin{bmatrix} \cos\theta_f & \cos\left(\theta_f - \frac{2\pi}{3}\right) & \cos\left(\theta_f + \frac{2\pi}{3}\right) \\ \sin\theta_f & \sin\left(\theta_f - \frac{2\pi}{3}\right) & \sin\left(\theta_f + \frac{2\pi}{3}\right) \end{bmatrix} \begin{bmatrix} i_{as} \\ i_{bs} \\ i_{cs} \end{bmatrix} \tag{8.48}$$

compactly expressed as

$$i_{qd} = [T][i_{abc}] \tag{8.49}$$

where

$$i_{qd} = [i_{qs}^e \quad i_{ds}^e]^t \tag{8.50}$$

$$i_{abc} = [i_{as} \quad i_{bs} \quad i_{cs}]^t \tag{8.51}$$

$$[T] = \frac{2}{3} \begin{bmatrix} \cos\theta_f & \cos\left(\theta_f - \frac{2\pi}{3}\right) & \cos\left(\theta_f + \frac{2\pi}{3}\right) \\ \sin\theta_f & \sin\left(\theta_f - \frac{2\pi}{3}\right) & \sin\left(\theta_f + \frac{2\pi}{3}\right) \end{bmatrix} \tag{8.52}$$

where i_{as}, i_{bs}, and i_{cs} are the three phase stator currents. Note that the elements in the T matrix are cosinusoidal functions of electrical angle, θ_f. The electrical field angle in this case is that of the rotor flux-linkages phasor and is obtained as the sum of the rotor and slip angles:

$$\theta_f = \theta_r + \theta_{sl} \tag{8.53}$$

and the slip angle is obtained by integrating the slip speed and is given as

$$\theta_{sl} = \int \omega_{sl} dt \tag{8.54}$$

From these derivations, a drive scheme and its phasor diagram are developed.

8.5 INDIRECT VECTOR-CONTROL SCHEME

A vector controller accepts the torque and flux requests and generates the torque- and flux-producing components of the stator-current phasor and the slip-angle, θ_{sl}, commands. The request/command values and the controller-instrumented parameters are denoted with asterisks throughout this text. From equations (8.39), (8.40), and (8.44), the commanded values of i_T, i_f, and ω_{sl} are

$$i_T^* = \frac{T_e^*}{K_{te}\lambda_r^*} = \frac{T_e^*}{\lambda_r^*}\frac{L_r^*}{L_m^*}\left(\frac{2}{3}\right)\left(\frac{2}{P}\right) = K_{it}\left(\frac{T_e^*}{\lambda_r^*}\right)\left(\frac{L_r^*}{L_m^*}\right) \tag{8.55}$$

$$i_f^* = (1 + T_r^* p)\frac{\lambda_r^*}{L_m^*} \tag{8.56}$$

$$\omega_{s1}^* = K_{it}\left[\frac{L_r^*}{T_r^*}\right]\left[\frac{T_e^*}{(\lambda_r^*)^2}\right] = K_{it}R_r^*\frac{T_e^*}{(\lambda_r^*)^2} = \frac{L_m^*}{T_r^*}\frac{i_T^*}{\lambda_r^*} \tag{8.57}$$

The command slip angle, θ_{s1}^* is generated by integrating ω_{s1}^*. The torque-angle command is obtained as the arctangent of i_T^* and i_f^*. The field angle is obtained by summing the command slip angle and rotor angle. With the torque- and flux-producing components of the stator-current commands and rotor field angle, the *qd* axes current commands (and, hence, *abc* phase-current commands) are obtained as follows. The transformation from the flux- and torque-producing currents to the *qd* axes current commands are derived from the phasor diagram shown in Figure 8.1. The relevant steps involved in the realization of the indirect vector controller are as follows:

$$\begin{bmatrix} i_{qs}^* \\ i_{ds}^* \end{bmatrix} = \begin{bmatrix} \cos\theta_f & \sin\theta_f \\ -\sin\theta_f & \cos\theta_f \end{bmatrix}\begin{bmatrix} i_T^* \\ i_f^* \end{bmatrix} \tag{8.58}$$

and

$$\begin{bmatrix} i_{as}^* \\ i_{bs}^* \\ i_{cs}^* \end{bmatrix} = [T^{-1}]\begin{bmatrix} i_{qs}^* \\ i_{ds}^* \\ 0 \end{bmatrix} \tag{8.59}$$

where

$$T^{-1} = \begin{bmatrix} 1 & 0 & 1 \\ -\frac{1}{2} & -\frac{\sqrt{3}}{2} & 1 \\ -\frac{1}{2} & \frac{\sqrt{3}}{2} & 1 \end{bmatrix} \tag{8.60}$$

By using equations (8.58) to (8.60), the stator *q* and *d* axes and *abc* current commands are derived as

$$\begin{aligned} i_{qs}^* &= |i_s^*|\sin\theta_s^* \\ i_{ds}^* &= |i_s^*|\cos\theta_s^* \\ i_{as}^* &= |i_s^*|\sin\theta_s^* \\ i_{bs}^* &= |i_s^*|\sin\left(\theta_s^* - \frac{2}{3}\pi\right) \\ i_{cs}^* &= |i_s^*|\sin\left(\theta_s^* + \frac{2}{3}\pi\right) \end{aligned} \tag{8.61}$$

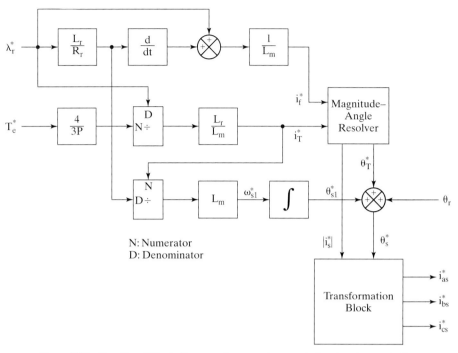

Figure 8.22 Functional block diagram of a current-source indirect vector controller

where

$$\theta_s^* = \theta_f + \theta_T^* = \theta_r + \theta_{s1}^* + \theta_T^* \qquad (8.62)$$

The current commands are enforced by the inverter with suitable control techniques outlined elsewhere. A realization of the scheme in block-diagram form is shown in Figure 8.22. The flowchart for the generation of control variables in the indirect vector-control scheme is given in Figure 8.23. An implementation of the indirect vector-control scheme on an inverter-fed induction motor is discussed in the next section.

8.6 AN IMPLEMENTATION OF AN INDIRECT VECTOR-CONTROL SCHEME

An implementation of the indirect vector-controlled induction motor is shown in Figure 8.24. The torque command is generated as a function of the speed error signal, generally processed through a PI controller. The flux command for a simple drive strategy is made to be a function of speed, defined by

$$\begin{aligned}\lambda_r^* &= \lambda_b; \quad 0 \leq \pm\omega_r \leq \pm\omega_{\text{rated}} \\ &= \frac{\omega_b}{|\omega_r|}\lambda_b; \quad \pm\omega_b \leq \pm\omega_r \leq \pm\omega_{r(\max)} \text{ and } |\omega_r| \geq \omega_b\end{aligned} \qquad (8.63)$$

Section 8.6 An Implementation of an Indirect Vector-Control Scheme 451

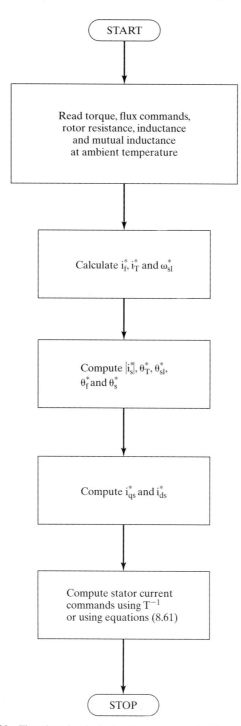

Figure 8.23 Flowchart for the indirect vector-controlled induction motor

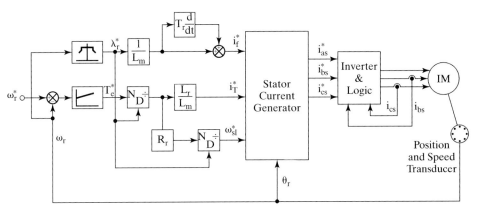

Figure 8.24 The implementation of the indirect vector-controlled induction motor servo

where λ_b and ω_b are the rated or base rotor flux linkages and rotor speed, respectively. The flux is kept at rated value up to rated speed; above that, the flux is weakened to maintain the power output at a constant, very much as in the dc motor drive. In such a case, the rotor flux-linkages programming is a complex task; it is considered in detail in section 8.10. By following the flow chart given in Figure 8.15, the three-phase stator-current commands are generated. These current commands are simplified through a power amplifier, which can be any standard converter–inverter arrangement discussed in the earlier chapters. For small motor drives (of less than 25-hp rating), the power amplifier is of the type shown in Figure 8.8. The regenerative energy is dumped in the brake resistor in this power circuit. The output of the inverter is connected to the induction-motor stator terminals. The third-phase current is reconstructed from the two measured phase currents if the system does not have a grounded neutral.

The rotor position, θ_r, is measured with an encoder/synchronous resolver and converted into necessary digital information for feedback. Some transducers are currently available to convert the rotor position information into velocity; they can be used to eliminate a tachogenerator to obtain the velocity information. The controllers are implemented with microprocessors; a vast amount of literature is available for reference on this topic. Next in order are the tuning considerations for the indirect vector controller.

Example 8.3

Prove that the transfer function between the rotor flux linkage and its command is unity; from that, prove that the transfer relationship between the torque and its command is unity in the indirect vector controller for the induction machine. Assume that the machine and controller parameters match and that there is no time delay between the currents and their commands.

Solution From the flux-linkage equation derived from the rotor equation, the rotor flux linkage is given as

Section 8.6 An Implementation of an Indirect Vector-Control Scheme

$$\lambda_r(s) = L_m \frac{i_f(s)}{(1 + sT_r)}$$

but the flux-producing component of the stator current, i_f, is equal to its command value, i_f^*, because there is no time delay and the vector controller enforces the command faithfully through the inverter, which then is written as

$$i_f = i_f^*$$

The flux-current command obtained from the vector controller is given by

$$i_f^*(s) = \frac{(1 + sT_r^*)}{L_m^*} \lambda_r^*(s)$$

From these equations, the flux-producing component of the stator-current phasor is substituted in terms of the flux linkage command and motor parameters as

$$\lambda_r(s) = \frac{L_m}{L_m^*} \frac{(1 + sT_r^*)}{(1 + sT_r)} \lambda_r^*(s)$$

As the motor and controller parameters match, the rotor flux linkages and its command become equal, giving the transfer function between the rotor flux linkage and its command as unity. Then the transfer relationship between the electromagnetic torque and its command is derived as follows:

$$T_e = K_{te}\lambda_r i_T = K_{te}\lambda_r i_T^* = K_{te}\lambda_r \frac{T_e^*}{K_{te}^* \lambda_r^*}$$

As the motor and controller parameters match, the torque constant for the controller and the machine become the same, and the ratio between the rotor flux linkage and its command is unity. That makes the torque equal to its command value, resulting in a unity transfer relationship between them.

Example 8.4

Derive the time constants associated with the flux- and torque-producing channels in an indirect vector-controlled induction motor drive.

Solution

(i) Flux channel: From the equation of the rotor flux linkages, it is seen that the time constant of the rotor flux-linkages channel is the rotor time constant, denoted by T_r. Its value is on the order of seconds, like the field constant in the separately-excited dc machine.

(ii) Torque channel: To derive the time constant in this channel, it is assumed that the rotor flux-linkages phasor is maintained constant. Then the torque is influenced only by the torque-producing component of the stator-current phasor, i_T. Therefore, the time constant associated with this channel will be the time constant involved in the transfer function between this current and the corresponding voltage, which is the quadrature axis voltage in synchronous reference frames. It is derived as follows.

The stator quadrature axis voltage in synchronous frames is

$$v_{qs}^e = (R_s + L_s p)i_{qs}^e + \omega_s L_s i_{ds}^e + L_m p i_{qr}^e + \omega_s L_m i_{dr}^e$$
$$= R_s i_{qs}^e + p\lambda_{qs}^e + \omega_s \lambda_{ds}^e$$

where

$$\lambda_{qs}^e = L_s i_{qs}^e + L_m i_{qr}^e$$
$$\lambda_{ds}^e = L_s i_{ds}^e + L_m i_{dr}^e$$

Substituting for the rotor q and d axes currents in terms of the stator q and d axes currents and rotor flux linkages, the d and q axes stator flux linkages are

$$\lambda_{qs}^e = \frac{L_s L_r - L_m^2}{L_r} i_{qs}^e = \sigma L_s i_{qs}^e$$

$$\lambda_{ds}^e = L_s i_{ds}^e + L_m \frac{\lambda_r - L_m i_{ds}^e}{L_r} = \frac{L_m}{L_r}\lambda_r + \frac{L_s L_r - L_m^2}{L_r} i_{ds}^e$$

where σ is the leakage coefficient of the induction machine, defined as

$$\sigma = 1 - \frac{L_m^2}{L_s L_r}$$

but in steady state, the rotor flux-linkages phasor is given by

$$\lambda_r = L_m i_{ds}^e$$

which, upon substitution, reduces the d axis stator flux linkages:

$$\lambda_{ds}^e = L_s i_{ds}^e$$

resulting in the q axis voltage:

$$v_{qs}^e = R_s i_{qs}^e + \sigma L_s p i_{qs}^e + \omega_s L_s i_{ds}^e$$

This gives the time constant in the q axis current path, i.e., i_T path, which is the torque channel in the induction machine:

$$\tau_T = \sigma \frac{L_s}{R_s} = \sigma \tau_s = \tau_s' = \text{stator–transient time constant}$$

Note that it is of the same order, a few ms, as the armature time constant in dc machines.

8.7 TUNING OF THE VECTOR CONTROLLER

The tuning of the vector controller requires the exact values of rotor resistance, mutual inductance, and rotor self-inductance of the induction machine. Implementing these values in the vector controller with proper normalization for the torque and flux commands completes the tuning task. The tuning task is simple if the motor parameters remain constant. The fact that the rotor resistance changes with temperature and frequency and the leakage inductance changes with the magnitude of the stator current complicates the tuning problem. The natural question is, then, what values are to be instrumented in the vector controller? Changes in the rotor resistance are of paramount importance. Hence, it should be possible to have the value of rotor resistance adjusted or to keep it as a nominal value (most usually a value corresponding to some steady-state operating point). The need for parameter compensation becomes apparent when the consequences of parameter changes are studied with the vector controller having fixed parameters. This is taken up in later sections.

Example 8.5

An induction motor with the following data is to be used with an indirect vector controller:

5 hp, Y − connected, 3 ϕ, 60 Hz, 4 poles, 200 V, $R_s = 0.277\ \Omega$, $R_r = 0.183\ \Omega$, $L_m = 0.0538$ H, $L_r = 0.05606$ H, $L_s = 0.0553$ H, $J = 0.01667$ kg-m^2, Rated speed = 1766.9 rpm.

Find the rated rotor flux linkages and torque commands and the corresponding flux- and torque-producing components of the stator-current command, the stator-current phasor command, the torque-angle command, and the slip-speed command. The drive is assumed to be a torque amplifier.

Solution The q and d axes voltages in synchronous reference frames are

$$v_{qs}^e = V_s \frac{\sqrt{2}}{\sqrt{3}} = 163.3\ \text{V and } v_{ds}^e = 0$$ as it is equal to the peak value of phase a voltage.

The rated speed is 1766.9 rpm; in mechanical radians, it is

$$\omega_m = \frac{2\pi N_r}{60} = 185.029\ \text{rad/sec}$$

The steady-state flux linkages are evaluated from the steady-state currents; they, in turn, are found by using synchronous reference frame equations with the substitution of p = 0 and with the slip speed being zero. (Refer to explanations in Chapter 5.) Because the slip speed is zero, the machine does not produce electromagnetic torque; thus the stator currents are utilized to produce solely the stator and rotor flux linkages.

$$\begin{bmatrix} I_{qs}^e \\ I_{ds}^e \\ I_{qr}^e \\ I_{dr}^e \end{bmatrix} = \begin{bmatrix} R_s & \omega_s L_s & 0 & \omega_s L_m \\ -\omega_s L_s & R_s & -\omega_s L_m & 0 \\ 0 & \omega_{sl} L_m & R_r & \omega_{sl} L_r \\ -\omega_{sl} L_m & 0 & -\omega_{sl} L_r & R_r \end{bmatrix}^{-1} \begin{bmatrix} v_{qs}^e \\ v_{ds}^e \\ 0 \\ 0 \end{bmatrix} = \begin{bmatrix} 0.104 \\ 7.832 \\ 0 \\ 0 \end{bmatrix}$$

Electromagnetic torque under this condition is,

$$T_e = \frac{3}{2} \cdot \frac{P}{2} \cdot L_m (I_{qs}^e \cdot I_{dr}^e - I_{ds}^e \cdot I_{qr}^e) = 0\ \text{N·m}$$

Rotor flux linkages are,

$$\lambda_{qr}^e = L_m I_{qs}^e + L_r I_{qr}^e = 0.056\ \text{Wb–Turn}$$
$$\lambda_{dr}^e = L_m I_{ds}^e + L_r I_{dr}^e = 0.4213\ \text{Wb–Turn}$$

The resultant rotor flux linkage is,

$$\lambda_r = \sqrt{(\lambda_{qr}^e)^2 + (\lambda_{dr}^e)^2} = 0.4214\ \text{Wb–Turn}$$

It is instructive to note the difference between the stator flux-linkage magnitude and that of its rotor counterpart from the following calculations:

$$\lambda_{qs}^e = L_s I_{qs}^e + L_m I_{qr}^e = 0.0058 \text{ Wb-Turn}$$
$$\lambda_{ds}^e = L_s I_{ds}^e + L_r I_{dr}^e = 0.4331 \text{ Wb-Turn}$$

and the stator flux linkage is

$$\lambda_s = \sqrt{(\lambda_{qs}^e)^2 + (\lambda_{ds}^e)^2} = 0.4331 \text{ Wb-Turn}$$

which is approximately 2.8% higher than the rotor flux linkages

The stator-current phasor magnitude is

$$I_s = \sqrt{(I_{ds}^e)^2 + (I_{qs}^e)^2} = \sqrt{(0.104)^2 + (7.832)^2} = 7.832 \text{ A} = I_f$$

which is equal to the flux-producing stator current in the machine and is denoted as I_f. Note that this is peak value and not rms value. The friction and windage losses are not given, so they can be neglected. Therefore, the electromagnetic torque is equal to shaft torque; its rated value is obtained as

$$T_e = \frac{P_o}{\omega_m} = \frac{5*745.6}{185.029} = 20.178 \text{ N·m}$$

The torque constant K_{te} is

$$K_{te} = \frac{3}{2} \cdot \frac{P}{2} \cdot \frac{L_m}{L_r} = 2.88$$

and, by using this, the torque-producing component of the stator current and the stator-current phasor are obtained as

$$I_t = \frac{T_e}{K_{te}\lambda_r} = \frac{20.178}{2.882 \times 0.4213} = 16.63 \text{ A}$$

$$I_s = \sqrt{I_f^2 + I_t^2} = 18.38 \text{ A}$$

The torque angle is

$$\theta_T = \tan^{-1}\left\{\frac{I_t}{I_f}\right\} = 1.13 \text{ rad}$$

The slip speed is verified from the above as

$$\omega_{sl} = \frac{R_r}{L_r} \cdot \frac{I_t}{I_f} = \frac{0.183}{0.056} \cdot \frac{16.63}{7.832} = 6.932 \text{ rad/sec}$$

The actual variables will equal their command values in steady state. Therefore, the computed variable values are the respective command values for rated operation.

Example 8.6

Consider Example 8.5. Compute the stator-voltage magnitude at rated operating conditions with rated rotor flux linkages, and comment on the finding.

Solution The steady-state conditions, determined in Example 8.5 to produce rated torque at rated speed and rated rotor flux linkages, are

$$I_T = I_{qs}^e = 16.63 \text{ A}$$

$$I_f = I_{ds}^e = 7.832 \text{ A}$$

$$\omega_{sl} = 6.932 \text{ rad/sec}$$

From the steady-state rotor equations, the rotor currents are found as follows:

$$\begin{bmatrix} I_{qr}^e \\ I_{dr}^e \end{bmatrix} = \omega_{sl} L_m \begin{bmatrix} R_r & \omega_{sl} L_r \\ -\omega_{sl} L_r & R_r \end{bmatrix}^{-1} \begin{bmatrix} -I_{ds}^e \\ I_{qs}^e \end{bmatrix}$$

$$I_{qr}^e = -15.96 \text{ A}$$

$$I_{dr}^e = 0$$

Then the stator voltages are computed from the stator steady-state equations as

$$V_{qs}^e = R_s I_{qs}^e + \omega_s L_s I_{ds}^e + \omega_s L_m I_{dr}^e = 167.85 \text{ V}$$

$$V_{ds}^e = R_s I_{ds}^e - \omega_s L_s I_{qs}^e - \omega_s L_m I_{qr}^e = -20.84 \text{ V}$$

The magnitude of stator voltage is

$$V_s = \sqrt{(V_{qs}^e)^2 + (V_{ds}^e)^2} = 169.18 \text{ V}$$

which, in terms of the line-to-line voltage, is given by

$$V = \frac{\sqrt{3}}{\sqrt{2}} V_s = 207.2 \text{ V}$$

It is 7.2 V higher than the rated value. Therefore, to operate within rated voltage and to deliver rated power, the rotor flux linkage has to be decreased. Flux-weakening is treated in a later section.

8.8 FLOWCHART FOR DYNAMICS COMPUTATION

It is useful to determine the transient response of the indirect vector-controlled induction motor, in order to evaluate its application to a specific requirement. Also, such a study is a vehicle for avoiding the costly process of prototype construction and testing. The guidelines adopted for the flowchart are that the machine and vector controller parameters are specified along with the inverter and switching logic. Note that any converter can be used in the simulation. The system equations are assembled by starting from the speed controller, vector controller, inverter, switching logic, and the machine with the currents, speed, and position feedbacks. The speed command and torque disturbance need to be specified to evaluate the response. The nonlinear equations of the machine are used in the simulation, and stator dynamics are included. The flowchart is given in Figure 8.25.

458 Chapter 8 Vector-Controlled Induction Motor Drives

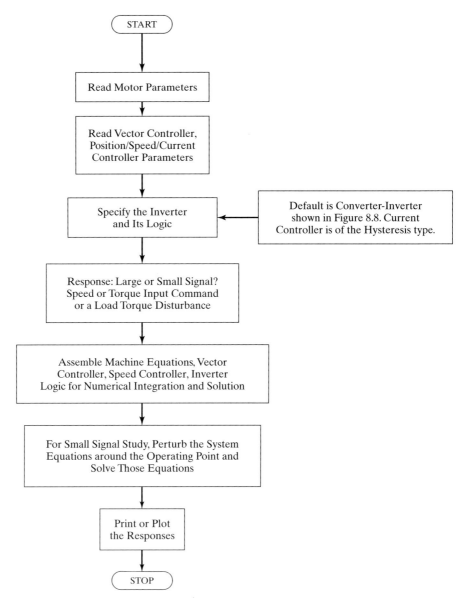

Figure 8.25 Flowchart for dynamic simulation of the vector controlled induction motor drive

8.9 DYNAMIC SIMULATION RESULTS

A functional block diagram of an indirect vector-controlled induction motor drive with dq axes current control is shown in Figure 8.26. The phase-current feedback loops are substituted with the qd axes current-feedback loops, thus enabling the individual design of the respective current controllers, considering their different time constants dominating the torque and flux responses. This is the distinct advantage of such a configuration.

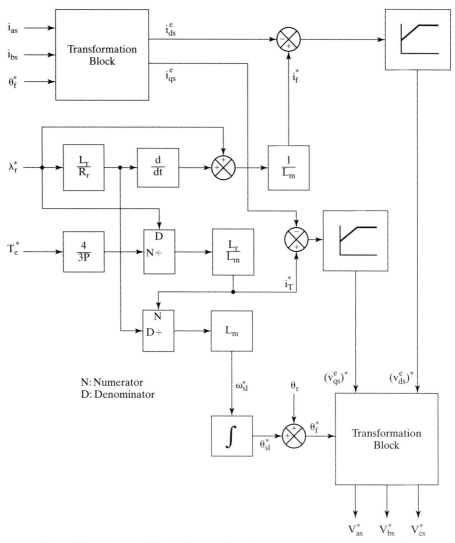

Figure 8.26 Functional block diagram of a voltage-source indirect vector controller

A 0.25-p.u. bidirectional-step speed reference is given; its response is shown in Figure 8.27. The electromagnetic torque command is limited to 2 p.u., and rotor flux linkage is at rated value at the time of application of the speed command. The response of the torque is almost instantaneous, and the speed response is very smooth without ripples. Further, the drive goes through the first-, fourth-, and third-quadrant operation without any current, flux, and torque oscillations. The drive uses PWM voltage control at the final stage of the controller.

The performance of the drive system for a sinusoidal speed reference is shown in Figure 8.28. Small-signal sinusoidal response is helpful to determine the speed loop bandwidth of the system. For a low-frequency sinusoidal command, note that the response is a true replica of the reference itself. For higher frequencies of the reference, both attenuation and phase shift occur in the responses.

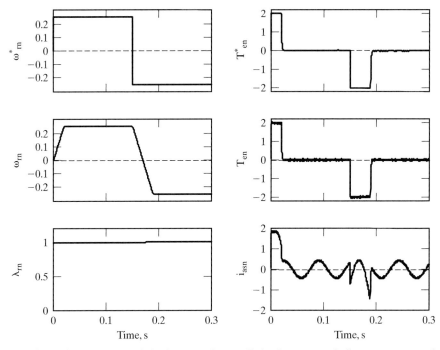

Figure 8.27 Output characteristics for a speed-controlled voltage-source indirect vector-control scheme in p.u.

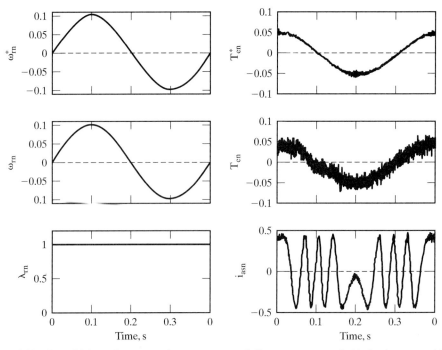

Figure 8.28 Sinusoidal speed response of a current source indirect vector-controlled induction motor drive in p.u.

8.10 PARAMETER SENSITIVITY OF THE INDIRECT VECTOR-CONTROLLED INDUCTION MOTOR DRIVE

A mismatch between the vector controller and induction motor occurs as a result either of the motor parameters changing with operating conditions such as temperature rise and saturation or of the wrong instrumentation of the parameters in the vector controller. The latter phenomenon is controllable, but the former is dependent on the operating conditions of the motor drive and hence is uncontrollable. The mismatch produces a coupling between the flux- and torque-producing channels in the machine. This has the following consequences:

(i) The rotor flux linkages deviate from the commanded value.
(ii) The electromagnetic torque, hence, deviates from its commanded value, producing a nonlinear relationship between the actual torque and its commanded value.
(iii) During torque transients, an oscillation is caused both in the rotor flux linkages and in torque responses, with a settling time equal to the rotor time constant. The rotor time constant is large: on the order of 0.5 second or greater.

In a torque drive, consequences (ii) and (iii) are most undesirable. Although, in the speed-controlled drive, the nonlinear torque-to-torque command characteristic will not have a detrimental effect on the steady-state operation, its effect is considerable during the transients. The load and motor inertia are required to smooth these torque excursions so that they do not appear as speed ripples. For the present, the parameter sensitivity effects are quantified in the steady state and the transient state, considering both open and closed outer speed loop.

8.10.1 Parameter Sensitivity Effects When the Outer Speed Loop Is Open

With the outer speed loop open in the vector-controlled induction motor drive, the external commands are the rotor flux linkages and electromagnetic torque. The actual slip speed and commanded value are equal in this mode of the drive. Therefore, the mismatch between the vector controller and induction motor induces deviations in the flux- and torque-producing stator-current components and hence in the torque angle. The net effects are the deviations in the magnitude of the rotor flux linkages and electromagnetic torque. They are derived in the following sections.

8.10.1.1 Expression for electromagnetic torque. In the steady state, an expression for the rotor flux linkages is obtained by substituting $p = 0$ in equation (8.56):

$$\lambda_r^* = L_m^* i_f^* \tag{8.64}$$

Substituting equation (8.64) into equation (8.57) gives the command value of the slip speed as

$$\omega_{sl}^* = \left[\frac{1}{T_r^*}\right]\left[\frac{i_T^*}{i_f^*}\right] \tag{8.65}$$

462   Chapter 8   Vector-Controlled Induction Motor Drives

The torque-angle command is known to be

$$\theta_T^* = \tan^{-1}\left[\frac{i_T^*}{i_f^*}\right] \tag{8.66}$$

Substituting equation (8.66) into (8.65) cause the torque-angle-to-slip-speed relationship to emerge as

$$\tan \theta_T^* = \omega_{s1}^* T_r^* \tag{8.67}$$

from which the sine and cosine of the torque angle are found as

$$\sin \theta_T^* = \frac{\omega_{s1}^* T_r^*}{\sqrt{1 + (\omega_{s1}^* T_r^*)^2}} \tag{8.68}$$

$$\cos \theta_T^* = \frac{1}{\sqrt{1 + (\omega_{s1}^* T_r^*)^2}} \tag{8.69}$$

The command value of the torque, from related equations, could be written as

$$T_e^* = K_{te}\lambda_r^* i_T^* = K_{te}L_m^* i_T^* i_f^* = \frac{1}{K_{it}}\frac{(L_m^*)^2}{L_r^*}(i_s^*)^2 \cos \theta_T^* \sin \theta_T^* \tag{8.70}$$

Likewise, the actual electromagnetic torque generated in the motor is expressed as

$$T_e = \frac{1}{K_{it}}\frac{L_m^2}{L_r} i_s^2 \cos \theta_T^* \sin \theta_T^* \tag{8.71}$$

because the command and actual torque angles are equal in steady state. In the torque mode, the constraints in force are

$$i_s = i_s^* \tag{8.72}$$

$$\omega_{s1} = \omega_{s1}^* \tag{8.73}$$

Substituting equations (8.67), (8.68), (8.72), and (8.73) into (8.70) and (8.71), the ratio of torque to its command value is obtained:

$$\frac{T_e}{T_e^*} = \left[\frac{L_m}{L_m^*}\right]^2\left[\frac{L_r^*}{L_r}\right]\left[\frac{T_r}{T_r^*}\right]\left[\frac{1 + (\omega_{s1}^* T_r^*)^2}{1 + (\omega_{s1}^* T_r)^2}\right] \tag{8.74}$$

Defining the nondimensional variables between the motor (actual) and controller-instrumented rotor time constants and magnetizing inductances to generalize the results and their interpretation, we get

$$\alpha = \frac{T_r}{T_r^*} \tag{8.75}$$

$$\beta = \frac{L_m}{L_m^*} \tag{8.76}$$

and making the approximation that the leakage inductance is negligible compared to the mutual inductance leads to the elimination of one more parameters, resulting in a simpler and compact representation of the parameter sensitivity effects:

Section 8.10 Parameter Sensitivity of the Indirect Vector-Controlled Induction

$$\frac{L_r^*}{L_r} \cong \frac{L_m^*}{L_m} = \frac{1}{\beta} \tag{8.77}$$

Substituting equations from (8.75) to (8.77) into equation (8.74) gives the actual-to-command torque value in terms of the normalized values α and β as

$$\frac{T_e}{T_e^*} = \alpha\beta \left[\frac{1 + (\omega_{sl}^* T_r^*)^2}{1 + (\alpha \omega_{sl}^* T_r^*)^2} \right] \tag{8.78}$$

Note that the factors α and β embody the aspects of controller mismatch, saturation, and machine temperature, as reflected in the rotor resistance.

8.10.1.2 Expression for the rotor flux linkages.
The parameter variations affect primarily the rotor flux linkages. The change in steady-state rotor flux linkages gives an indication of the change in the electromagnetic torque. In steady state, the actual and the commanded rotor flux linkages are

$$\lambda_r^* = L_m^* i_f^* \tag{8.79}$$

$$\lambda_r = L_m i_f \tag{8.80}$$

Substituting for i_f in terms of stator current and finding the ratio of the actual to command rotor flux linkages gives

$$\frac{\lambda_r}{\lambda_r^*} = \left[\frac{L_m}{L_m^*}\right]\left[\frac{\cos\theta_T}{\cos\theta_T^*}\right] = \left[\frac{L_m}{L_m^*}\right]\sqrt{\frac{1 + (\omega_{sl}^* T_r^*)^2}{1 + (\alpha\omega_{sl}^* T_r^*)^2}} = \beta\sqrt{\frac{1 + (\omega_{sl}^* T_r^*)^2}{1 + (\alpha\omega_{sl}^* T_r^*)^2}} \tag{8.81}$$

A set of normalized curves between the torque ratio, flux ratio, and α for various slip speeds and saturation levels β are presented in the following section.

8.10.1.3 Steady-state results.
The rationale for choosing the range of values for α and β is discussed before the results are presented.

Ranges of α and β The practical temperature excursion of the rotor is approximately 130°C above ambient. This increases the rotor resistance by 50% over its ambient or nominal value. Magnetic saturation can decrease the self-inductance to 80% of its nominal value. Hence, the lower limit of α can be obtained from the following:

$$T_r = \frac{0.8 L_m^*}{1.5 R_r^*} \cong 0.533 T_r^* \tag{8.82}$$

Since $\alpha = T_r/T_r^*$, the lowest value of α is approximately 0.5. The upper limit of α approaches 1.5, depending primarily on errors in the instrumented vector controller and an increase in rotor self-inductance because of unsaturated operating point on the BH material characteristics of the laminations during flux weakening. Hence, the range of value for α is chosen to be

$$0.5 < \alpha < 1.5 \tag{8.83}$$

A typical value of β in the magnetic saturation region is around 0.8, and operation of the induction motor drive in the linear portion of the iron B-H characteristics

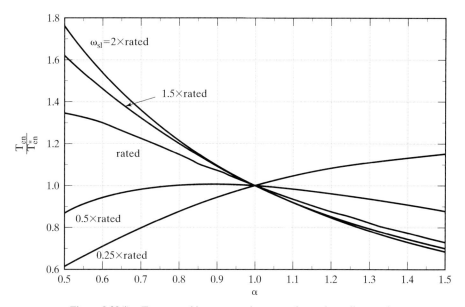

Figure 8.29(i) Torque and its command versus α for various slip speeds

increases β to 1.2. Note that these values are given for stator currents not exceeding twice the rated value. Hence, the range of values for β is chosen as

$$0.8 < \beta < 1.2 \tag{8.84}$$

Steady-State Torque and Flux The steady-state responses for commands of torque and rotor flux linkages as a function of α are shown in Figures 8.29(i) and 8.29(ii), respectively. In these figures, the saturation factor β is maintained at 1 while the slip speed is varied from rated to three times rated value. For α < 1, mainly signifying the increase in rotor temperature, the rotor flux linkages and the electromagnetic torque are greater than their command values. Though the rotor flux is shown to be increasing to value exceeding 1.2 p.u., it is difficult to maintain this high value, because of saturation. Increasing motor saturation at ambient temperature increases α; for such operating points, the torque and rotor flux are less than their commanded values. Note that α increases when the induction motor drive is started at a temperature less than the assumed ambient temperature that usually is used to set the nominal vector-controller values. The same figures also delineate the characteristics resulting from a mismatch between the instrumented-controller parameters and the actual value of the motor parameters.

The linearity of the input–output torque relationship has degraded, and this induction motor drive is unsuitable for use in torque-control applications. A parameter compensator is essential to achieve a linear torque amplifier. High-performance applications, such as robotic drives, require precise control of torque; the torque takes precedence over the speed loop and comes directly after the position loop. Those applications can elude the uncompensated induction motor drive. Saturation affects

Section 8.10   Parameter Sensitivity of the Indirect Vector-Controlled Induction   465

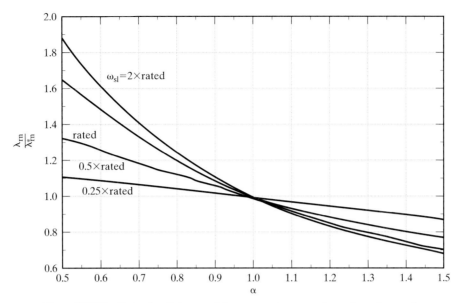

Figure 8.29(ii)   Rotor flux linkage and its command versus α for various slip speeds

the torque and flux proportionally, as is seen from equations (8.78) and (8.81); hence, it will not be elaborated any further.

In general, increasing temperature causes saturation and an increase in the output torque. However, for torque commands less than 0.5 p.u., output torque is reduced.

Example 8.7

The induction motor given in Example 8.4 was run in the torque mode. The ratio of rotor flux linkage to its command value was observed to be 1.1. The pertinent data available for the operating point are

$$\alpha = 0.6$$
$$\beta = 0.9$$

Find the ratio of actual torque to its command value, the slip-speed command, and the torque command, if $\lambda_r^* = 0.4$ Wb-turns.

Solution

$$\frac{\lambda_r}{\lambda_r^*} = 1.1 = \beta\sqrt{\frac{1 + (\omega_{sl}^* T_r^*)^2}{1 + (\alpha \omega_{sl}^* T_r^*)^2}} = 0.9\sqrt{\frac{1 + (0.3063\omega_{sl}^*)^2}{1 + (0.6 \times 0.3063\omega_{sl}^*)^2}}$$

Rearrange this and solve for ω_{sl}^*; $\omega_{sl}^* = 3.375$ rad/s and

$$\frac{T_e}{T_e^*} = \alpha\beta\left[\frac{1 + (\omega_{sl}^* T_r^*)^2}{1 + (\alpha \omega_{sl}^* T_r^*)^2}\right] = 0.6 \times 0.9\left[\frac{1 + (0.3063 \times 3.375)^2}{1 + (0.6 \times 0.3063 \times 3.375)^2}\right] = 0.806$$

$$\omega_{sl}^* = \frac{L_m^* i_T^*}{T_r^* \lambda_r^*}$$

from which the torque current and torque command are obtained as

$$i_T^* = \frac{\omega_{sl}^* \lambda_r^* T_r^*}{L_m^*} = \frac{3.375 \times 0.4 \times 0.3063}{0.0538} = 7.696 \text{ A}$$

$$T_e^* = K_t \lambda_r^* i_T^* = \frac{3}{2} \frac{P}{2} \frac{L_m^*}{L_r^*} \lambda_r^* i_T^* = \frac{3}{2} \times \frac{4}{2} \times \frac{0.0538}{0.05606} \times 0.4 \times 7.696 = 8.862 \text{ N·m}$$

8.10.1.4 Transient characteristics. Simulation results for the dynamic operation of the parameter-sensitive induction motor drive are obtained by solving the rotor equations given in (8.27) and (8.28) and the electromechanical equation given by

$$J \frac{d\omega_r}{dt} = \frac{P}{2}(T_e - T) \tag{8.85}$$

where J is the moment of inertia and T is the load torque.

The effect of switching delays in the control logic of the inverter and in the inverter itself is considered to be negligible. The stator dynamics are omitted, as they can be for a high-performance transistorized inverter with a 5 kHz current loop bandwidth. Experimental results have confirmed the accuracy of the above assumptions [31].

It can be shown that the flux and torque have a time constant equal to the rotor time constant when the rotor flux linkage is allowed to vary and the natural frequency of oscillation of the system corresponds with the slip speed. The system damping factors are determined by many factors, including α and β. Different behavior has been observed for the motoring and the regenerating operation of the drive on the open-speed loop. A sample simulation result is shown in Figure 8.30 for a stator temperature rise of 100°C. This figure contains the torque command, actual torque, and the rotor flux linkages for a shaft speed of 300 rpm. Note that the rotor flux linkages have not been modified for saturation. It is symbolic of the behavior for unsaturated conditions in the motor drive.

Although the developed flux and torque increase for rising temperature, the oscillatory responses are undesirable and make the induction motor drive a non-ideal torque amplifier. The transient performance of the indirect vector-controlled induction motor drive is derived analytically in the following for both the rotor flux linkages and the torque responses. These derivations demonstrate the validity of the statements in the preceding discussion and help us to understand the parameters influencing the transient responses. A time delay, τ_c, involved in the currents by stator dynamics, is also incorporated, along with the parameter mismatch between the controller and motor.

Rotor Flux Linkages Response Considering parameter sensitivity in the indirect vector-controlled induction motor drive and a time lag of τ_c between the currents and their commands, the transfer function between the rotor flux linkage and its command is derived. From the steady-state-parameter sensitivity effects on the rotor flux linkages, the ratio between the current and its command is derived as

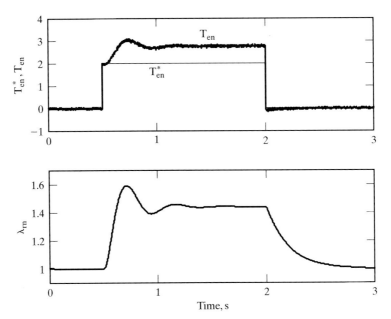

Figure 8.30 Simulation results for torque command at a stator temperature 100°C above ambient

$$\frac{i_f}{i_f^*} = \frac{i_s \cos \theta_T}{i_s^* \cos \theta_T^*} \quad (8.86)$$

The torque angle and slip speed are enforceable instantaneously, but not the current magnitude (because of the stator dynamics of the induction machine). The above relationship can be written in terms of rotor time constant and slip speed as

$$\frac{i_f}{i_f^*} = \frac{i_s}{i_s^*}\sqrt{\frac{1 + (\omega_{sl}T_r^*)^2}{1 + (\alpha\omega_{sl}T_r^*)^2}} \quad (8.87)$$

Considering the time lag between the stator current and its command, its transfer function is written as

$$\frac{I_f(s)}{I_f^*(s)} = \frac{1}{(1 + s\tau_c)}\sqrt{\frac{1 + (\omega_{sl}T_r^*)^2}{1 + (\alpha\omega_{sl}T_r^*)^2}} \quad (8.88)$$

from which the transfer function between the rotor flux linkage and its command is derived as

$$\frac{\lambda_r(s)}{\lambda_r^*(s)} = \beta\sqrt{\frac{1 + (\omega_{sl}T_r^*)^2}{1 + (\alpha\omega_{sl}T_r^*)^2}} \frac{(1 + sT_r^*)}{(1 + s\tau_c)(1 + s\alpha T_r^*)} \quad (8.89)$$

The rotor flux linkage, including the initial condition, is derived as

$$\lambda_r(s) = \beta\sqrt{\frac{1 + (\omega_{sl}T_r^*)^2}{1 + (\alpha\omega_{sl}T_r^*)^2}} \frac{\{(1 + sT_r^*)\lambda_r^*(s) - T_r^*\lambda_r(0)\}}{(1 + s\tau_c)(1 + s\alpha T_r^*)} + \frac{\alpha T_r^*}{(1 + s\alpha T_r^*)}\lambda_r(0) \quad (8.90)$$

For a step command of rotor flux linkages with a magnitude of λ_r^*, the time response is obtained from the above as

$$\lambda_r(t) = \beta\sqrt{\frac{1+(\omega_{s1}T_r^*)^2}{1+(\alpha\omega_{s1}T_r^*)^2}}\left[\lambda_r^* + \frac{1}{\tau_c - \alpha T_r^*}[(T_r^* - \tau_c)\lambda_r^* - T_r^*\lambda_r^*(0)]e^{-\frac{t}{\tau_c}}\right.$$
$$\left. + T_r^*[\lambda_r^*(0) + \lambda_r^*(\alpha - 1)]e^{-\frac{t}{\alpha T_r^*}}\right] + \lambda_r(0)e^{-\frac{t}{\alpha T_r^*}} \quad (8.91)$$

The derivative term in the flux command path causes a large flux-producing component of the stator-current command to be generated, which is usually limited to rated value or slightly above it. Hence, to correspond with this reality, the solution for the response has to be modified by dropping the derivative term in the flux command path, resulting in the following solution:

$$\lambda_r(t) = \beta\sqrt{\frac{1+(\omega_{s1}T_r^*)^2}{1+(\alpha\omega_{s1}T_r^*)^2}}\left[\begin{array}{l}\lambda_r^* + \frac{1}{\tau_c - \alpha T_r^*}[\tau_c\lambda_r^* - T_r^*\lambda_r^*(0)]e^{-\frac{t}{\tau_c}}\\ + T_r^*[\lambda_r^*(0) + \lambda_r^*]e^{-\frac{t}{\alpha T_r^*}}\end{array}\right] + \lambda_r(0)e^{-\frac{t}{\alpha T_r^*}} \quad (8.92)$$

Torque Response The torque-producing component of the stator-current response to a step command is evaluated; when combined with the rotor flux-linkages response, it gives the torque response. It is derived as follows. Like the transfer function between the rotor flux linkages and its command, the transfer function between the torque current and its command is derived as

$$\frac{I_T(s)}{I_T^*(s)} = \frac{\alpha}{(1+s\tau_c)}\sqrt{\frac{1+(\omega_{s1}T_r^*)^2}{1+(\alpha\omega_{s1}T_r^*)^2}} \quad (8.93)$$

The time response due to a step command of torque current with a magnitude of i_T^* including an initial condition is derived as

$$i_T(t) = \alpha i_T^*\sqrt{\frac{1+(\omega_{s1}T_r^*)^2}{1+(\alpha\omega_{s1}T_r^*)^2}}[1-e^{-\frac{t}{\tau_c}}] + i_T(0)e^{-\frac{t}{\tau_c}} \quad (8.94)$$

Then the torque response is written as

$$T_e(t) = \frac{3}{2}\frac{P}{2}\frac{L_m}{L_r}\lambda_r(t)i_T(t) \quad (8.95)$$

where the instantaneous values of rotor flux linkages and torque current are substituted from the equations derived in the above.

8.10.2 Parameter Sensitivity Effects on a Speed-Controlled Induction Motor Drive

The closure of an outer loop ensures that the electromagnetic torque, T_e^*, will be modified until the actual output torque is equal to the load torque in steady state, regardless of the parameter variations in the induction motor. The ratio of torque to its command value can be denoted by C. The relationship is expressed as

$$T_e = T_l = CT_e^* \quad (8.96)$$

Section 8.10   Parameter Sensitivity of the Indirect Vector-Controlled Induction   469

The manner in which this relationship is enforced in the induction motor drive has consequences for many key variables of the drive. Insight into the drive's performance is possible by fixing the value of C and then evaluating the stator current, slip speed, and flux for a given load torque. The key variable to be solved for is the torque command for a given load torque. The relevant equations are derived next.

Let ω_{s1}^* be written from equation (8.57) as

$$\omega_{s1}^* = mT_e^* \qquad (8.97)$$

where the following are defined:

$$m = K_{it}\frac{R_r^*}{(\lambda_r^*)^2} \qquad (8.98)$$

$$x = \omega_{s1}^* T_r^* = hT_e^* \qquad (8.99)$$

$$h = K_{it}\frac{R_r^* T_r^*}{(\lambda_r^*)^2} = \frac{4L_r^*}{3P(\lambda_r^*)^2} \qquad (8.100)$$

By substituting equation (8.99) into equation (8.78), the ratio between the torque and its command is derived as

$$\frac{T_e}{T_e^*} = \alpha\beta\left[\frac{1+x^2}{1+\alpha^2 x^2}\right] = \alpha\beta\left[\frac{1+(hT_e^*)^2}{1+(\alpha hT_e^*)^2}\right] \qquad (8.101)$$

Note that $T_e = T_l$ and that solving for T_e^* leads to the following cubic equation in T_e^*:

$$(T_e^*)^3 + (T_e^*)^2\left(-\frac{\alpha}{\beta}T_l\right) + T_e^*\left(\frac{1}{h^2}\right) + \left(-\frac{T_L}{\alpha\beta h^2}\right) = 0 \qquad (8.102)$$

The roots of the equation are obtained by using one of the standard numerical methods. One real root constitutes the solution of equation (8.102); complex roots have no physical identity for the system considered. From T_e^*, the solutions for ω_{s1}^* and the rotor flux linkages are obtained from equation (8.99) as

$$\omega_{s1}^* = \frac{hT_e^*}{T_r^*} \qquad (8.103)$$

and hence the ratio of rotor flux linkage to its command is given by

$$\frac{\lambda_r}{\lambda_r^*} = \beta\sqrt{\frac{1+(hT_e^*)^2}{1+(\alpha hT_e^*)^2}} \qquad (8.104)$$

Peak stator current can be computed from h, T_e^*, and ω_{s1}^*.

8.10.2.1 Steady-state characteristics. Since the key drive variables are not related by simple expressions, as they are in the case of the torque mode of operation, a set of graphs is drawn to illustrate the effects of parameter sensitivity on the operation of the induction motor drive. All the graphs are drawn with load torque as the *x*-axis and the variables of interest on the *y*-axis. The major variables considered for study are the actual/command torque, actual/commanded rotor flux linkages, slip speed, and peak value of the stator current.

470 Chapter 8 Vector-Controlled Induction Motor Drives

Figures 8.31 and 8.32 show the actual/commanded torque and actual/commanded rotor flux linkages for extreme values of α = 0.5 and 1.5 and with a nominal value of 1, maintaining β at a value of 1. The torque and flux follow their command values for α = 1.0 independent of operational speed. With decreasing α, the torque is less than its commanded value up to 0.75 p.u. of load torque, at which

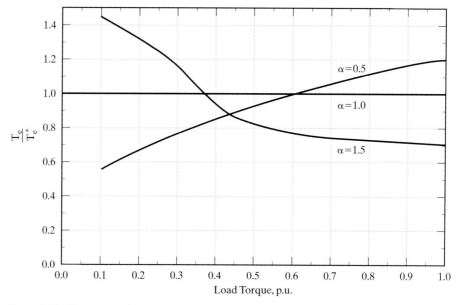

Figure 8.31 Torque over its commanded value vs. load torque for various values of α, with β = 1.0

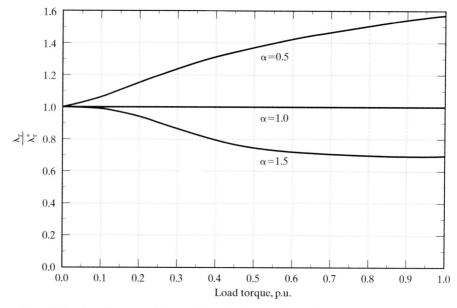

Figure 8.32 Actual/commanded rotor flux vs. load torque for various values of α, with β = 1.0

Section 8.10   Parameter Sensitivity of the Indirect Vector-Controlled Induction   471

point the actual torque increases over its commanded value. The opposite trend in the behavior of the output torque is noticed with α greater than 1. Increasing α to greater than one radically decreases actual torque up to 0.75 p.u. of load torque, after which it remains at a constant level. At α = 0.5, the actual flux increases monotonically with load torque. The saturation of the motor has a damping effect on this variable, and only modest increases in its magnitude result. The effect of saturation on torque and rotor flux is displayed in Figures 8.33 and 8.34. As expected, saturation (with β = 0.8) decreases the flux level by 20%, whereas the torque decreases with respect to its commanded value. The value of α is maintained at 0.5 in these graphs. Values of α greater than 1 have not been considered in these graphs. The reason lies in the fact that it is the lower values of α that are usually encountered in practice. Operation of the motor drive in the linear region of the B–H characteristics increases the value of β and, for a value of 1.2, the figures show the increase in both the flux and torque over their commanded values. The ratios of the actual to command values are plotted in order to appreciate the generalized nature of the characteristics. Values of 1.2 for the ratio of actual to commanded rotor flux at very low flux levels are a distinct possibility, but caution must be exercised in interpreting the flux, because the simulations are for the linear operating region.

Slip-speed characteristics are shown in Figures 8.35 and 8.36 for varying α and β. The slip speed is linear for α = 1 but exhibits other trends for α greater than and less than 1. Defining the nominal value of a variable as that for α = 1, if saturation is accounted for, in the case of α = 0.5, the slip speed increases over its nominal value. There is a slight decrease in the slip speed for operation on the linear region of the B–H characteristics of the iron. Overall, for α less than 1, the slip speed is higher than the nominal value up to a reasonable fraction of full load, at which point

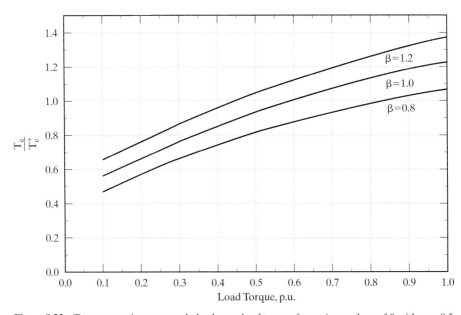

Figure 8.33   Torque over its commanded value vs. load torque for various values of β with α = 0.5

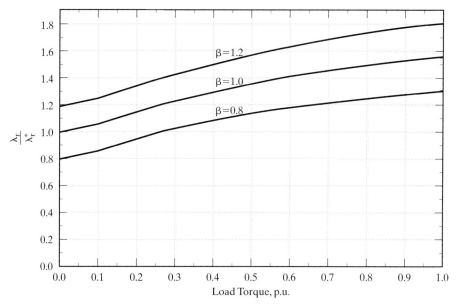

Figure 8.34 Actual/commanded rotor flux vs. load torque for various values of β, with $\alpha = 0.5$

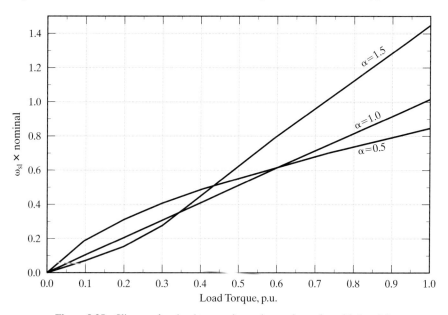

Figure 8.35 Slip speed vs. load torque for various values of α, with $\beta = 1.0$

it becomes slightly lower than the nominal value. This signifies that the rotor losses increase for load torques less than 0.8 p.u. In the field-weakening region of the drive, the losses increase for load torques higher than 0.4 p.u., as evidenced by the curve with α equal to 1.5. Notably, at $\alpha = 0.5$ and around 1 p.u. of load torque, the drive will have slightly lower losses than its nominal value and designed peak capacity, even when saturation is included in the computation.

Section 8.10 Parameter Sensitivity of the Indirect Vector-Controlled Induction 473

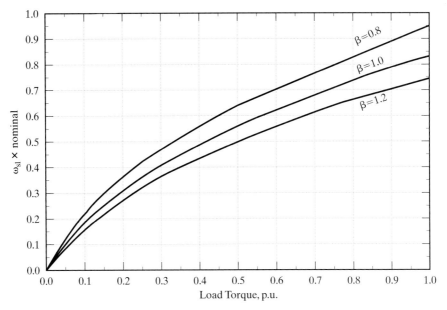

Figure 8.36 Slip speed vs. load torque for various values of β, with $\alpha = 0.5$

Peak values of the stator phase currents for discrete values of α and β are shown in Figures 8.37 and 8.38 against the load torque. A close resemblance to the slip-speed characteristics is evident. Saturation of the induction motor increases the input stator current as shown in Figure 8.38. The drive operation is usually confined to 1 p.u. torque and lower; within this region, the trend to have higher stator losses is evident for α less than 1. This is highly undesirable from the point of view of thermal rating and system efficiency.

8.10.2.2 Discussion on transient characteristics. It is well known that the stability of this drive system is not affected by machine-parameter sensitivity. Notably, the dampings of the oscillations in the flux and torque responses are dependent on the machine parameters. These oscillations in the flux and torque are not transmitted to the rotor shaft for two reasons:

1. the high bandwidth of the outer speed loop, which forces the torque demand to match the load torque in a very short time;
2. the filtering introduced by the moment of inertia of the induction motor and the load.

Saturation of the machine increases the field component of the stator current and decreases the torque component. Depending on the value of the magnetizing inductance, the field component of the stator current can vary from 0.25 to 0.7 p.u. Saturation curtails peak reserve electromagnetic torque, which, in turn, affects the dynamics of the drive, as reflected in increased acceleration and deceleration times.

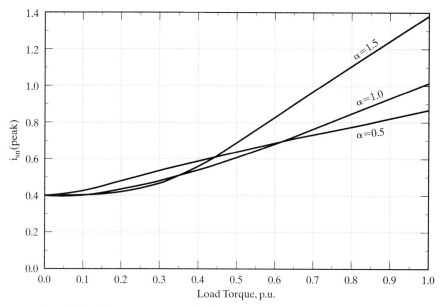

Figure 8.37 Peak stator current vs. load torque for various values of α, with $\beta = 1.0$

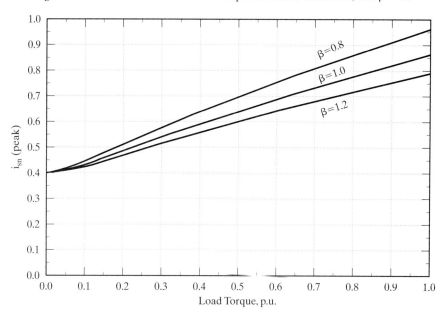

Figure 8.38 Peak stator current vs. load torque for various values of β, with $\alpha = 0.5$

8.10.2.3 Parameter sensitivity of other motor drives.
Ceramic permanent-magnet dc and synchronous motor drives have 20% less flux for a 100°C rise in temperature. A typical induction motor drive has almost the same sensitivity as a permanent-magnet motor drive, except that it is of positive sensitivity. The Samarium–Cobalt permanent-magnet machines have a sensitivity less by an order of magnitude compared to ceramic permanent-magnet motor drives.

Increasing temperature reduces the stator-current magnitude in the induction motor drive, whereas the drives with ceramic permanent magnets experience an increase in the armature/stator current; it is inversely proportional to the flux for a given torque. This factor not only contributes to an increase in the stator losses but also reduces the power output capability with a given inverter. For example, assuming a temperature rise of 100°C and an inverter rated for 2 p.u. power at ambient-temperature operating conditions, the output of a ceramic permanent magnet motor drive is reduced to 1.6 p.u. at 100°C because of a loss of 20% flux in the rotor. Consequently, there is a 20% reduction in the capacity of the motor drive. This is an interesting consequence; the induction motor drive has the opposite effect, enhancing power output under a similar situation. Including saturation effect, it could be conservatively stated that the capacity of the induction motor drive will not suffer from operation at temperatures higher than the ambient.

8.11 PARAMETER-SENSITIVITY COMPENSATION

The effects of mismatch between the induction motor and vector controller can be minimized by adapting the parameters in the vector controller to the actual motor parameters, at all times. This requires monitoring of the motor parameters, but direct monitoring is difficult while the drive is in operation. Several methods have come to the forefront to minimize the consequences of parameter sensitivity in indirect vector-control schemes. These are parameter-adaptation schemes based on one of the following strategies:

(i) direct monitoring of the alignment of the flux and the torque-producing stator-current component axes or real-time measurement of the instantaneous rotor resistance or both;

(ii) measurement of modified reactive power, measurement and estimation of rotor flux or the deviation of the field angle, or a combination of the rotor flux and the torque-producing component of the stator current. (An error in these measured variables is an indication of the amount of the parameter variation present in the induction motor.)

While strategy (i) can be classified as a direct scheme for parameter adaptation, strategy (ii) is an indirect parameter-adaptation scheme. Note that this classification is based on how the parameter-compensation signal is generated: by direct measurement of the parameter/alignment of the axes, or by indirectly monitoring the system for parameter variations as in strategy (ii). Most of the parameter-adaptation algorithms are themselves parameter-dependent. This particular aspect can cause significant error in the computation of the variables used in the parameter-compensation schemes. Besides such other critical factors as the speed of computation or method of measurement of the variables and complexity of the implementation, the error due to the parameter dependency of the algorithms could very well provide a yardstick to assess one scheme vis-à-vis other candidate schemes.

8.11.1 Modified Reactive-Power Compensation Scheme

A scheme using the modified reactive power, F, is discussed in this section, for parameter compensation of the indirect vector-controlled induction motor drive. The modified reactive power is defined by

$$F = -\frac{L_m}{L_r} \omega_s i_{ds}^e \lambda_r \quad (8.105)$$

and its command value is given as

$$F^* = -\frac{L_m^*}{L_r^*} \omega_s^* i_{ds}^{*e} \lambda_r^* \quad (8.106)$$

and they are estimated from terminal voltages, phase currents, and other command variables. The modified reactive power command, F^*, is obtained from the command variables of rotor speed, slip speed, flux-producing component of stator current, and rotor flux linkages. The computation of modified reactive power, F, can be shown to be, in terms of phase currents, their derivatives, and phase voltages,

$$F = \frac{2}{\sqrt{3}}\left[\left(v_{cs} - \frac{pi_{cs}}{K_{11}}\right)i_{as} - \left(v_{as} - \frac{pi_{as}}{K_{11}}\right)i_{cs}\right] \quad (8.107)$$

where

$$K_{11} = \frac{L_r}{(L_s L_r - L_m^2)} \quad (8.108)$$

The parameter K_{11} is dependent on motor inductances; hence, their variation will affect the computation of F and, in the final analysis, the parameter compensation. A change in the operating point will alter F^* and, accordingly, F. Note that any parameter variation in the induction motor will change F and make it deviate from its expected or commanded value F^*, indicating a change in the rotor time constant. The error between F^* and F is amplified through a controller and a correction signal is obtained to correct the rotor time constant in the vector controller. This correction signal is made equivalent to the incremental slip speed required to compensate for the parameter variations in this particular implementation.

The implementation of this parameter-compensation scheme is shown in Figure 8.39. The error in the modified reactive power is converted into an incremental slip-speed signal and added to the slip-speed command of the vector controller. Figure 8.40 shows an experimental result from a laboratory prototype when the stator temperature was at 30°C. The motor drive is originally tuned at 20°C, where the parameters in the controller match those of the induction motor. With the temperature change, the torque characteristic has become nonlinear, as expected. In that situation, modified reactive power F is diverging from its command value, as shown in the plot. The parameter adaptation is achieved by closing the modified-reactive-power loop shown in Figure 8.39 by using estimated phase voltages and measured stator currents. Even without much tuning

Section 8.11   Parameter-Sensitivity Compensation   477

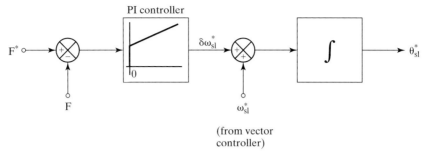

Figure 8.39 Block diagram of the parameter-adaptation with modified-reactive-power feedback control

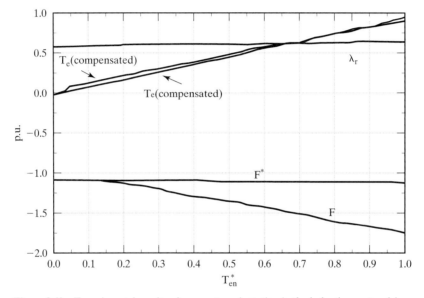

Figure 8.40 Experimental results of parameter adaptation in the induction motor drive

of the controller in the modified reactive power path, the torque characteristic has become linear. The parameter compensation is effective only from 2.5 N·m, i.e., 0.125 p.u. torque of the motor under parameter-compensated conditions, and that the estimated modified reactive power closely follows its command value is seen.

8.11.2 Parameter Compensation with Air Gap-Power Feedback Control

A parameter compensation scheme using air gap power is described in some detail in this section to overcome the dependence on inductances in the modified-reactive-power method. The basis of the scheme is that the air gap power in the machine is equal to its reference value when the controller and motor parameters match, but

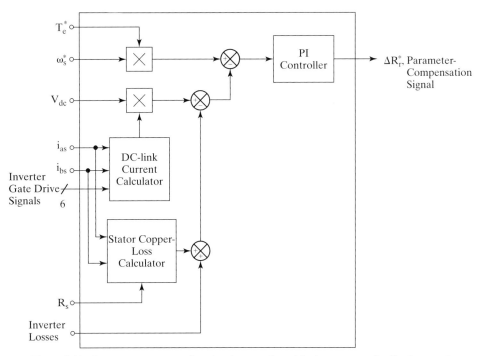

Figure 8.41 Parameter-compensation signal generation with air gap-power feedback control

develops an error in case of a divergence between the controller and motor parameters. This control scheme is shown in block-diagram form in Figure 8.41, with the dc-link voltage and stator currents as feedback variables. From the stator phase currents and inverter gate drive logic signals, the dc-link current can be reconstructed; when multiplied with the dc-link voltage, it gives the input power from the dc link. Subtracting the three-phase stator copper and inverter losses from the dc-link power provides the air gap power, P_a. The reference air gap power, P_a^*, is generated by the product of the torque and stator frequency references. The error between the reference and measured air gap power is amplified, then limited with a proportional-plus-integral controller to provide a parameter-compensation signal. Since this is a single signal and three parameters are involved in the vector controller, it is judicious to view this signal as an adjunct to the most sensitive parameter, i.e., rotor resistance in the controller. Hence, it is approximately identified as ΔR_r^*, which is summed with R_r^* in the block diagram of the slip-speed signal generator in the vector controller. The place of this parameter-compensation scheme in the overall indirect vector-controlled induction motor drive is shown in Figure 8.42 to appreciate the implementation aspect. Note that the parameter-compensation control is secondary to the indirect vector controller; accordingly, a lower priority is assigned during the execution of its control in software implementations.

8.11.2.1 Steady-state performance. The performance of this control scheme in steady state is explored for independent variations in rotor resistance and

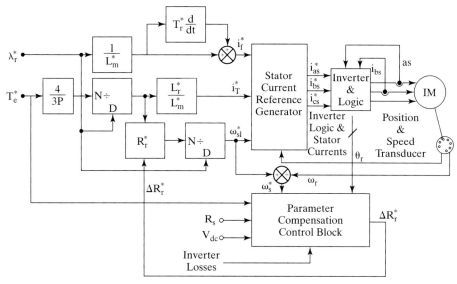

Figure 8.42 Parameter-compensated indirect vector-controlled induction motor drive

magnetizing inductance of the induction machine. The motor-phase rms current is obtained as

$$I_s^* = \frac{\sqrt{I_f^{*2} + I_T^{*2}}}{\sqrt{2}} \quad (8.109)$$

where

$$I_f^* = \frac{\lambda_r^*}{L_m^*} \quad (8.110)$$

$$I_T^* = \frac{4}{3P} \cdot \frac{L_r}{L_m} \cdot \frac{T_e^*}{\lambda_r^*} \quad (8.111)$$

This motor current, when applied with a stator frequency of ω_s given by

$$\omega_s = \omega_r + \omega_{sl}^* \quad (8.112)$$

where

$$\omega_{sl}^* = \frac{4}{3P} \cdot \frac{R_r^* T_e^*}{(\lambda_r^*)^2} \quad (8.113)$$

gives a rotor phase current, which is computed from the equivalent circuit as

$$I_r = \frac{\omega_s L_m}{\sqrt{\left(\frac{R_r}{s}\right)^2 + \omega_s^2 (L_m + L_{lr})^2}} \cdot I_s^* \quad (8.114)$$

480   Chapter 8   Vector-Controlled Induction Motor Drives

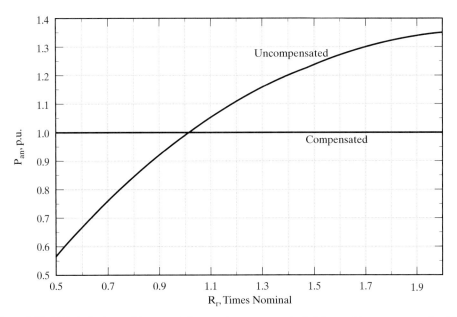

Figure 8.43 Effect of rotor-resistance variation on air-gap power at rated flux and torque commands with the rotor speed at rated value

The air gap power is then calculated as

$$P_a = 3I_r^2 \frac{R_r}{s} \qquad (8.115)$$

and the reference air gap power is given by

$$P_a^* = \frac{\omega_s T_e^*}{(P/2)} \qquad (8.116)$$

where the parameters with asterisk indicate that they are the instrumented values in the controller.

These equations enable the determination of parameter sensitivity on air-gap power.

The effect of rotor-resistance variation in the machine from 0.5 to 2 times its nominal value, with nominal value set in the controller, on air gap power is shown in Figure 8.43. The flux and torque commands are maintained at rated values with the speed at rated value. The parameters of the machine are given in Example 8.4. Invariably, the air gap power is at rated value for the nominal value of rotor resistance that matches with the controller instrumented value. If the rotor resistance in the controller is made equal to the actual value in the machine that amounts to exact compensation, the air gap power remains at rated value, as is shown in the figure.

The effect due to variation in magnetizing inductance from 0.8 to 1.2 times nominal value is shown in Figure 8.44 for the same conditions as for the above. Both the variations produce a distinct error in air gap power from their reference value of

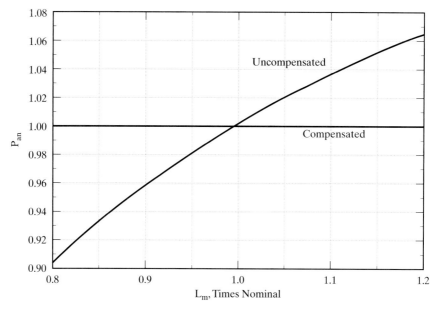

Figure 8.44 Effect of magnetizing-inductance variation on air gap power for rated flux and torque commands at rated speed

one p.u., thus clearly indicating that this air gap power error could be effectively employed to compensate for the parameter sensitivity of the indirect vector controller.

The magnetizing-inductance variation also is to be compensated in the vector controller, only in the slip-speed-signal generation, i.e., through the variation of the R_r^* value. In that case, the new value of R_r^* for variation of L_m can be calculated from the air gap equation with the following steps:

$$P_a = 3I_r^2 \frac{R_r}{s} = \frac{3\omega_{sl}^*(\omega_r + \omega_{sl}^*)(L_m I_s)^2 R_r}{R_r^2 + \omega_{sl}^{*2}(L_m + L_{1r})^2} \quad (8.117)$$

from which a quadratic equation in slip speed is obtained as

$$a\omega_{sl}^{*2} + b\omega_{sl}^* + c = 0 \quad (8.118)$$

where

$$a = 3(L_m I_s)^2 R_r - P_a(L_m + L_{1r})^2 \quad (8.119)$$
$$b = 3\omega_r R_r (L_m I_s)^2 \quad (8.120)$$
$$c = -P_a R_r^2 \quad (8.121)$$

Solving for the slip-speed command gives ω_{sl}^* as

$$\omega_{sl}^* = \frac{-b \pm \sqrt{b^2 - 4ac}}{2a} \quad (8.122)$$

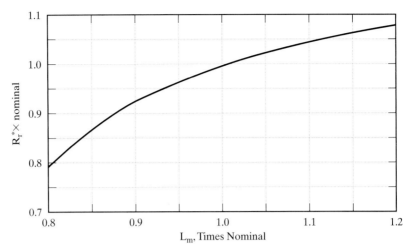

Figure 8.45 Variation of rotor resistance in the vector controller vs. magnetizing-inductance variation to deliver rated air gap power

from which the rotor resistance to be instrumented in the controller is obtained by

$$R_r^* = \left(\frac{3P}{4}\right)\frac{\omega_{sl}^*(\lambda_r^*)^2}{T_e^*} \quad (8.123)$$

The rotor-resistance variation in the controller required to overcome the magnetizing inductance variation from 0.8 to 1.2 times nominal value is shown in Figure 8.45 for the same conditions discussed for the previous two figures. This figure verifies that the parameter compensation can be achieved through the slip-speed variation alone for all parameter changes in the vector controller.

8.11.2.2 Dynamic performance. The dynamic performance of the parameter-compensated vector-controlled drive with a step change in the rotor resistance from nominal to 1.5 times the nominal value is shown in Figure 8.46; that for a linear variation in R_r from nominal to 2 times nominal is shown in Figure 8.47. The responses demonstrate that the air gap-power feedback control for parameter compensation works and that the error in air gap power reduces to zero. Even that it takes 0.75 seconds to reach steady state is acceptable, because changes in rotor resistance are slowed down by the thermal time constant of the rotor. The steady-state error is not exactly zero; there is a window in the implementation of the air gap power error. If the error is within this window, the controller does not generate a correction signal. Steady-state error can be reduced to zero with a zero window error size, leading to oscillations in the dynamic response of the drive system. For the linear change in rotor resistance, which is more realistic in the motor, the parameter compensation is fairly swift and smooth, with oscillations only in the beginning as is evident from Figure 8.47.

Section 8.11 Parameter-Sensitivity Compensation

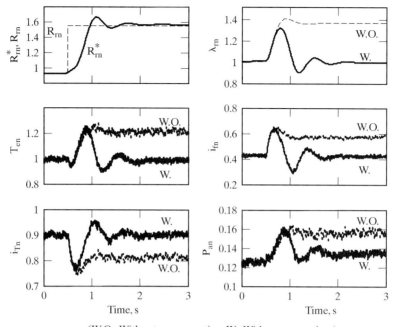

(W.O.–Without compensation, W.–With compensation,)

Figure 8.46 Plots for step change in rotor resistance

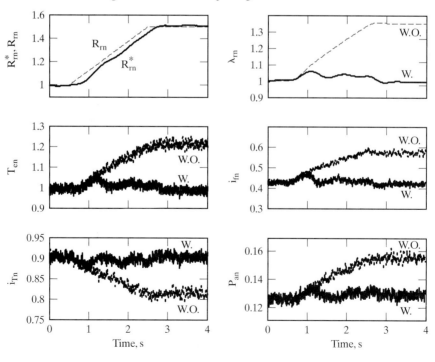

(W.O.–Without compensation, W.–With compensation,)

Figure 8.47 Plots for linear change in rotor resistance

484 Chapter 8 Vector-Controlled Induction Motor Drives

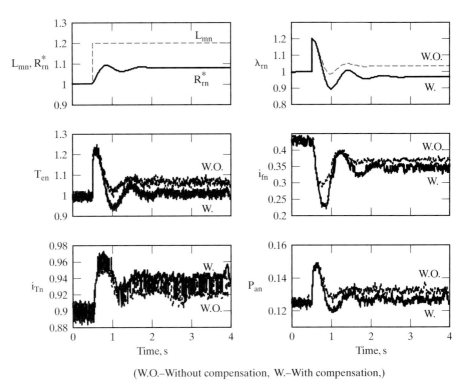

(W.O.–Without compensation, W.–With compensation.)

Figure 8.48 Effects of increase in mutual inductance

A step change in mutual inductance by 20% likewise is effectively handled by this scheme. It is shown in Figure 8.48. Note that the rotor flux linkage is not properly compensated; that is the penalty for using the same compensation signal to counter three different parameter variations. The torque linearity is maintained intact in this parameter-compensation scheme.

8.12 FLUX-WEAKENING OPERATION

Principle of Flux Weakening The motor drive is operated with rated flux linkages up to a speed where the ratio between the induced emf and stator frequency, known as volts/Hz, with neglected stator impedance, can be maintained constant, i.e., at rated value. After a certain stator frequency, known as the base frequency, the volts/Hz ratio, and hence the stator flux linkage, will become lower than its rated value that is due to the fixed dc-link-voltage magnitude. The operation above the base speed necessarily involves the weakening of the flux, resulting in the reduction of torque for the same torque-producing component of the stator current. By coordinating the torque level for each operating speed, the power output can be maintained constant in the flux-weakening region, much as in the dc motor drive case.

Section 8.12 Flux-Weakening Operation

This section covers the flux-weakening algorithms for stator (direct schemes) and rotor (indirect schemes) flux-linkage-controlled drive systems.

8.12.1 Flux Weakening in Stator-Flux-Linkages-Controlled Schemes

Consider the normalized stator voltage equations of the induction machine with flux linkages as variables. Neglecting the stator resistive-voltage drops and considering only the steady state, the equations are

$$v_{qsn}^e = R_s i_{qsn}^e + p\lambda_{qsn}^e + \omega_{sn}\lambda_{dsn}^e \cong p\lambda_{qsn}^e + \omega_s\lambda_{dsn}^e \quad (8.124)$$

$$v_{dsn}^e = R_s i_{dsn}^e + p\lambda_{dsn}^e - \omega_s\lambda_{qsn}^e \cong p\lambda_{dsn}^e - \omega_{sn}\lambda_{qsn}^e \quad (8.125)$$

$$v_{sn} = \sqrt{(v_{qsn}^e)^2 + (v_{dsn}^e)^2} = \omega_{sn}\sqrt{(\lambda_{qsn}^e)^2 + (\lambda_{dsn}^e)^2} = \omega_{sn}\lambda_{sn} \quad (8.126)$$

The stator flux-linkages phasor is derived as

$$\lambda_{sn} = \frac{v_{sn}}{\omega_{sn}} \quad (8.127)$$

In the case of the direct vector-control scheme, as with the indirect vector-control derivation, it may be assumed that the stator flux-linkages phasor is aligned with the d axis stator flux linkages. Thereby, the q axis stator flux linkages are forced to zero, with the resulting elegant torque expression in terms of the product of the stator flux-linkages phasor and the q axis current obtained as

$$\lambda_{qsn}^e = L_{sn}i_{qsn}^e + L_{mn}i_{qrn}^e = 0 \quad (8.128)$$

$$\lambda_{dsn}^e = L_{sn}i_{dsn}^e + L_{mn}i_{drn}^e = \lambda_{sn}^e \quad (8.129)$$

which, upon substitution in the torque expression, results as

$$T_{en} = \{i_{qsn}^e \lambda_{dsn}^e - i_{dsn}^e \lambda_{qsn}^e\} = i_{qsn}^e \lambda_{sn}^e \quad (8.130)$$

The air gap power is calculated as

$$P_{an} = T_{en}\omega_{sn} \quad (8.131)$$

By substituting for torque and flux linkages phasor from the previous equations, the normalized air gap power is derived as

$$P_{an} = T_{en}\omega_{sn} = i_{qsn}^e \lambda_{sn}^e \omega_{sn} = i_{qsn}^e \frac{v_{sn}^e}{\omega_{sn}}\omega_{sn} = v_{sn}^e i_{qsn}^e \quad (8.132)$$

If the q axis stator current is maintained constant, note that, in the flux-weakening region, the air gap power remains constant as the stator voltage phasor is maintained constant. The output power is the difference between the air gap power and the slip power. By keeping the slip constant, the output power becomes constant for a constant air gap power. Therefore, constant-power operation is obtained in the entire flux-weakening region of the stator-flux-linkages-controlled drive by forcing the stator flux linkages to be inversely proportional to stator frequency.

8.12.2 Flux Weakening in Rotor-Flux-Linkages-Controlled Schemes

To obtain the maximum speed of operation at rated power, a simple inverse variation of rotor flux linkages as a function of speed is not sufficient. During flux-weakening, the flux linkages are controlled in such a way that the stator-voltage phasor is within the admissible limit of the applied voltage. It is more natural, then, to control the stator flux linkages, which are directly related to the stator voltages, than the rotor flux linkages, which are indirectly related to the stator voltages. Note that high-performance drive systems are based on indirect vector control, in which the rotor flux-linkages phasor is varied as a function of the rotor speed. It is important, therefore, to have the flux weakening with the modified rotor flux linkages in indirect vector control, to effectively extend the range of speed of operation of the drive system.

Control of the rotor flux linkages affects the stator flux linkages, but not in a proportional manner. This results in the stator flux linkages being greater than the desired value, i.e., rated value, requiring stator voltages greater than rated values when the stator frequency exceeds the rated value. Since that could not be supplied from the limited dc link, constant-power operation is not possible to maintain in the flux-weakening mode. This is proven analytically in this section, to gain an insight into the process of the flux-weakening mode of operation in the indirect VCIM drives.

Consider the case where rotor flux is set inversely proportional to rotor speed. The relevant equations in steady state and in normalized units are

$$\lambda_{rn} = \frac{1}{\omega_{rn}} \tag{8.133}$$

$$I_{fn} = I_{frn}\lambda_{rn} \tag{8.134}$$

where λ_{rn} is rotor flux linkages in p.u., ω_{rn} is the rotor speed in p.u., I_{fn} is the field current in p.u., and I_{frn} is the rated field current in normalized units.

The stator equations in these variables are,

$$V_{qsn}^e = R_{sn}I_{Tn} + L_{sn}\omega_{sn}I_{fn} \tag{8.135}$$

$$V_{dsn}^e = R_{sn}I_{fn} - a_n\omega_{sn}I_{Tn} \tag{8.136}$$

where

$$a_n = \left(L_{sn} - \frac{L_{mn}^2}{L_m}\right) = \left(1 - \frac{L_{mn}^2}{L_{mn}L_{sn}}\right)L_{sn} = \sigma L_{sn} \tag{8.137}$$

Neglecting the stator resistive-voltage drops, we have the stator voltages in steady state as

$$V_{qsn}^e \approx L_{sn}\omega_{sn}I_{fn} \tag{8.138}$$

$$V_{dsn}^e \cong -a_n\omega_{sn}I_{Tn} \tag{8.139}$$

and the stator-voltage phasor as

$$V_{sn} = \sqrt{(V_{qsn}^e)^2 + (V_{dsn}^e)^2} = \frac{L_{sn}}{L_{mn}}\omega_{sn}\sqrt{\lambda_{rn}^2 + (\sigma L_{mn}I_{Tn})^2} \tag{8.140}$$

from which the rotor flux-linkages phasor is derived as

$$\lambda_{rn} = \sqrt{\left(\frac{L_{mn}}{L_{sn}} \frac{V_{sn}}{\omega_{sn}}\right)^2 - (\sigma L_{mn} I_{Tn})^2} \qquad (8.141)$$

This equation demonstrates that the rotor flux linkage is not simply inversely proportional to the stator frequency; it is also strongly influenced by the torque-producing component of the current. The contrast between the rotor flux linkages in the indirect vector-control scheme and the stator flux linkages in the direct vector-control scheme is significant. In the latter case, the product of the stator flux linkages and stator frequency is constant for a fixed stator-voltages phasor. The product of the rotor flux linkages and stator frequency is not a constant for a fixed stator-voltage phasor in the indirect vector-control scheme. This fact implies that the air gap power cannot be maintained constant even if the rotor flux linkage is forced to be inversely proportional to stator frequency. The stator flux linkages increase under this circumstance, thus demanding a higher stator-voltage phasor, which may not be possible with a fixed dc-link voltage. The stator flux linkages increase under this circumstance, as shown in the following by rearranging equation (8.141) in terms of stator flux linkages, rotor flux linkages, and torque as

$$\lambda_{sn} = \frac{L_{sn}}{(L_{mn}\lambda_{rn})}\sqrt{bT_{en}^2 + \lambda_{rn}^4} \qquad (8.142)$$

where

$$b = \sigma^2 L_{mn}^2 \qquad (8.143)$$

This equation indicates clearly that the stator flux linkages are influenced by the rotor flux linkages and electromagnetic torque. As the electromagnetic torque is generated, the linear relationship between rotor and stator flux linkages is lost. Figure 8.49 shows the rotor flux linkages, electromagnetic torque, and voltage phasor versus rotor speed for a 1-p.u. stator current. Note that the stator voltage requirement exceeds 1 p.u. as the stator flux linkage increases. The key, then, to the control of flux-weakening operation is to maintain the stator flux linkages inversely proportional to the stator frequency. The constraint of a 1-p.u. current in steady state has to be further incorporated, as is given in the next section.

8.12.3 Algorithm to Generate the Rotor Flux-Linkages Reference

For the sake of simplicity, the stator resistances are neglected in the following derivations. This will introduce an error of 2 to 3% in the estimation of such variables as the torque and power in small machines, but much less error in large machines. Note that the following development makes it simple to incorporate the effects of the stator resistance if required for final tuning of the solution. To modify the rotor flux linkages to keep the stator voltage requirement to 1 p.u., a factor x is introduced, leading to

$$\lambda_{rn} = \frac{x}{\omega_{rn}} \qquad (8.144)$$

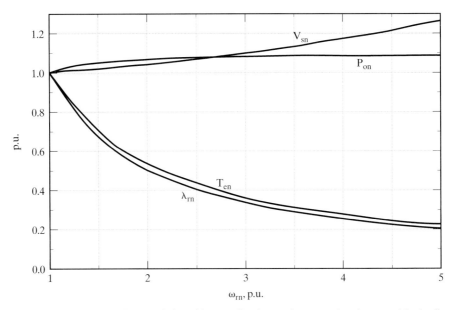

Figure 8.49 Performance characteristics with rotor flux inversely proportional to speed in the flux-weakening region

and

$$T_{en} = \frac{1}{\omega_{rn}} \qquad (8.145)$$

and the torque-producing component of the stator current in terms of the above equations and rated values in normalized units is

$$I_{Tn} = I_{Trn}\frac{T_{en}}{\lambda_{rn}} = I_{Trn}\frac{1/\omega_{rn}}{x/\omega_{rn}} = \frac{I_{Trn}}{x} \qquad (8.146)$$

where

$$I_{Trn} = \frac{I_{Tr}}{I_b} \qquad (8.147)$$

and the rotor flux linkages in p.u. is

$$\lambda_{rn} = \frac{\lambda_r}{\lambda_{rb}} = \frac{L_m i_f}{L_m i_{fr}} = \frac{I_{fn}}{I_{frn}} \qquad (8.148)$$

where

$$I_{frn} = \frac{I_{fr}}{I_b} \qquad (8.149)$$

from which the field current is obtained as

$$I_{fn} = I_{frn}\lambda_{rn} = I_{frn}\frac{x}{\omega_{rn}} \qquad (8.150)$$

Section 8.12   Flux-Weakening Operation   489

The torque component of the stator current is

$$I_{Tn} = \frac{I_{Trn}}{x} = \sqrt{I_{sn}^2 - I_{fn}^2} = \sqrt{I_{sn}^2 - \frac{(xI_{frn})^2}{\omega_{rn}^2}} = I_{sn}\sqrt{1 - \left(\frac{xI_{frn}}{I_{sn}\omega_{rn}}\right)^2} \quad (8.151)$$

where the permissible stator current phasor is I_{sn}.

The synchronous speed is written as the sum of the rotor speed and slip speed, in normalized units:

$$\omega_{sn} = \omega_{rn} + \omega_{sln} = \omega_{rn} + \frac{\omega_{sl}}{\omega_b} \quad (8.152)$$

By substituting for the slip speed in terms of I_{Tn} and I_{fn}, the stator angular frequency is obtained as

$$\omega_{sn} = \omega_{rn} + \frac{R_r I_T}{L_r I_f} \cdot \frac{1}{\omega_b} \quad (8.153)$$

Substituting for the torque- and flux-producing components of stator currents, in terms of the variable x from equations (8.146) and (8.150), into the stator angular frequency equation gives

$$\omega_{sn} = \omega_{rn}\left\{1 + \frac{K_s}{x^2}\right\} \quad (8.154)$$

where the constant K_s is given by

$$K_s = \frac{R_r I_{Trn}}{\omega_b L_r I_{frn}} \quad (8.155)$$

Then the stator voltages are obtained, by substituting for the stator frequency into stator-voltage equations in steady state, as

$$V_{qsn}^e \cong L_{sn}\omega_{sn}I_{fn} = \left(\frac{x^2 + K_s}{x}\right)L_{sn}I_{frn} \quad (8.156)$$

$$V_{dsn}^e = -a_n\omega_{sn}I_{Tn} = -a_n\omega_{rn}\left(1 + \frac{K_s}{x^2}\right)\sqrt{I_{sn}^2 - \frac{x^2 I_{frn}^2}{\omega_{rn}^2}} \quad (8.157)$$

The magnitude of the stator-voltage phasor is given by

$$V_{sn} = \sqrt{(V_{qsn}^e)^2 + (V_{dsn}^e)^2} \quad (8.158)$$

By the combining of the last three equations, a polynomial in variable x is obtained:

$$d_1 x^6 + d_2 x^4 + d_3 x^2 + d_4 = 0 \quad (8.159)$$

where

$$d_1 = b_2^2 - a_n^2 I_{frn}^2 \quad (8.160)$$

$$d_2 = -v_{sn}^2 + 2K_s b_2^2 - 2K_s a^2 I_{frn}^2 + a_n^2 \omega_{rn}^2 I_{sn}^2 \quad (8.161)$$

$$d_3 = b_2^2 K_s^2 - (aK_s I_{frn})^2 + 2K_s(a_n I_{sn}\omega_{rn})^2 \quad (8.162)$$

TABLE 8.5 Required rotor flux linkages vs. ω_{rn}

ω_{rn}, p.u.	x	$\lambda_{rn} = x/(1.03\omega_{rn})$ p.u.
1.0	1.030	1.000
1.5	1.013	0.656
2.0	0.990	0.480
2.5	0.958	0.372
3.0	0.919	0.297
3.5	0.870	0.241
4.0	0.809	0.196
4.5	0.733	0.158
5.0	0.636	0.123

$$d_4 = (a_n K_s \omega_{rn} I_{sn})^2 \qquad (8.163)$$

$$b_2 = L_{sn} I_{frn} \qquad (8.164)$$

Solving the polynomial equation for the roots of x will lead to the rotor flux linkages, and, together with the torque reference, the flux and torque components of the stator current and the slip speed can be evaluated. Note that the real root provides the solution for x; the other roots have no physical significance. The solution leads to the possible maximum torque and power output for a given stator voltage phasor and current phasor magnitude under the flux-weakening conditions. For the machine given in Example 8.4, the required rotor flux linkages to operate the drive system with 1-p.u. voltage and current magnitudes are given in Table 8.5 for various speeds.

At 1-p.u. speed, the rotor flux has to be 1 p.u., but x and λ_r have turned out to be 1.03 p.u. The additional 0.03 p.u. flux is due to the voltage that must have been dropped by stator resistance. Hence, to account for its neglect, the rotor flux linkage is modified by adjusting to 1 p.u. at 1-p.u. rotor speed. This is done by dividing all the ensuing rotor flux linkages by 1.03 in the flux-weakening region.

The rotor and stator flux linkages, torque, and power output versus rotor speed are shown in Figure 8.50 for this control with 1-p.u. voltage and current. The power output is not maintained at 1 p.u. beyond 2.5-p.u. speed. Increasing stator frequency rapidly increases the d axis stator voltage, necessitating a reduction in the q axis voltage by reducing the field current component of the stator current.

8.12.4 Constant-Power Operation

Constant-power operation is not obtainable over a wide speed range, such as 4 p.u. and above, in the illustrated case, but constant-power operation over a wide speed range is desirable in many applications, such as machine-tool spindles, electric vehicles, hand tools, and centrifuges. One of the solutions to this problem lies in having a reserve of dc-link voltage to meet the constant-power requirement. For example, the above illustration in Figure 8.49 shows that with 1.26-p.u. voltage, constant-power operation is achieved up to 5-p.u. speed with a 1-p.u. stator current. Several ways are available to increase the dc-link voltage or available voltage from the dc-link to the machine windings. Among them are increasing the dc-link voltage through a boost front-end rectifier, increasing the ac voltage by a step-up or tapped transformer, and changing the stator-

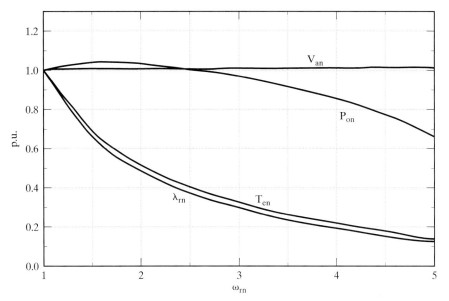

Figure 8.50 Performance characteristics of the vector-controlled induction motor drive with modified rotor flux weakening

windings connection from star to delta. While all of them are feasible, the solution using a step-up or tapped transformer is certainly not desirable; it increases the space requirement for the installation and also the cost. The other alternatives, using the boost front-end rectifier and changing the winding connections, are elegant solutions.

The boost front-end rectifier solves more than the one problem of varying the dc-link voltage; it also provides near-unity power factor while drawing near-sinusoidal currents, thus reducing the ac-line harmonic currents. Further, it makes the dc link independent of line variations compared to a diode-bridge rectifier in the front end. Such a solution is desired for emerging applications, to improve the line quality and to draw the minimum peak current and hence minimum kVA from the utility.

The star–delta change of winding connections enables higher voltage to be obtained for the machine phases, thus relieving the problem of insufficient voltage for flux weakening in the induction motor drives. Note that this method provides an inexpensive solution.

The solution methodologies suggested are suitable for general-purpose induction machines; no special design of the machines is required, and by providing a reserve of dc-link voltage, such as 20 to 30%, constant-power operation over a wide speed range could be obtained. The other approach is to coordinate the flux-weakening control with a motor of prespecified design parameters that would lend itself to working within the available dc-link voltage, yet be capable of providing the desired range of constant-power operation. This, then, would involve the coordination of motor design, which could be expensive for small-volume production but could turn out to be the most cost-effective for high-volume applications. In the light of increasing demand for ac drives as against dc drives, this solution becomes feasible for large-drive system manufacturers. The problem of determining the ideal machine parameters to provide constant-power operation over a wide range of speeds as a function of inverter variables along with other design constraints will be crucial in the product development of such drives in the future.

8.13 SPEED-CONTROLLER DESIGN FOR AN INDIRECT VECTOR-CONTROLLED INDUCTION MOTOR DRIVE

The principle, derivation, and implementation of decoupling nonlinear controllers for both the direct and indirect vector-control schemes made possible the independent control of flux and torque in the induction machine. In addition to torque control, speed control is required in a large number of applications. For speed regulation, usually an outer speed loop is closed in many applications. Then the design of the speed controller is of importance. An analytical approach using the transfer function is considered in the design of the speed controller. The vector controller transforms the induction motor drive into a linear system, even for large signals, when the flux linkages are maintained constant and hence resembles the separately-excited dc motor drive in all aspects, including in the development of the block diagram and hence in the synthesis of speed controller. This section contains the systematic development of the transfer-function derivation for the speed-controlled indirect vector-controlled induction motor drive. That similar derivations are possible for the direct vector controller goes without saying. Based on the transfer function, the speed controller is designed by using symmetric optimum method, but other methods could well be employed effectively, too. Use of symmetric optimum is made to maintain the uniformity of the speed-controller design for all ac and dc drive systems in this text.

8.13.1 Block Diagram Derivation

The block diagram of the indirect vector-controlled induction motor drive is derived in this section by developing the transfer functions of the various subsystems, such as the induction machine, inverter, speed controller, and feedback-transfer functions. By combining the subsystem block diagrams, the final block diagram of the induction motor drive is assembled, which has an overlap between the torque-current feedback loop and the induced-emf feedback loop. That overlap is overcome by block-diagram reduction techniques, making the inner current loop totally independent of the motor mechanical-transfer function; this approach lends itself to a simpler synthesis of the current controller, if need be.

8.13.1.1 Vector-controlled induction machine.
To design the speed controller for the indirect vector-controlled induction motor drive, the key assumption of constant rotor flux linkages is made. Symbolically, the assumption leads to

$$\lambda_r = \text{a constant} \tag{8.165}$$

$$p\lambda_r = 0 \tag{8.166}$$

The stator equations of the motor are

$$v_{qs}^e = (R_s + L_s p)i_{qs}^e + \omega_s L_s i_{ds}^e + L_m p i_{qr}^e + \omega_s L_m i_{dr}^e \tag{8.167}$$

$$v_{ds}^e = -\omega_s L_s i_{qs}^e + (R_s + L_s p)i_{ds}^e - \omega_s L_m i_{qr}^e + L_m p i_{dr}^e \tag{8.168}$$

Section 8.13 Speed-Controller Design for an Indirect Vector-Controlled Induction

but from the vector controller, the following relationships of the rotor q and d axes flux linkages are made use of to recast the stator voltage equations as

$$i_{qr}^e = -\frac{L_m}{L_r} i_{qs}^e \qquad (8.169)$$

$$i_{dr}^e = \frac{\lambda_r}{L_r} - \frac{L_m}{L_r} i_{ds}^e \qquad (8.170)$$

Substitution of the rotor currents into the stator-voltage equations results in

$$v_{qs}^e = (R_s + \sigma L_s p) i_{qs}^e + \sigma L_s \omega_s i_{ds}^e + \omega_s \frac{L_m}{L_r} \lambda_r \qquad (8.171)$$

$$v_{ds}^e = (R_s + \sigma L_s p) i_{ds}^e - \sigma L_s \omega_s i_{qs}^e + \frac{L_m}{L_r} p\lambda_r \qquad (8.172)$$

where σ is the leakage coefficient. It is known that the flux-producing component of the stator current is constant in steady state, and that is the d axis stator current in the synchronous frames. Its derivative is also zero, giving the following:

$$i_f = i_{ds}^e \qquad (8.173)$$

$$p i_{ds}^e = 0 \qquad (8.174)$$

The torque-producing component of the stator current is the q axis current in the synchronous frames, given by

$$i_T = i_{qs}^e \qquad (8.175)$$

Substituting these into the q axis voltage equation gives

$$v_{qs}^e = (R_s + L_a p) i_T + \omega_s L_a i_f + \omega_s \frac{L_m}{L_r} \lambda_r \qquad (8.176)$$

where L_a is given by

$$L_a = \sigma L_s = \left(L_s - \frac{L_m^2}{L_r} \right) \qquad (8.177)$$

Substituting for $\lambda_r = L_m i_f$ gives the q axis stator voltage in synchronous reference frames:

$$v_{qs}^e = (R_s + L_a p) i_T + \omega_s L_a i_f + \omega_s \frac{L_m^2}{L_r} i_f = R_s + L_a p i_T + \omega_s L_s i_f \qquad (8.178)$$

The second stator equation is not required; the solution of either will yield i_T, which is the variable under control in the system. Now, the stator frequency is represented as

$$\omega_s = \omega_r + \omega_{sl} = \omega_r + \frac{i_T}{i_f}\left(\frac{R_r}{L_r}\right) \qquad (8.179)$$

The electrical equation of the motor is obtained by substituting for ω_s from (8.179):

$$v_{qs}^e = (R_s + L_a p)i_T + \omega_r(L_s i_f) + \omega_{sl} L_s i_f = (R_s + L_a p)i_T + \omega_r(L_s i_f) + i_T \frac{R_r L_s}{L_r}$$

$$= \left(R_s + \frac{R_r L_s}{L_r} + L_a p\right) i_T + \omega_r L_s i_f \quad (8.180)$$

from which the torque-producing component of the stator current is derived as

$$i_T = \frac{v_{qs}^e - \omega_r L_s i_f}{R_s + \frac{R_r L_s}{L_r} + L_a p} = \frac{K_a}{(1 + sT_a)} \{v_{qs}^e - \omega_r L_s i_f\} \quad (8.181)$$

where

$$R_a = R_s + \frac{L_s}{L_r} R_r \quad (8.182)$$

$$K_a = \frac{1}{R_a} \quad (8.183)$$

$$T_a = \frac{L_a}{R_a} \quad (8.184)$$

From this block, which converts the voltage and speed feedback into the torque current, the electromagnetic torque is written as

$$T_e = K_t i_T \quad (8.185)$$

where the torque constant is defined as

$$K_t = \frac{3}{2} \frac{P}{2} \frac{L_m^2}{L_r} i_f \quad (8.186)$$

The load dynamics can be represented, given the electromagnetic torque and a load torque that is considered to be frictional for this particular case, as

$$J \frac{d\omega_m}{dt} + B\omega_m = T_e - T_l = K_t i_T - B_l \omega_m \quad (8.187)$$

which, in terms of the electrical rotor speed, is derived by multiplying both sides by the pair of poles:

$$J \frac{d\omega_r}{dt} + B\omega_r = \frac{P}{2} K_t i_T - B_l \omega_r \quad (8.188)$$

and hence the transfer function between the speed and the torque-producing current is derived as

$$\frac{I_T(s)}{\omega_r(s)} = \frac{K_m}{1 + sT_m} \quad (8.189)$$

where

$$K_m = \frac{P}{2}\frac{K_t}{B_t}; B_t = B + B_l; T_m = \frac{J}{B_t} \qquad (8.190)$$

8.13.1.2 Inverter. The stator q axis voltage is delivered by the inverter with a command input that is the error between the torque-current reference and the torque-current feedback. This current error could be amplified through a current controller. The gain of the current controller is considered unity here, but any other gain can easily be incorporated in the subsequent development. The inverter is modeled as a gain, K_{in}, with a time lag of T_{in}. The gain is obtained from the dc-link voltage to the inverter, V_{dc}, and maximum control voltage, V_{cm}, as

$$K_{in} = 0.65\frac{V_{dc}}{V_{cm}} \qquad (8.191)$$

The factor 0.65 here is introduced to account for the maximum peak fundamental voltage obtainable from the inverter with a given dc-link voltage. The interested reader is referred to Chapter 7 for details on this. The torque current error is restricted within the maximum control voltage, V_{cm}. The time lag in the inverter is equal to the average carrier switching-cycle time, i.e., half the period, and is expressed in terms of the PWM switching frequency as

$$T_{in} = \frac{1}{2f_c} \qquad (8.192)$$

8.13.1.3 Speed controller. A proportional-plus-integral (PI) controller is used to process the speed error between the speed-reference and filtered speed-feedback signals. The transfer function of the speed controller is given as

$$G_s(s) = \frac{K_s(1 + sT_s)}{sT_s} \qquad (8.193)$$

where K_s and T_s are the gain and time constants of the speed controller, respectively.

8.13.1.4 Feedback transfer functions. The feedback signals are current and speed, which are processed through first-order filters. They are given in the following.

(i) **Current Feedback Transfer Function:** Very little filtering is common in the current feedback signal; the signal gain is denoted by

$$G_c(s) = H_c \qquad (8.194)$$

(ii) **Speed-Feedback Transfer Function:** The speed-feedback signal is processed through a first-order filter given by

$$G_\omega(s) = \frac{\omega_{rm}(s)}{\omega_r(s)} = \frac{H_\omega}{1 + sT_\omega} \qquad (8.195)$$

where H_ω is the gain and T_ω is the time constant of the speed filter.

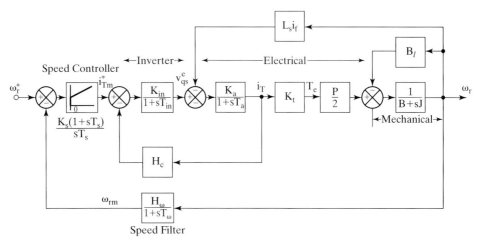

Figure 8.51 Block diagram of the vector-controlled induction motor with constant rotor flux linkages

The speed filter accepts the speed signal as input and produces a modified speed signal for comparison to the speed-reference signal, ω_r^*.

This completes the development of all the subsystems of the vector-controlled induction motor drive with constant rotor flux linkages. By incorporating equations (8.180), (8.181), (8.183), and from (8.191) to (8.195) with the mechanical impedance of the load, speed filter, speed controller, and i_T loop, the block diagram shown in Figure 8.51 is derived.

8.13.2 Block-Diagram Reduction

Further, the speed-feedback pickoff point for the electrical system can be moved to the i_T point resulting in the diagram shown in Figure 8.52(i) which can be further simplified as in Figure 8.52(ii), where the current closed loop transfer function

$$G_i(s) = \frac{K_a K_{in}(1 + sT_m)}{\{(1 + sT_{in})[(1 + sT_a)(1 + sT_m) + K_a K_b] + H_c K_a K_{in}(1 + sT_m)\}} \quad (8.196)$$

where the emf constant is given by

$$K_b = K_m L_s i_f \quad (8.197)$$

8.13.3 Simplified Current-Loop Transfer Function

This third-order current transfer function, $\dfrac{i_{Tm}^*}{H_c i_T^*}$, can be approximated to a first-order transfer function as follows. T_{in} is usually negligible compared to T_1, T_2, and

Section 8.13 Speed-Controller Design for an Indirect Vector-Controlled Induction

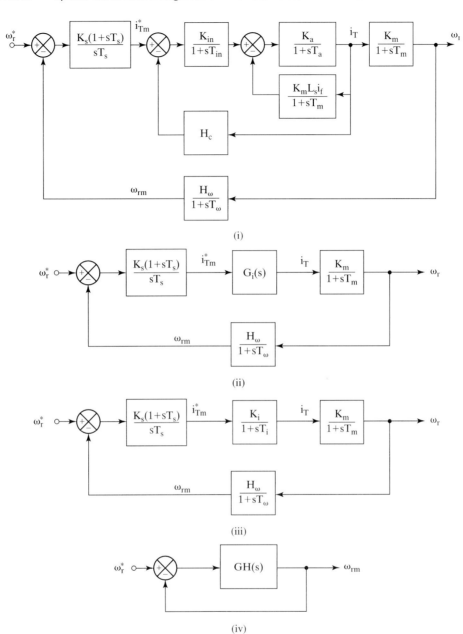

Figure 8.52 Block-diagram reduction of Figure 8.51

T_m, and, in the vicinity of the crossover frequency, the following approximations are valid:

$$1 + sT_{in} \cong 1 \qquad (8.197)$$

$$(1 + sT_a)(1 + sT_{in}) \cong 1 + s(T_a + T_{in}) \cong 1 + sT_{ar}$$

where $T_{ar} = T_a + T_{in}$.

Substitution of these into $G_i(s)$ results in

$$G_i(s) = \frac{K_a K_{in}(1 + sT_m)}{(1 + sT_{ar})(1 + sT_m) + K_a K_b + H_c K_a K_{in}(1 + sT_m)} \qquad (8.198)$$

which is written compactly as

$$G_i(s) = \frac{T_1 T_2 K_a K_{in}}{T_{ar} T_m} \cdot \frac{(1 + sT_m)}{(1 + sT_1)(1 + sT_2)} \qquad (8.199)$$

where

$$-\frac{1}{T_1}, -\frac{1}{T_2} = \frac{-b \pm \sqrt{b^2 - 4ac}}{2a} \qquad (8.200)$$

$$a = T_{ar} \cdot T_m$$
$$b = T_{ar} + T_m + H_c K_a K_{in} T_m$$
$$c = 1 + K_a K_b + H_c K_a K_{in}$$

The transfer function $G_i(s)$ is simplified by using the fact that $T_1 < T_2 < T_m$, and, near the vicinity of the crossover frequency, the following approximations are valid:

$$1 + sT_m \cong sT_m \qquad (8.201)$$
$$1 + sT_2 \cong sT_2 \qquad (8.202)$$

Substitution of these into $G_i(s)$ gives

$$G_i(s) = \frac{K_a K_{in} T_1}{T_{ar}} \cdot \frac{1}{(1 + sT_1)} = \frac{K_i}{(1 + sT_i)} \qquad (8.203)$$

where K_i and T_i are the gain and time constants of the simplified current-loop transfer function, given by

$$K_i = \frac{K_a K_{in} T_1}{T_{ar}} \qquad (8.204)$$

$$T_i = T_1 \qquad (8.205)$$

The model reduction of the current loop is necessary to synthesize the speed controller. The loop transfer function of the speed is given then by the substitution of this simplified transfer function of the current loop, as shown in Figure 8.52(iii), and by combining all the blocks to obtain the final block diagram, as shown in Figure 8.52(iv).

8.13.4 Speed-Controller Design

The loop transfer function of the speed loop is given by

$$GH(s) \cong \frac{K_s}{T_s} K_g \frac{1 + sT_s}{s^2(1 + sT_{\omega i})} \qquad (8.206)$$

where approximation $1 + sT_m \cong sT_m$ is made and the current-loop time constant and speed-filter time constant are combined into an equivalent time constant:

$$T_{\omega i} = T_\omega + T_i$$
$$K_g = K_i K_m \frac{H_\omega}{T_m} \qquad (8.207)$$

The transfer function of the speed to its command is derived as

$$\frac{\omega_r(s)}{\omega_r^*(s)} = \frac{1}{H_\omega} \left\{ \frac{1 + sT_s}{1 + sT_s + \frac{T_s}{K_g K_s} s^2 + \frac{T_s T_{\omega i}}{K_g K_s} s^3} \right\} \qquad (8.208)$$

and, by equating the coefficient of the denominator polynomial to the coefficient of the symmetric optimum function, K_s and T_s can be evaluated as given in Chapter 3. The symmetric optimum function for a damping ratio of 0.707 is given by

$$\frac{(1 + sT_s)}{1 + (T_s)s + \left(\frac{3}{8}T_s^2\right)s^2 + \left(\frac{1}{16}T_s^3\right)s^3} \qquad (8.209)$$

from which the speed-controller constants are derived as

$$T_s = 6T_{\omega i} \qquad (8.210)$$
$$K_s = \frac{4}{9} \frac{1}{K_g T_{\omega i}} \qquad (8.211)$$

The proportional and integral gains of the speed controller are, respectively, then obtained as

$$K_p = K_s = \frac{4}{9} \frac{1}{K_g T_{\omega i}} \qquad (8.212)$$
$$K_i = \frac{K_s}{T_s} = \frac{2}{27} \frac{1}{K_g T_{\omega i}^2} \qquad (8.213)$$

The overshoot can be suppressed by canceling the zero with the addition of a pole $(1 + sT_s)$ in the path of the speed command. Consider the following example to test the validity of the various assumptions made in the derivation of the speed-controller design.

Example 8.8

Consider the induction motor given in example in Section 8.4, with the inverter and load parameters given below:

$$I_f = 6 \text{ A}, f_c = 2000 \text{ Hz}, B_t = 0.05, H_\omega = 0.05, T_\omega = 0.002, V_{cm} = 10 \text{ V},$$
$$J = 0.0165 \text{ kg} - \text{m}^2, V_{dc} = 285 \text{ V}, H_c = 0.333 \text{ V/A}$$

Calculate the speed-controller constants, and verify the validity of the assumptions in design.

Solution

$$R_a = R_s + R_r \cdot \frac{L_s}{L_r} = 0.457 \, \Omega; \quad K_a = \frac{1}{R_a} = 2.1885; \quad L_a = L_s - \frac{L_m^2}{L_r} = 0.0047 \text{ H}$$

$$T_a = \frac{L_a}{R_a} = 0.0104 \text{ sec}; \quad T_m = \frac{J}{B_t} = 0.33 \text{ sec}; \quad K_m = \frac{P}{2} \frac{K_t}{B_t} = 36.224$$

$$K_t = \frac{3}{2} \cdot \frac{P}{2} \cdot \frac{L_m^2}{L_r} I_f = 0.9056; \text{ N·m/A} \quad K_b = \frac{P}{2} \cdot \frac{K_t}{B_t} \cdot L_s I_f = 11.9672 \text{ v/rad/sec};$$

$$T_{in} = \frac{T_c}{2} = \frac{1}{2f_c} = 0.00025 \text{ sec};$$

$$K_{in} = 0.65 \frac{V_{dc}}{V_{cm}} = 18.525 \text{ V/V}$$

$$T_{ar} = T_a + T_{in} = 0.01065 \text{ sec} \,; \quad T_1 = 0.00074 \text{ sec} \,; \quad T_2 = 0.1173 \text{ sec}$$

Approximated current loop:

$$K_i = \frac{K_{in}}{R_a} = 2.8708; \quad T_i = T_1 = 0.00074 \text{sec}$$

Speed controller:

$$K_g = \frac{K_i K_m H_\omega}{T_m} = 104.1; \quad T_{\omega i} = T_\omega + T_i = 0.00274 \text{ sec}; \quad T_s = 6T_{\omega i} = 0.0164 \text{ sec};$$

$$K_s = \frac{4}{9K_g T_{\omega i}} = 1.5605$$

Proportional gain, $K_{ps} = K_s = 1.5605$

Integral gain, $K_{is} = \frac{K_s}{T_s} = 95.0655$

With these calculated values, the following transfer functions are obtained:

Exact current loop transfer function, $G_i(s) = \dfrac{G_{in}(s)G_{12}(s)}{1 + H_c G_{in}(s) G_{12}(s)}$

where

$$G_{in}(s) = \frac{K_{in}}{1 + sT_{in}}; \quad G_{12}(s) = \frac{G_1(s)}{1 + G_1(s)G_2(s)}; \quad G_1(s) = \frac{K_a}{1 + sT_a}; \quad G_2(s) = \frac{K_b}{1 + sT_m}$$

Simplified current-loop transfer function, $G_{is}(s) = \dfrac{K_i}{1 + sT_i}$

Section 8.13 Speed-Controller Design for an Indirect Vector-Controlled Induction

Exact speed-loop transfer function, $G_{\omega e}(s) = \dfrac{\omega_r(s)}{\omega_r^*(s)} = \dfrac{G_{\omega f}(s)}{1 + G_{\omega f}(s)G_\omega(s)}$

where

$$G_{\omega f}(s) = \dfrac{K_m}{(1 + sT_m)} \cdot \dfrac{K_s(1 + sT_s)}{sT_s} \cdot G_i(s) \; ; \; G_\omega(s) = \dfrac{H_\omega}{1 + sT_\omega}$$

Simplified speed-loop transfer function: Obtained from simplified current loop-transfer function and given by

$$G_{\omega s}(s) = \dfrac{G_f(s)}{1 + G_f(s)G_\omega(s)}$$

where

$$G_f(s) = \dfrac{K_m}{(1 + sT_m)} \cdot \dfrac{K_s(1 + sT_s)}{sT_s} \cdot \dfrac{K_i}{(1 + sT_i)}$$

The smoothed speed-loop transfer function is obtained by introducing a pole of $-\dfrac{1}{T_s}$ in the forward path of the speed reference, resulting in approximate and exact smoothed speed-loop transfer functions:

Approximate: $G_{s\omega s}(s) = \dfrac{G_{\omega s}(s)}{1 + sT_s}$

Exact: $G_{s\omega e}(s) = \dfrac{G_{\omega e}(s)}{1 + sT_s}$

All these transfer functions are computed; their gain and phase plots are given in Figures 8.53(i), (ii), and (iii). Even though there seems to be a significant discrepancy between the gains of the simplified and the exact current-loop transfer functions, note that their phases are identical in the frequency range of interest. The discrepancy due to the approximation of the current loop from third order to first order has hardly affected the accuracy of the speed-loop transfer functions, as is clearly seen from the gain and phase plots. This justifies the assumptions made in various approximations.

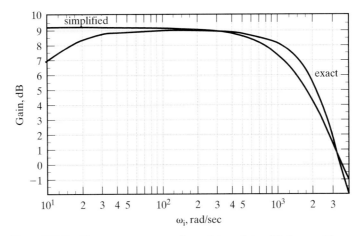

Figure 8.53(i) Frequency response of complete and simplified current loops

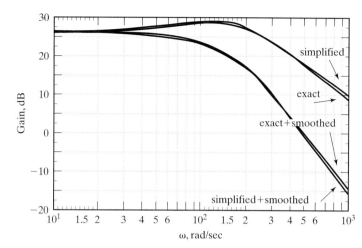

Figure 8.53(ii) Comparison of speed-loop frequency responses with and without smoothing with complete and simplified current loops

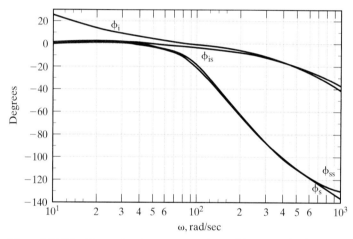

Figure 8.53(iii) Phase plots for current- and speed-loop transfer functions

Note that φ_{is} is the simplified current-loop phase, φ_i is its exact phase, φ_{ss} is the simplified, and φ_s is the exact phase of speed-loop transfer functions. The phase plot for the smoothed speed-loop transfer function is not shown; it is evident from φ_{is} and φ_s.

8.14 PERFORMANCE AND APPLICATIONS

The vector-controlled induction motor drive has delivered high performance. Some of its key performance achievements are given in the following:

(i) more than 6 kHz of torque bandwidth;
(ii) 200-Hz small-signal speed-loop bandwidth;
(iii) precise and smooth zero-speed operation;

(iv) trouble-free four-quadrant operation;

(v) fast torque and speed reversals.

Such performance figures have enabled these motor drives to find applications in

(i) machine-tool servos and spindles;

(ii) robotic actuators;

(iii) punch presses;

(iv) electric traction and propulsion;

(v) rolling mills;

(vi) paper and pulp drives;

(vii) centrifuges.

An illustrative example of the vector-controlled induction motor drive for centrifuges is given in the next section.

8.14.1 Application: Centrifuge Drive [58]

Centrifuges are used to separate the granular and crystalline parts from saturated solutions by spinning them at controlled speeds for set times. For example, a sugar centrifuge application is described with its speed and torque requirements over a cycle of operation. The sugar centrifuge has a basket, which is connected to the motor shaft through an elastomer coupling, as shown in Figure 8.54. The supersaturated sugar solution, with a thick suspension of sugar crystals, will be poured into the basket when the centrifuge is spinning at 250 rpm. After charging with the solution, the centrifuge basket is accelerated to a speed of 1200 rpm. This speed is maintained until the sugar solution is forced through the perforations in the bottom of the basket. This operation is known as the spin cycle. During spin, note that the motor is delivering friction and windage losses and hence requires the development of a very small electromagnetic torque. During the spin cycle, the sugar crystals stick to the walls of the basket and become dry. Before the removal of the sugar crystals, the basket has to be stopped. To stop it in a short time and to recover the kinetic energy stored in the basket, a maximum braking torque is applied to the centrifuge. The removal of dry sugar crystals is effected by spinning the basket in the reverse direction at 35 rpm and requires nearly 0.5 p.u. of torque to loosen the crystals. The sugar crystals are discharged through the bottom. A complete cycle of operation is shown in Figure 8.55. From the operational cycle it is evident that, to improve productivity, the acceleration and deceleration have to be swift, which requires peak torque capability from standstill to maximum speed. The deceleration can be smooth and fast only if it is done under regenerative conditions. Further, a controlled speed reversal and low-speed operation with 0.5 p.u. of torque to loosen the sugar crystals for final discharge to packaging requires a four-quadrant drive with very high torque and speed bandwidths. The power requirement can be up to 250 kW for the centrifuges: the baskets are 1219 mm (48 inches) in diameter and 762 mm (30 inches) deep and weigh considerably with the charge in them. From these requirements, the autosequentially commutated current-source inverter (ASCI)-driven induction motor emerges as the strong choice for this application, provided that the following modifications are incorporated in the design of the inverter control.

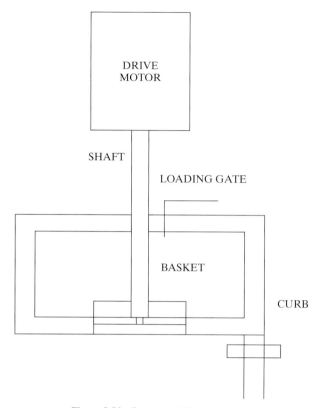

Figure 8.54 Sugar centrifuge schematic

- To reduce the torque pulsations at low speeds, PWM of the current pulses has to be provided for.
- For faster torque response, vector control is required to overcome the oscillatory responses and to develop the maximum torque per ampere input.

Note that the ASCI induction motor drive is superior to dc drives, because it can be operated under hazardous environment, having no commutator or brushes. Compared to other inverter drives, ASCI is cost effective, highly reliable, and simpler in construction.

8.15 RESEARCH STATUS

The present research is directed to the formulation and experimental verification of parameter-compensation schemes, speed and position sensorless schemes in the vector-controlled induction motor drives. Given the variety of parameter-compensation schemes available, effort is made to compare and contrast and to choose the best of them. All the compensation schemes require feedback of many machine variables. Inexpensive means of acquiring them remains to an extent a seri-

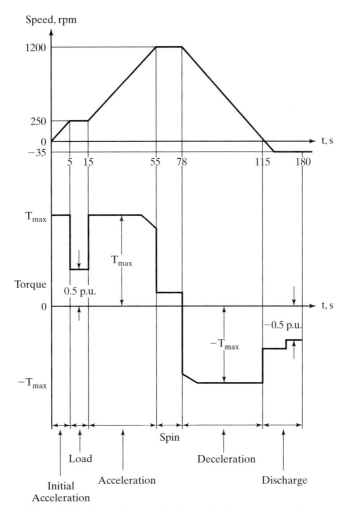

Figure 8.55 A complete cycle of operation in a sugar centrifuge

ous problem in the implementation of parameter-compensation schemes. This is being realized, and some efforts are being made in this direction.

Selftuning algorithms are emerging to match a vector controller to an induction motor with unknown parameters. This is the best approach to overcome the secondary problem of parameter compensation.

8.16 REFERENCES

1. E. Bassi, F. Benzi, S. Bolognani, and G.S. Buja, "A field orientation for current fed induction motor (CSIM) drive based on the torque-angle closed loop control," *Conf. Record of IEEE–Industry Applications Society Annual Meeting*, vol. 1, pp. 384–389, Oct. 1989.

2. K. H. Bayer and F. Blaschke, "Stability problems with the control of induction motors using the method of field orientation," *Conf. Record of IFAC Symposium on Power Electronics,* pp. 483–492, 1977.

3. A. Bellini, G. Gifalli, and G. Ulivi, "A microcomputer based direct field oriented control of induction motors," *Proc. ICEM '86 München, International Conference on Electrical Machines,* vol. 2, pp. 652–655, Sept. 1986.

4. F. Blaschke, "Das Verfahren der Feldorientierung zur Regelung der Drehfelmachine" ("The method of field orientation for control of three phase machines"), Ph.D. Dissertation, TU Braunschweig, 1973.

5. F. Blaschke, "The principle of field orientation as applied to the new transvektor closed-loop control system for rotating-field machines," *Siemens Review,* vol. 34, pp. 217–220, May 1972.

6. F. Blaschke, "A new method for the structural decoupling of ac induction machines," *Proc. IFAC Symposium,* Dusseldorf, pp. 1–15 (6.3.1), Oct. 1971.

7. V. F. Blaschke, H. Ripperger, and H. Steinkonig, "Regelung Umrichtergespeister Asynchronmachinnen mit Eingepragtem Standerstrom," *Siemens-Zeitschrift,* no. 42, pp. 773–777, 1968.

8. B. K. Bose, "Scalar decoupled control induction motor," *Trans. on IEEE–Industry Applications Society,* vol. IA-20, no. 1, pp. 216–225, Jan./Feb. 1984.

9. D. Dalal and R. Krishnan, "Parameter compensation for indirect vector controlled induction motor drive using estimated airgap power," *Conf. Record of IEEE–Industry Applications Society Annual Meeting,* pp. 170–176, Oct. 1987.

10. D. Dalal, "Novel parameter compensation scheme for indirect vector controlled induction motor drives," M.S. Thesis, The Bradley Department of Electrical Engineering, Virginia Polytechnic Institute & State University, Blacksburg, VA, July 1987.

11. M. Depenbrock, "Direct self-control (DSC) of inverter-fed induction machine," *Trans. on IEEE–Power Electronics,* vol. 3, no. 4, pp. 420–429, Oct. 1988.

12. W. Floter and H. Ripperger, "Field-oriented closed-loop control of an induction machine with transvektor control system," *Siemens Review,* vol. 39, no. 6, pp. 248–252, June 1972.

13. W. Floter and H. Ripperger, "Field-oriented closed-loop control of an induction machine with the new TRANSVECTOR control system," *Siemens—Z.45,* pp. 761–764, 1971.

14. A. Fratta, A. Vagati, and F. Villata, "Vector control of induction motors without shaft transducers," *Conf. Record of IEEE–Power Electronics Specialists' Conference,* vol. 2, pp. 839–846, April 1988.

15. R. Gabriel, W. Leonhard, and C. Nordby, "Microprocessor control of induction motors using field coordinates," *Proc. IEEE International Conference on Electrical Variable-Speed Drives,* pp. 146–150, Sept. 1979.

16. l. J. Garces, "Parameter adaptation for speed controlled static ac drive with squirrel cage induction motor," *Conf. Record of IEEE Industry Applications Society Annual Meeting,* pp. 843–850, Oct. 1979.

17. T. G. Habetler, F. Profumo, M. Pastorelli, and L. M. Tolbert, "Direct torque control of induction machines using space vector modulation," *IEEE Trans. on Industry Applications,* vol. 28, no. 5, pp. 1045–1053, Sept./Oct. 1992.

18. T. G. Habetler, F. Profumo, G. Griva, M. Pastorelli, and A. Bettini, "Stator resistance tuning in a stator flux field oriented drive using an instantaneous hybrid flux estimator," *Conf. Record, EPE Conf., Brighton, UK,* vol. 4, pp. 292–299, 1993.

19. K. Hasse, "Zur Dynamik Drehzahlgeregelter Antriebe mit Stomrichtergespeisten Asynchron-Kurzschlusslaufermachinen" ("On the dynamic of speed control of static ac drives with squirrel-cage induction machines"), Ph.D. Dissertation, TH Darmstadt, 1969.
20. J. Holtz and A. Khambadkone, "Vector controlled induction drive with a self-commissioning scheme," *Proc. IEEE Industrial Electronics Conference*, vol. II, pp. 927–932, Nov. 1990.
21. J. Holtz and T. Thimm, "Identification of the machine parameters in a vector controlled induction motor drive system," *Conf. Record of IEEE-Industry Applications Society Annual Meeting*, vol. 1, pp. 601–606, Oct. 1989.
22. J. Holtz and E. Bube, "Field-oriented asynchronous pulse width modulation for high performance ac machine drives operating at low switching frequency," *Conf. Record of IEEE Industry Applications Society Annual Meeting*, pp. 412–417, Oct. 1988.
23. R. Joetten and G. Maeder, "Control methods for good dynamic performance induction motor drives based on current and voltages as measured quantities," *Trans. on IEEE Industry Applications Society*, vol. IA-19, no. 3, pp. 356–363, May/June 1983.
24. M. P. Kazmierkowski and A. B. Kasprowicz, "Improved direct torque and flux vector control of PWM inverter-fed induction motor drives," *IEEE Trans. on Industrial Electronics*, vol. 42, no. 4, pp. 344–350, Aug. 1995.
25. R. J. Kerkman, B. J. Seibel, T. M. Rowan, and D. Schlegel, "A new flux and stator resistance identifier for ac drive systems," *Conf. Record, IEEE-IAS, Orlando, Florida*, pp. 310–318, Oct. 1995.
26. R. Krishnan and A. S. Bharadwaj, "A review of parameter sensitivity and adaptation in indirect vector controlled induction motor drive systems," *IEEE Trans. on Power Electronics*, vol. 6, no. 4, pp. 695–703, Oct. 1991.
27. R. Krishnan, A. S. Bharadwaj, and R. A. Bedingfield, "Computer aided simulation for the dynamic analysis of the indirect vector controlled induction motor drive systems," *Proc. Fifth IFAC/IMACS Symposium on Computer Aided Design in Control Systems*, pp. 583–588, University of Wales, Swansea, UK, July 1991.
28. R. Krishnan and P. Pillay, "Parameter sensitivity in vector controlled ac motor drives," *Proc. IEEE-Industrial Electronics Conference*, vol. 1, pp. 212–218, Nov. 1987.
29. R. Krishnan and F. C. Doran, "A method of sensing line voltages for parameter adaptation of inverter fed induction motor servo drives," *Trans. on IEEE-Industry Applications Society*, vol. IA-23, no. 4, pp. 617–622, July/Aug. 1987.
30. R. Krishnan, F. C. Doran, and T. S. Latos, "Identification of thermally safe load cycles for an induction position servo," *Trans. on IEEE Industry Applications Society*, vol. IA-23, no. 4, pp. 636–643, July/Aug. 1987.
31. R. Krishnan and F. C. Doran, "Study of parameter sensitivity in high performance inverter-fed induction motor drive systems," *Trans. on IEEE Industry Applications Society*, vol. IA-23, no. 4, pp. 623–635, July/Aug. 1987.
32. R. Krishnan, J. F. Lindsay, and V. R. Stefanovic, "Comparison of control schemes for inverter fed induction motor drives," *Conf. Record of IEE Conference on Power Electronics and Variable Speed Drives*, (Conf. Publ. no. 234), pp. 312–316, May 1984.
33. R. Krishnan, V. R. Stefanovic, and J. F. Lindsay, "Control characteristics of inverter-fed induction motor," *Conf. Record of IEEE-Industry Applications Society Annual Meeting*, pp. 548–561, Oct. 1981.

34. R. Krishnan, W. A. Maslowski, and V. R. Stefanovic, "Control principles in current source induction motor drives," *Conf. Record of IEEE-Industry Applications Society Annual Meeting*, pp. 605–617, Oct. 1980.

35. B. S. Lee and R. Krishnan, "Adaptive stator resistance compensator for high performance direct torque controlled induction motor drives," *Conf. Record of IEEE-Industry Applications Society Annual Meeting*, pp. 423–430, Oct. 1998.

36. W. Leonhard, *Control of electric drives*, Springer-Verlag, New York, 1985.

37. W. Leonhard, "Introduction to ac motor control using field coordinates," *Conf. Record of Simposia Sulla Evoluziona Nella Diamica Della Machine Elettriche Rotanti, Tirrania*, pp. 370–376, 1975.

38. R. D. Lorenz, "Tuning of field oriented induction motor controllers for high performance applications," *Conf. Record of IEEE Industry Applications Society Annual Meeting*, pp. 607–612, Oct. 1985.

39. R. Magureanu, l. Kreindler, D. Floricau, and C. Solacolu, "Current versus voltage control of the induction motor operating at constant rotor flux," *Proc. Third International Conference on Electrical Machines and Drive*, pp. 221–225, Nov. 1987.

40. T. Matsui, T. Okuyama, J. Takahashi, T. Sukegawa, and K. Kamiyama, "A high accuracy current component detection method for fully digital vector-controlled PWM VSI-fed ac drives," *Conf. Record of IEEE-Power Electronics Specialists' Conference*, pp. 877–884, April 1988.

41. S. Mir, M. E. Elbuluk, and D. S. Zinger, "PI and fuzzy estimators for tuning the stator resistance in direct torque control of induction machines," *IEEE Trans. on Power Electronics*, vol. 13, no. 2, pp. 279–287, March 1998.

42. A. Nabae, K. Otsuka, H. Uchino, and R. Kurosawa, "An approach to flux control of induction motors operated with variable frequency power supply," *Conf. Record of IEEE-Industry Applications Society Annual Meeting*, pp. 890–896, Oct. 1978.

43. D.Y. Ohm, Y. Khersonsky, and J. R. Kimzey, "Rotor time constant adaptation method for induction motors using dc link power measurements," *Conf. Record of IEEE-Industry Applications Society Annual Meeting*, vol. 1, pp. 588–593, Oct. 1989.

44. K. Ohnishi, H. Suzuki, K. Miyachi, and M. Terashima, "Decoupling control of secondary flux and secondary current in induction motor drive with controlled voltage source and its comparison with volts/hertz control," *Trans. on IEEE Industry Applications Society*, vol. IA-21, no. 1, pp. 241–247, Jan./Feb. 1985.

45. T. Ohtani, N. Takada, and K. Tanaka, "Vector control of induction motors without shaft encoders," *Conf. Record of IEEE-Industry Applications Society Annual Meeting*, Vol 1., pp. 500–507, Oct. 1989.

46. T. Ohtani, "Torque control using the flux derived from magnetic energy in induction motors driven by static converter," *Conf. Record of International Power Electronics Conference*, pp. 696–707, March 1983.

47. T. Okuyama, N. Hujimoto, and H. Hujii, "A simplified vector control system without speed and voltage sensors—Effect of saturating errors of control parameters and their compensation," *Electrical Engineering in Japan*, vol. 110, no. 4, pp. 129–139, 1990.

48. C. E. Rettig, "U.S. Patent # 962,614," June 1976.

49. K. Sattler, U. Sachaefer, and R. Gheysens, "Field oriented control of an induction motor with field weakening under consideration of saturation and rotor heating," *Fourth International Conference on Power Electronics and Variable-Speed Drives*, pp. 286–291, July 1990.

50. C. D. Schauder, "Adaptive speed identification for vector control of induction motors without rotational transducers," *Conf. Record of IEEE-Industry Applications Society Annual Meeting*, vol. 1, pp. 493–499, Oct. 1989.
51. S. Tadakuma, K. Tanaka, K. Miura, and H. Naito, "Vector-controlled induction motors under feedforward and feedback controls," *Electrical Engineering in Japan* (English Translation of Denki Gakkai Robunshi), vol. 110, no. 5, pp. 100–110, 1990.
52. I. Takahashi and Y. Ohmori, "High performance direct torque control of an induction motor," *Trans. on IEEE Industry Applications Society*, vol. IA-25, no. 2, pp. 257–264, Mar./Apr. 1989.
53. I. Takahashi and T. Noguchi, "A new quick-response and high efficiency control strategy of an induction motor," *Trans. on IEEE Industry Applications Society*, vol. IA-22, no. 5, pp. 820–827, Sept./Oct. 1986.
54. I. Takahashi and T. Noguchi, "Quick control of an induction motor by means of instantaneous slip frequency control," *Electrical Engineering in Japan*, vol. 106, no. 2, pp. 46–53, Mar./Apr. 1986.
55. C. R. Wasko, "500HP, 120Hz Current-fed field-oriented control inverter for fuel pump test stands," *Conf. Record of IEEE-Industry Applications Society Annual Meeting*, pp. 314–320, Sept. 1986.
56. C. R. Wasko, "AC vector control drives for process applications," *Conf. Record of IEEE-Industry Applications Society Annual Meeting*, pp. 134–137, Oct. 1985.
57. X. Xu, De Donker, and D. W. Novotny, "A stator flux oriented induction machine drive," *Conf. Record of IEEE-Power Electronics Specialists' Conference*, pp. 870–876, April 1988.
58. Richard H. Osman and Joseph B. Bange, "A regenerative centrifuge drive using a current-fed inverter with vector control," *IEEE Trans. on Industry Applications*, vol. 27, no. 6, pp. 1076–1080, Nov./Dec. 1991.

8.17 DISCUSSION QUESTIONS

1. The command variables i_f^* and i_T^* are in synchronously rotating reference frames in the vector control scheme. Is it possible to have them in rotor reference frames? Are there any special advantages in doing so?
2. The field angle θ_f could be obtained by using flux-sensing coils. Discuss the measurement scheme in a block-diagram form. Discuss the situation when the stator windings are fed dc currents.
3. Using flux-sensing coils and output signals, devise a scheme to measure the rotor flux linkages and electromagnetic torque. (Hint: Reference 39.)
4. Vector control is the instantaneous control of the stator-current phasor. This implies that the performance of the motor drive is very much dependent on the response of the inverter. Discuss the impact of switching speed on the transient response of the motor drive.
5. In high-performance drives, two types of current control are commonly devised. They are hysteresis control (or bang-bang control) and PWM control. Discuss the impact of these controllers on the performance of the motor drive.
6. Which is the most significantly changing parameter in the induction motor and why? In that case, identify a simple and inexpensive parameter-compensation scheme.

7. A flux or torque feedback control is sufficient to overcome the effects of parameter sensitivity of the induction motor drive. Justify this statement.
8. Is there a method to differentiate among the various parameter-compensation schemes? Explain the basic principle of such a method. Discuss the merits and demerits of the method. (Hint: Reference 28.)
9. Discuss the high performance of vector-controlled induction motor drives *vis-à-vis* dc motor drives and permanent-magnet synchronous motor drives.
10. There are many variations of vector control based on the same principle of instantaneous control of the stator-current phasor. Summarize the key variations, their merits, and demerits. (Hint: Reference 37.)
11. Prove that position or speed signal is not obtainable from the fundamental model with sinusoidally distributed windings and with no saliency in the rotor structure.
12. Saliency on the rotor can be introduced by any one of the following:
    (i) varying the rotor copper fill;
    (ii) varying the rotor slot opening;
    (iii) varying the rotor slot height;
    (iv) varying the tooth lip thickness.
    Will these methods lend themselves to realistic implementation in an actual drive system? Discuss the merits and demerits of use of saliency for position detection.
13. Rotor-slot harmonics can be determined from the stator-current spectrum. This then has indirectly the position and speed information. This signal is extracted by passing through a band-pass filter and extracted with a phase-locked loop. At low speeds, this signal has a large time constant and is not usable for high dynamic performance at zero and low speeds. Justify this statement.
14. Direct self-control does not rely on position sensor feedback. Is this correct?
15. For high performance, stator or rotor flux-linkages-position information is required in the vector controller. Explain why?
16. How does direct self control get its position information?
17. Stator resistance is the only machine parameter that direct self control is dependent on for its implementation. Stator-current error can be used to adapt the stator resistance. What other signals can be used for stator-resistance adaptation?
18. Air gap power feedback is ineffective for tracking stator-resistance variations. Explain why.
19. A stator-flux-linkages-based vector controller does not give a simple decoupling of flux and torque channels, as in the indirect vector controller. Prove this statement.
20. When the stator flux-linkages position is unavailable or has significant error at zero and low speeds, how can the direct-self-control-based induction motor operate? Will there be a degradation of performance, and, if so, how can it be averted?
21. In the torque and flux feedback loops, the fundamental components of currents and voltages or their actual values, including the harmonics, can be used. Discuss the merits and demerits of the two options, and, from the discussion, explain which one is appropriate for high dynamic performance.
22. How can fundamental components be extracted from the switched machine input voltages? Will this affect the dynamic performance of the drive system?
23. The flux component of stator-current computation usually neglects the derivative term because it can produce a large flux current command in the indirect vector con-

troller for variations in the flux-linkages command. Is it safe to neglect it for small variations in the flux-linkages command?

24. Indirect vector controller is dependent on three motor parameters. Direct online measurement of these parameters is not possible. Indirect methods to detect all the parameter variations with one variable, such as reactive or real power errors, are easier. These variables are then used to adjust the slip-speed reference, for example. Justify with reasons how this can be adequate in parameter compensation, although individual parameter update is required in the indirect vector controller.

25. Enumerate all the possible signals that can be measured or extracted for use in the parameter adaptation of the indirect vector controller.

26. By adapting parameters in the vector controller, tuning is automatic, but compensating the parameter variations with only a slip-speed signal is not exact; it leaves out other input variables, such as i_f and i_T. Comment on the quality of the tuning of the vector controller achieved in this case.

27. Air gap power feedback control for parameter adaptation is better than modified reactive-power-feedback control. Justify it with reasons.

28. Will switching strategies affect the performance of the vector-controlled induction motor drive, particularly from the point of view of torque, speed, and position responses?

29. Develop a procedure to synthesize the current controllers, assuming that each phase current is individually controlled in the indirect vector-controlled induction motor drive.

30. Develop a procedure to synthesize the torque- and flux-component current controllers for a direct vector-controlled induction motor drive. What are the salient differences between these and the controllers discussed in question 29?

31. A flux-weakening controller is least parameter-sensitive in the stator and rotor flux-linkages-based vector controllers. Comment on the veracity of this statement.

8.18 EXERCISE PROBLEMS

1. Consider an induction motor whose parameters are given in worked Example 5.1 with $J = 0.01667$ kg-m^2 and $B = 0.0$ N·m/rad/sec. It is being driven by a current-source PWM inverter. It has a flux and torque processor for estimating the torque, the rotor flux linkage, and its angle. Assume that the control strategy used is the direct vector control (per Figure 8.2) and the drive is in the torque mode with no dynamometer attached.

   (i) Develop a subsystem of the inverter with three inner current feedback loops. The inverter is operating in the PWM mode with a switching frequency of 6 kHz. The current error is amplified by a PI current controller (or, for ease, use a high-gain P controller) and limited to the rated current. Then this control signal is compared with the triangular PWM carrier signal to find the switching signals from which the applied line-to-line and hence phase voltages are found. Assume that the system is balanced.

   (ii) Use the algorithm given in the text and the block diagram shown in Figure 8.3 to realize the flux and torque processor.

   (iii) Once these subsystems are developed, close the torque and the rotor flux-linkages feedback loops to control torque and rotor flux-linkages. The flux-linkages command is maintained at 0.405 Wb–turn and the torque command is set at 20.178 N·m. The PI (or P) limiters are set at 7.49 and 17.27 A for the flux- and torque-producing-component command currents, respectively. Then this gives rise to a peak stator-phase current of 18.84 A. Note that these are not rms values but peak values.

(iv) Test each subsystem before integrating the system completely. Plot the torque and its command, rotor flux linkage and its command, field and torque components of the stator current and their commands, stator-current phasor magnitude and torque angle, field angle, phase a current and its command, phase a voltage and rotor speed. Exercise care when you find the field angle by properly implementing the arctangent function for all quadrants.

2. Consider an indirect vector-controlled induction motor drive with rotor-position feedback. The inverter has three inner current feedback loops and works with a PWM frequency of 20 kHz. The parameters of the motor and drive are from the above problem. The field and torque current commands are limited to their rated and twice their rated values, respectively. The dc-link voltage can be considered to be 285 V, and the parameters of the controller match those of the machine. The load torque is zero throughout this problem. The following dynamic simulation results are required to enhance your understanding of the indirect vector-controlled induction motor drive:

   (i) With torque command kept zero, apply a rated rotor flux-linkages command. (The drive is in torque mode with speed loop opened.)

   (ii) When the steady state is reached in (i), apply a step torque command of 1 p.u. for 20 ms, then a step of -1 p.u. for 10 ms, and then 0 p.u. for 5 ms. (The drive is in torque mode with open speed loop.)

   (iii) Close the speed feedback loop with a high proportional gain in the speed controller and with the rated rotor flux linkages in steady state in the machine; then apply a step speed command of 0.75 p.u. for 100 ms, then apply a -0.75 p.u. for 50 ms, and then 0 p.u. for 10 ms.

   Plot all the key variables, such as the speed, torque and rotor flux-linkages commands and their responses, flux and torque current commands and their responses, phase a voltage and phase a current and its command. Plot these results in p.u.

3. Realize the indirect vector controller with discrete analog and digital integrated chips. Give only a block-diagram schematic of the realization. Minimize the number of components in the realization.

4. Stability is taken for granted, in general, in vector-controlled induction motor drives. Instability can arise when the controller and machine parameters do not match, as is shown in the case of the direct self control. Examine the stability of the indirect vector-controlled induction motor drive.

5. A perfect matching of machine and controller parameters is an ideal case that hardly ever arises in practice. Explore control strategies that use modern control theory to make the vector controller highly insensitive to machine-parameter variations.

6. Most of the parameter-compensation schemes are themselves parameter-dependent. Consider an example, and illustrate the effects on the performance of the indirect and direct vector-controlled motor drives.

7. Devise a control scheme to maximize the efficiency of the motor drive operating with a vector-control strategy.

CHAPTER 9

Permanent-Magnet Synchronous and Brushless DC Motor Drives

9.1 INTRODUCTION

The availability of modern permanent magnets (PM) with considerable energy density led to the development of dc machines with PM field excitation in the 1950s. Introduction of PM to replace electromagnets, which have windings and require an external electric energy source, resulted in compact dc machines. The synchronous machine, with its conventional field excitation in the rotor, is replaced by the PM excitation; the slip rings and brush assembly are dispensed with. With the advent of switching power transistor and silicon-controlled-rectifier devices in later part of 1950s, the replacement of the mechanical commutator with an electronic commutator in the form of an inverter was achieved. These two developments contributed to the development of PM synchronous and brushless dc machines. The armature of the dc machine need not be on the rotor if the mechanical commutator is replaced by its electronic version. Therefore, the armature of the machine can be on the stator, enabling better cooling and allowing higher voltages to be achieved: significant clearance space is available for insulation in the stator. The excitation field that used to be on the stator is transferred to the rotor with the PM poles. These machines are nothing but 'an inside out dc machine' with the field and armature interchanged from the stator to rotor and rotor to stator, respectively.

This chapter contains the description of the PM synchronous and brushless dc machines, derivation of their respective dynamic models, principle of control, analysis of the drive system, various control strategies, design of speed controller, flux-weakening operation, and position-sensorless operation. Some converter topologies for low-cost operation are also included at the end of the chapter.

9.2 PERMANENT MAGNETS AND CHARACTERISTICS

Permanent-magnet characteristics and their operation within a magnetic circuit are described. The air gap or load line to determine the operating point of the magnet is derived. The energy-density definition and magnet-volume calculation are given to clarify the rudiments of application considerations for the magnet in machine design. The design computation for the rotor magnet in ac machines is outside the scope of this book; the interested reader is directed to the cited references at the end of this chapter.

9.2.1 Permanent Magnets

Materials to retain magnetism were introduced in electrical-machine research in 1950's. There has been a rapid progress in materials in the interim. Materials that retain magnetism are known as *hard magnet* materials. Various materials, such as Alnico-5, ferrites, samarium–cobalt, and neodymium–boron–iron are available as permanent magnets for use in machines. The B–H demagnetization characteristics of these materials are shown in Figure 9.1—for second quadrant only, because the magnets do not have an external excitation once they are magnetized and therefore have to operate with a negative magnetic field strength. That includes the second and third quadrants of the B–H characteristics. It is not usual to encounter the third quadrant in operation; hence, only the second quadrant is considered here. The last three listed materials have straight-line B vs. H characteristics, whereas Alnico-5 has the highest remnant flux density but has a nonlinear B vs. H.

The flux density at zero excitation is known as remnant flux density, B_r, shown in Figure 9.2 for a generic hard permanent magnet. The low-grade magnet has a curve, usually termed a *knee*, at low flux density, and at around this point it sharply dives down to zero flux density and reaches the magnetic field strength of H_{cl} known as coercivity. The magnetic field strength at the knee point is H_k. If the external excitation acting against the magnet is removed, then it recovers magnetism along a line parallel to the original B-H characteristic. In that process, it reaches a new remnant flux density, B_{rr}, considerably lower than the original remnant flux density. The magnet has lost this difference in flux density irretrievably. Even though the recoil line is shown as a straight line, it usually is a loop, and the average of the loop flux densities is represented by the straight line. In the case of a high-grade magnet, the B–H characteristic is a straight line, and its coercivity is indicated by H_{ch}. The line along which the magnet can be demagnetized and restored to magnetism is known as the *recoil line*. The slope of this line is equal to $\mu_0\mu_{rm}$, as derived from the relationship between the flux density and magnetic field intensity, where μ_{rm} is relative recoil permeability. For samarium–cobalt and neodymium–boron magnets, the recoil permeability, μ_{rm}, nearly has a value between 1.03 to 1.1.

9.2.2 Air Gap Line

To find the operating point on the demagnetization characteristic of the magnet, consider the flux path in the machine. The flux crosses from a north pole of the rotor magnet to the stator across an air gap and then closes the flux path from the stator to the rotor south pole via an air gap. In the process, the flux crosses two times the

Section 9.2 Permanent Magnets and Characteristics

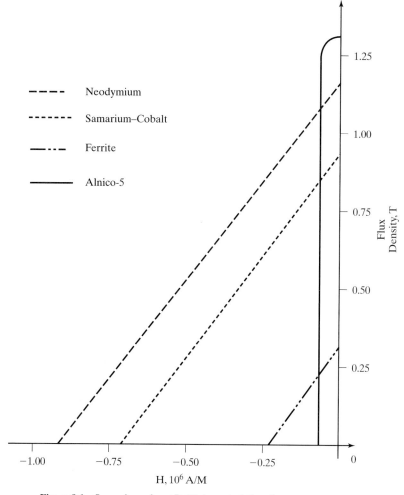

Figure 9.1 Second-quadrant B–H characteristics of permanent magnets

magnet length and two times the air gap, as shown in Figure 9.3. The mmf provided by magnets is equal to the mmf received by the air gap if the mmf requirement of stator and rotor iron is considered negligible. Then,

$$H_m l_m + H_g l_g = 0 \tag{9.1}$$

where H_m and H_g are magnetic field strengths in magnet and air, respectively, and, l_m and l_g are the length of the magnet and air gap, respectively. The operating flux density on the demagnetization characteristic can be modeled, assuming that it is a straight line, as

$$B_m = B_r + \mu_0 \mu_{rm} H_m \tag{9.2}$$

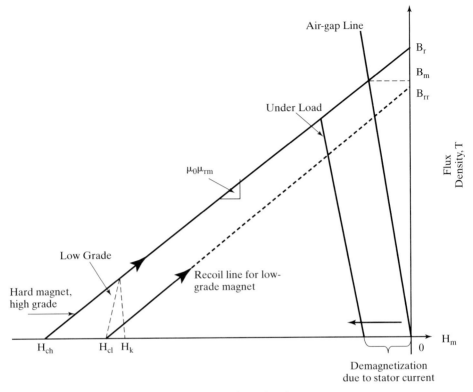

Figure 9.2 Operating point of magnets

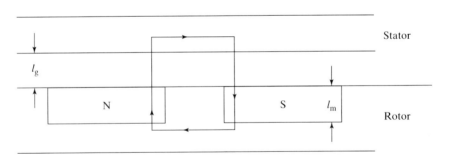

Figure 9.3 Simple layout of the stator and rotor of a machine

Substituting for H_m from (9.1) into (9.2) in terms of H_g, and then writing it in terms of the air gap flux density, which is equal to the magnet flux density, gives the operating magnet flux density:

$$B_m = \frac{B_r}{\left(1 + \dfrac{\mu_{rm} l_g}{l_m}\right)} \qquad (9.3)$$

This clearly indicates that the operating flux density is always less than the remnant flux density, because of the air gap excitation requirement. Note that the excitation requirements of iron and leakage flux are neglected in this conceptual derivation.

The operating point is shown in Figure 9.2, and the line connecting the operating flux density B_m and origin is known as the air gap line or *load line*. The slope of this line is equal to a fictitious permeance coefficient μ_c times the permeance of the air. If the stator is electrically excited, producing demagnetization, then the load line moves toward the left but remains parallel to the original load line, as shown in the figure. The operating flux density is further reduced from B_m. Note that the permeance coefficient is derived for an operating point defined by B_m and H_m as

$$B_m = B_r + \mu_0 \mu_{rm} H_m = -\mu_0 \mu_c H_m \qquad (9.4)$$

from which the permeance coefficient is derived as

$$\mu_c = \frac{B_r}{-\mu_0 H_m} - \mu_{rm} = \frac{-\mu_0 \mu_{re} H_m}{-\mu_0 H_m} - \mu_{rm} = \mu_{re} - \mu_{rm} \qquad (9.5)$$

where μ_{re} can be considered as external permeability. The variations in the remnant flux density are due to temperature changes as well as to the impact of the applied magnetic field intensity, both of which are induced by external operating conditions, as is clearly seen from this formulation of μ_c, and therefore it is appropriately labeled as external permeability. As the demagnetizing field is introduced by external operating conditions, it is seen that the permeance coefficient will decrease as the external permeability also decreases for that operating point. In hard permanent magnets, the external permeability is on the order of from 1 to 10 in the nominal operating region.

9.2.3 Energy Density

The energy density of the magnet is found as the product of its magnetic field strength and its operating flux density. This measure serves to differentiate magnets for use in machines from those having higher values preferred for high-power-density machines. The peak-energy operating point is optimal from the point of view of magnet utilization. It is found by differentiating the energy density with respect to magnetic field strength and equating to zero to find the magnetic field strength at which it is the maximum. The maximum energy density for a hard high-grade permanent magnet shown in Figure 9.4 is

$$E_{max} = -\frac{B_r^2}{4\mu_0 \mu_{rm}} \qquad (9.6)$$

and the flux density at which the maximum energy density available is at $0.5 B_r$. The operating line for this flux density is shown as giving the required magnetic field strength. Note that this operating point for maximum energy density requires a considerable amount of demagnetizing field strength from the stator excitation of the machine. Moreover, it will not be possible to maintain this operating point in a variable-speed machine drive; the stator currents will be varying widely over the entire torque–speed region.

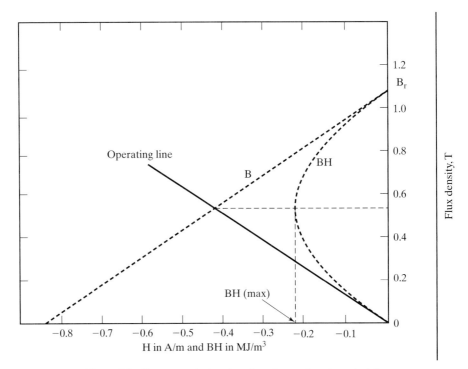

Figure 9.4 Energy product and preferred operating characteristics

9.2.4 Magnet Volume

The magnet volume is found in terms of operating point and air gap volume, as follows:

$$B_g l_g = \mu_0 H_m l_m \tag{9.7}$$

$$B_m A_m = B_g A_g \tag{9.8}$$

From these ideal relationships, magnet volume is

$$V_m = A_m l_m = \left(\frac{B_g A_g}{B_m}\right)\left(\frac{B_g l_g}{\mu_0 |H_m|}\right) = \frac{B_g^2 (A_g l_g)}{\mu_0 |B_m H_m|} = \frac{B_g^2 V_g}{\mu_0 |E_m|} \tag{9.9}$$

where V_g is the air gap volume, E_m is the magnet operating energy density, and A_m and A_g are the magnet and air gap area.

From this relationship, it is inferred that the maximum operating energy density point of the magnet will yield the magnet with the minimum volume and, hence, cost.

9.3 SYNCHRONOUS MACHINES WITH PMs

Prevalent rotor configurations of PM synchronous machines and their impact on the direct and quadrature axis inductances and the distinct differences between the PM brushless dc and synchronous machine are discussed in this section.

9.3.1 Machine Configurations

The permanent magnet (PM) synchronous machines can be broadly classified on the basis of the direction of field flux, as follows:

1. Radial field: the flux direction is along the radius of the machine.
2. Axial field: the flux direction is parallel to the rotor shaft.

The radial-field PM machines are common; the axial-field machines are coming into prominence in a small number of applications because of their higher power density and acceleration. Note that these are very desirable features in high-performance applications.

The magnets can be placed in many ways on the rotor. The radial-field versions are shown in Figure 9.5. The high-power-density synchronous machines have surface PMs with radial orientation intended generally for low speed applications, whereas the interior-magnet version is intended for high-speed applications. Regardless of the manner of mounting the PMs, the basic principle of operation is the same. An important consequence of the method of mounting the rotor magnets is the difference in direct and quadrature axes inductance values. It is explained as follows. The rotor magnetic axis is called direct axis and the principal path of the flux is through the magnets. The permeability of high-flux-density permanent magnets is almost that of the air. This results in the magnet thickness becoming an extension of air gap by that amount. The stator inductance when the direct axis or magnets are aligned with the stator winding is known as *direct axis inductance*. By rotating the magnets from the aligned position by 90 degrees, the stator flux sees the interpolar area of the rotor, containing only the iron path, and the inductance measured in this position is referred to as *quadrature axis inductance*. The direct-axis reluctance is greater than the quadrature-axis reluctance, because the effective air gap of the direct axis is multiple times that of the actual air gap seen by the quadrature axis. The consequence of such an unequal reluctance is that

$$L_q > L_d \quad (9.10)$$

where L_d is the inductance along the magnet axis (i.e., direct axis) and L_q is the inductance along an axis in quadrature to the magnet axis.

This is quite contrary to the wound-rotor salient-pole synchronous machine, where the quadrature-axis inductance is always greater than the direct-axis inductance. Note that, in the wound-rotor salient-pole synchronous machine, the direct axis, having the excitation coils, has a small air gap, whereas the quadrature axis has the large air gap.

Figure 9.5(i) shows the magnets mounted on the surface of the outer periphery of rotor laminations. This arrangement provides the highest air gap flux density, but it has the drawback of lower structural integrity and mechanical robustness. Machines with this arrangement of magnets are known as *surface mount PMSMs*. They are not preferred for high-speed applications, generally greater than 3,000 rpm. There is very little (less than 10%) variation between the quadrature- and direct-axis inductances in this machine. This particular fact has consequences for the control, operation, and characteristics of the surface-mount PMSM drives. The details are deferred until later sections.

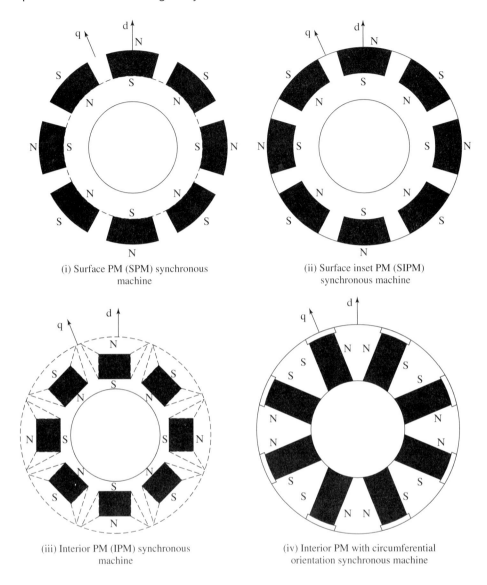

Figure 9.5 (i) Surface-PM (SPM) synchronous machine

Figure 9.5(ii) shows the magnets placed in the grooves of the outer periphery of the rotor laminations, providing a uniform cyclindrical surface of the rotor. In addition, this arrangement is much more robust mechanically as compared to surface-mount machines. The ratio between the quadrature- and direct-axis inductances can be as high as 2 to 2.5 in this machine. This construction is known as inset PM synchronous machine.

Figure 9.5(iii) and (iv) show the placement of magnets in the middle of the rotor laminations in radial and circumferential orientations, respectively. This construction is mechanically robust and therefore suited for high-speed applications.

The manufacturing of this arrangement is more complex than for the surface-mount or inset-magnet rotors. Note that the ratio between the quadrature- and direct-axis inductances can be higher than that of the inset-magnet rotor but generally does not exceed three in value. This type of machine construction is generally referred to as *interior PMSM*.

The inset-magnet construction has the advantages of both the surface- and interior-magnet arrangements: easier construction and mechanical robustness, with a high ratio between the quadrature- and direct-axis inductances, respectively. Many more arrangements of the magnets on the rotor are possible, but they are very rarely used in general industrial practice. Flux-reversal machines with magnets and armature windings on the salient stator poles and salient rotors with no windings or magnets are another possible construction. As they are in the early stages of development, they are not considered any further in this book.

9.3.2 Flux-Density Distribution

The flux plot and flux density vs. rotor position of a surface PM with radial orientation are shown in Figures 9.6 and 9.7, respectively. The dips in the flux density at various points occur because of the slot opening of the stator lamination where the

Figure 9.6 Flux distribution in the machine at no-load

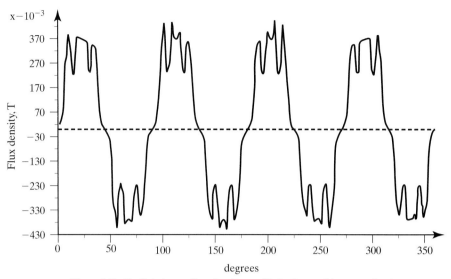

Figure 9.7 Radial air-gap flux-density profile in the machine at no-load

reluctance is much higher and, hence, the flux and its density are lower. The slotting effect also affects the induced emfs in the machine armature, clearly raising design and practical concerns about possible ripple effects. The slotting effect exists regardless of whether the machine is designed with sinusoidal or trapezoidal flux-density distribution. For these distributions, predetermined sinusoidal or rectangular currents are injected to produce the torque. Invariably, that results in ripple air gap torque capable of causing undesirable effects at low speed.

9.3.3 Line-Start PM Synchronous Machines

Some PM synchronous machines are intended and designed for constant-speed applications, to improve efficiency and power factor in comparison to induction and wound-rotor synchronous motors. Such machines have a squirrel-cage winding to provide the torque from standstill to near-synchronous speed. The same cage windings also serve to damp rotor oscillations. Once the motor pulls into synchronism, the cage windings do not contribute to electrical torque, because there are no induced voltages and hence no currents in them at zero slip.

Variable-speed PM synchronous motor drives have no need for the damper windings to offset hunting and oscillation. The damping is provided by properly controlling the input currents from the inverter. This results in a compact and a smaller rotor than that of the machine with damper windings. The way damping is produced in the PM synchronous motor with and without damper windings deserves a comment. The machine with the damper windings operates to suppress the oscillations with no external feedback. The feedback comes internally through the induced emf due to the slip speed in the cage windings. In the inverter-controlled PM drives, the control has to be initiated by an external signal or feedback variable to counter the oscillation. Its dependence on an external feedback loop compromises reliability. Wherever reliable operation regardless of the accuracy in speed control is a major

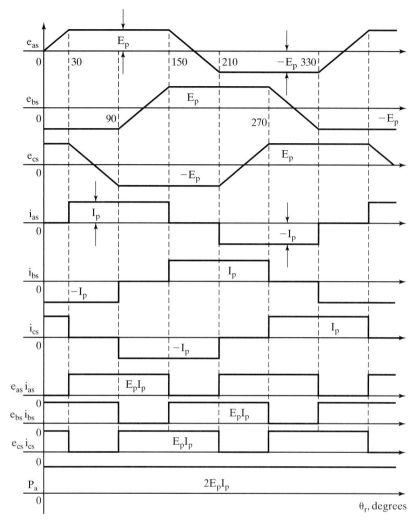

Figure 9.8 PM dc brushless-motor waveforms

concern or requirement, the synchronous motor with damper windings might prove to be an intelligent choice.

9.3.4 Types of PM Synchronous Machines

The PM synchronous motors are classified on the basis of the wave shape of their induced emf, i.e., sinusoidal and trapezoidal. The sinusoidal type is known as PM synchronous motor; the trapezoidal type goes under the name of PM dc brushless machine.

Even though the trapezoidal type of induced emfs have constant magnitude for 120 electrical degrees both in the positive and negative half-cycles, as shown in Figure 9.8, the power output can be uniform by exciting the rotor phases with 120 degrees (electrical) wide currents. The currents cannot rise and fall in the motor windings in zero time; hence, in actual operation, there are power pulsations during the turn-on and turn-off of the currents for each half-cycle. The severity of such pulsations is absent in the sinusoidal-emf-type motors.

PM dc brushless machines have 15% more power density than PM synchronous machines. This can be attributed to the fact that the ratio of the rms value to peak value of the flux density in the PM dc brushless machine is higher than in the sinusoidal PM machine. The ratio of the power outputs of these two machines is derived in the following, based on equal copper losses in their stators.

Let I_{ps} and I_p be the peak values of the stator currents in the synchronous and dc brushless machine. The rms values of these currents are

$$I_{sy} = \frac{I_{ps}}{\sqrt{2}} \tag{9.10}$$

$$I_d = I_p \sqrt{\frac{2}{3}} \tag{9.11}$$

Equating the copper losses and substituting for the currents in terms of their peak currents gives

$$3I_{sy}^2 R_a = 3I_d^2 R_a \tag{9.12}$$

i.e.,

$$3\left(\frac{I_{ps}}{\sqrt{2}}\right)^2 \cdot R_a = 3\left(\sqrt{\frac{2}{3}} \cdot I_p\right)^2 \cdot R_a \tag{9.13}$$

from which the relationship between the peak currents of the PM synchronous and brushless dc machine is obtained as

$$I_p = \frac{\sqrt{3}}{2} \cdot I_{ps} \tag{9.14}$$

The ratio of their power outputs is obtained from these relationships as follows:

$$\text{Power output ratio} = \frac{\text{PM dc brushless power}}{\text{PM synchronous power}} = \frac{2 \times E_p \times I_p}{3 \times \frac{E_p}{\sqrt{2}} \times \frac{I_{ps}}{\sqrt{2}}} = \frac{2 \times E_p \times \frac{\sqrt{3}}{2} \times I_{ps}}{3 \times \frac{E_p \times I_{ps}}{2}} = 1.1547 \tag{9.15}$$

Note that the power output ratio has been derived under the assumption that the power factor of the PM synchronous motor is unity.

From the waveforms of the PM dc brushless motor, it is seen that its control is simple if the absolute position of the rotor is known. Knowing the rotor position amounts to certain knowledge of the rotor field and the induced emf and, hence, instances of applying the appropriate stator currents for control.

9.4 VECTOR CONTROL OF PM SYNCHRONOUS MACHINE (PMSM)

A dynamic model of the PMSM is required to derive the vector-control algorithm to decouple the air gap-flux and torque channels in the drive system. The derivation of the dynamic model is easily made from the dynamic model of the induction machine in flux linkages from Chapter 5.

9.4.1 Model of the PMSM

The two-axes PMSM stator windings can be considered to have equal turns per phase. The rotor flux can be assumed to be concentrated along the d axis while there is zero flux along the q axis, an assumption similarly made in the derivation of indirect vector-controlled induction motor drives. Further, it is assumed that the machine core losses are negligible. Also, rotor flux is assumed to be constant at a given operating point. Variations in rotor temperature alter the magnet flux, but its variation with time is considered to be negligible. There is no need to include the rotor voltage equations as in the induction motor, since there is no external source connected to the rotor magnets, and variation in the rotor flux with respect to time is negligible.

The stator equations of the induction machine in the rotor reference frames using flux linkages are taken to derive the model of the PMSM. The rotor frame of reference is chosen because the position of the rotor magnets determines, independently of the stator voltages and currents, the instantaneous induced emfs and subsequently the stator currents and torque of the machine. Again, this is not the case in the induction machine; there, the rotor fluxes are not independent variables, they are influenced by the stator voltages and currents, and that is why any frame of reference is suitable for the dynamic modeling of the induction machine. When rotor reference frames are considered, it means the equivalent q and d axis stator windings are transformed to the reference frames that are revolving at rotor speed. The consequence is that there is zero speed differential between the rotor and stator magnetic fields and the stator q and d axis windings have a fixed phase relationship with the rotor magnet axis, which is the d axis in the modeling.

The stator flux-linkage equations are

$$v_{qs}^r = R_q i_{qs}^r + p\lambda_{qs}^r + \omega_r \lambda_{ds}^r \tag{9.16}$$

$$v_{ds}^r = R_d i_{ds}^r + p\lambda_{ds}^r - \omega_r \lambda_{qs}^r \tag{9.17}$$

where R_q and R_d are the quadrature- and direct-axis winding resistances, which are equal (and hereafter referred to as R_s), and the q and d axes stator flux linkages in the rotor reference frames are

$$\lambda_{qs}^r = L_s i_{qs}^r + L_m i_{qr}^r \tag{9.18}$$

$$\lambda_{ds}^r = L_s i_{ds}^r + L_m i_{dr}^r \tag{9.19}$$

but the self-inductances of the stator q and d axes windings are equal to L_s only when the rotor magnets have an arc of electrical 180°. That hardly ever is the case in practice. This has the implication that the reluctances along the magnet axis and the interpolar axis are different. When a stator winding (say d axis) is in alignment with the rotor magnet axis, the reluctance of the path is maximum; the magnet reluctance is almost the same as the air gap reluctance, and hence its inductance is the lowest at this time. The inductance then is referred to as the direct-axis inductance, L_d. At this time, the q axis winding faces the interpolar path in the rotor, where the flux path encounters no magnet but only the air gaps and iron in the rotor, resulting in lower reluctance and higher inductance. The inductance of the q axis winding is L_q at this time. As the rotor magnets and the stator q and d axis windings are fixed in space, that the winding inductances do not change in rotor reference frames is to be noted. In order, then, to compute the stator flux linkages in the q and d axes, the currents in the rotor and stator are required. The permanent-magnet excitation can be modeled as a constant current source, i_{fr}. The rotor flux is along the d axis, so the d axis rotor current is i_{fr}. The q axis current in the rotor is zero, because there is no flux along this axis in the rotor, by assumption. Then the flux linkages are written as

$$\lambda_{qs}^r = L_q i_{qs}^r \tag{9.20}$$

$$\lambda_{ds}^r = L_d i_{ds}^r + L_m i_{fr} \tag{9.21}$$

where L_m is the mutual inductance between the stator winding and rotor magnets.

Substituting these flux linkages into the stator voltage equations gives the stator equations:

$$\begin{bmatrix} v_{qs}^r \\ v_{ds}^r \end{bmatrix} = \begin{bmatrix} R_q + L_q p & \omega_r L_d \\ -\omega_r L_q & R_d + L_d p \end{bmatrix} \begin{bmatrix} i_{qs}^r \\ i_{ds}^r \end{bmatrix} + \begin{bmatrix} \omega_r L_m i_{fr} \\ 0 \end{bmatrix} \tag{9.22}$$

The electromagnetic torque is given by

$$T_e = \frac{3}{2} \frac{P}{2} \{\lambda_{ds}^r i_{qs}^r - \lambda_{qs}^r i_{ds}^r\} \tag{9.23}$$

which, upon substitution of the flux linkages in terms of the inductances and currents, yields

$$T_e = \frac{3}{2} \frac{P}{2} \{\lambda_{af} i_{qs}^r + (L_d - L_q) i_{qs}^r i_{ds}^r\} \tag{9.24}$$

where the rotor flux linkages that link the stator are

$$\lambda_{af} = L_m i_{fr} \tag{9.25}$$

The rotor flux linkages are considered constant except for temperature effects. The temperature sensitivity of the magnets reduces the residual flux density and, hence, the flux linkages with increasing temperature. The samarium–cobalt magnets have the least amount of temperature sensitivity: -2 to -3 % per 100°C rise in temperature. Neodymimum magnets have -12 to -13% per 100°C rise in temperature sensitivity; the ceramic magnets have -19% per 100°C rise in temperature sensitivity. Therefore, the temperature sensitivity of the magnets has to be included in the dynamic simulation by appropriately correcting for the rotor flux linkages from their nominal values.

9.4.2 Vector Control

The polyphase PMSM control is rendered equivalent to that of the dc machine by a decoupling control known as vector control. The vector control separates the torque and flux channels in the machine through its stator-excitation inputs. The vector control for PMSM is very similar to that derived in Chapter 8 on vector-controlled induction motor drives. Many variations of vector control similar to that of the induction motor are possible. In this section, the vector control of the PM synchronous machine is derived from its dynamic model. Considering the currents as inputs, the three phase currents are

$$i_{as} = i_s \sin(\omega_r t + \delta) \tag{9.26}$$

$$i_{bs} = i_s \sin\left(\omega_r t + \delta - \frac{2\pi}{3}\right) \tag{9.27}$$

$$i_{cs} = i_s \sin\left(\omega_r t + \delta + \frac{2\pi}{3}\right) \tag{9.28}$$

where ω_r is the electrical rotor speed and δ is the angle between the rotor field and stator current phasor, known as the torque angle.

The rotor field is traveling at a speed of ω_r rad/sec; hence, the q and d axes stator currents in the rotor reference frame for a balanced three-phase operation are given by

$$\begin{bmatrix} i_{qs}^r \\ i_{ds}^r \end{bmatrix} = \frac{2}{3} \begin{bmatrix} \cos \omega_r t & \cos\left(\omega_r t - \frac{2\pi}{3}\right) & \cos\left(\omega_r t + \frac{2\pi}{3}\right) \\ \sin \omega_r t & \sin\left(\omega_r t - \frac{2\pi}{3}\right) & \sin\left(\omega_r t + \frac{2\pi}{3}\right) \end{bmatrix} \begin{bmatrix} i_{as} \\ i_{bs} \\ i_{cs} \end{bmatrix} \tag{9.29}$$

Substituting the equations from (9.26) in (9.28) into (9.29) gives the stator currents in the rotor reference frames:

$$\begin{bmatrix} i_{qs}^r \\ i_{ds}^r \end{bmatrix} = i_s \begin{bmatrix} \sin \delta \\ \cos \delta \end{bmatrix} \tag{9.30}$$

The q and d axes currents are constants in rotor reference frames, since δ is a constant for a given load torque. As these are constants, they are very similar to armature and field currents in the separately-excited dc machine. The q axis current is distinctly equivalent to the armature current of the dc machine; the d axis current is field current, but not in its entirety. It is only a partial field current; the other part is contributed by the equivalent current source representing the permanent magnet field. It is discussed in detail in the section on flux-weakening operation of the PMSM.

Substituting this equation into the electromagnetic torque expression gives the torque:

$$T_e = \frac{3}{2} \cdot \frac{P}{2} \left[\frac{1}{2}(L_d - L_q)i_s^2 \sin 2\delta + \lambda_{af} i_s \sin \delta \right] \tag{9.31}$$

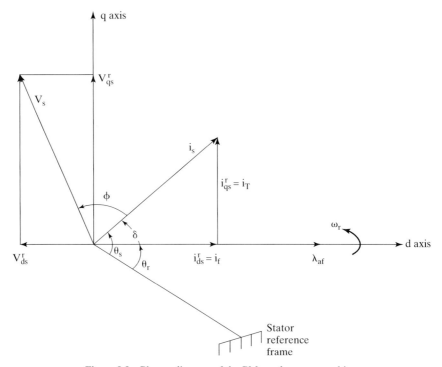

Figure 9.9 Phasor diagram of the PM synchronous machine

For $\delta = \pi/2$,

$$T_e = \frac{3}{2} \cdot \frac{P}{2} \lambda_{af} i_s = K_1 \lambda_{af} i_s, \text{N·m} \qquad (9.32)$$

where

$$K_1 = \frac{3}{2} \cdot \frac{P}{2} \qquad (9.33)$$

Note that the equation (9.24) is similar to that of the torque generated in the dc motor and vector-controlled induction motor. If the torque angle is maintained at 90° and flux is kept constant, then the torque is controlled by the stator-current magnitude, giving an operation very similar to that of the armature-controlled separately-excited dc motor.

The electromagnetic torque is positive for the motoring action, if δ is positive. Note that the rotor flux linkages λ_{af} are positive. Then the phasor diagram for an arbitrary torque angle δ is shown in Figure 9.9. Note that

$$i_{qs}^r = \text{Torque-producing component of stator current} = i_T \qquad (9.34)$$

$$i_{ds}^r = \text{Flux-producing component of stator current} = i_f \qquad (9.35)$$

and the torque angle is given by,

$$\theta_T = \delta \qquad (9.36)$$

Section 9.4 Vector Control of PM Synchronous Machine (PMSM)

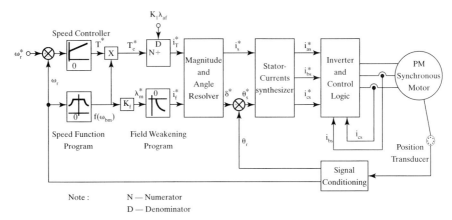

Figure 9.10 Vector-controlled PM synchronous motor drive

The equations (9.34) to (9.36) complete the similarity of the vector-controlled PM synchronous and induction motors. Note that the mutual flux linkage is the resultant of the rotor flux linkages and stator flux linkages. It is then given as

$$\lambda_m = \sqrt{(\lambda_{af} + L_d i^r_{ds})^2 + (L_q i^r_{qs})^2} \text{ (Wb} - \text{Turn)} \qquad (9.37)$$

If δ is greater than $\pi/2$, i^r_{ds} becomes negative. Hence, the resultant mutual flux linkages decrease. This is the key to flux-weakening in the PM synchronous motor drives. If δ is negative with respect to the rotor or mutual flux linkages, the machine becomes a generator.

9.4.3 Drive-System Schematic

From the previous derivations and the understanding gained from them, the vector-controlled PM synchronous motor drive schematic is obtained and is shown in Figure 9.10. The torque reference is a function of the speed error, and the speed controller is usually of PI type. For fast response of the speed, a PID controller is appropriate. The product of torque reference and air gap flux linkage λ_m^*, in p.u., generates the torque-producing component, i_T^*, of the stator current. The reason for this block is to adjust the torque-producing component of the stator current both in the constant-torque and in the constant-power regions of operation. It is proved as follows: The function generator on the speed has the following characteristics:

$$\left. \begin{array}{ll} f(\omega_{bm}) = \dfrac{\omega_b}{\omega_m}; & \pm\omega_b < \omega_m < \pm\omega_{max} \\ \quad\quad\quad = 1; & 0 < \omega_m < \pm\omega_b \end{array} \right\} \qquad (9.38)$$

and

$$P_a = \omega_m T_e^* = \omega_m f(\omega_{bm}) T^* \qquad (9.39)$$

Substituting the function from (9.38) into (9.39) gives the air gap power for the constant-torque region:

$$P_a = \omega_m T^* \qquad (9.40)$$

and, for the constant-power operation region, the air gap power is similarly obtained:

$$P_a = \omega_b T^* \qquad (9.41)$$

The base speed and T^* are constant; hence, the constant-torque and -power modes of operation are implemented with this function-generator block, as seen from the equations (9.40) and (9.41), respectively. Note that subtracting shaft losses from air gap power gives the shaft power output.

The function generator, $f(\omega_{bm})$, sets the reference for the resultant mutual flux linkages, involving a constant K_f. The output of the function generator being in p.u., the value of K_f is unity. The flux-linkages reference, λ_m^*, determines the field current reference, i_f^*, required to counter the flux linkages due to the rotor magnet. Note that, in the field-weakening region, i_f^* is negative; by making this positive, the mutual flux linkages could be strengthened, even though it may lead to saturation and therefore is not usually resorted to in practice. The flux-producing component of the stator-current phasor, i_f^*, can be predetermined and programmed in a function generator, such as in a Read-Only Memory (ROM).

The stator-current phasor magnitude i_s^* and the torque angle δ^* commands are evaluated by

$$i_s^* = \sqrt{(i_f^*)^2 + (i_T^*)^2} \qquad (9.42)$$

$$\delta^* = \tan^{-1}\left[\frac{i_T^*}{i_f^*}\right] \qquad (9.43)$$

The instantaneous position of the stator-current phasor command is given by

$$\theta_s^* = \theta_r + \delta^* = \omega_r t + \delta^* \qquad (9.44)$$

from which the stator-phase current commands are obtained for a balanced three-phase operation by substituting for i_T^* and i_f^* from (9.24):

$$\begin{bmatrix} i_{as}^* \\ i_{bs}^* \\ i_{cs}^* \end{bmatrix} = \begin{bmatrix} \cos\theta_r & \sin\theta_r \\ \cos\left(\theta_r - \frac{2\pi}{3}\right) & \sin\left(\theta_r - \frac{2\pi}{3}\right) \\ \cos\left(\theta_r + \frac{2\pi}{3}\right) & \sin\left(\theta_r + \frac{2\pi}{3}\right) \end{bmatrix} \begin{bmatrix} i_T^* \\ i_f^* \end{bmatrix} = i_s^* \begin{bmatrix} \sin(\theta_r + \delta^*) \\ \sin\left(\theta_r + \delta^* - \frac{2\pi}{3}\right) \\ \sin\left(\theta_r + \delta^* + \frac{2\pi}{3}\right) \end{bmatrix} \qquad (9.45)$$

These stator-phase current commands are amplified by the inverter and its logic and fed to the PM synchronous machine. The rotor position is obtained with a position encoder or synchronous resolver. The velocity signal, ω_r, is extracted from the rotor position by using signal processors available commercially at present. In the derivation of stator-current commands, it was assumed that the zero-sequence current is zero, but the derivation is not limited by this fact; it can be included easily by an additional first-order differential equation similar to the stator zero-sequence equation in the induction-machine model.

9.5 CONTROL STRATEGIES

The torque-angle control provides a wide variety of control choices in the PMSM drive system. Some key control strategies are the following:

   (i) constant torque-angle control or zero-direct-axis-current control;
   (ii) unity power-factor control;
   (iii) constant mutual air gap flux-linkages control;
   (iv) optimum-torque-per-ampere control;
   (v) flux-weakening control.

These control strategies are derived systematically and analyzed in the following, but illustrated for steady-state operation only.

9.5.1 Constant ($\delta = 90°$) Torque-Angle Control

In this control, the torque angle δ is maintained at 90 degrees; hence, the field or direct-axis current is made to be zero, leaving only the torque or quadrature-axis current in place. This is the mode of operation for speeds lower than the base speed. Such a strategy is commonly used in many of the drive systems. The relevant equations of performance in this mode of operation are

$$T_e = \frac{3}{2} \cdot \frac{P}{2} \lambda_{af} \cdot i_{qs}^r = \frac{3}{2} \cdot \frac{P}{2} \lambda_{af} \cdot I_s \tag{9.46}$$

and torque per unit of stator current is constant, given by

$$\frac{T_e}{I_s} = \frac{3}{2} \cdot \frac{P}{2} \lambda_{af} \tag{9.47}$$

and normalized electromagnetic torque is expressed as

$$T_{en} = \frac{T_e}{T_b} = \frac{\frac{3}{2} \cdot \frac{P}{2} \lambda_{af} \cdot I_s}{\frac{3}{2} \cdot \frac{P}{2} \lambda_{af} \cdot I_b} = I_{sn}(\text{p.u.}) \tag{9.48}$$

indicating that torque equals the stator current in p.u, which gives the simplest control for implementation in the PMSM drives. Note that I_{sn} is the normalized stator-current phasor magnitude. Relevant equations to determine the steady-state performance of the PMSM drive with this control strategy are derived in the following. The q and d axes voltages, in steady state, are

$$v_{qs}^r = (R_s + L_q p) I_s + \omega_r \lambda_{af} = R_s I_s + \omega_r \lambda_{af} \tag{9.49}$$

$$v_{ds}^r = -\omega_r L_q I_s \tag{9.50}$$

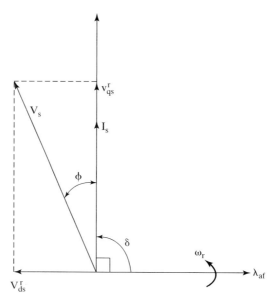

Figure 9.11 Constant-torque-angle control

Note that the rate of change of currents is zero in the rotor reference frames, because the currents are constants in steady state. The magnitude of the voltage phasor is given by

$$V_s = \sqrt{(v_{qs}^r)^2 + (v_{ds}^r)^2} \qquad (9.51)$$

and V_s in normalized units is obtained from combining equations (9.49) to (9.51):

$$V_{sn} = \frac{V_s}{V_b} = \frac{V_s}{\omega_b \lambda_{af}} = \sqrt{(\omega_{rn} + R_{sn}I_{sn})^2 + (L_{qn}I_{sn}\omega_{rn})^2} \qquad (9.52)$$

From the phasor diagram shown in Figure 9.11 and the axis voltages, power factor is obtained as

$$\cos\phi = \frac{v_{qs}^r}{V_s} = \frac{v_{qs}^r}{\sqrt{(v_{qs}^r)^2 + (v_{ds}^r)^2}} = \frac{1}{\sqrt{1 + \dfrac{(L_{qn}I_{sn})^2}{\left(1 + \dfrac{R_{sn}I_{sn}}{\omega_{rn}}\right)^2}}} \qquad (9.53)$$

This equation implies that the power factor deteriorates with increasing rotor speed as well as with increasing stator current. The maximum rotor speed with this control strategy for a given stator current (and neglecting stator resistive drop) is obtained from the voltage magnitude expression (9.52) as follows:

$$\omega_{rn}(\text{max}) \cong \frac{V_{sn}(\text{max})}{\sqrt{1 + L_{qn}^2 I_{sn}^2}} \qquad (9.54)$$

where $V_{sn(\text{max})}$ is obtained from the dc-link voltage, V_{dc}, approximately, as

$$V_{sn(\text{max})} = \sqrt{2} \times 0.45 V_{dc} \qquad (9.55)$$

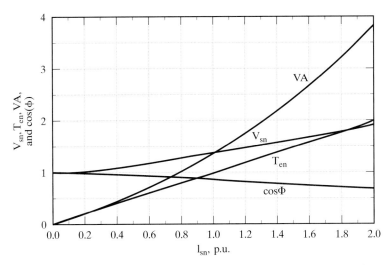

Figure 9.12 Performance characteristics for constant-torque-angle control

assuming that six-step switching is performed in the inverter and neglecting device and cable voltage drops. It is realistic to consider a PWM-based inverter, in which case the available voltage is further reduced by a factor of K_{dr}, usually in the range of from 0.85 to 0.95, giving the voltage phasor as

$$V_{sn(max)} \cong 0.636 K_{dr} V_{dc} \tag{9.56}$$

A PMSM with

$$R_{sn} = 0.1729 \text{ p.u.}, L_b = 0.0129 \text{ H}, L_{qn} = 0.6986 \text{ p.u.}, L_{dn} = 0.4347 \text{ p.u.},$$
$$\omega_b = 628.6 \text{ rad/sec}, V_b = 97.138 \text{V}, I_b = 12 \text{ A}$$

is considered for the plotting of performance characteristics utilizing the constant-torque-angle control strategy, as shown in Figure 9.12 for a 1-p.u. rotor speed. Note that the volt-ampere (VA) in p.u. is also plotted as a function of stator current, to assess the VA rating required of the inverter and for the purpose of comparing various control strategies under discussion. The change in power factor from 1 to 0.859 over a zero-to-1 p.u. change in stator current is significant: this creates a need for reactive VA, thus demanding a higher VA from the inverter.

The dc-link voltage required to operate this PMSM at 1-p.u. speed and current is approximately estimated by using equation (9.56). Consider the derating factor to be 0.8, to allow for a margin of voltage for current control. V_{sn} required for this operating point is computed as 1.365 p.u. From this, the required dc-link voltage is obtained as

$$V_{dc} = \frac{1.365}{0.636 \times 0.8} = 2.68 \text{ p.u.} = 2.68 \times 97.138 \text{ V} = 260.6 \text{ V}$$

If current control is not required and six-step voltage-source operation is resorted to in the inverter, the derating factor K_{dr} will go up to from 0.92 to 0.95. This would generate peaky currents, resulting in higher copper losses in the machine, in addition to increased peak inverter-current rating, but the six-step operation will contribute to significant enhancement of the electromagnetic torque and power output, particularly in the flux-weakening region. The maximum speed for this drive (neglecting stator-resistive voltage drop), from equation (9.54), is 1.18 p.u. for 1-p.u. current.

9.5.2 Unity-Power-Factor Control

Unity-power-factor (UPF) control implies that the VA rating of the inverter is fully utilized for real power input to the PMSM. UPF control is enforced by controlling the torque angle as a function of motor variables. The performance equations in this mode of operation are derived and given below.

The q and d axes currents are

$$I_{qs}^r = I_s \sin \delta \quad (9.57)$$

$$I_{ds}^r = I_s \cos \delta \quad (9.58)$$

and normalized torque is obtained as

$$T_{en} = I_{sn} \left\{ I_{sn} \cdot \frac{(L_{dn} - L_{qn})}{2} \cdot \sin 2\delta + \sin \delta \right\} \text{ (p.u.)} \quad (9.59)$$

The q and d axes normalized stator voltages are

$$v_{qsn}^r = \omega_{rn} \left\{ 1 + L_{dn} I_{sn} \cos \delta + \frac{R_{sn} I_{sn}}{\omega_{rn}} \cdot \sin \delta \right\} \text{ (p.u.)} \quad (9.60)$$

$$v_{dsn}^r = \omega_{rn} I_{sn} \left\{ \frac{R_{sn}}{\omega_{rn}} \cdot \cos \delta - L_{qn} \sin \delta \right\} \text{ (p.u.)} \quad (9.61)$$

from which the stator-voltage phasor magnitude is obtained as

$$V_{sn} = \sqrt{(v_{qsn}^r)^2 + (v_{dsn}^r)^2} \text{ (p.u.)} \quad (9.62)$$

The angle between d axis and the resultant voltage V_{sn} is

$$\tan(\delta + \phi) = \frac{v_{qsn}^r}{v_{dsn}^r} \quad (9.63)$$

Since the power factor has to be zero in this control,

$$\phi = 0 \quad (9.64)$$

which gives the following relationship for the torque angle:

$$\tan \delta = \frac{v_{qsn}^r}{v_{dsn}^r} \quad (9.65)$$

Upon substitution of equations (9.60) and (9.61) into (9.65),

$$\frac{\sin \delta}{\cos \delta} = \frac{1 + L_{dn} I_{sn} \cos \delta + \dfrac{R_{sn} I_{sn}}{\omega_{rn}} \cdot \sin \delta}{\dfrac{R_{sn} I_{sn} \cos \delta}{\omega_{rn}} - L_{qn} I_{sn} \sin \delta} \quad (9.66)$$

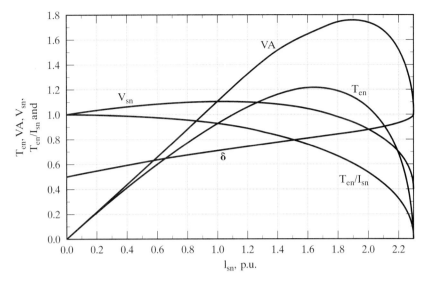

Figure 9.13 Performance characteristics for UPF control

which, upon simplification, gives

$$I_{sn}(L_{qn} \sin^2 \delta + L_{dn} \cos^2 \delta) = -\cos \delta \qquad (9.67)$$

from which torque angle is computed as

$$\delta = \cos^{-1}\left\{\frac{-1 \pm \sqrt{1 - 4L_{qn}I_{sn}^2(L_{dn} - L_{qn})}}{2I_{sn}(L_{dn} - L_{qn})}\right\} \text{ (rad)} \qquad (9.68)$$

Note that torque angle has to be greater than 90°; hence, depending on the value of the denominator, a positive or negative sign in the above equation has to be employed. Torque angles less than 90° will result in an increase of flux linkages contributing to saturation in the machine; that is undesirable from point of view of losses. This expression provides the enforcement of UPF control; its implementation requires the motor phase current magnitude and motor parameters L_{qn} and L_{dn}. Note that it is independent of rotor speed.

For the same machine cited in the illustration of constant-torque-angle control, the performance characteristics with UPF control are shown in Figure 9.13. The torque per unit current is less than one, indicating that this control strategy is not optimal in terms of torque generation; its efficiency will be reduced by increased copper losses for the generation of same torque. Its overall volt–ampere requirement is only 1.09 p.u., as against 1.355 p.u. for the constant-torque-angle control. The motor-voltage phasor requirement is 1.098 p.u. against 1.365 p.u. for the constant-torque-angle control. This demonstrates that the reserve voltage availability with UPF control would extend the constant torque region, resulting in higher output of the PMSM drive. This feature is very desirable in many applications requiring extended speed range.

9.5.3 Constant-Mutual-Flux-Linkages Control

In this control strategy, the resultant flux linkage of the stator q and d axes and rotor, known as the mutual flux linkage, is maintained constant, most usually at a value equal to the rotor flux linkages, λ_{af}. Its main advantage is that, by the limiting of the mutual flux linkages, the stator-voltage requirement is kept comparably low. In addition, varying the mutual flux linkages provides a simple and straightforward flux-weakening for operation at speeds higher than the base speed. Hence, mutual flux-linkages control is one of the powerful techniques useful in the entire speed range, as against other schemes that are limited to operation lower than the base speed only. The mutual flux linkage is expressed as follows:

$$\lambda_m = \sqrt{(\lambda_{af} + L_d I_{ds}^r)^2 + (L_q I_{qs}^r)^2} \tag{9.69}$$

and, for example, equating it to the rotor flux linkages as

$$\lambda_m = \lambda_{af} \tag{9.70}$$

and substituting for the direct- and quadrature-axis currents in terms of the stator-current phasor and torque angle in the above equations results in

$$I_s = -\frac{2\lambda_{af}}{L_d}\left[\frac{\cos\delta}{\cos^2\delta + \rho^2 \sin^2\delta}\right] \tag{9.71}$$

where

$$\rho = \frac{L_q}{L_d} \tag{9.72}$$

Two distinct cases arise here, depending on the saliency ratio ρ. For surface-mounted magnets, ρ is around unity; for buried or interior PM rotor construction it could have values as high as 3. These two cases are analyzed separately in the following.

Case (i): $\rho = 1$

This amounts to a value of torque angle, δ, given by

$$\delta = \cos^{-1}\left\{\frac{-L_d I_s}{2\lambda_{af}}\right\} (\text{rad}) \tag{9.73}$$

Note that the base voltage is defined as

$$V_b = \omega_b \lambda_{af} \ (V) \tag{9.74}$$

and the base impedance is

$$Z_b = \omega_b L_b = \frac{V_b}{I_b} \ (\Omega) \tag{9.75}$$

Using these for normalization of the motor current given in (9.71) yields the torque angle:

$$\delta = \cos^{-1}\left\{-\frac{L_d I_b \cdot I_s/I_b}{2\lambda_{af}}\right\} = \cos^{-1}\left\{-\frac{I_{sn}L_{dn}}{2}\right\} (\text{rad}) \tag{9.76}$$

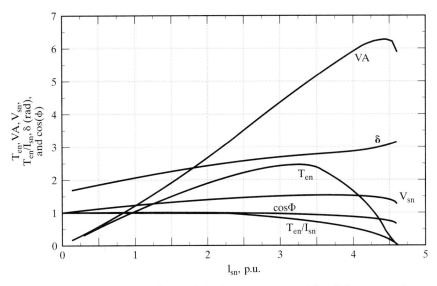

Figure 9.14 Performance characteristics for constant-mutual-flux-linkages control

Case (ii) : $\rho \neq 1$

This yields the torque angle as

$$\delta = \cos^{-1}\left\{\frac{1}{L_{dn}I_{sn}(1-\rho^2)} \pm \sqrt{\left\{\frac{1}{L_{dn}(1-\rho^2)I_{sn}}\right\}^2 - \frac{1}{(1-\rho^2)}}\right\} \text{ (rad)} \quad (9.77)$$

The minimum of the two possible values of δ is chosen, so that the demagnetizing current is small. Note also that δ has to be greater than 90°. Otherwise, the *d* axis stator current will strengthen the mutual flux over and above the rotor flux linkages, resulting in saturation of the stator core. The other performance equations are as given in the previous section, and so also are the machine parameters.

The performance characteristics are shown in Figure 9.14. The torque per ampere is slightly less than unity, and stator voltage is approximately 1.17 p.u. (less than is required for constant-torque-angle control), and the VA rating is also less for this control. The power factor is close to unity, up to 1 p.u. of stator current, indicating that this scheme is much closer to UPF control than to constant-torque-angle control.

9.5.4 Optimum-Torque-Per-Ampere Control

A control strategy to maximize electromagnetic torque for unit stator current is valuable from the optimum-machine and inverter-utilization points of view. As in other control strategies, this control strategy is enforced with torque-angle control. The theoretical basis for this control strategy is derived as follows. The electromagnetic torque is

$$T_e = \frac{3}{2} \cdot \frac{P}{2}\left[\lambda_{af}i_s \sin \delta + \frac{1}{2}(L_d - L_q)i_s^2 \sin 2\delta\right] \quad (9.78)$$

which, in terms of normalized units, is

$$T_{en} = i_{sn}\left[\sin\delta + \frac{1}{2}(L_{dn} - L_{qn})i_{sn}\sin 2\delta\right] \text{ (p.u.)} \quad (9.79)$$

where the base torque is defined as

$$T_b = \frac{3}{2}\cdot\frac{P}{2}\lambda_{af}I_b \text{ (N·m)} \quad (9.80)$$

The torque per ampere is given as

$$\left(\frac{T_{en}}{i_{sn}}\right) = \left[\sin\delta + \frac{1}{2}(L_{dn} - L_{qn})i_{sn}\sin 2\delta\right] \quad (9.81)$$

Its maximum is found by differentiating with respect to δ and equating to zero to obtain the condition for δ as

$$\delta = \cos^{-1}\left\{-\frac{1}{4a_1 i_{sn}} - \sqrt{\left(\frac{1}{4a_1 i_{sn}}\right)^2 + \frac{1}{2}}\right\} \text{ (rad)} \quad (9.82)$$

where

$$a_1 = (L_{dn} - L_{qn}) = L_{dn}(1 - \rho) \quad (9.83)$$

Only the negative sign in the argument is considered, because δ has to be greater than 90° to reduce the flux in the air gap. To illustrate the superior performance of this control as against constant-torque-angle control, the torque vs. stator current is shown in Figure 9.15. It has 3.2% and 11.05% torque enhancement for 1- and 2-p.u. stator currents, respectively, compared to the constant-torque-angle operation, and this control strategy requires stator voltages of 1.286 and 1.736 p.u., respectively. This is the

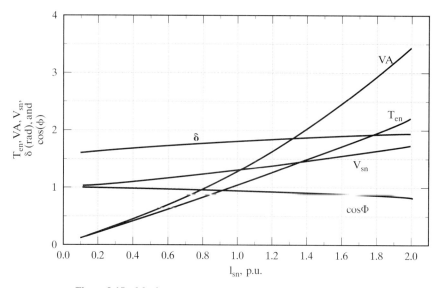

Figure 9.15 Maximum-torque-per-ampere control performance

penalty paid for using this control strategy in terms of the poor inverter utilization, which needs a 3.472-p.u. VA rating at 2-p.u. stator current, as against 2.8 p.u. for constant-mutual-flux-linkages control with a torque derating of 0.065 p.u. only. Note that the torque enhancement could be nearly 100% for $\rho = 3.5$ at a 2-p.u. stator current with this control strategy. This control strategy is, in general, preferred for highly salient machines with $\rho > 2$.

9.6 FLUX-WEAKENING OPERATION

The upper limits placed on the available dc-link voltage and current rating of a given inverter cause the maximum motor-input voltage and current to be limited. The voltage and current limits affect the maximum-speed-with-rated-torque capability and the maximum torque-producing capability of the motor drive system, respectively. It is required and desirable to produce the rated power with the highest attainable speed for many applications, such as electric vehicles, people-carriers in airport lobbies, forklifts, and machine-tool spindle drives. Corresponding to the maximum dc-link voltage and hence to the maximum input machine voltage and rated torque, the machine attains a speed known as the rated speed. Above this speed, the induced emf will exceed the maximum input voltage, making the flow of current into machine phases impractical. To overcome this situation, the induced emf is constrained to be less than the applied voltage by weakening the air gap flux linkages. The flux-weakening is made to be inversely proportional to the stator frequency, so that the induced emf is a constant and will not increase with the increasing speed.

This section considers the operation of the permanent-magnet synchronous motor drives when they are constrained to be within the permissible envelope of the maximum inverter voltage and current to produce the rated power and to provide this at the highest attainable rotor speed. The rated power is intended for steady-state operation; multiple times that is preferred for fast accelerations and decelerations during transient operation. Effective current control during the flux-weakening operation, with high transient capability, is preferable to a saturation of the current loop, resulting in significant harmonic content in the stator currents, resulting in higher torque ripples and higher losses.

Two approaches are considered in this section: (i) The demagnetizing current is predetermined only by rotor speed; this approach is called *direct approach;* (ii) The demagnetizing stator current is derived as a function of not only rotor speed but also the electromagnetic torque. This method is referred to as the *indirect approach*.

9.6.1 Maximum Speed

To understand the scope of the flux-weakening of the PMSM drive, it is essential to find the maximum speed. The maximum speed for a given set of stator voltages and currents is obtained analytically for the purpose of design calculations. The maximum operating speed with zero torque is calculated from the steady-state stator voltage equations as follows. The normalized stator equations in the rotor reference frames are given as

$$v_{qsn}^r = (R_{sn} + L_{qn}p)i_{qsn}^r + \omega_{rn}(L_{dn}i_{dsn}^r + 1) \qquad (9.84)$$
$$v_{dsn}^r = -\omega_{rn}L_{qn}i_{qsn}^r + (R_{sn} + L_{dn}p)i_{dsn}^r$$

where the *abc*-to-*qd* transformation valid for voltages, currents, and flux linkages is

$$\begin{bmatrix} v_{qsn}^r \\ v_{dsn}^r \end{bmatrix} = \frac{2}{3} \begin{bmatrix} \cos\theta_r & \cos\left(\theta_r - \frac{2\pi}{3}\right) & \cos\left(\theta_r + \frac{2\pi}{3}\right) \\ \sin\theta_r & \sin\left(\theta_r - \frac{2\pi}{3}\right) & \sin\left(\theta_r + \frac{2\pi}{3}\right) \end{bmatrix} \begin{bmatrix} v_{asn} \\ v_{bsn} \\ v_{csn} \end{bmatrix} \quad (9.85)$$

The steady-state stator-voltage equations are obtained by setting the derivative of the current variables to zero in equation (9.84):

$$\begin{aligned} v_{qsn}^r &= R_{sn} i_{qsn}^r + \omega_{rn}(L_{dn} i_{dsn}^r + 1) \\ v_{dsn}^r &= -\omega_{rn} L_{qn} i_{qsn}^r + R_{sn} i_{dsn}^r \end{aligned} \quad (9.86)$$

when $i_{qsn}^r = 0$, and the stator-voltage phasor is given as

$$v_{sn}^2 = v_{dsn}^{r\,2} + v_{qsn}^{r\,2} = \omega_{rn}^2(1 + L_{dn} i_{dsn}^r)^2 + R_{sn}^2 i_{dsn}^{r\,2} \quad (9.87)$$

from which the maximum speed for a given stator-current magnitude of i_{dsn}^r is

$$\omega_{rn}(\max) = \frac{\sqrt{v_{sn}^2 - R_{sn}^2 i_{dsn}^{r\,2}}}{(1 + L_{dn} i_{dsn}^r)} \quad (9.88)$$

Note that the denominator of equation (9.87) has to be positive, giving the condition that the maximum stator current to be applied to counter the magnet flux linkages is

$$i_{dsn}^r(\max) < -\frac{1}{L_{dn}} \quad (9.89)$$

9.6.2 Direct-Flux-Weakening Algorithm

The direct-flux-weakening algorithm finds the demagnetizing component of stator current satisfying the maximum current and voltage limits only. It is very similar to the field control of a separately-excited dc machine, where the field current is determined usually by the speed alone. Such a method has the advantage of simplicity, but it has the disadvantage of not optimizing the stator current by considering the operating torque in the machine. Such an optimization is possible with a torque-request feedforward. The PMSM drive-system control with both constant-torque and constant-power operation is presented in this section. Flux-weakening algorithm, control scheme, controller realization, simulation, and performance are described in this section.

By considering the steady state, the voltage phasor is written (neglecting the resistive terms) as

$$v_{sn}^2 = \omega_{rn}^2\{(1 + L_{dn} i_{dsn}^r)^2 + (L_{qn} i_{qsn}^r)^2\} \text{ (p.u.)} \quad (9.90)$$

where the voltage phasor, v_{sn}, is defined as

$$v_{sn} = \sqrt{(v_{qsn}^{r\,2} + v_{dsn}^{r\,2})} \text{ (p.u.)} \quad (9.91)$$

The quadrature current i_{qsn}^r can be written in terms of the stator-current phasor and the direct-axis current as

$$i_{qsn}^r = \sqrt{i_{sn}^2 - i_{dsn}^{r\,2}} \text{ (p.u.)} \quad (9.92)$$

Substituting equation (9.90) into (9.89) gives the following equation relating the voltage phasor, current phasor, and rotor speed.

$$v_{sn}^2 = \omega_{rn}^2 \{L_{qn}^2(i_{sn}^2 - i_{dsn}^{r\,2}) + (1 + L_{dn}i_{dsn}^r)^2\} \text{ (p.u.)} \tag{9.93}$$

Note that the voltage phasor and current phasor (v_{sn} and i_{sn}, respectively) are the maximum values that could be obtained from the inverter operation. Hence, for the flux-weakening operation, these are considered to be constant. That leads to the appreciation that the equation contains only two variables, ω_{rn} and i_{dsn}^r. Therefore, from one of these two variables the other could be computed analytically. This is the key to the control and operation of the PMSM drive in the flux-weakening region. Further, equation (9.93) is written in terms of i_{dsn}^r and ω_{rn} as

$$v_{sn} = \omega_{rn}\sqrt{(ai_{dsn}^{r\,2} + bi_{dsn}^r + c)} \text{ (p.u.)} \tag{9.94}$$

where the constants are

$$a = L_{dn}^2 - L_{qn}^2 \tag{9.95}$$

$$b = 2L_{dn} \tag{9.96}$$

$$c = 1 + L_{qn}^2 i_{sn}^2 \tag{9.97}$$

Assuming the rotor speed is available for feedback control and using equations (9.93) to (9.97) gives the d axis stator current, which would automatically satisfy the constraints of maximum stator current, i_{sn}, and stator voltage, v_{sn}. From the stator-current magnitude and the d axis stator current, by using equation (9.92), the maximum permitted q axis stator current could be calculated. The d and q axes currents then determine the stator-phase currents obtained by using the inverse transformation from equation (9.85):

$$\begin{bmatrix} i_{asn} \\ i_{bsn} \\ i_{csn} \end{bmatrix} = \begin{bmatrix} \cos\theta_r & \sin\theta_r \\ \cos\left(\theta_r - \dfrac{2\pi}{3}\right) & \sin\left(\theta_r - \dfrac{2\pi}{3}\right) \\ \cos\left(\theta_r + \dfrac{2\pi}{3}\right) & \sin\left(\theta_r + \dfrac{2\pi}{3}\right) \end{bmatrix} \begin{bmatrix} i_{qsn}^r \\ i_{dsn}^r \end{bmatrix} \tag{9.98}$$

Combining equations $i_{qsn}^r = i_{sn}\sin\delta$ and $i_{dsn}^r = i_{sn}\cos\delta$ with (9.98) yields the normalized stator phase currents via the following relationship:

$$\begin{bmatrix} i_{asn} \\ i_{bsn} \\ i_{csn} \end{bmatrix} = \begin{bmatrix} \sin(\theta_r + \delta) \\ \sin\left(\theta_r + \delta - \dfrac{2\pi}{3}\right) \\ \sin\left(\theta_r + \delta + \dfrac{2\pi}{3}\right) \end{bmatrix} i_{sn} \tag{9.99}$$

The torque angle is obtained as,

$$\delta = \tan^{-1}\left(\dfrac{i_{qsn}^r}{i_{dsn}^r}\right) \tag{9.100}$$

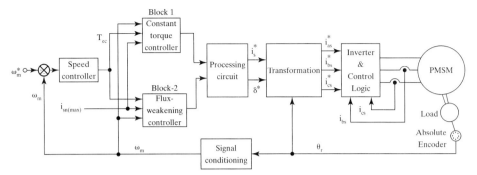

Figure 9.16 Schematic of the PMSM-drive control strategy

Note that the calculated q axis stator current, together with d axis stator current, determines the torque, T_{ef}, that could be produced, and it is then used to modify the torque command, T_{ec}, generated from the speed error in a drive system.

If the torque request, T_{ec}, is more than T_{ef}, the torque calculated by the flux-weakening module, then the final torque request is made equal to the calculated one. If T_{ec} is less than T_{ef}, then the final torque command is maintained at T_{ec}. This final torque request is denoted as T_e^*. From this T_e^*, the required q axis current in the machine could be calculated from the torque equation, since i_{dsn}^r request is known:

$$i_{qsn}^r = \frac{T_e^*}{\{1 + (L_{dn} - L_{qn})i_{dsn}^r\}} \text{ (p.u.)} \tag{9.101}$$

Although the above relationships were derived for steady-state performance, it is to be noted that some voltage reserve is to be available for dynamic control of currents. With a smaller voltage reserve, the current loops will become sluggish; beyond their limit points, they will no longer control the current, thus making the applied voltages six-step. This results in currents rich in harmonics and hence in higher air gap-torque pulsations in the machine.

9.6.2.1 Control scheme. The control scheme for the PMSM drive both in the region of constant torque and in that of flux-weakening could be formulated from the derivations and understanding provided in previous section. Schematically, the control scheme is shown in Figure 9.16. Assuming a speed-controlled drive system, the torque command T_{ec} is generated by the speed error. Depending on the mode of operation, this torque command is processed by Block 1 or Block 2. Block 1 corresponds to the constant-torque-mode controller; Block 2 corresponds to the flux-weakening-mode controller. These blocks are explained in the subsequent paragraphs. The outputs of these controllers are the stator-current-magnitude command and the torque-angle command. They, together with the electrical rotor position, provide the phase-current commands through the transformation block. The current commands are enforced with an inverter by current feedback control, with any one of the current control schemes available. For illustration, pulse-width mod-

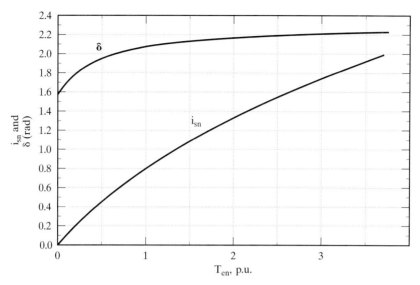

Figure 9.17 Stator-current magnitude and torque angle vs. electromagnetic torque

ulation for the current control is chosen. The rotor position and rotor speed are obtained with an encoder and a signal conditioner, respectively.

9.6.2.2 Constant-torque-mode controller. Block 1 contains the constant-torque-mode controller, with maximum torque per ampere, as explained in 9.4.4. The torque-vs.-stator-current magnitude and torque angle are computed from equations (9.81) and (9.82). Note that, for easier computation, the stator-current magnitude is varied and the torque angle and torque are evaluated. For implementation, note that the inverse is needed, i.e., given the torque request, the stator-current magnitude and torque angle are to be made available. That could be accomplished by curve-fitting the characteristics and programming it in memory for retrieval and use. The characteristics of stator-current magnitude and torque angle vs. torque are shown in Figure 9.17.

These characteristics, for the drive system under illustration, are curve-fitted by the following expressions with minimum error:

$$i_{sn} = 0.01 + 0.954 T_{en} - 0.189 T_{en}^2 + 0.02 T_{en}^3 \tag{9.102}$$

$$\delta = 1.62 + 0.715 T_{en} - 0.3 T_{en}^2 + 0.04 T_{en}^3 \tag{9.103}$$

Algorithm and procedure for obtaining a curve fit can be obtained from a standard textbook on numerical analysis. The detailed schematic of the torque-mode controller, then, takes the form shown in Figure 9.18. The equations (9.102) and (9.103) can be realized in the form of tables to realize Block 1. The speed signal determines the mode of operation of the drive system. In the torque-control mode of operation, if the rotor speed is less than the base speed, then it enables Block 1 in the form of onward transmission of the torque-command signal, T_{ec}. The torque signal is limited

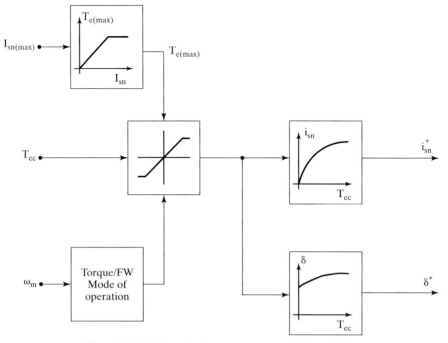

Figure 9.18 Schematic of constant-torque-mode controller

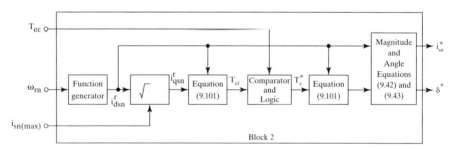

Figure 9.19 Schematic realization of the flux-weakening controller

by the maximum torque that could be generated with the maximum permissible stator-current phasor. Its magnitude can be a variable, depending on whether the drive is in steady state or in intermittent peak operation. Then the resulting torque signal provides the stator-current-magnitude and torque-angle commands from the memory that may have characteristics given in equations (9.102) and (9.103). For regenerative action, the torque angle is negative.

9.6.2.3 Flux-weakening controller. From the flux-weakening algorithm, the flux-weakening controller emerges, which could be schematically captured as shown in Figure 9.19. The inputs to this module are three variables: the torque request, rotor speed, and maximum permissible stator current. The module's outputs are the stator-current-magnitude request and the torque-angle request.

The rotor speed determines the d axis stator current request through the equation (9.94) in a slightly modified form:

$$i_{dsn}^r = \frac{\left(-b + \sqrt{\left\{b^2 - 4a\left(c - \frac{v_{sn}^2}{\omega_{rn}^2}\right)\right\}}\right)}{2a} \quad (9.104)$$

where a, b, and c are given in equations (9.95) to (9.97).

Note that constant c is dependent on the maximum stator-current limit. This could either be programmed or be captured in a tabular form for retrieval and use in the feedback-control computations. This d axis current request, indicated with an asterisk in Figure 9.19, then, along with the maximum stator current, determines the permissible quadrature-axis current, i_{qsn}^r.

This q axis current, with the d axis current, determines the maximum electromagnetic torque allowed, T_{ef}, within maximum voltage and current constraints. This is compared with the torque request T_{ec}, generated by the speed error. Then the following logic is applied to find the torque-command input to compute the command q axis current:

$$\begin{aligned} \text{If} \quad & T_{ec} > T_{ef}, \text{ then } T_e^* = T_{ef} \\ \text{If} \quad & T_{ec} < T_{ef}, \text{ then } T_e^* = T_{ec} \end{aligned} \quad (9.105)$$

This logic submodule adjusts the torque request depending on the load and maximum capability of the motor drive system as a function of the rotor speed. From this final torque request, T_e^*, the stator q axis current is computed by using equation (9.93). The direct- and quadrature-axis stator-current requests are then used to calculate the stator-current phasor magnitude and torque-angle requests. The torque-mode or flux-weakening controller module is chosen, based on the rotor speed being lower or higher than its base value, respectively.

9.6.2.4 System performance.
The drive-system performance with the strategy incorporating both the optimized constant-torque mode and flux-weakening controllers is modeled and simulated to evaluate its performance. The PWM frequency of the inverter is set at 5 kHz. The dc-link voltage is 280 V, and the load torque is zero. To prove the operation in the four-quadrant operation, a step speed command from 7 p.u. to −7 p.u. is given, and various machine and control variables are viewed. The machine variables of interest are the torque and speed. Likewise, the control variables, such as speed and torque request are monitored. The simulation results are shown in Figure 9.20. The torque command follows the maximum trajectory as a function of speed, and the actual torque very closely follows its command. The power trajectory is maintained at its set maximum in the flux-weakening mode, and the drive speed envelope is smooth during the entire speed of operation. Note that the base speed is 1 p.u.; beyond it, flux-weakening is exercised in the drive system.

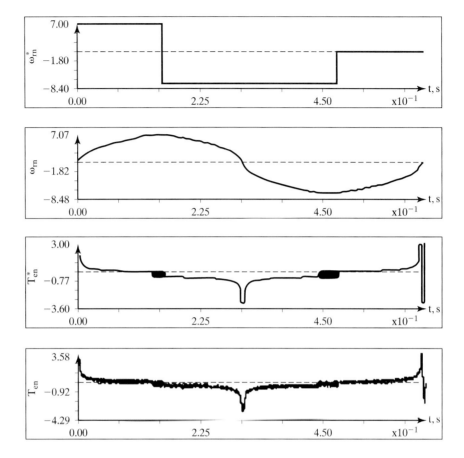

Figure 9.20 Simulation results

Example 9.1

A PMSM has the following parameters:

$$R_{sn} = 0.173 \text{ p.u}, L_{dn} = 0.435 \text{ p.u}, L_{qn} = 0.699 \text{ p.u}, V_{sn} = 1.45 \text{ p.u}, I_{sn} = 1 \text{ p.u.}$$

Find (i) the maximum speed with and without neglecting stator resistances and (ii) the steady-state characteristics in the flux-weakening region.

Solution

Part (i): Maximum Speed

$$\text{Maximum speed} = \frac{\sqrt{v_{sn}^2 - (R_{sn}I_{sn})^2}}{(1 + L_{dn}I_{dsn}^r)} \text{ (p.u.)}$$

Note that $I_{dsn}^r = -I_{sn}$; hence,

$$\omega_{rn}(\text{max}) = \frac{\sqrt{1.45^2 - 0.173^2}}{1 - 0.435} = 2.547 \text{ p.u.}$$

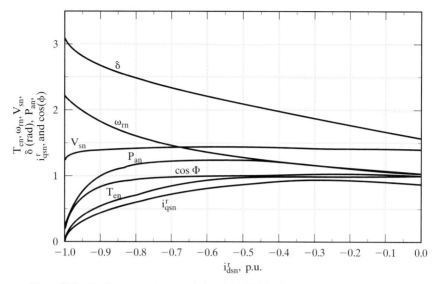

Figure 9.21 Performance characteristics of the PMSM in the flux-weakening region

Neglecting stator resistance gives the maximum speed of the PMSM:

$$\omega_{rn}(\max) = \frac{V_{sn}}{(1 - L_{dn}I_{sn})} = \frac{1.45}{1 - 0.435} = 2.565 \text{ (p.u.)}$$

The effect of the stator resistance is usually negligible in the maximum-speed prediction, but not for small machines, because they have proportionally larger stator resistances compared to large machines.

Part (ii): Steady-State Performance Computation in the Flux-Weakening Operation

The steady-state flux-weakening performance is computed from the following steps.

Step 1: For computing I_{dsn}^r during field-weakening when $I_{qsn}^r = 0$, the effect of resistive drops is, in general, as much as $I_{sn}R_{sn}$ (approximately); hence, reduce V_{sn} to approximately 1.25 p.u.

Step 2: With $V_{sn} = 1.25$ p.u., compute I_{dsn}^r from

$$V_{sn} = \omega_{rn}\sqrt{aI_{dsn}^{r\,2} + bI_{dsn}^r + c}$$

for each value of increasing ω_{rn}. Select the smaller of the roots for I_{dsn}^r. From this, compute I_{qsn}^r with I_{sn} set at 1 p.u.

Step 3: Compute V_{qsn}^r, V_{dsn}^r, and V_{sn} to check on whether it is equal to or less than the original assigned value of 1.45 p.u. Otherwise, go to step 2 to give a smaller value of V_{sn}.

Step 4: Compute voltage-phasor angle and current-phasor angle, hence power-factor angle, torque, and power output, all in p.u.

The performance characteristics are plotted and shown in Figure 9.21.

9.6.3 Indirect Flux-Weakening

An alternative flux-weakening control strategy is based on controlling directly the mutual flux linkage, which is inversely proportional to electrical rotor speed. This is very similar to controlling the field flux of a separately-excited dc machine. The mutual flux linkage is defined as

$$\lambda_{mn} = \sqrt{(L_{dn}i_{dsn} + 1)^2 + (L_{qn}i_{qsn})^2} \qquad (9.106)$$

with the assumption that the base flux linkage is equal to the rotor flux linkages. Neglecting the stator resistance voltage drops gives the stator-voltage phasor in terms of the mutual flux linkages and electrical rotor speed by using the equation (9.90) as

$$v_{sn} = \omega_{rn}\lambda_{mn} \qquad (9.107)$$

This implies that, during flux-weakening, when the dc link voltage and hence the stator-voltage-phasor magnitude is constant, the mutual flux linkage has to be decreased inversely proportional to the rotor electrical speed. Such control is straightforward and simple in the high-speed operational region, when compared to other schemes in the literature. To implement this, the independent references to be commanded are the mutual flux linkages and electromagnetic torque. In order to incorporate any control strategy (such as the maximum torque per ampere, constant air gap flux linkages, unity power factor, or constant torque angle) below the base speed region, the corresponding mutual flux linkages are calculated and preprogrammed in the mutual-flux-linkages controller. For the region above the base speed, the mutual flux linkage is made to be inversely proportional to the rotor electrical speed in order to work within the limited dc-link voltage. The machine will be operated within its maximum torque capability for a given mutual-flux-linkages command.

9.6.3.1 Maximum permissible torque.

The maximum electromagnetic torque for a given mutual flux linkage is found as follows. The torque in terms of motor parameters, mutual flux linkages, and d axis stator current is given by

$$T_{en} = \frac{(L_{dn} - L_{qn})i^r_{dsn} + 1}{L_{qn}}\sqrt{\lambda_{mn}^2 - (1 + L_{dn}i^r_{dsn})^2} \qquad (9.108)$$

By differentiating this with respect to the d axis stator current and equating to zero, the condition for maximum permissible electromagnetic torque for a given mutual flux linkage is found. Substituting this condition for the d axis stator current in (9.108) yields the maximum permissible electromagnetic torque for the given mutual flux linkages.

Figure 9.22 shows such a relationship for $L_{qn} = 0.699$ and $L_{dn} = 0.434$. It can be seen that, for the given system parameters, the T^*_{enm} vs. λ^*_{mn} relationship is nearly linear. This means that a simple first- or second-order polynomial can be used to implement this relationship. In general, the function T^*_{enm}-vs.-λ^*_{mn} can be computed off-line for the appropriate range of λ^*_{mn} and subsequently incorporated into the system as a simple lookup table.

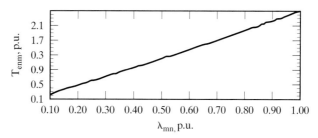

Figure 9.22 Maximum permissible torque command as a function of flux command

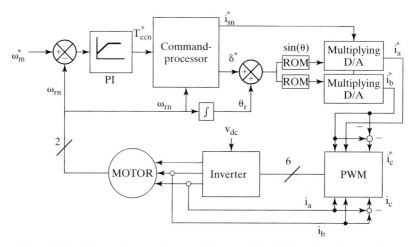

Figure 9.23 The speed-control system with mutual-flux-linkages-based controller

9.6.3.2 Speed-control scheme. The mutual-flux-linkage-weakening and torque controls are incorporated in a speed-control system as shown in Figure 9.23. T^*_{ecn} is the output of the speed PI controller. The command-processor weakens T^*_{ecn} in inverse proportion to speed when the speed is higher than 1 p.u., in order to limit the machine's power. Note that the torque command is limited to its maximum permissible value, which depends on the mutual-flux-linkage command. The command-processor also generates the mutual-flux-linkage command. The mutual-flux-weakening operation begins at base speed, which could be less than or equal to one p.u.. The outcomes, T^*_{en}, along with the commanded mutual flux linkage, λ^*_{mn}, are provided to the current and angle resolver to generate the appropriate commands for the stator current, i^*_{sn}, and its angle, δ^*. The angle δ^* is added to the rotor's absolute angle to provide the desired current angle with respect to the stator's reference frame. The $\{i^*_{sn}, \delta^*\}$ pair is itself a command input to a PWM current controller. Figure 9.24 shows the operations performed by the command processor.

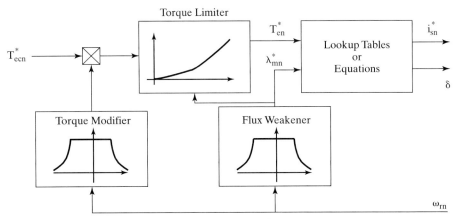

Figure 9.24 The command processor

9.6.3.3 Implementation strategy. In this section, an implementation strategy for the on-line computation of $\{i_{sn}^*, \delta^*\}$ from $\{T_{en}^*, \lambda_{mn}^*\}$ is discussed. The strategy follows this general format:

$$i_{sn}^* = \Omega(T_{en}^*, \lambda_{mn}^*) \tag{9.109}$$

$$\delta^* = \Lambda(T_{en}^*, \lambda_{mn}^*) \tag{9.110}$$

A method for realizing the functions $\Omega(.,.)$ and $\Lambda(.,.)$ based on lookup tables is presented. The method is illustrated via an example using the model parameters of a PMSM. These values are

$L_{dn} = 0.434$ p.u., $L_{qn} = 0.699$ p.u. and $R_{sn} = 0.1729$ p.u. The base values are $V_b = 97.138$ V, $I_b = 12$ A, $L_b = 0.0129$ H and $\omega_b = 628.6$ rad/sec.

Lookup Tables Realization Equations (9.109) and (9.110) can be realized by using separate three-dimensional lookup tables. These lookup tables are generated off-line by solving system equations numerically. Two of the independent axes of each table are assigned to T_{en}^* and λ_{mn}^*. Off the third dimension, the respective i_{sn}^* or δ^* is read. Note that the tables need only provide the data for positive torque commands. For negative torque commands, the respective angle from the table must be applied with negative sign. λ_{mn}^* is limited to values between 0 and 1, since the only interest is in weakening the mutual flux linkage. The numerical solution to the system variables for the following range of the normalized mutual-flux-linkage command:

$$0.2 \leq \lambda_{mn}^* \leq 1 \text{ (p.u.)}$$

and for the motor parameters given earlier, yields the viable ranges for the other three variables as

$$0 \leq T_{en}^* \leq 2.44 \text{ (p.u.)}$$
$$0 \leq i_{sn} \leq 3.3 \text{ (p.u.)}$$
$$1.57 \leq \delta \leq 3.14 \text{ (rad)}$$

Section 9.6   Flux-Weakening Operation   551

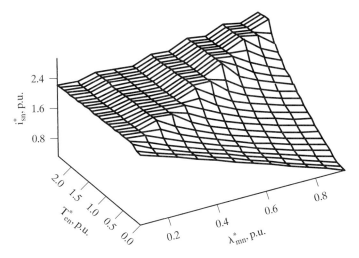

Figure 9.25   Current command as a function of torque and mutual-flux-linkage commands

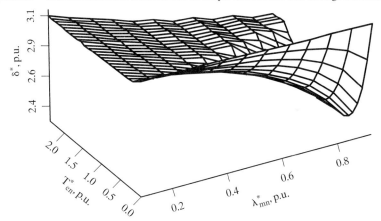

Figure 9.26   Angle command as a function of the torque and mutual-flux-linkage commands

Figures 9.25 and 9.26 depict the three-dimensional tables for this system. The variables are digitized under the following constraints,

$$T_{en}^*: 9 \text{ bits}; \lambda_{mn}^*: 7 \text{ bits}; i_{sn}^* \text{ and } \delta^*: 8 \text{ bits each}$$

Once the torque command reaches its maximum permissible value, the current and angle commands are frozen. The fact that each graph has two distinct areas is due to this limitation. In each graph, the border between these two surfaces appears to be a straight line, which can be attributed to the fact that the T_{enm}^*-vs.-λ_{mn}^* relationship is nearly linear. The T_{enm}^*-vs.-λ_{mn}^* relationship suggests that, for the given parameters, the relationship can adequately be approximated with a first- or at most a second-order polynomial. Thus, either of the following two polynomials can be used for on-line calculation of T_{enm}^* from λ_{mn}^* in this example.

$$T_{enm}^* = 0.21\lambda_{mn}^{*2} + 2.23\lambda_{mn}^* + .0059$$

or, more simply,

$$T^*_{enm} = 2.44\lambda^*_{mn}$$

The polynomials are both derived by using the least-squares-fit method.

9.6.3.4 System performance. Utilizing these tables results in the accurate enforcement of the commanded torque and the commanded mutual flux linkage. As a result, the voltage requirement is closely maintained at a maximum level of 1 p.u. Figure 9.27 shows the simulation results for a speed-control system that utilizes the two tables for a speed-controlled operation with $+3$-p.u. and -3-p.u. speed commands. The dotted lines represent commanded values, the solid lines actual values. The PWM is operating at 20 kHz, and the current error is amplified by a factor of 200. The stator resistance is taken into account in the simulation; therefore, the mutual flux-weakening is initiated at a speed of .53 p.u. instead of 1 p.u. As a result, the voltage requirement is successfully limited to 1 p.u. The limited bandwidth of the current controller decreases the ability of the system to enforce desired currents as speed increases. As a consequence, the mutual-flux-linkage error increases with speed. Note that memory chips are required to store the relatively large look-up tables. Each table has 2^{16} entries. Therefore, each table requires 64 kilobytes of storage capacity.

The table-look-up approach provides a high level of accuracy in terms of enforcing the torque and mutual-flux-linkage commands. Utilizing the look-up-table method results in meeting the specified voltage requirements of the drive system, which is one of the key factors in the drive's performance. The system is also relatively fast: implementing it mainly involves reading data from a memory chip, plus the performance of minimal calculations. However, implementing this system requires a considerable amount of digital memory. For the example considered, 128 kilobytes of ROM is required to store the tables.

9.6.3.5 Parameter sensitivity. Parameter variations are a major source of error in model-based motor controllers, such as the one discussed. The effects of the variations of three of the PMSM parameters (namely, stator resistance, q axis inductance, and rotor flux linkages) on the input-voltage requirement of the simulated example are briefly discussed in this section.

(i) Stator Resistance The stator resistance can vary in the range of approximately 1 to 2 times its nominal value. As the rotor resistance increases, the input-voltage requirement also increases. Figure 9.28 shows the effect of a stator resistance increase of 100% on the voltage requirements of the simulated system. If maintaining a 1-p.u. input voltage is the main objective, the solution is to initiate the mutual-flux-weakening operation at lower speeds. Note that, as the mutual flux linkage is further weakened, the maximum permissible torque also decreases. This, in turn, increases the response time of the motor.

(ii) Rotor Flux Linkage The rotor flux linkage of a PMSM can decrease as much as 20 percent depending on the kind of magnet used. Figure 9.29 shows the effect of a 20-percent reduction of λ_{af} on the simulated example. It can be seen that the overall voltage requirement is higher than the case when the rotor flux linkage is at its

Section 9.6 Flux-Weakening Operation 553

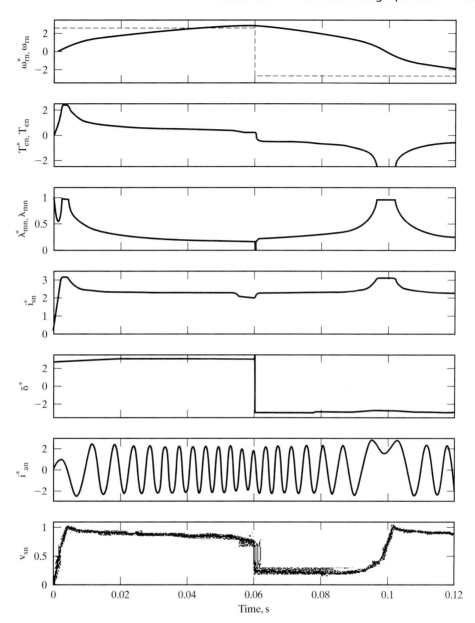

Figure 9.27 Simulation results for a ±3-p.u.-step speed command

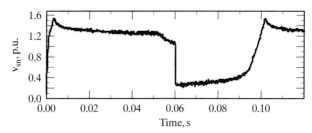

Figure 9.28 Input-voltage requirement with double the stator resistance

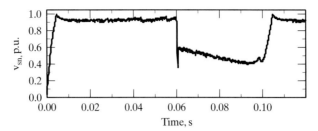

Figure 9.29 Input-stator phasor-voltage requirement when λ_{af} decreases by 20% from its nominal value

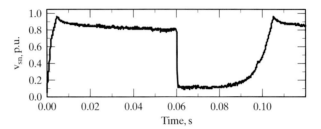

Figure 9.30 Input-voltage requirement for a 20-percent reduction in L_q

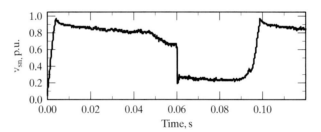

Figure 9.31 Input-voltage requirement for a 10-percent increase in L_q

nominal value. Under identical torque requirements, as the rotor flux linkage decreases, the current requirements increase and, therefore, the input-voltage requirements increase.

(iii) *q* Axis Self-Inductance The q axis self-inductance, L_q, can vary in the range of approximately from 0.8 to 1.1 times its nominal value. Figures 9.30 and 9.31 show the effect of L_q variations on the input-voltage requirement. The increase of L_q over its nominal value results in higher voltage requirements. The voltage requirement exceeds the 1-p.u. limit at some points. This indicates that a good reserve margin in the dc-link voltage has to be given to handle the parameter variations that are bound to affect the performance. The extreme cases need to be taken into account in the design process. Alternatively, estimating the individual parameters directly and using them in the controller can compensate for the effects of parameter sensitivity.

9.7 SPEED-CONTROLLER DESIGN

The design of the speed-controller is important from the point of view of imparting desired transient and steady-state characteristics to the speed-controlled PMSM drive system. A proportional-plus-integral controller is sufficient for many industrial applications; hence, it is considered in this section. Selection of the gain and time constants of such a controller by using the symmetric-optimum principle is straightforward if the d axis stator current is assumed to be zero. In the presence of a d axis stator current, the *d* and *q* current channels are cross-coupled, and the model is nonlinear, as a result of the torque term.

Under the assumption that $i_{ds}^r = 0$, the system becomes linear and resembles that of a separately-excited dc motor with constant excitation. From then on, the block-diagram derivation, current-loop approximation, speed-loop approximation, and derivation of the speed-controller by using symmetric optimum are identical to those for a dc or vector-controlled induction-motor-drive speed-controller design.

9.7.1 Block-Diagram Derivation

The motor q axis voltage equation with the d axis current being zero becomes

$$v_{qs}^r = (R_s + L_q p)i_{qs}^r + \omega_r \lambda_{af} \tag{9.111}$$

and the electromechanical equation is

$$\frac{P}{2}(T_e - T_l) = J p \omega_r + B_1 \omega_r \tag{9.112}$$

where the electromagnetic torque is given by

$$T_e = \frac{3}{2} \cdot \frac{P}{2} \lambda_{af} i_{qs}^r \tag{9.113}$$

and, if the load is assumed to be frictional, then

$$T_l = B_l \omega_m \tag{9.114}$$

which, upon substitution, gives the electromechanical equation as

$$(Jp + B_t)\omega_r = \left\{\frac{3}{2}\left(\frac{P}{2}\right)^2 \cdot \lambda_{af}\right\} i_{qs}^r = K_t \cdot i_{qs}^r \tag{9.115}$$

where

$$B_t = \frac{P}{2} B_l + B_1 \tag{9.116}$$

$$K_t = \frac{3}{2}\left(\frac{P}{2}\right)^2 \cdot \lambda_{af} \tag{9.117}$$

The equations (9.111) and (9.115), when combined into a block diagram with the current- and speed-feedback loops added, are shown in Figure 9.32.

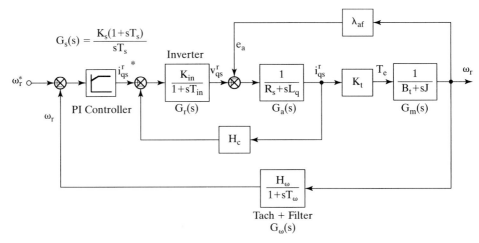

Figure 9.32 Block diagram of the speed-controlled PMSM drive

The inverter is modeled as a gain with a time lag by

$$G_r(s) = \frac{K_{in}}{1 + sT_{in}} \quad (9.118)$$

where

$$\left.\begin{array}{l} K_{in} = 0.65 \dfrac{V_{dc}}{V_{cm}} \\ T_{in} = \dfrac{1}{2f_c} \end{array}\right\} \quad (9.119)$$

where V_{dc} is the dc-link voltage input to the inverter, V_{cm} is the maximum control voltage, and f_c is the switching (carrier) frequency of the inverter.

The induced emf due to rotor flux linkages, e_a, is

$$e_a = \lambda_{af}\omega_r \text{ (V)} \quad (9.120)$$

9.7.2 Current Loop

This induced-emf loop crosses the q axis current loop, and it could be simplified by moving the pick-off point for the induced-emf loop from speed to current output point. This gives the current-loop transfer function from Figure 9.33 as

$$\frac{i_{qs}^r(s)}{i_{qs}^{r*}(s)} = \frac{K_{in}K_a(1 + sT_m)}{H_cK_aK_{in}(1 + sT_m) + (1 + sT_{in})\{K_aK_b + (1 + sT_a)(1 + sT_m)\}} \quad (9.121)$$

where

$$K_a = \frac{1}{R_s};\ T_a = \frac{L_q}{R_s};\ K_m = \frac{1}{B_t};\ T_m = \frac{J}{B_t};\ K_b = K_tK_m\lambda_{af} \quad (9.122)$$

Section 9.7 Speed-Controller Design

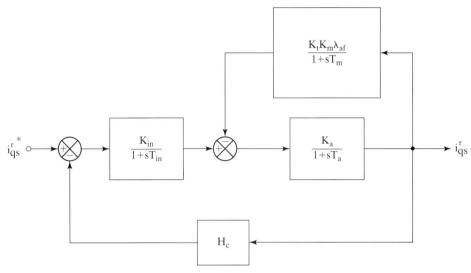

Figure 9.33 Current controller

The following approximations are valid near the vicinity of crossover frequency:

$$1 + sT_r \cong 1 \tag{9.123}$$

$$1 + sT_m \cong sT_m \tag{9.124}$$

$$(1 + sT_a)(1 + sT_{in}) \cong 1 + s(T_a + T_{in}) \cong 1 + sT_{ar} \tag{9.125}$$

where

$$T_{ar} = T_a + T_{in} \tag{9.126}$$

With this, the current-loop transfer function is approximated as

$$\frac{i_{qs}^r(s)}{i_{qs}^{r*}(s)} \cong \frac{(K_a K_{in} T_m)s}{K_a K_b + (T_m + K_a K_{in} T_m H_c)s + (T_m T_{ar})s^2} \cong \left(\frac{T_m K_{in}}{K_b}\right) \frac{s}{(1 + sT_1)(1 + sT_2)} \tag{9.127}$$

It is found that $T_1 < T_2 < T_m$; hence, on further approximation, $(1 + sT_2) \cong sT_2$. The approximate current-loop transfer function is then given by

$$\frac{i_{qs}^r(s)}{i_{qs}^{r*}(s)} \cong \frac{K_i}{(1 + sT_i)} \tag{9.128}$$

where

$$K_i = \frac{T_m K_{in}}{T_2 K_b} \tag{9.129}$$

$$T_i = T_1 \tag{9.130}$$

This simplified current-loop transfer function is substituted in the design of the speed controller as follows.

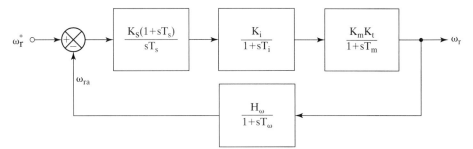

Figure 9.34 Simplified speed-control loop

9.7.3 Speed-Controller

The speed loop with the simplified current loop is shown in Figure 9.34.

Near the vicinity of the crossover frequency, the following approximations are valid:

$$(1 + sT_m) \cong sT_m \tag{9.131}$$

$$(1 + sT_i)(1 + sT_\omega) \cong 1 + sT_{\omega i} \tag{9.132}$$

$$1 + sT_\omega \cong 1 \tag{9.133}$$

where

$$T_{\omega i} = T_\omega + T_i \tag{9.134}$$

The speed-loop transfer function, with these approximations, is given by

$$GH(s) \cong \frac{K_i K_m K_t H_\omega}{T_m} \cdot \frac{K_s}{T_s} \cdot \frac{(1 + sT_s)}{s^2 (1 + sT_{\omega i})} \tag{9.135}$$

from which the closed-loop speed transfer function is obtained as

$$\frac{\omega_r(s)}{\omega_r^*(s)} \cong \frac{1}{H_\omega} \left\{ \frac{K_g \dfrac{K_s}{T_s}(1 + sT_s)}{s^3 T_{\omega i} + s^2 + K_g \dfrac{K_s}{T_s}(1 + sT_s)} \right\} \tag{9.136}$$

where

$$K_g = \frac{K_i K_m K_t H_\omega}{T_m} \tag{9.137}$$

Equating this transfer function to a symmetric-optimum function with a damping ratio of 0.707 gives the closed-loop-speed transfer function as

$$\frac{\omega_r(s)}{\omega_r^*(s)} \cong \frac{1}{H_\omega} \cdot \frac{(1 + sT_s)}{1 + (T_s)s + \left(\dfrac{3}{8}T_s^2\right)s^2 + \left(\dfrac{1}{16}T_s^3\right)s^3} \tag{9.138}$$

Equating the coefficients of equations (9.136) and (9.138) and solving for the constants yields the time and gain constants of the speed controller as

Section 9.7 Speed-Controller Design

$$T_s = 6T_{\omega i} \tag{9.139}$$

$$K_s = \frac{4}{9K_g T_{\omega i}} \tag{9.140}$$

Hence, the proportional gain, K_{ps}, and integral gain, K_{is}, of the speed controller are derived as

$$K_{ps} = K_s = \frac{4}{9K_g T_{\omega i}} \tag{9.141}$$

$$K_{is} = \frac{K_s}{T_s} = \frac{1}{27 K_g T_{\omega i}^2} \tag{9.142}$$

The validity of various approximations is verified through a worked example.

Example 9.2

The PMSM drive system parameters are as follows:

$R_s = 1.4\ \Omega, L_d = 0.0056\ H, L_q = 0.009\ H, \lambda_{af} = 0.1546\ Wb\text{-Turn}, B_t = 0.01\ N\cdot m/rad/sec,$
$J = 0.006\ kg - m^2, P = 6, f_c = 2\ kHz, V_{cm} = 10\ V, H_\omega = 0.05\ V/V, H_c = 0.8\ V/A, V_{dc} = 285\ V.$

Design a symmetric-optimum-based speed-controller, and verify the validity of assumptions made in its derivation. The damping ratio required is 0.707.

Solution

Inverter: Gain, $K_{in} = 0.65 \dfrac{V_{dc}}{V_{cm}} = 18.525\ V/V$

Time constant, $T_{in} = \dfrac{1}{2f_c} = 0.00025\ s$

$$G_r(s) = \frac{K_{in}}{1 + sT_{in}} = \frac{18.525}{(1 + 0.00025\ s)}$$

Motor (electrical): Gain, $K_a = \dfrac{1}{R_s} = 0.7143$; Time constant, $T_a = \dfrac{L_q}{R_s} = 0.0064\ s$

$$G_a(s) = \frac{K_a}{1 + sT_a} = \frac{0.7143}{(1 + 0.0064\ s)}$$

Induced emf loop: Torque constant, $K_t = \dfrac{3}{2}\left(\dfrac{P}{2}\right)^2 \cdot \lambda_{af} = 2.087\ N\cdot m/A$

Mechanical gain, $K_m = \dfrac{1}{B_t} = 100\ rad/s/Nm$

$$G_b(s) = \frac{K_t K_m \lambda_{af}}{(1 + sT_m)} = \frac{32.26}{(1 + 0.6s)}$$

where the mechanical time constant is

$$T_m = \frac{J}{B_t} = 0.6\ s$$

Motor (mechanical): $G_m(s) = \dfrac{K_m K_t}{(1 + sT_m)} = \dfrac{208.7}{(1 + 0.6\,s)}$

Equivalent electrical time constants of the motor: Solve for the roots of $as^2 + bs + c = 0$, where

$$a = T_m T_{ar}$$
$$b = T_m + K_a K_{in} T_m H_c$$
$$c = K_a K_b$$

where

$$K_b = K_t K_m \lambda_{af} = 32.26$$

Then the inverse of the roots will give T_1 and T_2 as

$$T_1 = 0.0005775 \text{ (sec)}$$
$$T_2 = 0.301 \text{ (sec)}$$

Simplified current-loop transfer function: $G_{is}(s) = \dfrac{K_i}{1 + sT_i}$

$$T_i = T_1 = 0.0005775 \text{ (s)}$$
$$K_i = \dfrac{T_m K_r}{T_2 K_b} = 1.1443$$

Exact current-loop transfer function: $G_i(s) = \dfrac{G_r(s) \cdot G_a(s)/[1 + G_a(s) \cdot G_b(s)]}{1 + H_c \cdot G_a(s) \cdot G(s)/[1 + G_a(s) \cdot G_b(s)]}$

Speed controller:

$$K_g = K_i K_m K_t \dfrac{H_\omega}{T_m} = 19.90$$
$$T_{\omega i} = T_\omega + T_i = 0.0025775 \text{ (s)}$$
$$T_s = 6T_{\omega i} = 0.0155 \text{ (s)}$$
$$K_s = \dfrac{4}{9 K_g T_{\omega i}} = 8.6638$$

Simplified speed-loop transfer function: $G_{ss}(s) \cong \dfrac{1}{H_\omega} \left\{ \dfrac{K_g \dfrac{K_s}{T_s}(1 + sT_s)}{s^3(T_{\omega i}) + s^2 + K_g \dfrac{K_s}{T_s}(1 + sT_s)} \right\}$

Exact speed-loop transfer function: $G_{se}(s) = \dfrac{G_m(s) \cdot G_i(s) \cdot G_s(s)}{1 + G_\omega(s) \cdot G_m(s) \cdot G_i(s) \cdot G_s(s)}$

where

$$G_s(s) = \dfrac{K_s}{T_s} \cdot \dfrac{(1 + sT_s)}{s} = (560.2)\dfrac{(1 + 0.0155\,s)}{2}$$

$$G_\omega(s) = \dfrac{H_\omega}{1 + sT_\omega} = \dfrac{0.05}{(1 + 0.002\,s)}$$

Smoothing: It is achieved by canceling the zero, $(1+sT_s)$, with a pole inserted in series with the speed reference. Note that this leaves the final speed-loop transfer function with only poles. The smoothing can also be thought of as the soft-start controller employed in almost all of the drive systems in practice.

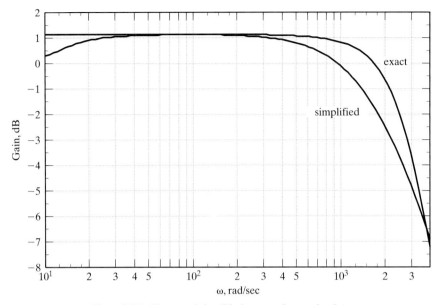

Figure 9.35 Exact and simplified current-loop-gain plots

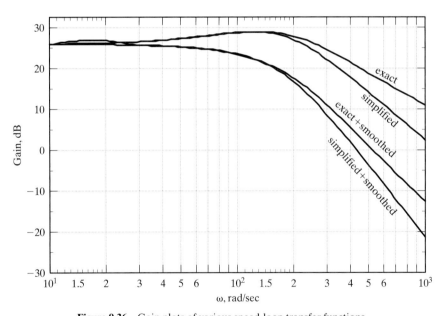

Figure 9.36 Gain plots of various speed-loop transfer functions

All the transfer-function gains and phases are plotted in Figures 9.35, 9.36, and 9.37. In the frequency regions of interest, note that the approximations hold good both in magnitude and in phase, in spite of the reduction of the fifth-order system to an equivalent third-order in the case of the speed loop and of a third to a first in the current loop.

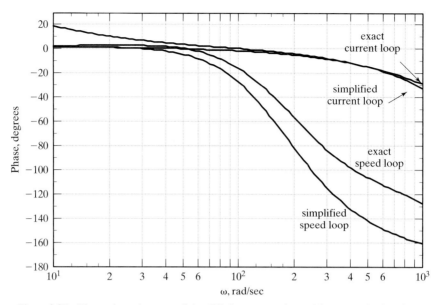

Figure 9.37 Phase plots of exact and simplified current and speed-loop transfer functions

9.8 SENSORLESS CONTROL

The PMSM drive requires two current sensors and an absolute-rotor-position sensor for the implementation of any control strategy discussed in the previous section. The rotor position is sensed with an optical encoder or a resolver for high-performance applications. The position sensors compare to the cost of the low-power motor, thus making the total system cost very noncompetitive compared to other types of motor drives. As for the current sensors, they are not as expensive as the rotor-position sensor; note that other types of drives also require their use in feedback control. Hence, the control and operation of PMSM drive without a rotor-position sensor would enhance its applicability to many cost-sensitive applications and provide a back-up control in sensor-based drives during sensor failures. One method of rotor-position sensorless control strategy is discussed in this section.

The basis for this control strategy is that the error between measured and calculated currents from the machine model gives the difference between the assumed rotor speed and the actual rotor speed of the motor drive. Nulling this current error results in the synchronous operation of the motor drive by estimating its rotor position accurately. The relevant algorithm for sensorless control is developed next.

The following assumptions are made to develop this control algorithm.

(i) Motor parameters and rotor PM flux are constant.
(ii) Induced emfs in the machine are sinusoidal.
(iii) The drive operates in the constant-torque region, and flux-weakening operation is not considered.

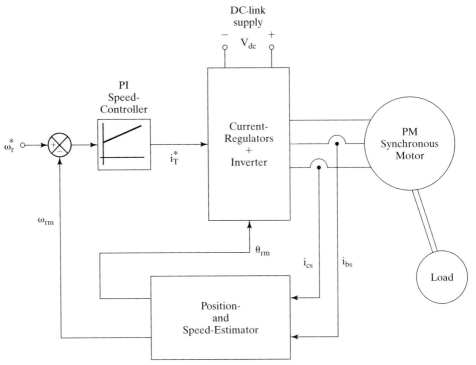

Figure 9.38 Control schematic diagram of the sensorless PMSM drive

The basic control schematic diagram is shown in Figure 9.38. Two phase currents constitute the inputs to the electrical rotor position and speed estimator. The error between the reference speed, ω_r^*, and the estimated rotor speed, ω_{rm}, is amplified and limited to provide the torque-producing component of the stator current, i_T^*, which is the q axis stator current in the rotor reference frames. The estimated rotor position, together with i_T^*, provides the stator current commands, which are enforced by a three-phase inverter feeding the PMSM. The position and speed estimator is derived from the machine equations.

Consider that the machine is running at a speed ω_r, whereas the model starts with an assumed rotor speed ω_{rm}. This is shown in the phasor diagram given in Figure 9.39. The assumed rotor position θ_{rm} lags behind the actual rotor position θ_r by $\delta\theta$ radians. They are related to the actual and assumed or model speed as follows:

$$\theta_r = \int \omega_r \, dt \tag{9.143}$$

$$\theta_{rm} = \int \omega_{rm} \, dt \tag{9.144}$$

$$\delta\theta = \theta_r - \theta_{rm} = \int (\omega_r - \omega_{rm}) \, dt \tag{9.145}$$

The machine model is utilized to compute the stator currents; note that it is carried out in a reference frame at an assumed rotor speed. That implies the reference frames are α and β axes and not d and q axes, which are the usual rotor reference

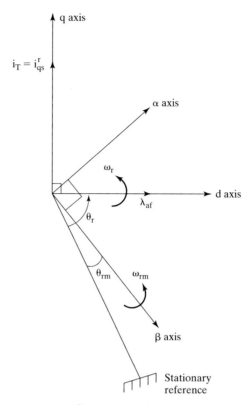

Figure 9.39 Phasor diagram corresponding to an error between the actual and assumed rotor position

frames. Therefore, the machine equations in the assumed rotor-speed reference frames are

$$\begin{bmatrix} pi_{\alpha m} \\ pi_{\beta m} \end{bmatrix} = \begin{bmatrix} -\dfrac{R_s}{L_q} & -\omega_{rm}\dfrac{L_d}{L_q} \\ \omega_{rm}\dfrac{L_q}{L_d} & -\dfrac{R_s}{L_q} \end{bmatrix} \begin{bmatrix} i_{\alpha m} \\ i_{\beta m} \end{bmatrix} + \begin{bmatrix} -\dfrac{\omega_{rm}\lambda_{af}}{L_q} \\ 0 \end{bmatrix} + \begin{bmatrix} \dfrac{v_\alpha}{L_q} \\ \dfrac{v_\beta}{L_d} \end{bmatrix} \quad (9.146)$$

In the model α and β reference frames, the actual machine equations are written from the d and q axes as

$$\begin{bmatrix} pi_\alpha \\ pi_\beta \end{bmatrix} = \begin{bmatrix} -\dfrac{R_s}{L_q} & -\omega_{rm}\dfrac{L_d}{L_q} \\ \omega_{rm}\dfrac{L_q}{L_d} & -\dfrac{R_s}{L_q} \end{bmatrix} \begin{bmatrix} i_\alpha \\ i_\beta \end{bmatrix} + \begin{bmatrix} -\dfrac{\omega_r\lambda_{af}}{L_q}\cos\delta\theta \\ \dfrac{\omega_r\lambda_{af}}{L_d}\sin\delta\theta \end{bmatrix} + \begin{bmatrix} \dfrac{v_\alpha}{L_q} \\ \dfrac{v_\beta}{L_d} \end{bmatrix} \quad (9.147)$$

The variables without the second subscript indicate that they are actual machine variables; the machine model (or estimated) variables end with subscript m. The actual machine equations are derived on the understanding that α–β axes are the considered reference axes and hence the rotor flux linkages have components on them from the d axis given as a function of the error in rotor position, $\delta\theta$. It is assumed that the entire rotor field is aligned on the d axis.

Discretize the two sets of model and actual current equations, respectively, as

$$\begin{bmatrix} i_{\alpha m}(kT) \\ i_{\beta m}(kT) \end{bmatrix} = \begin{bmatrix} i_{\alpha m}(\overline{k-1}T) \\ i_{\beta m}(\overline{k-1}T) \end{bmatrix} + \begin{bmatrix} pi_{\alpha m}(\overline{k-1}T) \\ pi_{\beta m}(\overline{k-1}T) \end{bmatrix} T \qquad (9.148)$$

$$\begin{bmatrix} i_{\alpha}(kT) \\ i_{\beta}(kT) \end{bmatrix} = \begin{bmatrix} i_{\alpha}(\overline{k-1}T) \\ i_{\beta}(\overline{k-1}T) \end{bmatrix} + \begin{bmatrix} pi_{\alpha}(\overline{k-1}T) \\ pi_{\beta}(\overline{k-1}T) \end{bmatrix} T \qquad (9.149)$$

where T is the sampling time and k is the present sampling instant. Substituting for the derivative terms in terms of the currents, induced emfs, and input voltages from equations (9.146) and (9.147) into (9.148) and (9.149) and finding their respective current errors yields the following:

$$\begin{bmatrix} \delta i_{\alpha}(kT) \\ \delta i_{\beta}(kT) \end{bmatrix} = \begin{bmatrix} i_{\alpha}(kT) - i_{\alpha m}(kT) \\ i_{\beta}(kT) - i_{\beta m}(kT) \end{bmatrix} = T \begin{bmatrix} -\dfrac{\lambda_{af}}{L_q}(\omega_r \cos\delta\theta - \omega_{rm}) \\ \dfrac{\omega_r \lambda_{af}}{L_d}\sin\delta\theta \end{bmatrix} \qquad (9.150)$$

A number of assumptions have been made to derive (9.150): (i) The difference between the model and actual currents, when multiplied by T, becomes negligible; (ii) The sampling time is very small compared to the electrical and mechanical time constants of the drive system.

If $\delta\theta$ is small, then the following approximations are valid for use in interpreting the above results:

$$\sin\delta\theta \cong \delta\theta \qquad (9.151)$$

$$\cos\delta\theta \cong 1 \qquad (9.152)$$

Hence, substituting these into error currents results in

$$\delta i_{\alpha}(kT) = -\dfrac{\lambda_{af}}{L_q} T(-\omega_{rm} + \omega_r) \qquad (9.153)$$

$$\delta i_{\beta}(kT) = \omega_r \dfrac{\lambda_{af}}{L_d} \delta\theta \qquad (9.154)$$

from which the actual rotor speed is obtained as

$$\omega_r = -\dfrac{L_q}{\lambda_{af}} \dfrac{1}{T} \delta i_{\alpha}(kT) + \omega_{rm} \qquad (9.155)$$

and the error in estimated rotor position is

$$\delta\theta = \dfrac{L_d}{\lambda_{af}} \dfrac{1}{T} \dfrac{\delta i_{\beta}(kT)}{\omega_r} \qquad (9.156)$$

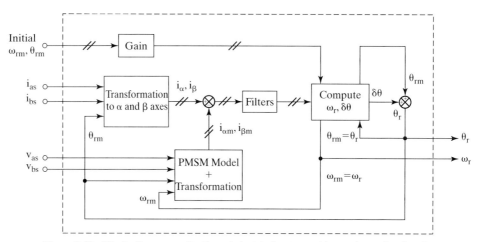

Figure 9.40 Block-diagram realization of electrical rotor position and speed estimation

Substitution for the rotor speed in the error rotor-position equation gives $\delta\theta$ as

$$\delta\theta = \frac{\left(\dfrac{L_d}{T\lambda_{af}}\right)\delta i_\beta(kT)}{\left[\omega_{rm} - \dfrac{L_q}{T\lambda_{af}}\delta i_\alpha(kT)\right]} \tag{9.157}$$

Note that $\delta\theta$, ω_{rm}, and ω_r are for the sampling instant of kT also. They have to be evaluated for each sampling instant to follow the rotor position closely. The rotor position then is

$$\theta_r = \theta_{rm} + \delta\theta \tag{9.158}$$

which becomes the estimate for θ_{rm} in the next sampling interval for feeding into the controller to compute stator-current commands. Filters are required to smooth ripples in the current errors due to PWM voltages fed to the machine. The realization of the position and speed estimator is shown in block-diagram form in Figure 9.40.

Starting from standstill is achieved by bringing the rotor to a particular position by energizing the stator phases accordingly. Alternatively, a sequence of pulse patterns is issued to the inverter, which would enable the motor to achieve a significant but small rotor speed to start the estimation. During this time, the estimator is kept inactive; it is brought into the control process by switching off the starting process. The input of initial values of speed and rotor position serves to smooth the transition process from initial starting to estimation for continued operation and control. Low-speed operation continues to be a challenge with this technique.

Several other methods exist to estimate the rotor position. For one, the induced emf of the stator phases could be estimated from the measured stator currents and voltages. This method has the problem of finding the position at zero speed; there are no induced emfs at that point, and at low speeds they would be difficult to estimate accurately because of small magnitude. Hence, this method has to incorporate an initial-starting process, as discussed above.

Ideally, rotor position can be obtained from stator self-inductances. The reluctance variation between, say, q and d axes in the machine occurs when the magnets have a pole arc less than 180 electrical degrees, regardless of the methods of magnet placement in the rotor. This variation in the reluctance can be measured by noninvasive measurement techniques, and then they can be used to extract the rotor position information. This reluctance variation is prominent in machines with high saliency. Even the surface-magnet machines exhibit a 10% variation in their reluctance between the d and q axes. References contain research papers discussing alternative methods of estimating rotor position.

9.9 PARAMETER SENSITIVITY

Temperature variation changes the stator resistance and flux remanence in the permanent magnets. The loss of magnetism for a 100°C rise in temperature in ceramic, neodymium, and samarium–cobalt magnets is 19%, 11%, and 4%, respectively, from their nominal values. The effect due to the loss of magnetism with temperature variations is predominant compared to the effect of stator-resistance variations on the performance of the drive system. Further, the stator-resistance sensitivity is overcome in current-regulated drives by the nature of closed-loop control. That the current-control has no impact on the drive system is due to temperature sensitivity of the magnets. A closed-loop speed-controlled drive system will minimize the effects due to temperature sensitivity of the magnets. To have an understanding of this operation, consider that the drive system is in steady state and that the magnet flux density has decreased in a step fashion (not usually the case, but taken up here as an extreme case for illustration). The torque will decrease instantaneously, because the current-magnitude command is a constant in steady state. With the decrease in the torque, the rotor will slow down, resulting in higher speed error, higher torque command, and hence higher current command. The torque then will rise and hence the rotor speed, and this cycle of events will go on, depending on the dynamics of the system, until the system reaches steady state again. In the wake of such a disturbance in the magnet flux linkages, the drive system will encounter electromagnetic torque oscillations as explained, and that may not be very desirable in high-performance drives.

Saturation usually will affect the quadrature-axis inductance rather than the direct-axis inductance. It is due to the fact that the direct axis, with its magnets, presents a very high-reluctance path; the magnets have a relative permeability comparable to that of air. In the quadrature axis, the reluctance is lower, because most of the flux path is through the iron.

By denoting the temperature-variation effects by α and the saturation effects by β, the relationship of the torque to its reference and of the mutual flux linkage to its reference are derived with the speed loop open for the PMSM drive. It is further assumed that the drive has inner current loops and that the stator currents equal their references.

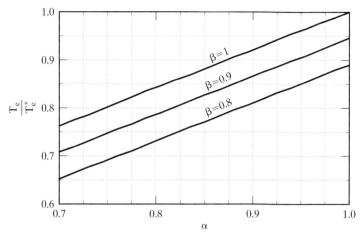

Figure 9.41 Ratio of electromagnetic torque to its command for various saturation levels, with speed loop open

9.9.1 Ratio of Torque to Its Reference

Ratio of torque to its reference is derived as

$$\frac{T_e}{T_e^*} = \frac{\alpha \lambda_{af}^* i_{qs}^{r*} + (L_d - \beta L_q) i_{ds}^{r*} i_{qs}^{r*}}{\lambda_{af}^* i_{qs}^{r*} + (L_d - L_q) i_{ds}^r i_{qs}^{r*}} = \frac{\alpha \lambda_{af}^* + L_d(1 - \beta\rho) i_{ds}^{r*}}{\lambda_{af}^* + L_d(1 - \rho) i_{ds}^{r*}} \quad (9.159)$$

Note that the ratio of torque to its reference is independent of the q axis stator current. By dividing the numerator and denominator by the base flux linkages, $L_b I_b$, the torque-to-reference is derived as

$$\frac{T_e}{T_e^*} = \frac{\alpha \lambda_{afn} + (1 - \beta\rho) L_{dn} i_{dsn}^r}{\lambda_{afn} + (1 - \rho) L_{dn} i_{dsn}^r} \quad (9.160)$$

where the normalized rotor flux linkage is given as

$$\lambda_{afn} = \frac{\lambda_{af}}{L_b I_b} = \frac{\lambda_{af}}{\lambda_b}, \text{ p.u.} \quad (9.161)$$

Note that

$$i_{dsn}^r = (i_{dsn}^r)^* \quad (9.162)$$

For the same machine details used in the previous illustrations, the ratio of torque to its reference, plotted against α, is shown in Figure 9.41 for various values of β ranging from 80% to 100% of the nominal value of L_q. Note that $i_{dn}^r = -1$ p.u., $\rho = 1.607$, and $L_{dn} = 0.435$ p.u.. α is varied from 0.7 to 1. Lower values of α indicate increased rotor temperature; the value of 1 corresponds to the operation at ambient temperature. Higher rotor temperature reduces the output torque for the same stator current, and saturation further decreases it from the nominal value. The variation is linear, unlike in the case of the indirect vector controlled induction motor drive described in Chapter 8.

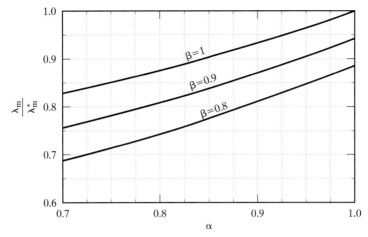

Figure 9.42 Ratio of mutual flux linkage to its reference value for various saturation levels, with speed loop open

Note that, when torque angle is set at $\pi/2$, $i_{dsn}^r = 0$, and hence the torque-to-reference equals α. Then it becomes independent of saturation. This is contrary to the performance of an indirect vector-controlled induction motor.

9.9.2 Ratio of Mutual Flux Linkage to Its Reference

The ratio of mutual flux linkage to its reference is derived as

$$\frac{\lambda_m}{\lambda_m^*} = \frac{\sqrt{(\alpha\lambda_{afn} + L_{dn}i_{dsn}^r)^2 + (\beta L_{qn}i_{qsn}^r)^2}}{\sqrt{(\lambda_{afn} + L_{dn}i_{dsn}^r)^2 + (L_{qn}i_{qsn}^r)^2}} = \frac{\sqrt{(\alpha\lambda_{afn} + L_{dn}i_{dsn}^r)^2 + (\beta\rho L_{dn}i_{qsn}^r)^2}}{\sqrt{(\lambda_{afn} + L_{dn}i_{dsn}^r)^2 + (\rho L_{dn}i_{qsn}^r)^2}} \quad (9.163)$$

where ρ is the saliency factor given by the ratio between the quadrature- and direct-axis inductances. For the same values considered in the previous illustration, the ratio of the mutual flux linkage to its reference vs. α for various β values is shown in Figure 9.42. They follow the trend of the torque-to-reference-vs.-α characteristics, except that they are not linear any more.

Temperature and saturation variations produce a nonideal torque and mutual-flux-linkages amplifier of the PMSM drive, with the consequence that the motor drive is not suitable for precision torque and speed-control applications. The changes in the parameter variations could be detected and compensated in a manner similar to the methods described for vector-controlled induction motor drives. One such method is given in the following section.

9.9.3 Parameter Compensation Through Air Gap Power Feedback Control

Air gap power is a variable that is a clear indicator of the variations in rotor flux linkages and saturation in the q axis of the PMSM. Air gap power is computed from

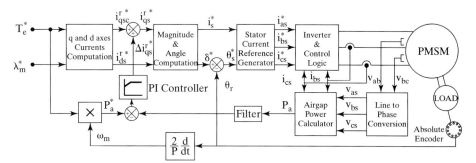

Figure 9.43 Parameter compensation of PMSM drive with air-gap-power feedback control

the input power of the machine by subtracting the stator resistance losses. If this air gap power is filtered to remove the switching ripple due to stator currents and transient magnetic energy, then it is a dc signal for a given operating point. This is denoted as

$$P_a = P_i - \frac{3}{2}\{(i_{qs}^r)^2 + (i_{ds}^r)^2\}R_s \qquad (9.164)$$

where input power, P_i, is

$$P_i = \frac{3}{2}\{v_{qs}^r i_{qs}^r + v_{ds}^r i_{ds}^r\} = v_{as}i_{as} + v_{bs}i_{bs} + v_{cs}i_{cs} \qquad (9.165)$$

The variations in the machine rotor flux linkages are embodied in the q axis voltage, and saturation effect in the form of L_q variation is contained in the d axis voltage. Hence, air gap power indicates the effects of major parameter variations, including that of R_s. The closure of air gap power feedback control requires the determination of reference air gap power.

The reference air gap power is obtained from the reference or actual speed and the torque reference in the drive system as

$$P_a^* = \frac{\omega_r T_e^*}{(P/2)} \qquad (9.166)$$

The error between the reference and the computed/measured air gap power is amplified and then limited to provide a q axis stator-current correction signal to compensate for the parameter variations, as is shown in Figure 9.43. This signal, denoted as Δi_{qs}^*, is summed with the calculated q axis current, $(i_{qsc}^r)^*$, to provide the compensated q axis current reference, $(i_{qs}^r)^*$.

The working of this feedback loop is explained as follows. Assuming a reduction in the rotor flux linkage from its nominal value, it is deduced that q axis voltage will be reduced; hence, a reduction in the measured/computed air gap power, P_a, sets in. This will result in a positive and a larger Δi_{qs}^{r*}, thus increasing the stator q axis current and so resulting in increasing air gap power to equal its reference, P_a^*. Similar reasoning would show that compensation for saturation effect also is achieved with this control strategy.

9.9.3.1 Algorithm.
The correction signal is denoted as Δi_{qs}^{r*} and is given by

$$\Delta i_{qs}^{r*} = K_p(P_a^* - P_a) + K_i \int (P_a^* - P_a) \quad (9.167)$$

where K_p and K_i are the proportional and integral gains of the PI controller.

Assuming no d axis stator current, the q axis current command in rotor reference frame is obtained as

$$i_{qs}^{r*} = \frac{T_e^*}{\frac{3}{2}\frac{P}{2}\lambda_{af}^*} \quad (9.168)$$

The reference torque component of the current is augmented by the correction signal as

$$i_T^* = i_{qs}^{r*} + \Delta i_{qs}^{r*} \quad (9.169)$$

from which the magnitude of the reference stator current is obtained as follows:

$$i_s^* = \sqrt{i_T^{*2} + i_f^{*2}} \quad (9.170)$$

The flux-producing component of the stator-current command, i_f^*, is zero for $\delta = 90°$, but it is nonzero for δ other than $90°$. i_f^* is nonzero when flux-weakening is resorted to, as incorporated later in this section.

The phase angle of the reference stator current is given by

$$\theta_s^* = \delta^* + \theta_r \quad (9.171)$$

where the torque angle is calculated as

$$\delta^* = \operatorname{atan}\left(\frac{i_T^*}{i_f^*}\right) \quad (9.172)$$

The reference stator-phase currents can be generated from i_s^* and θ_s^* by using the transformation equations (9.45):

$$i_{as}^* = i_s^* \sin(\theta_s^*) \quad (9.173)$$

$$i_{bs}^* = i_s^* \sin\left(\theta_s^* - \frac{2\pi}{3}\right) \quad (9.174)$$

$$i_{cs}^* = i_s^* \sin\left(\theta_s^* + \frac{2\pi}{3}\right) \quad (9.175)$$

The reference currents are fed to the inverter, in this particular case illustrated with a hysteresis controller, which makes the actual motor currents follow the commanded values at all times. Current feedback is required for the hysteresis controller to achieve this. A description of the hysteresis current controller can be found in Chapter 4. The reference phase currents are compared to the actual line currents by the hysteresis controller, which in turn controls the switching of the inverter and hence controls the average phase voltages supplied to the PMSM. To operate above the rated speed, flux-weakening is applied. The block labeled as q and d axis currents in Figure 9.43 is elaborated in Figure 9.44 by using a simple flux-weakening control

572 Chapter 9 Permanent-Magnet Synchronous and Brushless DC Motor Drives

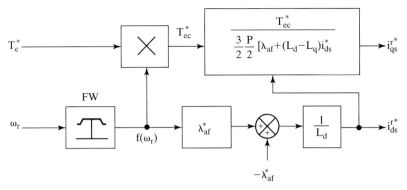

Figure 9.44 Schematic of the q and d axis command current generator of Figure 9.43

strategy. In Figure 9.44, the block labeled as FW (for flux-weakening unit) has the following output:

$$f(\omega_r) = \frac{\omega_b}{\omega_r}; \quad \omega_b < \omega_r < \omega_{max} \quad (9.176)$$

$$= 1; \quad 0 < \omega_r < \omega_b$$

If the speed is less than the rated or base speed, the output of block FW is 1, and hence $i_{ds}^* = i_f^* = 0$. If the speed is greater than the rated speed, the flux-weakening component of the stator current is

$$i_{ds}^* = \frac{(f(\omega_r) - 1)\lambda_{af}^*}{L_d} \quad (9.177)$$

Note that $i_{ds}^* < 0$ for flux-weakening. The q axis reference current, i_{qs}^r*, is obtained as follows:

$$T_{ec}^* = T_e^* . f(\omega_r) \quad (9.178)$$

$$i_{qs}^r* = \frac{T_{ec}^*}{\frac{3}{2}\frac{P}{2}[\lambda_{af} + (L_d - L_q)i_{ds}^*]} \quad (9.179)$$

This completes the algorithm for the q and d axis command current generator blocks shown in Figure 9.43.

9.9.3.2 Performance. Dynamic simulation results of the drive system with parameter compensation shown in Figure 9.43 are presented in this section.

Torque-Drive Performance: Figure 9.45 shows the simulations for a step change of rotor flux linkages in the uncompensated and compensated torque-drive system. The system starts with nominal rotor flux linkages; after t = 0.02 s, λ_{af} is changed to 85% of its nominal value and the corresponding effects are studied. It is observed that, in the uncompensated system, since $\Delta i_{qs}^r* = 0$, i_s^* does not change, thereby

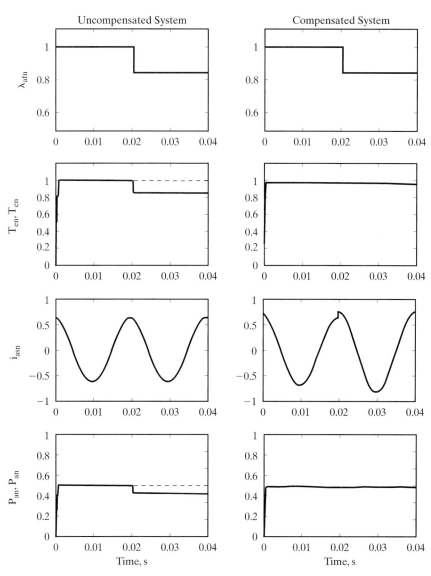

Figure 9.45 Simulation results of the constant-torque-angle-controlled torque drive system with and without parameter compensation when a step change in the rotor flux linkages is applied

reflecting a lack of changes in phase currents due to this step change in rotor flux linkages. Therefore, a decrease in the electromagnetic torque from its reference value is seen, resulting in a corresponding decrease in the actual air gap power. In the compensated system, it is noticed that torque rises to match its reference after initially dropping to a lower value. A corresponding rise in the value of i_{as}, the phase a current, is seen. In this torque-drive system, the rotor speed is assumed to be constant at $\omega_r = 0.5$ p.u. which is the rated speed for the machine under consideration.

574　Chapter 9　Permanent-Magnet Synchronous and Brushless DC Motor Drives

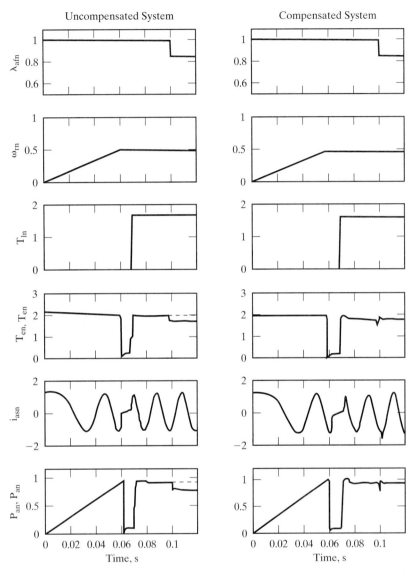

Figure 9.46　Simulation results of the speed-controlled drive system with constant torque angle for a step change in λ_{af}

Speed-Controlled-Drive Performance: The second set of simulations is for a speed-controlled drive system with a step change in its speed from 0 to 0.5 p.u applied initially; then, a step change of load torque from 0 to 1.65 p.u. is applied when the speed equals its command. Finally, λ_{af} is changed from nominal to 0.85 times the nominal value. The results for both the uncompensated and compensated speed-controlled drive systems are shown in Figure 9.46. The reference torque T_e^* is limited to 2 p.u. In the uncompensated system, the change in rotor flux linkages causes a reduction in torque and hence in the speed, momentarily. This increases the torque reference through the action of the outer speed-feedback loop, resulting in an increase in the air gap power reference (as is shown by dotted lines). The q axis stator current increases to meet the load torque resulting in the speed's being maintained equal to its reference in steady state. In the compensated system, the torque reference and air gap power reference do not change, but the stator q axis current is increased until it delivers torque equal to its reference. Resulting compensation has the advantage in that it does not change torque reference and hence will not lead to a speed reduction, as it can in the uncompensated system. For example, consider that the torque reference is altered by the speed feedback loop in the uncompensated system from 2 to 2.5 p.u., with the torque limiter having a maximum of 2 p.u. That means the stator q axis current will be generated only for 2 p.u. torque reference, hence generating less than 2 p.u. torque with the reduction in rotor flux linkages. If load torque is 2 p.u., then the speed will have to come down. That this is not the case in the compensated system is obvious.

Flux-Weakening-Drive Performance: The third set of simulations is for the speed-controlled drive system operating in the flux-weakening region. The condition for the simulation is to have a step change of λ_{af} from nominal to 0.85 times the nominal value with the drive system operating in the flux-weakening mode. The simulation results are shown in Figure 9.47. The flux-weakening process is initiated in this drive when the speed is greater than 0.5 p.u., where the reference torque T_e^* and hence the actual torque start falling as expected until the commanded speed is reached. When a step change is applied to λ_{af}, a rise in reference torque is observed, but the actual torque does not rise to match the reference torque, because of a drop in rotor flux linkages. The compensated system, however, shows a perfect matching of the torque and its reference, via almost instantaneous stator current compensation.

Effect of Quadrature-Axis Inductance Variation: The final set of simulations shows the step change of L_q from nominal to 0.85 times the nominal value in Figure 9.48, including the operation in the flux-weakening mode. The variation of L_q would not have any effect on a constant-torque-angle-controlled system where $\delta = 90°$, because the controller parameters and the output torque are not dependent on the quadrature-axis inductance, but it does affect the system in the flux-weakening mode: both the controller parameters and the torque output are dependent on the quadrature-axis inductance. The compensated system, however, corrects for changes in L_q, as is shown in Figure 9.48. The effects are similar to the case where λ_{af} changes.

Effect of Stator-Resistance Variation: Since the air gap power calculator is dependent on the stator resistance, which changes with temperature, simulations for a step change of stator resistance are necessary to verify the adequacy of the air gap-power-compensation scheme to neutralize the variations in stator resistance. The effects were seen to be negligible on the compensated system. The stator resistance

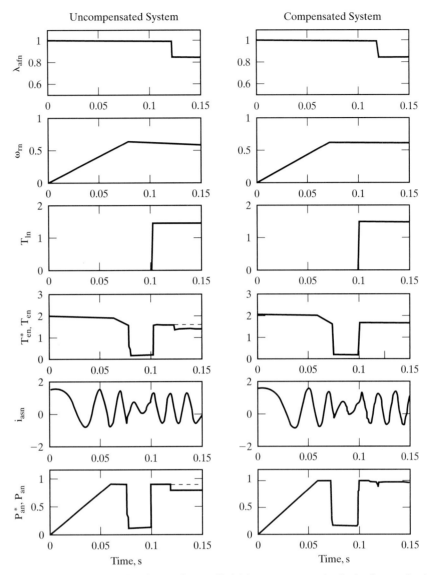

Figure 9.47 Simulation results of a speed-controlled drive system operating in the flux-weakening mode with a step change in λ_{af}

variations can also be compensated for via direct temperature measurement. The quadrature-axis inductance variation can be compensated with the q axis current magnitude. No direct monitoring of the rotor temperature or rotor flux linkages is possible to compensate for rotor-flux-linkages variation. Hence, schemes such as the air gap-power feedback control are necessary to overcome the parameter sensitivity of the rotor-flux-linkages variation, and the same then could be used to compensate for other motor-parameter variations.

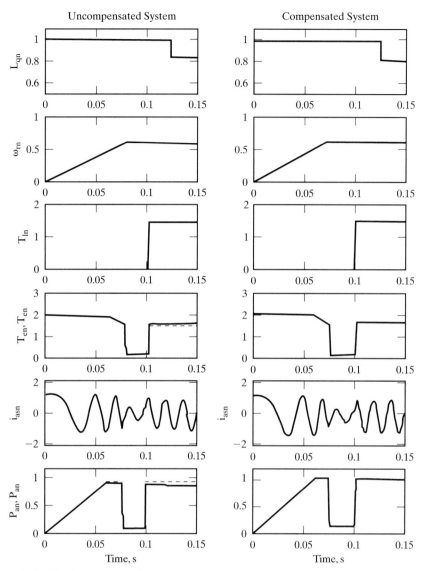

Figure 9.48 Simulation results of the effect of variation of L_q on the speed-controlled drive system, including the operation in the flux-weakening mode

9.10 PM BRUSHLESS DC MOTOR (PMBDCM)

PM synchronous machines having trapezoidal induced emf are known as PM brushless dc machines. The advantage of such a machine in comparison to the PMSM was discussed in the early part of this chapter. The major reason for the popularity of these machines over their counterparts is control simplicity. To initiate the onset and commutation of current in the phase of a machine, the beginning and end of the constant portion of the induced emf have to be tracked. That amounts to only six discrete positions for a three-phase machine in each electrical cycle. These signals could easily be generated with three Hall sensors displaced from each other by 120 electrical degrees. The Hall sensors are mounted facing a small magnet wheel fixed to the rotor and having the same number of poles as the rotor of the PMBDCM, or the extra magnet wheel may be dispensed with by extending the rotor beyond the stack length of the stator and using the rotor magnets to provide the position information. Such an arrangement tracks the absolute position of the rotor magnets and hence the shape and position of the induced emfs in all the machine phases. In contrast to the PMSM, which requires continuous and instantaneous absolute rotor position, the PMBDCM position-feedback requirement is much simpler: it requires only six discrete absolute positions for a three-phase machine, resulting in a major cost saving in the feedback sensor. Further, the control involves significant vector operations in the PMSM drive, whereas such operations are not required for operation of the PMBDCM drive.

9.10.1 Modeling of PM Brushless DC Motor

The flux distribution in a PM brushless dc motor is trapezoidal; therefore, the d–q rotor reference frames model developed for the PM synchronous motor is not applicable. Given the nonsinusoidal flux distribution, it is prudent to derive a model of the PMBDCM in phase variables. The derivation of this model is based on the assumptions that the induced currents in the rotor due to stator harmonic fields are neglected and that iron and stray losses are also neglected. Damper windings are not usually a part of the PMBDCM; damping is provided by the inverter control. The motor is considered to have three phases, even though the derivation procedure is valid for any number of phases.

The coupled circuit equations of the stator windings in terms of motor electrical constants are

$$\begin{bmatrix} v_{as} \\ v_{bs} \\ v_{cs} \end{bmatrix} = \begin{bmatrix} R_s & 0 & 0 \\ 0 & R_s & 0 \\ 0 & 0 & R_s \end{bmatrix} \begin{bmatrix} i_{as} \\ i_{bs} \\ i_{cs} \end{bmatrix} + p \begin{bmatrix} L_{aa} & L_{ab} & L_{ac} \\ L_{ba} & L_{bb} & L_{bc} \\ L_{ca} & L_{cb} & L_{cc} \end{bmatrix} \begin{bmatrix} i_a \\ i_b \\ i_c \end{bmatrix} + \begin{bmatrix} e_{as} \\ e_{bs} \\ e_{cs} \end{bmatrix} \quad (9.180)$$

where R_s is the stator resistance per phase, assumed to be equal for all three phases. The induced emfs e_{as}, e_{bs}, and e_{cs} are all assumed to be trapezoidal, as shown in Figure 9.8, where E_p is the peak value, derived as

$$E_p = (Blv)N = N(Blr\omega_m) = N\phi_a\omega_m = \lambda_p\omega_m \quad (9.181)$$

where N is the number of conductors in series per phase, v is the velocity, l is the length of the conductor, r is the radius of the rotor bore, ω_m is the angular velocity, and B is the flux density of the field in which the conductors are placed. This flux density is solely due to the rotor magnets. The product (Blr), denoted as ϕ_a, has the dimensions of flux and is directly proportional to the air gap flux, ϕ_g:

$$\phi_a = Blr = \frac{1}{\pi} B\pi lr = \frac{1}{\pi} \phi_g \quad (9.182)$$

Note that the product of flux and number of conductors in series has the dimension of flux linkages and is denoted by λ_p. Since this is proportional to phase *a* flux linkages by a factor of $\frac{1}{\pi}$, it is hereafter referred to as modified flux linkages.

Section 9.10 PM Brushless DC Motor (PMBDCM)

If there is no change in the rotor reluctance with angle because of a nonsalient rotor, and assuming three symmetric phases, the following are obtained:

$$L_{aa} = L_{bb} = L_{cc} = L \, ; \text{ and } L_{ab} = L_{ba} = L_{ac} = L_{ca} = L_{bc} = L_{cb} = M \text{ (H)} \quad (9.183)$$

Substituting equations (9.182) and (9.183) in equation (9.180) gives the PMBDCM model as

$$\begin{bmatrix} v_{as} \\ v_{bs} \\ v_{cs} \end{bmatrix} = R_s \begin{bmatrix} 1 & 0 & 0 \\ 0 & 1 & 0 \\ 0 & 0 & 1 \end{bmatrix} \begin{bmatrix} i_{as} \\ i_{bs} \\ i_{cs} \end{bmatrix} + \begin{bmatrix} L & M & M \\ M & L & M \\ M & M & L \end{bmatrix} p \begin{bmatrix} i_a \\ i_b \\ i_c \end{bmatrix} + \begin{bmatrix} e_{as} \\ e_{bs} \\ e_{cs} \end{bmatrix} \quad (9.184)$$

The stator phase currents are constrained to be balanced, i.e., $i_{as} + i_{bs} + i_{cs} = 0$, which leads to the simplification of the inductance matrix in the model as

$$\begin{bmatrix} v_{as} \\ v_{bs} \\ v_{cs} \end{bmatrix} = \begin{bmatrix} R_s & 0 & 0 \\ 0 & R_s & 0 \\ 0 & 0 & R_s \end{bmatrix} \begin{bmatrix} i_{as} \\ i_{bs} \\ i_{cs} \end{bmatrix} + \begin{bmatrix} (L-M) & 0 & 0 \\ 0 & (L-M) & 0 \\ 0 & 0 & (L-M) \end{bmatrix} p \begin{bmatrix} i_a \\ i_b \\ i_c \end{bmatrix} + \begin{bmatrix} e_{as} \\ e_{bs} \\ e_{cs} \end{bmatrix} \quad (9.185)$$

The electromagnetic torque is given by

$$T_e = [e_{as} i_{as} + e_{bs} i_{bs} + e_{cs} i_{cs}] \frac{1}{\omega_m} \text{ (N·m)} \quad (9.186)$$

The instantaneous induced emfs can be written from Figure 9.8 and equation (9.181) as

$$e_{as} = f_{as}(\theta_r) \lambda_p \omega_m \quad (9.187)$$
$$e_{bs} = f_{bs}(\theta_r) \lambda_p \omega_m \quad (9.188)$$
$$e_{cs} = f_{cs}(\theta_r) \lambda_p \omega_m \quad (9.189)$$

where the functions $f_{as}(\theta_r)$, $f_{bs}(\theta_r)$, and $f_{cs}(\theta_r)$ have the same shape as e_{as}, e_{bs}, and e_{cs}, with a maximum magnitude of ± 1. The induced emfs do not have sharp corners, as is shown in trapezoidal functions, but rounded edges. The emfs are the result of the flux-linkages derivatives, and the flux linkages are continuous functions. Fringing also makes the flux density functions smooth with no abrupt edges. The electromagnetic torque then is

$$T_e = \lambda_p [f_{as}(\theta_r) i_{as} + f_{bs}(\theta_r) i_{bs} + f_{cs}(\theta_r) i_{cs}] \text{ (N·m)} \quad (9.190)$$

It is significant to observe that the phase-voltage equation is identical to the armature-voltage equation of a dc machine. That is one of the reasons for naming this machine the *PM brushless* dc machine. The equation of motion for a simple system with inertia J, friction coefficient B, and load torque T_l is

$$J \frac{d\omega_m}{dt} + B \omega_m = (T_e - T_l) \quad (9.191)$$

and electrical rotor speed and position are related by

$$\frac{d\theta_r}{dt} = \frac{P}{2} \omega_m \quad (9.192)$$

Combining all the relevant equations, the system in state-space form is

$$\dot{x} = Ax + Bu \qquad (9.193)$$

where

$$x = \begin{bmatrix} i_{as} & i_{bs} & i_{cs} & \omega_m & \theta_r \end{bmatrix}^t \qquad (9.194)$$

$$A = \begin{bmatrix} -\dfrac{R_s}{L_1} & 0 & 0 & -\dfrac{\lambda_p}{L_1} f_{as}(\theta_r) & 0 \\ 0 & -\dfrac{R_s}{L_1} & 0 & -\dfrac{\lambda_p}{L_1} f_{bs}(\theta_r) & 0 \\ 0 & 0 & -\dfrac{R_s}{L_1} & -\dfrac{\lambda_p}{L_1} f_{cs}(\theta_r) & 0 \\ \dfrac{\lambda_p}{J} f_{as}(\theta_r) & \dfrac{\lambda_p}{J} f_{bs}(\theta_r) & \dfrac{\lambda_p}{J} f_{cs}(\theta_r) & -B/J & 0 \\ 0 & 0 & 0 & \dfrac{P}{2} & 0 \end{bmatrix} \qquad (9.195)$$

$$B = \begin{bmatrix} \dfrac{1}{L_1} & 0 & 0 & 0 \\ 0 & \dfrac{1}{L_1} & 0 & 0 \\ 0 & 0 & \dfrac{1}{L_1} & 0 \\ 0 & 0 & 0 & -\dfrac{1}{J} \\ 0 & 0 & 0 & 0 \end{bmatrix} \qquad (9.196)$$

$$L_1 = L - M \qquad (9.197)$$

$$u = \begin{bmatrix} v_{as} & v_{bs} & v_{cs} & T_l \end{bmatrix}^t \qquad (9.198)$$

The state variable θ_r, rotor position, is required so as to have the functions $f_{as}(\theta_r)$, $f_{bs}(\theta_r)$, and $f_{cs}(\theta_r)$ which can be realized from a stored table. This completes the modeling of the PMBDCM.

9.10.2 The PMBDCM Drive Scheme

For constant-torque operation at speeds lower than the base speed, this drive requires six discrete pieces of position information. They correspond to every 60 electrical degrees for energizing the three stator phases, as shown in Figure 9.8.

The flux-weakening is slightly different for this motor and is discussed later. The control scheme for the PMBDM drive is simple and is shown in Figure 9.49. The resolver gives absolute rotor position; it is converted into rotor speed through the signal processor. The rotor speed is compared to its reference, and the rotor speed error is amplified through the speed controller. The output of the speed controller

Section 9.10 PM Brushless DC Motor (PMBDCM)

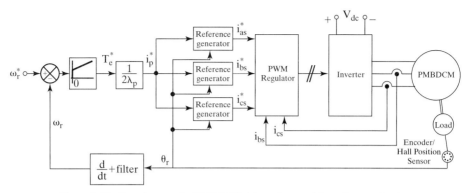

Figure 9.49 Speed-controlled PMBDM drive scheme without flux-weakening

provides the reference torque, T_e^*. The current-magnitude command, I_p^*, is obtained from the torque expression as

$$T_e^* = \lambda_p[f_{as}(\theta_r)i_{as}^* + f_{bs}(\theta_r)i_{bs}^* + f_{cs}(\theta_r)i_{cs}^*] \quad (9.199)$$

Only two machine phases conduct current at any time, with the two phases being in series for full-wave inverter operation, so the phase currents are equal in magnitude but opposite in sign. The rotor-position-dependent functions have the same signs as the stator phase currents in the motoring mode, but opposite signs in the regeneration mode. The result of such sign relationships is simplification of torque command as,

$$T_e^* = 2\lambda_p i_p^* \quad (9.200)$$

The stator-current command is derived from (9.200) as

$$i_p^* = \frac{T_e^*}{2\lambda_p} \quad (9.201)$$

The individual stator-phase current commands are generated from the current-magnitude command and absolute rotor position. These current commands are amplified through the inverter by comparing them with their respective currents in the stator phases. Only two phase currents are necessary in the balanced three-phase system to obtain the third phase current, since the sum of the three phase currents is zero.

The current errors are amplified and used with pulse-width modulation or hysteresis logic to produce the switching-logic signals for the inverter switches, as explained in Chapter 4.

9.10.3 Dynamic Simulation

The simulation results for a step speed reference input of from 0 to 1 p.u. are shown in Figure 9.50 for the PWM current controllers. The rotor is at standstill at time zero. With the onset of speed reference, the speed error and torque references attain a maximum value, which is limited to 2.0 p.u in this case. The current is made to follow the reference by the current controller. Therefore, the electromagnetic torque follows its

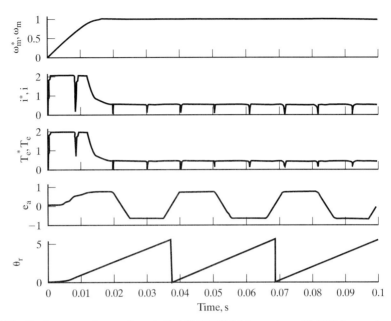

Figure 9.50 Performance of a speed-controlled PMBDCM drive system with PWM current controllers

reference very closely. The ripples on the torque are due to current ripples produced by the switching. The PWM control considered in this case is operating at 2 kHz, and the current controller has an amplification of 100.

9.10.4 Commutation-Torque Ripple

The desired current waveform is rectangular and 120° wide in each half-cycle for a three-phase PMBDCM drive. The leakage inductance, L_1, causes the stator currents to take a finite time to rise and fall, thus distorting the ideal waveform into a trapezoidal shape. The effect of this is the torque ripple generated at the current transitions. For a three-phase machine, there will be six torque ripples for every 360° electrical, as the six current transitions occur. They will also reduce the average torque if the conduction time is maintained at 120° electrical, whereby the constant-current region is reduced below 120° electrical. The consequences of a set of practical currents on the performance of the PMBDCM drive can be analyzed by using a Fourier-series approach as follows. The torque expressions are given considering only a two-pole machine; for a P-pole machine, the expressions have to be multiplied by the number of pole pairs.

The phase current is generalized, as shown in Figure 9.51. The phase a current can be resolved into Fourier series as

$$i_{as}(\theta_r) = \frac{4I_p}{\pi(\theta_2 - \theta_1)}\left[(\sin\theta_2 - \sin\theta_1)\sin\theta_r + \frac{1}{3^2}(\sin 3\theta_2 - \sin 3\theta_1)\sin 3\theta_r + \ldots\right]$$

(9.202)

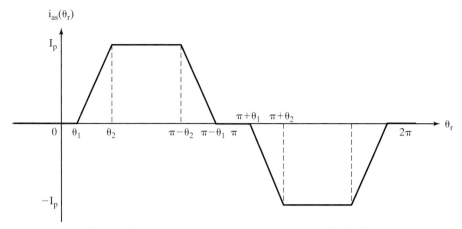

Figure 9.51 General phase-current waveform of the PMBDCM

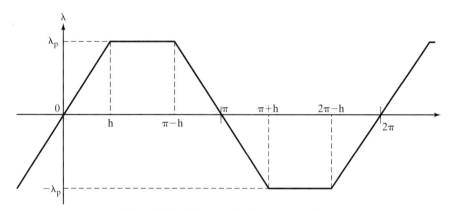

Figure 9.52 The rotor-flux-linkage waveform

Similarly, the Fourier series of the flux linkages of phase *a*, assuming a trapezoidal waveform and constancy for $(\pi - 2h)$ electrical degrees in each half-cycle, is

$$\lambda_{af}(\theta_r) = \frac{4\lambda_p}{\pi h}\left[\sin h \sin \theta_r + \frac{1}{3^2}(\sin 3h \sin 3\theta_r) + \frac{1}{5^2}(\sin 5h \sin 5\theta_r) + ...\right] \quad (9.203)$$

where λ_p is the peak value of the modified flux linkages and the flux linkage waveform is shown in Figure 9.52.

Similarly, the *b* and *c* phase currents and their modified flux linkages can be derived.

The fundamental electromagnetic torque is computed by considering the product of fundamental terms in the air gap flux linkages and respective stator currents for a 2-pole machine:

$$T_{e1} = \lambda_{af1}(\theta_r)i_{as1}(\theta_r) + \lambda_{bf1}(\theta_r)i_{bs1}(\theta_r) + \lambda_{cf1}(\theta_r)i_{cs1}(\theta_r) \text{ (N·m)} \quad (9.204)$$

Substituting the fundamental terms and expanding the expressions yields

$$T_{e1} = \frac{16 I_p \lambda_p}{\pi^2 h(\theta_2 - \theta_1)} [\sin h(\sin \theta_2 - \sin \theta_1)\sin^2 \theta_r + \sin h(\sin \theta_2 - \sin \theta_1)\sin^2(\theta_r - 2\pi/3)$$
$$+ \sin h(\sin \theta_2 - \sin \theta_1)\sin^2(\theta_r + 2\pi/3)] \quad (9.205)$$

For $h = \pi/6$, the electromagnetic torque for three phases is evaluated as

$$T_{e1} = [I_p \lambda_p]\left\{\frac{48}{\pi^3(\theta_2 - \theta_1)}(\sin \theta_2 - \sin \theta_1)\right\}\left[\frac{3}{2}\right] = 2.3193\left[\frac{\sin \theta_2 - \sin \theta_1}{\theta_2 - \theta_1}\right][I_p \lambda_p] \text{ (N·m)} \quad (9.206)$$

The normalized fundamental torque in p.u. as a function of θ_2 for θ_1 and h equal to 30° is shown in Figure 9.53, for a 1-p.u current. This shows that increasing the rise time of the current decreases the fundamental torque. At higher speeds, for the same rise time of the current, note that θ_2 increases; hence, there will be a greater reduction in the fundamental torque of the motor drive.

For 120° electrical rectangular current, the fundamental torque is

$$T_{e1} = 2.011 I_p \lambda_p \quad (9.207)$$

which closely approximates the available torque calculated from Figure 9.8.

The commutation torque is at six times the fundamental frequency. It can be seen as the result of the sum of the fundamental rotor flux linkages interacting with the fifth- and seventh-current harmonics and the fundamental of the current interacting with the fifth- and seventh-harmonic rotor flux linkages. It is derived as follows.

$$T_{e6} = \frac{4\lambda_p I_p}{\pi(\theta_2 - \theta_1)}\left[(\sin \theta_2 - \sin \theta_1)\left\{-\frac{1}{5^2}(\sin 5h) + \frac{1}{7^2}(\sin 7h)\right\}\right]\frac{4}{\pi h}$$
$$+ \frac{4}{\pi h} \cdot \frac{4}{\pi(\theta_2 - \theta_1)}\left[\sin h\left\{-\frac{1}{5^2}(\sin 5\theta_2 - \sin 5\theta_1) + \frac{1}{7^2}(\sin 7\theta_2 - \sin 7\theta_1)\right\}\right]$$

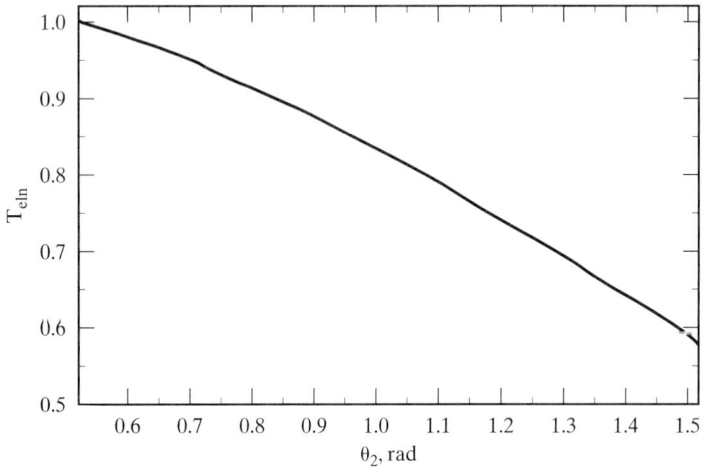

Figure 9.53 The fundamental torque vs. θ_2 for $\theta_1 = 0.524$ (rad) and $h = 0.524$ (rad) for $I_p = 1$ p.u.

$$= \frac{16\lambda_p I_p}{\pi^2 h(\theta_2 - \theta_1)} \left[(\sin\theta_2 - \sin\theta_1) \left\{ -\frac{1}{5^2}\sin 5h + \frac{1}{7^2}\sin 7h \right\} \right. \quad (9.208)$$

$$\left. + \sin h \left\{ -\frac{1}{5^2}(\sin 5\theta_2 - 5\theta_1) + \frac{1}{7^2}(\sin 7\theta_2 - \sin 7\theta_1) \right\} \right]$$

In general, the harmonic torque of frequency m times the fundamental is given by

$$T_{em} = \frac{16\lambda_p I_p}{\pi^2 h(\theta_2 - \theta_1)} \left[(\sin\theta_2 - \sin\theta_1) \left\{ -\frac{1}{(m-1)^2}\sin\overline{m-1}h + \frac{1}{(m+1)^2}\sin\overline{m+1}h \right\} \right.$$

$$\left. + \sin h \left\{ -\frac{1}{(m-1)^2}(\sin\overline{m-1}\theta_2 - \sin\overline{m-1}\theta_1) + \frac{1}{(m+1)^2}(\sin\overline{m+1}\theta_2 - \sin\overline{m+1}\theta_1) \right\} \right]$$

$$m = 6, 12, 18, 24, \ldots \quad (9.209)$$

Note that this is for one phase only. The negative sign attributed to the fifth-harmonic rotor flux linkage and current is due to the fact that they are revolving against the fundamental rotor flux linkage and current; hence, their torque contributions are considered negative. The detailed calculation for the sixth- and twelfth-harmonic torques are given in Tables 9.1 and 9.2, respectively.

The harmonic torques are normalized on the basis of the fundamental torque for a perfect rectangular current. It is rational from the point of view of

TABLE 9.1 Sixth-harmonic torque for various θ_2 with $\theta_1 = 30° = h$

Harmonic Number		Sixth-Harmonic Torque Component					
Rotor Flux Linkages	Stator Current	$\theta_2 = 32.5$	35°	37.5°	40°	45°	50°
1	5	0.424	0.439	0.447	0.448	0.428	0.383
1	7	−0.258	−0.221	−0.178	−0.132	−0.037	0.044
5	1	−0.079	−0.078	−0.077	−0.076	−0.075	−0.071
7	1	−0.041	−0.040	−0.039	−0.039	−0.037	−0.036
Total T_{e6n}		0.047	0.101	0.153	0.202	0.280	0.320

TABLE 9.2 Twelfth-harmonic torque for $\theta_1 = 30° = h$

Harmonic Number		Sixth-Harmonic Torque Component					
Rotor Flux Linkages	Stator Current	$\theta_2 = 32.5°$	35°	37.5°	40°	45°	50°
1	11	−0.201	−0.203	−0.190	−0.163	−0.088	−0.018
1	13	0.122	0.078	0.031	−0.012	−0.063	−0.057
11	1	0.016	0.016	0.016	0.016	0.015	0.015
13	1	0.012	0.012	0.011	0.011	0.011	0.010
Total T_{e12n}		−0.051	−0.097	−0.131	−0.149	−0.126	−0.05

application considerations. Hence, the normalized harmonic torques for the three phases are

$$T_{emn} = \frac{3T_{em}}{T_{e1}} = \frac{24}{\pi^2 h(\theta_2 - \theta_1)} \left[(\sin\theta_2 - \sin\theta_1) \left\{ -\frac{1}{(m-1)^2} \sin\overline{m-1}h + \frac{1}{(m+1)^2} \sin\overline{m+1}h \right\} \right.$$
$$\left. + \sin h \left\{ -\frac{1}{(m-1)^2} (\sin\overline{m-1}\theta_2 - \sin\overline{m-1}\theta_1) + \frac{1}{(m+1)^2} (\sin\overline{m+1}\theta_2 - \sin\overline{m+1}\theta_1) \right\} \right]$$

for m = 6, 12, 18, ... (9.210)

9.10.5 Phase Advancing

The magnitude of the induced emfs increases with increasing speed. When the magnitude of the line-to-line induced emf comes close to the magnitude of the dc-link voltage, it will not be possible to maintain the currents at desired level. Then, the currents can be advanced in phase so that the line-to-line induced emf is lower than the dc-link voltage, facilitating the establishment of the desired current magnitude. As the line-to-line induced emf becomes greater than the dc-link voltage, the current cannot flow from the dc link to the motor windings. The energy stored in the leakage inductance, (L-M), will keep the current circulating in the windings through the freewheeling diodes of the inverter. During this interval, the currents will decrease. This translates into a reduction of torque. If the speed is increased beyond rated value, then the power output can be maintained at rated value for a small range even with decreasing torque production. Note that this mode of operation is usually attempted for a very small speed range. The phase-advancing effect is quantified as follows.

Considering an ideal current with 120° constancy and an ideal flux linkage waveform, they can be resolved into harmonic components and written as

$$i_{as}(\theta_r) = \frac{4\sqrt{3}I_p}{2\pi} \left[\sin\theta_r + \frac{1}{5}\sin 5\theta_r + \ldots \right] \quad (9.211)$$

$$\lambda_{af1}(\theta_r) = \frac{24\lambda_p}{\pi^2} \left[\frac{1}{2}\sin\theta_r + \frac{1}{9}\sin 3\theta_r + \frac{1}{2}\cdot\frac{1}{25}\sin 5\theta_r + \ldots \right] \quad (9.212)$$

Advancing the current by an angle θ_a can be represented as

$$i_{as}(\theta_r + \theta_a) = \frac{4\sqrt{3}}{2\pi}I_p \left[\sin(\theta_r + \theta_a) + \frac{1}{5}\sin\{5(\theta_a + \theta_r)\} + \ldots \right] \quad (9.213)$$

Similarly, expressions for the other phase currents, $i_{bs}(\theta_r + \theta_a)$ and $i_{cs}(\theta_r + \theta_a)$, can be written. Substituting these into the torque expression gives the fundamental torque:

$$T_{e1} = \lambda_{af1}(\theta_r)i_{as1}(\theta_r + \theta_a) + \lambda_{bf1}(\theta_r)i_{bs1}(\theta_r + \theta_a) + \lambda_{cf1}(\theta_r)i_{cs1}(\theta_r + \theta_a)$$
$$= \frac{96\sqrt{3}}{4\pi^3}\lambda_p I_p[\sin\theta_r \cdot \sin(\theta_r + \theta_a) + \sin(\theta_r - 2\pi/3)\cdot\sin(\theta_r - 2\pi/3 + \theta_a) \quad (9.214)$$
$$+ \sin(\theta_r + 2\pi/3)\cdot\sin(\theta_r + 2\pi/3 + \theta_a)]$$
$$= 2.0085\lambda_p I_p \cos\theta_a$$

For speeds less than rated value, the advance angle is zero, giving the fundamental torque:

$$\therefore T_{e1} = 2.0085 \lambda_p I_p = T_{er} \quad (9.215)$$

where T_{er} is the rated value of the electromagnetic torque.

For speeds higher than rated value, the phase-advance angle, θ_a, is nonzero. Then, the torque can be represented in terms of the rated value and advance angle from equations (9.213) and (9.214) as

$$T_{e1} = T_{er} \cos \theta_a \quad (9.216)$$

Assuming that the base flux linkage is equal to λ_p and the base current is I_b, then the torque in normalized unit is

$$T_{en1} = I_{pn} \cos \theta_a \quad (9.217)$$

where I_{pn} is the normalized peak current in a machine phase.

It is to be noted that these expressions are valid only for ideal current waveforms; in a practical situation, they will deviate from the ideal considerably. The phase-advancing of currents not only decreases the available torque but also drastically increases the harmonic torque. As the current is advanced in one phase, initially the air gap flux linkage is not a constant but a ramp. The result of the interaction is a ramp-shaped torque, giving way to a significant pulsating-torque component. Their quantification can be achieved by following a procedure similar to the one given in the commutation-torque-ripple section.

9.10.6 Normalized System Equations

The equations of the PMBDCM can be normalized by using base voltage V_b, base current I_b, base flux linkages λ_b, base power P_b, and base frequency ω_b. Considering only phase a for normalization, it is achieved as follows.

$$v_{asn} = \frac{v_{as}}{V_b} = \frac{1}{V_b}[Ri_{as} + (L - M)pi_{as} + \lambda_p \omega_m f_{as}(\theta_r)] \quad (9.218)$$

but the base voltage can be written as

$$V_b = I_b \cdot Z_b = \lambda_b \cdot \omega_b \quad (9.219)$$

By substituting this in the voltage equation, the phase a normalized voltage expression is obtained as

$$v_{asn} = R_{an} i_{asn} + (X_{Ln} - X_{Mn}) \frac{p}{\omega_b} i_{asn} + \lambda_n \omega_{mn} f_{as}(\theta_r) \quad (9.220)$$

where

$$R_{an} = \frac{R}{Z_b} \text{ (p.u.)} \quad (9.221)$$

$$X_{Ln} = \frac{\omega_b L}{Z_b} \text{ (p.u.)} \quad (9.222)$$

$$X_{Mn} = \frac{\omega_b M}{Z_b} \text{ (p.u.)} \qquad (9.223)$$

$$\lambda_n = \frac{\lambda_p}{\lambda_b} \text{ (p.u.)} \qquad (9.224)$$

$$\omega_{mn} = \frac{\omega_m}{\omega_b} \text{ (p.u.)} \qquad (9.225)$$

$$T_{eb} = 2\lambda_b I_b \text{ (Nm)} \qquad (9.226)$$

$$Z_b = \frac{V_b}{I_b} \text{ (}\Omega\text{)} \qquad (9.227)$$

Similarly, the other two phase equations can be derived. The electromechanical equation is derived as

$$T_e = T_l + B\omega_m + J\frac{d\omega_m}{dt} \qquad (9.228)$$

and, in normalized form, the electromagnetic torque is given as

$$T_{en} = \frac{T_e}{T_b} = T_{ln} + B_n \omega_{mn} + 2Hp\omega_{mn} \qquad (9.229)$$

where

$$T_{en} = \frac{T_e}{T_b} \text{ (p.u.)} \qquad (9.230)$$

$$T_{ln} = \frac{T_l}{T_b} \text{ (p.u.)} \qquad (9.231)$$

$$B_n \frac{B\omega_b^2}{P_b} \text{ (p.u.)} \qquad (9.232)$$

$$H = \frac{1}{2} \frac{J\omega_b^2}{P_b(P/2)^2} \text{ (s)} \qquad (9.233)$$

The electromagnetic torque in terms of the flux linkages and motor currents is derived as

$$T_e = \frac{(e_{as}i_{as} + e_{bs}i_{bs} + e_{cs}i_{cs})}{\omega_m} = \lambda_p[i_{as}f_{as}(\theta_r) + i_{bs}f_{bs}(\theta_r) + i_{cs}f_{cs}(\theta_r)] \qquad (9.234)$$

and in normalized units it is given as

$$T_{en} = \frac{P}{4}[i_{asn}f_{as}(\theta_r) + i_{bsn}f_{bs}(\theta_r) + i_{csn}f_{cs}(\theta_r)] \text{(p.u.)}$$

$$T_b = 2\lambda_p I_p \qquad (9.235)$$

9.10.7 Half-Wave PMBDCM Drives

Several converter topologies emerge if the PMBDCM is operated in the half-wave mode, i.e., each phase is operated for 120 electrical degrees instead of an alternating current injection for 120 electrical degrees in each half-cycle in a three-phase machine, which is known as the full-wave operational mode. This full-wave operation has been described earlier in previous subsections. This mode of operation provides very limited opportunities for innovation in the inverter topology; six-switch full-bridge topology has stayed on as the most nearly optimal topology thus far. Cost minimization of the PMBDCM drives is of immense interest to the industry at present, with the opening up of a large number of applications to variable-speed operation. Such applications are to be found in HVAC, fans, pumps, washers, dryers, treadmills and other exercise equipment, wheel chairs, people carriers in airport lobbies, golf carts, freezers, refrigerators, automotive, handtools, and small-process drives with velocity control for packaging, bottling, and food-process applications. With the high-volume nature of these applications, cost minimization is of paramount importance, not only to save materials and labor (and possibly, by parts reduction, to enhance the reliability of the product) but also for the fact that without such a cost minimization many of these applications with variable-speed drives can not be realized, almost certainly in the present and probably in the future.

The cost of the motor and controller with single chips has been optimized; the only other subsystem available for optimization is the power converter. Half-wave operation is a major asset in this aspect, because many power-converter topologies are possible with minimum number of switches. Such a reduction in number of power devices has an impact on the reduction in logic power supplies, heat-sink volume, packaging size, enclosure size, and hence the overall cost of the drive system. Three power topologies for half-wave operation of the PMBDCM drive are presented and discussed in this section.

9.10.7.1 Split-supply-converter topology.
The power-converter topology with a single switch per phase and minimum number of diodes in the rectifying section, ideal for fractional-horsepower (fhp) PMBDC motor drives, is shown in Figure 9.54. This topology is similar to the split-voltage, single-switch-per-phase topology for the switched-reluctance motor drive, except that in switched-reluctance motor drives

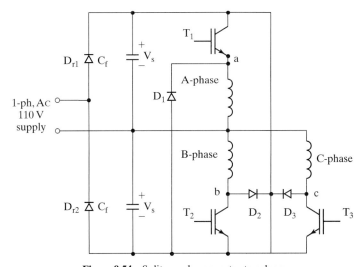

Figure 9.54 Split-supply converter topology

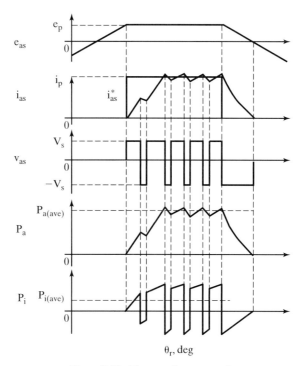

Figure 9.55 First-quadrant operation

it is advocated usually for machines with an even number of phases. The unidirectional current-handling capability of the converter is ideal for switched-reluctance motor drives, but a restriction in the form of half-wave operation is imposed in the PMBDC motor drive with this converter. While such a restriction results in the underutilization of the motor, it leads to other advantages and very welcome features.

By using bifilar windings for the machine, the emitters of the power switches can be tied together and connected to the negative of the dc-link voltage. This simplifies the gate circuitry and further can do away with the split supply (and hence one power capacitor in the dc link), but the bifilar windings would have leakage inductances due to the nonideal coupling of the bifilar windings, resulting in high switching turn-off voltages. The bifilar windings, in addition, also take up slot volume, thus decreasing the area for the main winding, resulting in lower power and torque densities. Before a discussion of merits and demerits of the split-supply converter topology is considered, it is instructive to see the operational modes of this converter with the PMBDCM drive.

A. Operation of the PMBDC Motor with the Split-Supply Converter

Consider the four-quadrant operation of the PMBDCM with the half-wave converter with a restriction of unidirectional current capability. The operation would be taken on a quadrant-by-quadrant basis.

The first quadrant operation is shown in Figure 9.55, where the phase sequence is maintained *abc* with the current injected into the windings when their respective induced emfs experience flat region and for 120° electrical.

Assuming that phase *A* operation is followed, for example, switch T_1 is turned on at the instant when the stator *A*-phase induced emf is 30° from its positive–zero crossover point. Turn-on of the switch T_1 results in the application of voltage V_s to phase winding A, assuming that the switches are ideal. If the phase current exceeds the reference current for that phase, the switch T_1 is turned off, depending on the chosen switching strategy, such as PWM or hysteresis.

During the turn-off of T_1, the phase current is routed through the diode D_1, A-phase winding, and the bottom capacitor in the dc link, resulting in a voltage of $-V_s$ applied to the phase A winding. The negative voltage application to the phase reduces the current swiftly by transferring energy from machine phase to the dc link. Thus, the current is controlled between desired limits in the machine phase with only one switch operation. When phase A has to be commutated, note that the switch T_1 is permanently turned off. The average power flow from the dc-link source to the machine winding is positive, indicating that the machine is in motoring mode, i.e., first quadrant. Air gap power P_a and instantaneous input power P_i are shown in Figure 9.55, indicating the same.

To brake when the motor drive is in the forward direction of motion, the electromagnetic torque has to be reversed from positive to negative polarity; thereby, the drive system will be in the fourth quadrant. Since the current cannot be reversed in this converter, the only alternative to get into this mode is to delay the onset of current in machine phases until their induced emfs have negative polarity and 30° from their negative–zero crossover point, as shown in Figure 9.56. This results in

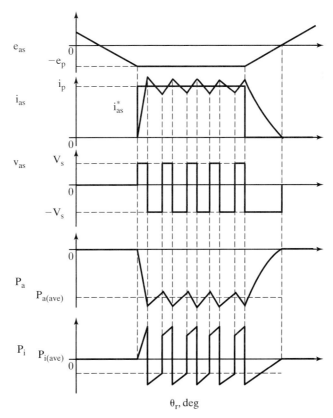

Figure 9.56 Fourth-quadrant operation

TABLE 9.3 Modes of the proposed converter

Mode	T_1	T_2	D_1	D_2	i_{as}	i_{bs}	V_{as}	V_{bs}
1	on	off	off	off	>0	0	V_s	0
2	off	off	on	off	>0	0	$-V_s$	0
3	on	on	on	off	>0	>0	$-V_s$	V_s
4	off	on	on	on	>0	>0	$-V_s$	$-V_s$

negative torque and air gap power in the machine, with the consequence that the drive system is in the fourth quadrant. The current control is very similar to the technique described in the first quadrant.

The third-quadrant operation is reverse motoring; it is obtained by changing the phase sequence to *acb* to obtain the reversal of rotation direction. Other than that, this mode is very similar to the first quadrant. Similar also is the second-quadrant operation to the fourth quadrant, but in the reverse rotational direction. For the sake of brevity, these two quadrants of operation are not considered in detail, even though they are dealt with in the modeling, simulation, and analysis later.

B. Operational Modes of the Converter

The operational modes of the converter are obtained from the above discussion and summarized in Table 9.3. Current conduction in one phase, which usually is the case during most of the operational time, and conduction in two phases, which occurs during the commutation of one phase and energization of the incoming phase, are considered here.

The distinct difference between the full-wave inverter modes and the proposed converter modes is that the proposed converter applies either a positive or a negative voltage across the phase winding when there is a current in it. This has the drawback of higher energy circulation in machine phases, resulting in a small reduction in efficiency and creating torque ripples higher than those in the full-wave-inverter-based PMBDC system that could result in higher acoustic noise in this drive system.

C. Merits and Demerits of the PMBDC Drive with the Split-Supply Converter

The PMBDC motor drive with this converter topology has the following advantages:

(i) One switch and one diode per phase halves the power-stage requirement of the switches and diodes compared to that of the full-wave inverter topology, resulting in low cost and high compactness in packaging.

(ii) Reduced gating driver circuits and logic power supplies leads to cost reduction and high compactness in packaging.

(iii) Full four-quadrant operational capability provides the attendant possibility for high-performance applications, such as in small-process drives. Note that the regenerative brake is to be added, which is common even for the full-wave inverter topology.

(iv) High reliability is due to the switch being always in series with the machine phase winding, thus preventing a shoot-through fault.

- **(v)** It is possible to operate with one switch or phase-winding failure, whereas it is not possible under these conditions in the PMBDC drive with full-wave inverter. This is an important and an attractive feature for applications such as wheelchair drives and some process drives.
- **(vi)** Lower conduction losses are due to one switch and diode per phase as compared to two switches and two diodes being in operation for the full-wave-inverter-fed PMBDC motor drive, resulting in half the losses compared to the full-wave drive, with the current being the same in both the drives. Further, the turn-on losses are reduced by the higher machine inductance in the PMBDC motor for operation with the proposed converter topology, because this allows for a soft turn-on of the power devices.
- **(vii)** This topology allows for sensing of the induced emf across the machine phases by looking at the switch voltages during their turn-off intervals, and mechanical or optical sensorless operation becomes possible by using these signals to generate the control signals. Note that two-phase switches have a common return, thus doing away with isolation requirement for the transducer signal, and also the fact that two phase voltages are sufficient for generation of the control signals for a three-phase PMBDC motor drive.

The disadvantages of this are as follows:

- **(i)** Poorer utilization of the motor is due to the half-wave operation. The torque density in terms of the torque per unit stator copper losses is lower in the half-wave-controlled machine by nearly 30% compared to the full-wave-controlled machine.
- **(ii)** Extra power capacitor is required in the dc link, due to split supply.
- **(iii)** Larger self-inductance for this PMBDC motor results in large electrical time constant, leading to a slow response in current and hence in the torque compared to the full-wave-inverter-fed PMBDC motor drive.
- **(iv)** The commutation-torque ripple frequency is halved in that drive compared to that of the full-wave-inverter drive. The commutation torque ripple can be attenuated, as in the full-wave-inverter drive, by coordinating the current in the rising phase with the outgoing phase winding current to yield a constant torque during the phase commutation.

Because of these disadvantages, this topology might not be useful and appropriate for integral-hp drive systems.

D. Design Considerations for the PMBDC Motor

This section contains the design considerations of the PMBDC motor required for use with all the half-wave converter topologies. Wherever possible, all these important factors are contrasted with those of the H-bridge inverter (full-wave)-operated PMBDC motor. For the sake of comparison, the full-wave operated motor is considered as the base.

The following relationships are made on the basis of equal copper volume in the slot of the PMBDC motor. The subscripts 1 and b correspond to the motor with this converter and the motor with the full-wave converter, respectively.

The ratio of the induced emfs is

$$\frac{e_1}{e_b} = \left(\frac{k_1}{k_b}\right)\left(\frac{\omega_1}{\omega_b}\right) \qquad (9.236)$$

where k is the emf constant and ω is the rotor electrical speed.

The air gap power ratio is given by

$$\frac{P_1}{P_b} = \frac{1}{2}\left(\frac{k_1}{k_b}\right)\left(\frac{\omega_1}{\omega_b}\right)\left(\frac{I_1}{I_b}\right) \qquad (9.237)$$

where I is the current in the winding.

The ratio of emf constants in terms of the number of turns per phase and conductor cross sections is

$$\frac{k_1}{k_b} = \frac{N_1}{N_b} = \frac{a_b}{a_1} \qquad (9.238)$$

where N denotes the number of turns/phase and a is the area of cross section of the conductor in general.

The ratio of the copper losses is given as

$$\frac{P_{c1}}{P_{cb}} = \frac{1}{2}\left[\frac{I_1 N_1}{I_b N_b}\right]^2 = 2\left[\frac{T_{e1}}{T_{eb}}\right]^2 \qquad (9.239)$$

where T_e corresponds to electromagnetic torque in general.

The ratio of the stator resistances is

$$\frac{R_1}{R_b} = \frac{a_b N_1}{a_1 N_b} = \left[\frac{N_1}{N_b}\right]^2 \qquad (9.240)$$

From these relationships, it is possible to find the number of turns per phase, electromagnetic torque, induced emf, air gap power, and stator copper losses in the PMBDC motor for the proposed converter topology, depending on the choice of such criteria as equality of torque, equality of copper losses, and equality of air gap power.

A comparison of PMBDC machines for use with half-wave and full-wave inverters is made here. The basis of the comparison is restricted to equal stator phase currents and equal dc-link voltage. Then, three distinct options emerge as follows:

(i) unequal copper losses, equal volume of copper and, hence, copper fill, and equal maximum speeds;
(ii) equal copper losses, equal volume of copper and, hence, copper fill, and unequal maximum speeds; and
(iii) equal copper losses, unequal volume of copper (and, hence, unequal copper fill factors), but such an option increases volume of the copper and possibly stator lamination size, resulting in a large machine size. This option also needs to be considered in applications where the size is not very critical and reliability is of high concern.

From Table 9.4, it is seen that the option (i) has the certain drawback of having twice the copper losses for the half-wave-inverter-based PMBDC machine compared to the full-wave-inverter-based PMBDC machine. This drawback could be interpreted as the result of moving the switch conduction and switching losses from the converter to the

TABLE 9.4 Comparison of the PMBDC machine variables based on the half-wave and the full-wave inverter

Ratio Between the Half-Wave-Inverter-Based and Full-Wave-Inverter-Based PMBDC Machine Variables	(i)	(ii)	(iii)
Maximum speed ratio	1	1.414	1
Electromagnetic torque ratio	1	0.707	1
Resistance ratio	4	2	2
Stator copper losses ratio	2	1	1
Size and cost ratio	1	1	>1

machine. It is easier to cool fractional-horsepower machines without significant additional resources; the thermal mass and surface area per unit output watt are higher in fhp machines compared to the integral-hp machines. Therefore, half-wave converter drives might be ideally suitable in fhp sizes. Further, the increase in cost to handle the thermal effects of higher copper losses in the machine has to be viewed from the overall perspective of the total cost of the motor drive system. The option (ii) is very preferable, to inherently exploit the full dc-link voltage in the half-wave-converter-based PMBDCM drive as compared to its counterpart and therefore to achieve a higher speed. That such a characteristic might be possible to exploit in many pump and fan drive applications is to be noted, thus making the half-wave-based PMBDCM drive an attractive technical solution. If the route through option (ii) is taken for comparison, a number of choices come into play, but they are not looked into for lack of space. Option (iii) might be the least desirable prima facie but has to be viewed in terms of the overall cost of the PMBDCM drive system to assess its suitability for a given application.

E. Impact of the Motor Inductance on the Dynamic Performance

From the machine equations, it is seen that the self-inductance of the phase winding plays a crucial role in the dynamics of the current loop and hence in the torque generation.

In the case of the full-wave-operated PMBDC machine, the electrical time constant is given by

$$\tau_{fw} = \frac{(L_s - M)}{R_s} \qquad (9.241)$$

where M is the mutual inductance.

Consider a PMBDCM with twice the number of turns per phase as compared to the full-wave-operated machine, for operation with the half-wave converter and its self-inductance in terms of L_s is equal to $4L_s$ and its resistance for equal copper losses is $2R_s$, thus giving its electrical time constant as

$$\tau_{hw} = \frac{4L_s}{2R_s} = \frac{2L_s}{R_s} \qquad (9.242)$$

To find the ratio between these two time constants, it is necessary to express M in terms of L_s, from which we get

$$\frac{\tau_{hw}}{\tau_{fw}} = \frac{2L_s}{(L_s - M)} = \frac{2L_s}{L_s(1 - k_m)} = \frac{2}{(1 - k_m)} \qquad (9.243)$$

TABLE 9.5 Ratio of electrical time constants for various designs of the PMBDCM

Stator Slots per Pole per Phase	Ratio of Mutual to Self-Inductance, k_m	Ratio of Time Constants τ_{hw}/τ_{fw} for Turns per Phase		
		$2N_b$	$\sqrt{2}\,N_b$	N_b
1	−0.333	1.500	1.050	0.750
2	−0.400	1.428	1.010	0.714
3	−0.415	1.413	0.999	0.706

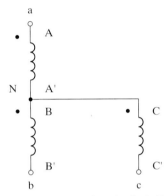

Figure 9.57 New machine winding connections for the half-wave converter topology

where

$$M = k_m L_s \qquad (9.244)$$

and k_m is used to evaluate the ratio of the time constants, given in Table 9.5 for a PMBDCM with turns per phase twice, $\sqrt{2}$, and equal to those of the full-wave-operated PMBDCM denoted by N_b. From the table, it is seen that the equivalent PMBDCM for operation with half-wave converter has the same electrical time constant compared to the full-wave PMBDCM drive, thus making the proposed drive suitable for very high-performance applications.

F. Winding Connections

The motor windings have to be connected as shown in Figure 9.57 for the split-supply converter as opposed to the half-wave converter connection, with all the phase wires forming the neutral having the same polarity for each winding. These connections will not create any additional manufacturing or design burden, thus having no impact on the cost.

G. Drive-System Description

The speed-controlled PMBDCM drive system schematic is shown in Figure 9.58. The feedback signals available for control are the phase currents, discrete rotor position signals from the Hall sensors to generate the gating instances for the phase switches, and rotor speed signal from a tachogenerator or from the position signal itself. The inner current loops enforce current commands, and the outer speed loop enforces the speed command. The speed signal is passed through a filter, and the resulting modified speed signal is compared with the speed reference to produce the speed-error signal. The

Section 9.10 PM Brushless DC Motor (PMBDCM)

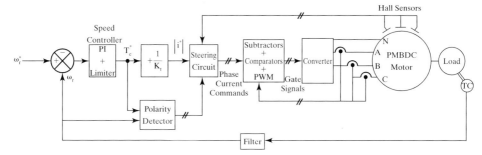

Figure 9.58 Schematic of the speed-controlled PMBDC motor drive system

torque-command signal is obtained from the speed-error signal through a speed controller that is a proportional-plus-integral (PI) type. The current magnitude reference is derived from the torque reference by a divider circuit, and the phase current commands are generated in combination with respective Hall position sensor signals through a steering circuit. The gating signals are generated for each phase switch by obtaining the phase current errors and processing these errors with PI current controllers and then combining them with carrier frequency to generate the pulse-width modulated signals.

H. Modeling, Simulation, and Analysis of the PMBDC Drive System

The various subsystems shown in Figure 9.58 are taken up for modeling, simulation, and analysis in this section.

Modeling of the PMBDCM with Converter Modes: PMBDCM model in *abc* phase variables is used in this simulation. Further, an ideal model with zero conduction voltage drop and zero switching times is utilized in this simulation for the switches and diodes. The operational modes determine whether one phase or two phases conduct at a given time; accordingly, the system equations emerge. To model when only phase A is conducting, the system equation is given by

$$R_s i_{as} + L_s p i_{as} + e_{as} = V_s \tag{9.245}$$

where R_s is stator phase resistance, L_s is self-inductance of a phase, e_{as} is a phase-induced emf, p is the derivative operator, and V_s is the dc-link voltage of the top half. The induced emf is given by

$$e_{as} = K_b f_{as}(\theta_r) \omega_r \ (V) \tag{9.246}$$

where f_{as} is a unit function generator to correspond to the trapezoidal induced emf of the PMBDCM as a function of θ_r (which is the rotor electrical position), K_b is the emf constant, and ω_r is the rotor electrical speed. f_{as} is given by

$$\begin{aligned}
f_{as}(\theta_r) &= (\theta_r)\frac{6}{\pi}, & 0 &< \theta_r < \frac{\pi}{6} \\
&= 1, & \frac{\pi}{6} &< \theta_r < \frac{5\pi}{6} \\
&= (\pi - \theta_r)\frac{6}{\pi}, & \frac{5\pi}{6} &< \theta_r < \frac{7\pi}{6} \\
&= -1, & \frac{7\pi}{6} &< \theta_r < \frac{11\pi}{6} \\
&= (\theta_r - 2\pi)\frac{6}{\pi}, & \frac{11\pi}{6} &< \theta_r < 2\pi
\end{aligned} \tag{9.247}$$

For mode 3, when the phase *a* is being commutated and phase *b* is energized, the following equations apply.

$$R_s i_{as} + L_s p i_{as} + M p i_{bs} + e_{as} = -V_s \quad (9.248)$$

$$R_s i_{bs} + L_s p i_{bs} + M p i_{as} + e_{bs} = V_s \quad (9.249)$$

where the subscript *b* corresponds to respective phase *b* variables and parameters defined above. The f_{bs} is similar to f_{as} but phase-displaced by 120° electrical. Similarly, the equations for other modes can be derived from (9.245) to (9.249).

The electromechanical equation with the load is given by

$$J \frac{d\omega_m}{dt} + B\omega_m = T_e - T_l \quad (9.250)$$

where J is the moment of inertia, B is the friction coefficient, T_l is the load torque and T_e is the electromagnetic torque, given by

$$T_e = \lambda_p \{f_{as}(\theta_r) i_{as} + f_{bs}(\theta_r) i_{bs} + f_{cs}(\theta_r) i_{cs}\} \quad (9.251)$$

The rotor position is derived for simulation from the following equation:

$$p\theta_r = \omega_r \quad (9.252)$$

Modeling of the Speed Controller: The speed controller is modeled as a PI type:

$$G_s(s) = K_{ps} + K_{is}/s \quad (9.253)$$

where *s* is Laplace operator, and from this equation the torque reference and current magnitude reference may be derived. This equation is written in time domain for simulation.

Steering Circuit: The steering circuit consists of three inputs, of which one is the current magnitude reference and the other two are polarity signals of the rotor speed and torque reference, denoted as |i*|, ω_{rp}, T_{ep}, respectively. Depending on the polarity of the rotor speed and torque reference, the quadrant of operation and the phase sequence are determined as given in the Table 9.6.

Depending on the quadrant of operation and the position of the rotor, which determines induced emfs in the machine phases and hence the emf functions, the phase current commands are determined as follows.

Quadrant I	Quadrant IV	
$f_{as}(\theta_r) \geq 1, i^*_{as} = \|i^*\|$	$f_{as}(\theta_r) \leq -1, i^*_{as} = \|i^*\|$	
$f_{bs}(\theta_r) \geq 1, i^*_{bs} = \|i^*\|$	$f_{bs}(\theta_r) \leq -1, i^*_{bs} = \|i^*\|$	(9.254)
$f_{cs}(\theta_r) \geq 1, i^*_{cs} = \|i^*\|$	$f_{cs}(\theta_r) \leq -1, i^*_{cs} = \|i^*\|$	

Similarly for the third and fourth quadrants, the phase current commands are derived. This completes the modeling of the steering circuit.

Current-Loop Modeling: For phase A, the current-loop model including the pulse-width modulation is derived in the following, but that the same algorithm is

Section 9.10 PM Brushless DC Motor (PMBDCM)

TABLE 9.6 Operational quadrant relationships

ω_{rp}	T_{ep}	Quadrant	Phase Seq.
≥ 0	≥ 0	I	abc
≥ 0	< 0	IV	abc
< 0	≥ 0	II	acb
< 0	< 0	III	acb

applicable to all other phases is to be noted. The current error, say for phase A, i_{aer}, which is the difference between the current reference and actual current, is amplified and processed through a proportional-plus-integral (PI) controller very similar to that of the speed controller. The output of the current controller is limited to its maximum value, given as i_{max}. The duty cycle of the switch for one period of the PWM cycle is then given by

$$d = \frac{i_{aer}}{i_{max}}, \quad i_{aer} > 0$$
$$= 0, \quad i_{aer} < 0 \qquad (9.255)$$

The on-time of the switch, T_{on}, is given by

$$T_{on} = dT_c = \frac{d}{f_c} \qquad (9.256)$$

where f_c is the carrier frequency of the PWM and T_c is the period of the PWM cycle.

The switch conduction time is updated only once in a PWM cycle, in order to avoid multiple switching in a PWM cycle. The off-time for the phase switches is obtained from the difference between T_c and T_{on}.

Simulation and Analysis: The parameters of the PMBDCM drive system for dynamic simulation are given next.

Poles = 4   hp = 0.5
$R_s = 0.7 \, \Omega/\text{ph}$ L = 2.72 mH   M = 1.5 mH
$K_b = 0.5128$ V/(rad/sec)(Mech.) $K_t = 0.049$ N·m/A
$J = 0.0002$ kg-m/s² $B = 0.002$ N·m/rad/sec.
$V_{s1} = V_{s2} = V_s = 160$ V; $f_c = 2$ kHz; Speed (base) = 4,000 rpm,
Current (base) = 17.35 A, V (base) = 40 V, Torque (base) = 0.89 N·m
Torque (max) = 2 * Torque(base); I_{max} = 2 * Current (base)
Speed Controller: Proportional gain $K_{ps} = 20$, Integral gain $K_{is} = 1$
Current Controllers: Proportional gain $K_{pi} = 50$, Integral gain $K_{ii} = 5$

A full four-quadrant operation is simulated by giving a two-directional step speed reference of rated value. The key responses of the actual speed, torque reference, actual torque, phase a current and its reference, phase b current, and rotor position are shown in Figure 9.59, together with the speed reference, all in normalized units. Although no

600 Chapter 9 Permanent-Magnet Synchronous and Brushless DC Motor Drives

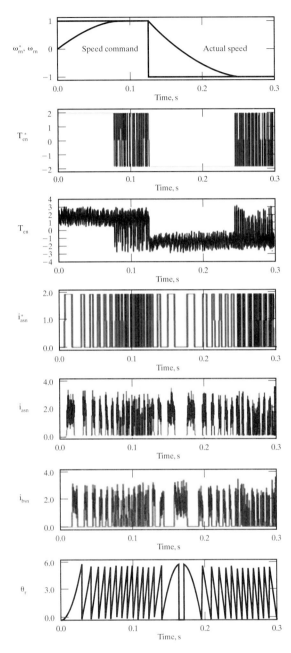

Figure 9.59 Dynamic simulation results of a four-quadrant PMBDC motor with the split supply converter topology

TABLE 9.7 Comparison of the half-wave converter with the full-wave converter-based PMBDCM system

Aspects	Proposed Converter-based PMBDCM	Full-wave Converter-based PMBDCM
Number of power switches (all phases)	3	6
Number of diodes	3	6
Switch voltage (min)	$2V_s$	V_s
Switch current (peak)	I_p	I_p
Switch current (rms)	$I_p/\sqrt{3}$	$I_p/\sqrt{3}$
Motor ph. current (rms)	$I_p/\sqrt{3}$	$\sqrt{2}I_p/\sqrt{3}$
Number of capacitors	2	1
Capacitor voltage	V_p	V_p
Min. number of diodes in the front-end rectifier	2	4
Number of logic power supplies for isolated operation (min.)	2	4
Number of gate drivers	3	6
Switch conduction losses	$hv_{sw}I_p$*	$2hv_{sw}I_p$
Diode conduction losses	$(1-h)v_dI_p$	$2(1-h)v_dI_p$
VA rating (peak)	$6V_sI_p$	$6V_sI_p$
VA rating (rms)	$2\sqrt{3}V_sI_p$	$2\sqrt{3}V_sI_p$

*Note: This current may be higher because of circulation energy.

steps have been taken to optimize the speed and current controllers, it becomes clear from the simulation results that the drive system is distinctly capable of four-quadrant operation. Further, it shows that the current response is fast enough for consideration of this drive in high-performance applications.

I. Comparison of the Half-Wave and Full-Wave Inverter-Based PMBDCM Drives

Consider the average duty cycle of the switch during a phase conduction as h, for deriving a comparison between the half-wave and the full-wave inverter-based PMBDCM drive system given in Table 9.7. Even though the VA ratings of the converters are the same, note that fewer devices with higher voltages usually cost less with the additional advantages of lower conduction and switching losses in the split-supply converter topology. The comparison does not consider the regenerative brake, which is required for both but with different VA ratings.

9.10.7.2 C-dump topology. The half-wave converter topology with one switch per phase has the disadvantage of utilizing only half of the available dc-link voltage. This fact precludes its use where high performance is a requirement. This could be resolved if topologies with more than one switch per phase but less than two switches per phase are resorted to. Such topologies have been developed for switched-reluctance motor drives with considerable success. One such topology is the C-dump converter with n+1 switches for an *n*-phase machine. The C-dump

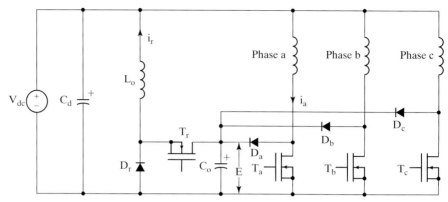

Figure 9.60 C-dump converter topology

topology is shown in Figure 9.60 for a three-phase machine. The principle of operation, analysis, and design of such a converter topology for a four-quadrant PMBDCM drive system are presented in this section. Design considerations for the PMBDCM to work with the C-dump power converter, based on power and torque equality with full-wave operated PMBDCM, are developed. A comparison of the C-dump and full-wave PMBDCM drives is derived to highlight key advantages and disadvantages of the C-dump-operated PMBDCM drive system.

A. Principle of Operation of the C-Dump PMBDCM Drive System

The C-dump converter for a three-phase system shown in Figure 9.60 is considered. It has four power switches and four power diodes with one of each for each phase winding and one set for energy recovery from the capacitor, C_o. Since the phase has only one switch, the current in it could only be unidirectional; hence, it is very similar to the half-wave converter-driven PMBDCM in operation. The motoring (I-quadrant) and regenerative (IV-quadrant) control of the C-dump-based PMBDCM are briefly described in the following.

Motoring Operation Assume that the direction of the motor is clockwise, which may be considered positive with the phase sequence *abc* of the motor phase windings for this discussion. The motoring operation is initiated when the phase voltage is in the flat region, i.e., at constant magnitude for a fixed speed and with the duration of 120° electrical. The phase *a* is energized when the phase current is commanded by turning on switch T_a; the equivalent circuit is shown in Figure 9.61(i). When the current error is negative, switch T_a is turned off, and the current in the phase *a* winding is routed through the diode D_a to the energy recovery capacitor, C_o shown in Figure 9.61(ii). During this time, a negative voltage of magnitude $(E - V_{dc})$ is applied across the machine winding, thus reducing the current and bringing the current error to positive. The average air gap power and the input power are positive, giving a positive electromagnetic torque, thus indicating the operation is firmly in the first quadrant of the torque-vs.-speed region.

The motoring operation in the counterclockwise (reverse) direction of rotation of the motor is similar, except that the phase sequence will be *acb* in the energization of the motor phase windings. That corresponds to the III quadrant of the torque–speed characteristics.

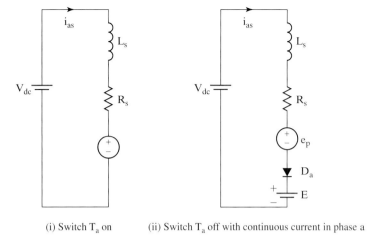

(i) Switch T_a on (ii) Switch T_a off with continuous current in phase a

Figure 9.61 First-quadrant motoring operation with phase *a* of the PMBDCM drive

Regenerative Operation Whenever the energy has to be transferred from load to supply, the PMBDCM is to be operated as a generator, i.e., by generating a negative torque in the machine as against the positive torque for the motoring operation. It is usual to provide a current of opposite polarity to that of the induced emf in the full-wave converter-operated PMBDCM to generate a negative torque. It is not feasible in this C-dump-operated PMBDCM: it has unidirectional current feature for the positive half-cycle of the induced emfs. Then, the only alternative is to exploit the negative cycles of the induced emf, where only positive currents are required to obtain the negative torque. Such an operation for phase *a* involves the turn-on of T_a during negative constant emf period, and then, when the error current becomes negative, turning off T_a, enabling the conduction of D_a, resulting in the energy transfer from the machine phase *a* to the energy recovery capacitor. These operations are shown in Figure 9.62(i) and (ii). Note that the air gap power and the average input power to phase *a* are negative, indicating that the power has been transferred from the machine to the energy-recovery capacitor, C_o. The energy from C_o is recovered by a step-down chopper, using switch T_r and diode D_r shown in Figure 9.60. Note that this regenerative operation corresponds to the IV quadrant for a phase sequence of *abc*; similar is the regenerative operation for reverse rotational direction of the PMBDCM, corresponding to the II quadrant.

B. Analysis of the C-Dump PMBDCM Drive

The analysis of the drive system with the C-dump topology is presented in this section. Effort is primarily made to obtain the maximum speed of the motor in terms of the duty cycle of the phase switches and the energy transferred to the energy-recovery capacitor, and hence an estimate of the power to be handled by the recovery chopper for a given motor rating. It is assumed that the commutation pulses are available through Hall sensors or encoders or resolvers or by estimation. Design guidelines and modeling can be found in reference [6].

Maximum Speed: Consider the machine voltage equation in steady state for rated stator current given by I_b; it is given as follows:

$$V_{as} = R_s I_b + K_b \omega_r \ (V) \tag{9.257}$$

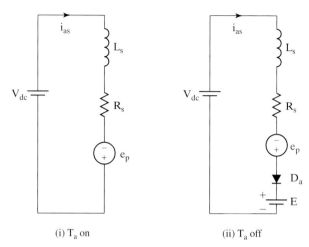

(i) T_a on   (ii) T_a off

Figure 9.62 Operational mode and waveforms of variables for the IV quadrant regenerative operation of PMBDCM drive system

and the electrical rotor speed is obtained from this as

$$\omega_r = \frac{V_{as} - R_s I_b}{K_b} \text{ (rad/sec)} \qquad (9.258)$$

For faster current-loop and hence torque and speed response, it is necessary to set aside some voltage, which is a fraction of the rated stator voltage, given as $k_a V_{as}$. Including this factor, the rotor speed is modified to

$$\omega_r = \frac{1}{K_b}[V_{as}(1 - k_a) - R_s I_b] \qquad (9.259)$$

If an average duty cycle of the phase switches is denoted as h, then the stator phase voltage in terms of the dc-link voltage is written as

$$V_{as} = h V_{dc} \qquad (9.260)$$

which, when combined with the rotor speed equation and normalization, yields the normalized rotor speed as

$$\omega_{rn} = h(1 - k_a)V_{dcn} - R_{sn} \text{ (p.u.)} \qquad (9.261)$$

where the additional subscript n denotes the normalized values of the variables and parameters.

Typically k_a is in the range from 0.2 to 0.4, and h is varied from nearly zero to one. This relationship explicitly gives speed in terms of the duty cycle, dc-link voltage, stator resistance, and dynamic voltage reserve. This expression allows the determination of range of h variation for the desired variation of speed range. Determination of h is crucial to the evaluation of average energy-recovery current and hence in the rating of that circuit.

Peak-Recovery Current: The energy transferred to the energy-storage capacitor, C_o, during the turn-off intervals of phase switches has to be recovered through the energy-recovery circuit if losses are neglected. The average duty cycle of energy transfer from

Section 9.10   PM Brushless DC Motor (PMBDCM)   605

the dc link and machine phase into capacitor C_o is $(1-h)$. Assuming that this stored energy is recovered through the chopper in a duty cycle of h, as this is essential to keep the separation of energy storage and recovery circuits, the powers can be equated as

$$E(1 - h)I_p = EhI_r \qquad (9.262)$$

where I_r is the peak recovery current through the chopper and could be written as

$$I_r = \frac{(1 - h)}{h} \cdot I_p \qquad (9.263)$$

As h increases, note that the energy recovered through the chopper reduces as I_r goes down, which in turn reduces the volt–ampere rating of the energy-recovery chopper circuit.

C. Comparison with Full-Wave Inverter-Controlled PMBDCM Drive

This section compares the C-dump-topology-based PMBDCM with that of the H-bridge inverter-fed PMBDCM drive system from the points of view of number of switches, passive components, their ratings, number of power supplies for gate drives, number of gate drives for isolated drive systems, converter losses, thermal management, and packaging requirements.

Some salient aspects and their comparison are given in Table 9.8. The average duty cycle for the phase switches is represented as h, and k_1 denotes the fraction of voltage for safety margin in the operation of the drive. Let k_2 be the fraction to give the regeneration brake current and

$$k_3 = \frac{E}{V_{dc}} \qquad (9.264)$$

TABLE 9.8 Comparison of the C-dump and full-wave-based PMBDCM drives

Aspects	C-Dump-Based PMBDC	Full-Wave-Based PMBDC
Number of switch devices	4	7 (including regenerative brake)
Number of diodes	4	7
Switch voltage	E	V_{dc}
Switch peak current	I_p	I_p
RMS switch current	$\frac{I_p}{\sqrt{3}}$	$\frac{I_p}{\sqrt{3}}$
Motor phase rms current	$\frac{I_p}{\sqrt{3}}$	$\sqrt{\frac{2}{3}}I_p$
Power capacitors	2	1
Capacitor voltages	V_{dc} and E	V_{dc}
Number of logic power supplies (minimum)	2	4
Inductor	1	0
Number of gate drivers	4	7
Turn-off snubbers (if needed)	0	6
Switch losses	$v_{sw}I_p\left[h + \frac{1-h}{h}\right]$	$2hv_{sw}I_p$
Diode losses	$v_dI_p(1 - h)$	$2v_dI_p(1 - h)$
Total peak switch VA rating	$EI_p\left[3 + \frac{1-h}{h}\right]$	$6v_{dc}(1 + k_1)I_p + v_{dc}(1 + k_1)I_pk_2$

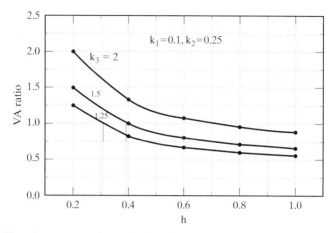

Figure 9.63 VA ratio vs. average duty cycle for various values of energy-recovery-capacitor voltages

This leads to the ratio of the VA rating as

$$\frac{VA_{cd}}{VA_{fw}} = \left[\frac{k_3}{1+k_1}\right] \frac{\left(3 + \dfrac{1-h}{h}\right)}{(6+k_2)} \quad (9.265)$$

For nominal values of $k_1 = 0.1$ and $k_2 = 0.25$, the relationship between the ratio of VA rating vs. the average duty cycle, h, for various values of k_3 ranging from 1.25 to 2 is shown in Figure 9.63. The break-even point for equal VA rating is shown with bold lines. Note that the duty cycle h is proportional to the normalized speed.

The converter switch losses are smaller for $h > 0.5$ in the case of the C-dump, and the diode losses are half those of the full-wave converter. This reduction in losses translates into heat sink and thermal management reduction by the same measure, resulting in a sizable reduction in packaging size. Further, it is helped by the requirement of smaller numbers of logic power supplies, snubbers, and gate drivers.

D. System Performance

The simulation takes the same drive-system parameters as given for the split-supply converter-operated drive system. The target value for the C-dump capacitor voltage is set to 175 V. The C-dump capacitor is initially charged to 100 V, the magnitude of the supply voltage. Figure 9.64 shows the commanded speed (ω_r^*), actual speed (ω_r), induced phase emf of phase a, phase-a current, air gap torque, and the voltage of the C-dump capacitor for the simulated speed loop with a speed command of ± 1000 r/min. The dump capacitor is charged to its target value of 175 V in less than 0.05 s. This voltage is then maintained in the proximity of 175 V. The transition from -1000 to $+1000$ r/min takes 0.9 seconds. As long as the commanded and actual speed are significantly different, the current command is at its maximum value of 20 A. Once the desired speed is achieved, the current drops to 8 A in order to support the motor load of 0.48 N·m. While the direction of rotation and the sign of the air gap torque are different, the load torque acts in unison with the air gap

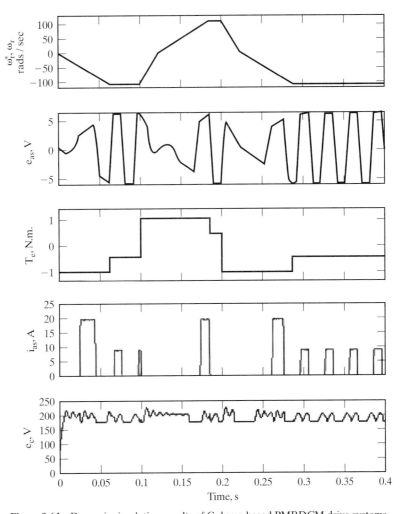

Figure 9.64 Dynamic simulation results of C-dump based PMBDCM drive systems

torque to decelerate the rotor. This results in faster deceleration of the rotor. The situation is different during acceleration, when the load opposes the air gap torque, resulting in slower acceleration than deceleration.

9.10.7.3 Variable-dc-link converter topology. One of the converter topologies with the advantage of varying the dc input voltage to the machine but with lower switch voltage not exceeding that of the dc source has other significant advantages also. Such a circuit is shown in Figure 9.65. In addition, this power converter topology has the advantages of the C-dump and split-supply converter topologies.

A. Principle of Operation

The converter circuit for a three-phase output has four switches and diodes, with the additional capacitor and inductor for its operation. The converter has two stages.

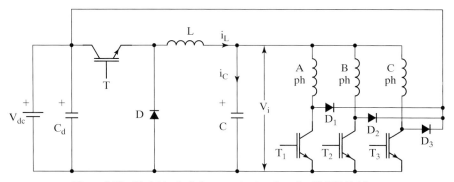

Figure 9.65 Variable-dc-link converter topology for PMBDC drives

The first stage is the chopper, which allows the variation of the input voltage to the machine. Switch T, diode D, inductor L, and capacitor C form a step-down-chopper power stage. The input voltage applied to the phases, V_i, is regulated by the operation of the chopper switch T. The inductor L and the capacitor C reduce the ripple content of the voltage V_i.

The second stage of the converter is the machine side, for handling the energy from the dc link to the machine and from the machine to the source. The chopper switch can be coordinated with the phase switches to regulate the current without having to switch the phase windings at carrier frequency. Because of the coordination option between the chopper switch and the phase switches, many modes of operation are possible in this drive system. The motoring (I quadrant) and regenerative (IV quadrant) control of the PMBDCM are briefly described with this converter.

Motoring Assume the direction of the motor is clockwise, which may be considered as positive with a phase sequence of *abc* of motor phase windings. The motoring operation is initiated when the phase voltage is constant positive for a fixed speed and with the duration of 120° electrical. Phase *a* is energized when the switch T_1 is turned on. To regulate current, T_1 is turned off, which initiates routing of the current through the freewheeling diode D_1, source voltage V_{dc}, and capacitor C, applying a voltage of $(V_i - V_{dc})$ across the machine phases. The motoring operation is similar in the reverse direction, except that the phase energization sequence will be *acb* in the motor phase windings. This corresponds to quadrant III operation.

Regenerative Operation To transfer energy from the load to the source, the PMBDCM has to be operated as a generator, i.e., by providing negative torque to the machine. Negative torque is achieved by injecting a positive current during the negative constant-emf period.

On the basis of the rotor-position information and the polarity of i*, the appropriate machine phase is turned on. The phase switch is turned on and modulated only if the current in the phase increases beyond a current window over the reference current, by hysteresis control. Normally, there is precise control of the machine input voltage through the chopper switch T, so the phase switch is rarely modulated to regulate the phase current. The phase switch is turned off only during commutation of the phase.

Section 9.10 PM Brushless DC Motor (PMBDCM)

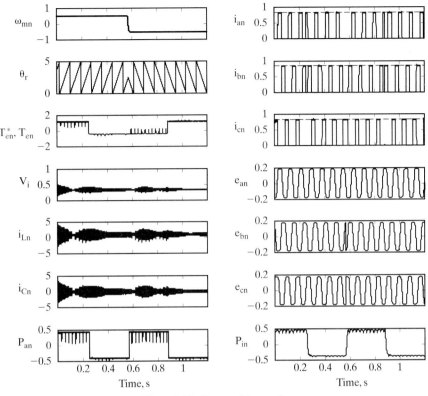

Figure 9.66 Torque-drive performance

B. System Performance

Consider the drive-system configuration as shown in Figure 9.58. For the torque-drive system, the speed-feedback loop is open. Both the torque- and speed-controlled drive systems are simulated, and results are given below.

Torque-Drive Performance Figure 9.66 shows the simulation results of torque-drive system performance, i.e., a system operating with only inner current loops. The machine is operating at 50% of the rated speed. The reference torque T_e^* is set at 1 p.u. by setting i* to 1 p.u.. The phase currents i_{as}, i_{bs}, and i_{cs} are seen to reach the same value as the reference current i* in the machine. It can be observed that the actual torque developed in the machine T_e has dips when there are transitions from one phase to another. Coordinating the currents in the incoming and outgoing phases can solve this problem. After 0.03 seconds, the commanded torque is switched from 1 p.u. to −1 p.u. while the speed remains the same, i.e., the system is operating in the IV quadrant. It can be observed that the voltage applied to the machine phases is about 0.5 p.u., unlike in other converter topologies, where a voltage switching between 0 and 1 p.u. is normally applied to the machine phases.

Speed-Controlled Drive Performance Figure 9.67 shows the simulation of a speed-controlled system operating in both forward and reverse directions. The current and torque are limited to 1.4 p.u. The current in the machine phases is unidirectional, but

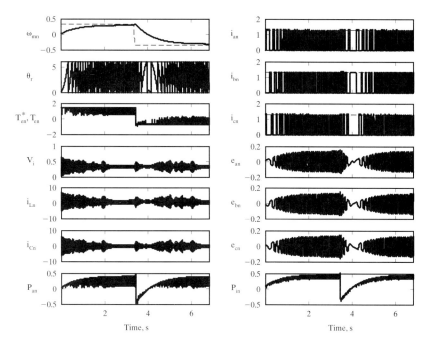

Figure 9.67 Speed-controlled-drive system performance

the torque developed by the machine is bidirectional. The input voltage to the phases, V_i, is varied with the speed. The current through the chopper inductor i_L has a high ripple content, but the currents in the machine phases have very little ripple. The variation in the ripple is due to machine inductances; additionally, it occurs because no subsequent switching occurs in the machine phases.

C. Merits and Demerits
The merits of the proposed topology are briefly listed next.

(i) Only four switches and diodes are required for four-quadrant operation with a three-phase PMBDCM.
(ii) Reducing gating driver circuits and logic power supplies saves money.
(iii) Full four-quadrant operational capability is available.
(iv) There is reduced possibility of shoot-through fault; switch is in series with the machine phase winding.
(v) Capability of operating with one switch failure or one phase-winding failure raises reliability.
(vi) Lower phase-switch losses raise efficiency.
(vii) High-frequency ripple in steady-state operation can be considerably lower than that of the fixed-dc-link-based converters; this is a distinct advantage for high-performance drive systems. Unlike the C-dump converter, this con-

verter has no circulating energy, resulting in better efficiency and very low torque ripple.

(viii) The power switches have a voltage rating equal to that of the source voltage, which is much lower than that of the C-dump and single-switch-per-phase converter.

The topology has the demerits associated with any half-wave converter topology, such as poorer utilization of the machine and a larger electrical time constant. In addition, this converter has two-stage power conversion, resulting in a slightly lower efficiency compared to single-stage power-converter topologies. Unlike the C-dump converter, this converter has no circulating energy, resulting in better efficiency and very low torque ripple.

9.10.8 Sensorless Control of PMBDCM Drive

The drive system is dependent on the position and current sensors for control. Elimination of both types of sensors is desirable in many applications, particularly in low-cost but high-volume applications, for cost and packaging considerations. Between the two sensors, the current sensor is easier to accommodate in the electronic part of the system; the position sensor requires a considerable labor and volume in the motor for its mounting. That makes it all the more important to do without the position sensor for the control of the PMBDCM drive system.

Current Sensing At least two phase currents are required for the current control of a three-phase machine. The phase currents can be sensed from the dc link current; hence, one sensor is sufficient for current control of the machine. The current sensors are relatively expensive if galvanic isolation is required. If isolation is not necessary, then the currents can be sensed inexpensively with precision resistors by measuring the voltage drops across them. The latter solution is used widely in low-cost motor drives. Another approach is to use MOSFET devices, with in-built current-sensing capability, to measure the currents. Alternatively, the MOSFET device itself serves as a sensing resistor during its conduction. The use of the drain-source voltage drop to estimate currents is fraught with inaccuracies due to temperature effect, and, for precise current control, the feedback from this voltage drop is not a viable method. Hall-effect current sensors are ideal for sensing the currents with galvanic isolation. At this stage, it is very nearly impossible to do away with current feedbacks for the control of the PMBDC machine to deliver high performance.

If precise torque and speed controls are not required, current feedback control and hence current sensing can be dispensed with. Then, a simple duty cycle using an open-loop PWM voltage controller is sufficient. However, the steering of the current to the appropriate machine phases requires the rotor-position information. A number of methods have come into practice to estimate rotor position without an externally mounted sensor.

Position Estimation Position can be sensed by Hall sensors overlooking a magnet wheel mounted on the shaft of the rotor extension with the magnets. This will provide just sufficient commutation signals, i.e., six per electrical cycle for a three-phase machine. Such a low discrete pulse count is not suitable for high-performance applications. Optical encoders and resolvers provide the rotor position with high resolution, but they are expensive. Further, the position sensors require extensive mounting arrangements. High-volume applications demand that they be dispensed with, on account of the cost and manufacturing burdens. Many methods are possible to estimate the commutation signals; they are briefly described here.

(i) Estimation by using machine model: The induced emf can be sensed from the machine model by using the applied currents and voltages and machine parameters of resistance, self-inductance, and mutual inductance. The advantage of this method is that an isolated signal can be extracted, because the input currents and voltages are themselves isolated signals. The voltages can be extracted from the base or gate drive signals and the dc-link voltage. The variations in the dc-link voltage can be estimated from the dc-link filter parameters and the dc-link current. Parameter sensitivity, particularly that of the stator resistance, will introduce an error in the induced emf estimation, resulting in inaccurate commutation signals to the inverter.

(ii) Induced emf from sensing coils: Sensing coils in the machine can be installed inexpensively to obtain induced-emf signals. The advantages of this method are that the signals are fairly clean, parameter-insensitive, and galvanically isolated. The disadvantages are in the additional manufacturing process and additional wire harness from the machine. The latter is not acceptable in refrigerator compressor motor drives, because of hermetic sealing requirements.

(iii) Sensing emfs from inactive phases: One of the most commonly used methods for acquiring position information is to monitor the induced emf of the machine phases when they are not being energized. Note that a machine phase is inactive for 33.33% of the time and that only two phases conduct at any given time. During the inactive time, an induced emf appears across the machine winding, which can be sensed. The induced emf of the phase yields the information on zero crossing and on when the emf reaches the constant region, indicating when that phase has to be energized. The polarity of the induced emf determines the appropriate polarity of the current to be injected into that machine phase. Instead of waiting for the constant region of the induced emf for energizing a machine phase, the induced emf on integration from its zero crossing will attain a particular value corresponding to thirty degrees from the zero crossing instant. The integrator output corresponding to thirty degrees from the positive zero crossing could be termed the *threshold value* used in energizing a phase. This threshold is independent of the rotor speed, as is shown below.

Assuming a trapezoidal induced emf whose peak is E_p at the rotor electrical speed of ω_b, the slope of the rising portion of the induced emf for any speed ω_s is given by dividing the peak value of the voltage at that speed by the time interval corresponding to thirty electrical degrees. Then, the instantaneous value of the induced emf during its rising instant is given by

$$e_{as}(t) = \frac{\left(\dfrac{E_p}{\omega_b}\right)\omega_s t}{(\pi/6\omega_s)} \qquad (9.266)$$

which, upon integration from 0 to $\pi/6\omega_s$, yields the sensor output voltage, V_{vs}:

$$V_{vs} = \int_0^{\pi/6\omega_s} e_{as}(t)dt = \frac{\pi}{12}\frac{E_p}{\omega_b} \qquad (9.267)$$

It can accordingly be proven that this algorithm also works for machines with sinusoidal induced emf, even though the sensor output voltage will be different. Note that the sensor output voltage is a constant, independent of the motor stator parameters, and that its magnitude is always the same for any operating speed of the machine. The only thing that could adversely affect the sensor output voltage is the induced-emf peak's decreasing with partial loss of rotor flux due to temperature sensitivity of the rotor magnets. This will clearly introduce errors, in that energization may not be exactly at thirty electrical degrees from the zero crossing as desired. Optimal utilization of the machine might not be possible in this condition, unless other corrective measures are taken.

(iv) **Third-harmonic induced emf:** An alternative method is to detect the third-harmonic induced emf in the machine windings and use them to generate the control signals. A three-phase, star-connected, four-wire system will allow the collection of the third-harmonic induced emf, and this can be inexpensively instrumented with four resistors.

All the methods that rely on the induced emf have the disadvantage that, at standstill, the position information is not available, as there is no induced emf at zero speed. Even at very low speeds, the induced emf might not be easily detectable. Therefore, a method to generate the control signals at and around zero speed has to be incorporated for successful starting of the machine and up to a speed at which the induced-emf methods can come in to generate the position information reliably. Therefore, a starting procedure at standstill is required. This procedure can consist of two steps:

Step (i): Exciting one or two phases, the rotor can be aligned to a predetermined rotor position. This way, the starting position is known; hence, correct starting control signals are generated. When the rotor starts moving at slow speed, the induced emf is so small that it cannot be used for generating the commutation pulses until the rotor speed reaches a certain level. This fact necessitates a second step to complete the starting process.

Step (ii): Once the rotor starts moving, the stator phases are energized at a slowly varying frequency, keeping the stator currents constant. The rate of frequency variation is kept low so that synchronism is maintained and can be controlled modestly if the load is known *a priori*. If not, the stator frequency is altered by trial and error until it reaches the minimum speed at which the induced emfs are of sufficient magnitude to render them useful for control. This constitutes the second step in the starting process. The problem with this approach is that this is not precise; some jitter and vibrations can be felt during the starting, which may not be significantly adverse in many applications. In many cases, step (i) is skipped, and only step (ii) is used for starting of the machine.

A method based on the saliency of the rotor is another alternative, but caution must be used here: the saliency in the PMBDCMs is not very significant. This requires a detection of the machine inductance and its profile, from which the rotor position can be extracted.

9.10.9 Torque-Smoothing

It is not possible to generate ideal rectangular currents, because of the time delay introduced by the machine inductance. Therefore, the currents become more or less trapezoidal and produce a large commutation-torque ripple, as much as 10 to 15% of the rated torque. Further, the induced emfs are not exact trapezoids, because of significant slot harmonics. They, in turn, will generate harmonic ripple torques, resulting in poorer torque performance. The quality of the induced emfs is further affected by the type of armature winding. Windings are chosen for low-cost manufacturing for high-volume applications, and they invariably cause a greater deviation from the ideal waveforms. The cumulative effect of all these imperfections leads to a drive with uneven torque over an electrical cycle of its operation. That makes the drive highly unsuitable for high-performance applications. To overcome these disadvantages, methods based on current-shaping to counter the ill effects of the flux distribution are successful.

To overcome the unevenness in the flux-density distribution, it is measured or computed, and the current is continuously adjusted accordingly, to generate a constant torque. To counter the commutation-torque pulsation, the incoming-phase and outgoing-phase currents are coordinated in such a manner that the sum of the torque produced by the two phases is kept constant. All of the algorithms require a set of fast-acting current-control loops to shape the current, with no deviation either in magnitude or phase from their references.

9.10.10 Design of Current and Speed Controllers

Since the PMBDCM is similar to the separately-excited armature-controlled dc machine, as can be seen from its model, the methods that are appropriate to the design of the current and speed controllers for that can be gainfully employed here. The design of current and speed controllers directly relevant to this motor drive can be obtained from Chapters 3 and 4.

9.10.11 Parameter Sensitivity of the PMBDCM Drive

The motor parameters that are sensitive to variations in temperature are stator resistances and rotor magnets. The use of inner current loops overcomes the effect of stator-resistance variations. The use of speed-control loop counters the rotor flux-linkages variation. In that process, the linearity of the torque with its reference might be lost. In order to preserve the torque linearity in the drive system, methods similar to the air gap-power feedback control have to be resorted to.

The inductance variation is a function of saturation and hence of the exciting current. Therefore, it is easy to counter the inductance variations if the excitation current is measured and made available.

9.11 REFERENCES

1. *DC Motors, Speed Controls and Servo Systems*, An Engineering Handbook by Electro-Craft Corporation, Fifth Edition, August, 1980, pp. 6–12.
2. R. Krishnan and P. N. Materu, "Analysis and design of a low cost converter for switched reluctance motor drives," *Conf. Record, IEEE–IAS Annual Meeting*, San Diego, CA, October 1989, pp. 561–567.
3. J. Bass, M. Ehsani, T. J. E. Miller, and R. L. Steigerwald, "Development of a unipolar converter for variable reluctance motor drives," *IEEE–IAS Annual Meeting*, Toronto, Canada, Oct. 1985, pp. 1062–1068.
4. R. Krishnan and P. N. Materu, "Design of a single-switch-per-phase converter for switched reluctance motor drives," *Conf. Record of IEEE IECON 1988*, pp. 773–779.
5. T. J. E. Miller, *Brushless Permanent-Magnet and Reluctance Motor Drives*, Clarendon Press, Oxford, 1989, pp. 78–80.
6. R. Krishnan and S. Lee, "PM brushless dc motor drive with a new power converter topology," *Conf. Record, IEEE–IAS Annual Meeting*, Orlando, FL, pp. 380–387, Oct. 1995.
7. P. J. Lawrenson, J. M. Stephenson, P. T. Blenkinsop, J. Corda, and N. N. Fulton, "Variable-speed switched reluctance motors," *IEE Proc. B, Power Applications*, vol. 127, no. 4, pp. 253–265, 1980.
8. C. Pollock and B. W. Williams, "Power converter circuits for wwitched reluctance motors with minimum number of switches," *Proc. IEE, London, Electric Power Applications*, vol. 137, pt. B, no. 6, pp. 373–384, Nov. 1990.
9. R. Krishnan and P. N. Materu, "Analysis and design of a low cost converter for switched reluctance motor drives," *Conf. Record, IEEE–IAS Annual Meeting*, San Diego, CA, October 1989, pp. 561–567.
10. R. Krishnan, "Analysis and design of switched-reluctance motor drives," Course Notes for EE 6444, MCSRG, Virginia Tech., pp. 150, Aug. 1995.
11. R. Krishnan, "Modeling, simulation, and analysis of permanent-magnet motor drives, Part II: The Brushless DC Motor Drive," P. Pillay and R. Krishnan, *IEEE Trans. on Industry Applications*, vol. 25, no. 2, pp. 274–279, March/April 1989.
12. R. Krishnan, "Control and operation of PM synchronous motor drives in the field weakening region," *Conf. Record, IEEE Ind. Electronics Conf.*, Invited Paper, pp. 745–750, Nov. 1993.
13. S. Morimoto et al, "Servo drive system and control characteristics of salient pole permanent magnet synchronous motor," *IEEE Trans. on Industry Applications*, vol. 29, no. 2, pp. 338–343, March/April 1993.
14. Marco Bilewski, A. Fretta, L. Giordano, A. Vagati, and F. Villata, "Control of high performance interior PM synchronous drives," *IEEE Trans. on Industry Applications*, vol. 29, no. 2, pp. 328–337, March/April 1993.
15. P. Pillay and R. Krishnan, "Modeling, Simulation, and Analysis of Permanent Magnet Motor Drives, Part I: The Permanent Magnet Synchronous Drives," *IEEE Trans. on Industry Applications*, vol. 25, no. 2, pp. 265–273, March/April 1989.
16. P. Pillay and R. Krishnan, "Application characteristics of PM synchronous and BLDC motor servo drives," *Conf. Record, IEEE IAS Annual Meeting*, Atlanta, pp. 380–390, Oct. 1987.
17. T. M. Jahns, "Flux-weakening regime operation of an interior PMSM drive," *Conf. Record, IEEE–IAS Annual Meeting*, pp. 814–823, Oct. 1986.

18. R. Monajemy and R. Krishnan, "Implementation strategies for concurrent flux weakening and torque control of the PM synchronous motor," *IEEE–IAS Conference*, pp. 238–245, vol. 1, Oct. 1995.
19. R. Krishnan, N. Tripathi, and R. Monajemy, "Neural control of high performance drives: An application to the PM synchronous motor drive," Invited Paper, *IEEE Ind. Electronics Conf.*, pp. 38–43, vol. 1, Nov. 1995.
20. R. Krishnan, S. Lee, and R. Monajemy, "Modeling, dynamic simulation and analysis of a C-dump brushless dc motor drive," *Proceedings of Applied Power Electronics Conference*, pp. 745–750, vol. 2, March 1996.
21. M. J. Corley and R.D. Lorenz, "Rotor position and velocity estimation for a salient pole PMSM," *IEEE Trans. In IA*, vol. 34, no. 4, pp. 784–789, July/Aug. 1998.
22. S. Ogasawara and H. Akagi, "Implementation and position control performance of a position-sensorless IPM motor drive system based on magnetic saliency," *IEEE Trans. In IA*, vol. 34, no. 4, pp. 806–812, July/Aug. 1998.
23. R. Mizutani, T. Takashita, and N. Matsui, "Current model-based sensorless drives of salient-pole PMSM at low speed and standstill," *IEEE Trans. In IA*, vol. 34, no. 4, pp. 841–846, July/Aug. 1998.
24. R. Krishnan and P. Vijayraghavan, "A new power converter topology for PM brushless dc motor drives," *IEEE Ind. Electronics Conf.*, Sept. 1998.

9.12 DISCUSSION QUESTIONS

1. A PM synchronous motor can also be realized with permanent magnets on the stator and armature windings on the rotor. Discuss its merits and demerits compared to a PMSM with magnets on the rotor.
2. A cage rotor with PMs (called a *line-start PM synchronous motor*) can be used to start the motor as an induction motor and run it as a synchronous machine at utility frequency. Such an arrangement will improve the efficiency of the motor in comparison to induction motors. Discuss its construction, detailed operation, and possible applications.
3. The line-start PMSMs can also be operated from an inverter. In that case, the cage rotor will act as damper windings to damp out oscillations and, in case of loss of synchronization, the machine can run as an induction motor without disrupting the motion. Discuss an application scenario for such a motor drive. (Hint: High-reliability propulsion applications.)
4. Synchronous machines with surface-mount magnets have very little difference between direct-axis and quadrature-axis inductances. Explain why.
5. Interior PMSMs are preferred for high L_q/L_d ratios. Where would such a feature find application?
6. If the rotor is made salient with neither windings nor PMs, then the machine is a synchronous-reluctance motor. Its stator is that of the conventional PM synchronous motor. Due to its difference in quadrature- and direct-axis path reluctances, a torque is produced for an armature excitation. Since its field excitation is not from PMs, it must come from stator excitation. Since the inverter gets only active power from the dc link, where will the reactive power be generated for the excitation of the machine?
7. For equal power rating of the synchronous reluctance and PM synchronous motor, will the ratings of their inverters be equal? Explain.

8. Is L_q greater than L_d in the wound-rotor salient-pole synchronous machine?
9. What is the consequence of $L_q > L_d$ in control of the PMSM? (Hint: Consider maximum torque generated per unit input current.)
10. High-speed PMSMs have the rotor enclosed in a stainless sleeve to restrain the magnets against centrifugal forces. Ideally, will there be losses in the sleeve?
11. If the stator currents are sinusoids at fundamental frequency superposed with harmonics, will there be losses in the magnet sleeves? If there are losses, will they increase with rotor speed?
12. Injecting ac rectangular currents into PMSM is used in low-performance applications. Give reasons for such a control strategy and discuss the disadvantages of this control.
13. Torque pulsation is one of the measures for evaluating the suitability of a motor drive for an application. For a critical application requiring minimum torque pulsation, which is a suitable candidate between PMSM and PMBDCM drive?
14. A large number of control strategies have been developed and discussed for PMSM drives. Consider a four-quadrant application requiring speed variation to a maximum of base speed only. Which control strategy will be ideal, based on each one of the following considerations?
    (i) Simplicity in implementation.
    (ii) Maximum utilization of the inverter or minimum rating of the inverter.
    (iii) Optimal output of the machine.
    (iv) Optimal voltage utilization of the inverter and machine.
15. For implementation of any control strategy, a mapping from the torque and flux commands to the stator current and torque angle has to be performed. Is it a good choice to recommend on-line computation for this mapping? If so, justify it in the context of present computational capability available with processors.
16. A look-up-table implementation is considered for the control of a PMSM and PMBDCM drive system. Discuss the merits and demerits of the performance obtained with this implementation. Relate the bit resolution and accuracy in stator current and torque with the memory requirement for the implementation for each motor drive.
17. The inner current-control loops can use stator currents in stator or rotor reference frames. Discuss the merits and demerits of using each of the reference frame currents for current-feedback control.
18. An interior PMSM is completely demagnetized by injecting a negative d axis current. What is the magnitude of stator current to achieve demagnetization?
19. An interior PMSM is completely demagnetized by injecting a negative d axis current. Assume also a q axis stator current in the machine during this operation. Determine the electromagnetic torque generated in the machine.
20. Can magnet flux linkages be reduced by stator currents?
21. Can magnet flux linkages be increased by stator currents?
22. Air gap flux linkage is varied for flux-weakening. Will this affect the stator flux linkages?
23. Discuss the effects of losing control over d axis current in the high-speed operational region in a PMSM.
24. What are the salient differences between the maximum-torque-per-ampere strategy and the constant-mutual-flux-linkages strategy?
25. Including the effects of core losses in the formulation of control strategy is desirable. How can it be achieved?

26. Core losses are constant in one of the following schemes for constant-speed but variable-torque operation in a PMSM drive:

    (i) maximum-torque-per-ampere control;
    (ii) constant-flux-linkages control;
    (iii) constant-torque-angle control.

    Identify the control strategy that gives constant core losses.

27. Propose a method of measurement for L_d and L_q of a PMSM, using the model developed in the text. {Hint: (i) Connect two phases together in a star-connected machine. (ii) Lock the rotor for each measurement.}

28. Almost all flux-weakening control schemes are dependent on machine parameters. Is there a control method to weaken the flux independent of machine parameters? If so, how can it be achieved?

29. Is flux-weakening possible in surface-mount-magnet machines?

30. Discuss a discrete-IC-chip-based implementation of a PMSM-drive constant-torque-angle control strategy.

31. Is it possible to implement other control strategies with discrete IC chips?

32. Why are processor-based implementations popular, and what are their advantages over discrete-IC-chip-based implementations?

33. Parameter sensitivity of PMSM has been discussed in this chapter. Can the state of the rotor magnet flux linkages be used to approximately predict stator temperature? If so, justify it with reasoning.

34. How critical is it to have parameter adaptation for a torque-controlled PMSM drive?

35. In the flux-weakening region, eventually the current loops saturate and six-step voltage operation will result. Discuss the effects of such an operation.

36. "Resorting to six-step operation in the flux-weakening region will give an enhanced torque-vs.-speed characteristic compared to the current-controlled operating region." Is this true? Develop a justification.

37. Sensorless operation is desirable. Enumerate the reasons and explain them.

38. Starting with precision from any rotor position without position transducers is difficult with many of the sensorless control algorithms. Explain the underlying reason for this statement.

39. The direct-axis self-inductance of the PMSM in stator reference frames can be modeled as $L_d^s = L_1 + L_2 \cos 2\theta_r$. Could this information be used to identify the rotor position, θ_r? {Hint: Ref. [21].}

40. A voltage-source PWM inverter applies a fundamental and a number of higher-order harmonics into a PMSM. The dominant-harmonic current can be detected, from which the inductance can be calculated. This is achievable, because the harmonic inductive voltage drops are dominant compared to the resistive voltage drops. From inductances, the instantaneous rotor position can be extracted for control. The following steps are involved in the detection algorithm:

$$v_h = L \frac{di_h}{dt}$$

$$L = \begin{bmatrix} L_0 + L_1 \cos 2\theta_r & L_1 \sin 2\theta_r \\ L_1 \sin 2\theta_r & L_0 - L_1 \cos 2\theta_r \end{bmatrix}$$

where $v_h = [v_{qsh} \; v_{dsh}]^T$ and $i_h = [i_{qsh} \; i_{dsh}]^T$.

v_h is the stator qd axis harmonic-voltages vector, and i_h is the stator qd axis harmonic-current vector. By measuring v_h and i_h and by using the inductances of the machine, θ_r can be estimated. What is the difference between this method and the one given in discussion question 39? {Hint: Ref. [22].}

41. A sensorless-control algorithm injects a voltage signal at high frequency into the d axis of the PMSM. The current response is correlated with the rotor position by using machine-model-based current estimation. From the rotor position, rotor speed is derived. Compare this method with the method described in discussion questions 39 and 40. {Hint: Ref. [23].}

42. What is the effect of parameter sensitivity on the sensorless methods described in discussion questions 39, 40 and 41?

43. The mutual flux linkage increases with stator current. Will this saturate the stator core?

44. In PMBDCM, the induced emfs might not be exactly trapezoidal. What can cause their distortion?

45. "PMBDCMs have surface-mounted magnets." Is this true?

46. Considering fundamentals of voltages and currents only, the operation of PMBDCM and PMSM is similar. Is this true? If so, justify.

47. Commutation-torque ripple poses a serious problem (i) only at low speeds, (ii) only at high speeds, or (iii) at all speeds. Which is true?

48. Phase advancing in PMBDCM is equivalent to flux-weakening in PMSM to enable high-speed operation. What are the factors limiting the advance angle?

49. Torque-smoothing is feasible if the flux-density waveforms are known as a function of rotor position. If they are not known, how can torque-smoothing be accomplished?

50. In order to increase the high-speed operational region in PMBDCM, phase voltages are applied up until 150 degrees. How does this enable higher speed of operation?

51. Sensorless-control methods in PMBDCM are mainly induced-emf-based. Are the sensorless methods of PMSM applicable to PMBDCMs?

52. Rotor magnet sensitivity to temperature is significant in PMBDCMs. How can it be compensated for? Why does it need to be compensated for?

53. "Half-wave controlled PMBDCMs can deliver 4-quadrant operation." Is it true?

54. Why consider half-wave operation of PMBDCMs as against full-wave-controlled PMBDCMs?

55. "Half-wave operation of PMBDCMs transfers the power-device losses to the machine-armature losses." Is this statement, in general, true?

56. Half-wave operation of PMBDCMs underutilizes the machine. How can that be corrected?

57. In cost considerations of the PMBDCM drive, which one of the following alternatives is to be aimed for?
    (i) Lowering motor cost only;
    (ii) Lowering converter and controller cost only;
    (iii) Lowering the total system cost, i.e., (i) and (ii) combined.

58. Some of the PMBDCM half-wave converters are in use with switched-reluctance motor drives. A certain industrial firm manufacturing both these motor drives can achieve cost efficiency by using the same half-wave converter for both. How can this statement be justified?

59. All the half-wave converters for PMBDCMs are shoot-through-failure-proof. Justify this statement.
60. For better utilization of the PMBDCM with half-wave operation, the windings will have a larger number of turns. If the fill factor of the slots is a constant for both the half-wave and full-wave PMBDCMs, enumerate the consequences of increasing the number of turns in the half-wave PMBDCM.

9.13 EXERCISE PROBLEMS

1. (i) For a PMSM, derive the normalized machine equations assuming

$$\lambda_b = \lambda_{af}, \quad L_b = L_d, \quad I_b = \frac{\lambda_b}{L_b}, \quad \rho = \frac{L_q}{L_d}$$

   (ii) What is the advantage of this normalization basis?

2. Prove that maximizing the mutual flux linkages with respect to stator current phasor results in

$$I_{sn} = \frac{-\cos\delta}{\cos^2\delta + \rho^2 \sin^2\delta}$$

   and that electromagnetic torque is,

$$T_{en} = \frac{-\rho^2 \cos\delta \sin\delta(\cos^2\delta + \rho \sin^2\delta)}{(\cos^2\delta + \rho^2 \sin^2\delta)^2}$$

   Use the normalized model derived in problem 1(i).

3. Will the strategy given in problem 2 yield a better performance than the maximum-torque-per-ampere strategy? Compare the strategies by using the machine parameters used throughout the text. {Hint: Compute δ from the torque equation, and then I_{sn} in the above.}

4. Using the model in problem 1, find the torque and mutual flux linkages developed by using the maximum-torque-per-unit-current and torque-for-constant-mutual-flux-linkages strategy with mutual flux linkages fixed at 1 p.u. The parameter of the machine: $P = 2$, and resistance can be neglected. What is the base speed in each case for these operating points? Assume base voltage is 1 p.u.

5. The parameters of a star-connected, 6-pole, 1.5-kW, 9.2-A, 1500-rpm, 9.55-N·m, 3-phase PMSM are as follows:

   $R_s = 0.513\ \Omega$, $L_d = 4.74$ mH, $L_q = 9.51$ mH, $B_1 = 9.36 \times 10^{-4}$ N·m/(rad/sec), $J = 0.01$ kg-m², Emf constant = 0.0669 V/rpm, Inverter input voltage = 285 V.

   (i) Determine the maximum speed of the PMSM drive system.
   (ii) Without exceeding the stator rated current and inverter input voltage, what is the maximum speed at which rated power is delivered?

   The stator resistive drop can be neglected in the calculations for (ii), and the maximum stator-phase peak voltage obtained through the inverter is 55% of the dc-link voltage.

Index

Air gap, 10
Air gap power, 10
Air gap torque, 10
Apparent power, 92
Applications, 115
    chopped controlled dc drives, 167
    frequency controlled induction drives, 405
    phase controlled dc drives, 115
    phase controlled induction drives, 282
    slip energy recovery drives, 305
Arbitrary reference frames, 209
Armature control, 38
Armature flux, 18
Auto sequentially commutated inverter, 382

Boundary matching condition, 66
    phase controlled dc motor drive, 66
    chopper controlled dc drive, 138
    frequency controlled induction drive, 342
Brush dc machine, 8, 18

Chopper
    time delay, 133
    duty cycle, 124
    four quadrant circuit, 126
    gain, 133
    hysteresis control, 155
    inversion, 132
    one quadrant, 135
    two quadrant, 135
    operation, 124
    pulse width modulation control, 152
    transfer function, 133
    with regeneration capability, 133
Chopper controlled dc motor drives, 124
    Applications, 167
    Armature losses including harmonics, 148

Chopper ratings, 143
Chopper operation, 126
Closed loop drive system, 151
Continuous current conduction, 137
Critical duty cycle, 139
Current controller, 151
    model, 156
    design, 157
Current loop, 151
Discontinuous current conduction, 140
Dynamic model, 160
Four quadrant drive, 126
Four quadrant operation, 126
Harmonic ripple torque, 147
Harmonics and its effects, 148
Speed controller, 158
    design, 158
    anti-windup circuit, 159
Speed loop, 158
Steady state analysis, 136
    average analysis, 136
    instantaneous analysis, 137
Symmetric optimum, 158
System simulation, 163
Commutation, 46
Commutation overlap, 50
Control strategies, 38
    Armature control, 38
    Constant air gap flux linkages, 536
    Direct torque or self control, 426
    Direct vector control, 414, 415
    Field control, 37
    Flux weakening, 37
    Indirect vector control, 446
    Optimum-torque-per-Ampere, 537
    Slip speed control, 346
    Unity power factor, 534

Vector control, 411, 527
Volts/Hz, 325
Converters, 12
Crossover frequency, 77
Current command, 72
Current control loop, 76

Dq axis model, 197
   Induction machine, 197
   PM Synchronous machine, 525
Damping ratio, 77
Dc link, 134
Dc link filter, 134, 314
Dc machines, 8, 18
   Armature, 18
   Armature control, 38
   Armature copper loss, 21
   Armature inductance, 21, 32
   Armature resistance, 21, 31
   Block diagram, 23
   Commutator, 18
   Compound excited, 30
   Eigen values, 23
   Emf constant, 20, 32
   Equivalent circuit, 21
   Field control, 37
   Field excitation, 18, 24
   Field winding, 18
   Friction constant, 22
   Induced emf, 20
   Load torque, 22
   Moment of inertia, 22
   Mutual inductance, 20
   Parameter measurement, 31
   Permanent magnet, 30
   Separately excited, 24
   Series excited, 27
   Shunt excited, 27
   Stability, 23
   State space model, 22
   Torque, 22
   Torque constant, 22
   Transfer functions, 23
   Winding arrangement, 20
Diode, 2
Diode bridge rectifier, 133
Direct self control, 426
Direct vector control, 414
Duty cycle, 124
Dynamic models, 21
   Dc machine, 21, 22
   Induction machine, 194
   PM brushless dc machine, 580
   PM synchronous machine, 525

Effective torque, 11
Efficiency, 9
Electromagnetic torque, 10
Emf constant, 20, 32

Field control, 37
Flux producing stator current, 413, 528
Flux-weakening, 37, 88, 377, 484, 539, 586
Flying shear application, 115
Fork lift application, 167
Form wound coils, 177
Forward regeneration, 44
Forward motoring, 45
Four quadrant operation, 43
Frame matrix, 212
Frequency controlled induction motor drives, 313
   Applications, 405
   Control of harmonics, 362
   Constant air gap flux control, 350
      principle of operation, 350
      drive strategy, 350
   Constant slip-speed control, 346
      drive strategy, 346
      steady-state analysis, 347
   Constant Volt/Hz control, 325
      boundary matching condition, 342
      dc link voltage, 328
      direct steady-state evaluation, 340
      dynamic simulartion, 330
      implementation, 328
      initial steady state vector, 342
      offset voltage, 327
      small signal responses, 338
      stator voltage, 326
      steady-state currents, 340
      steady-state performance, 330
   Current source induction motor drives, 381
      ASCI, 382
      steady-state performance, 385
      equivalent circuit approach, 386
      direct steady state evaluation, 389
      closed-loop CSIM drive system, 396
      dynamic simulation, 398
   Effects of time harmonics, 360
   Flux weakening, 377
   Harmonic flux linkages, 356
   Harmonic slips, 356
   Modulation ratio, 367
   Phase-shifting control, 362
   Phasor diagram, 357
   Pulse-width modulation (PWM), 365
   Reactive power, 323
   Real power, 323
   Sixth harmonic torque, 358
   Speed control, 324
   Steady state with PWM voltages, 369
   Static frequency changers, 313
      current regulated inverter drive, 316
      cycloconverter drive, 314
      full bridge inverter, 319
      inverter phase output voltage, 321
      modified McMurray inverter, 317
      PWM inverter drive, 314
      variable voltage variable frequency, 314

voltage source inverter, 317
Torque pulsations, 354

GTO device, 5

Harmonic control, 362
Harmonic resonance in power system supply, 98
Hysteresis control, 155
IGBT, 5
Indirect vector control, 446
Induction machines, 8, 174
   Air gap power, 185
   Air gap torque, 186
   Bearings, 179
   Breakdown torque, 191
   Cooling, 179
   Control principle, 254
   Deep bar rotor, 175
   Distribution factor, 176
   Dynamic model, 197
      arbitrary reference frames, 209
      normalized small signal, 235
      rotor reference frames, 214
      small signal, 233
      stator reference frames, 213
      synchronous reference frames, 215
   Dynamic simulation, 223
   Electromagnetic torque, 186, 212, 249
   Enclosures, 179
   Equivalent circuit, 181
   Form-wound windings, 177
   Frequency response, 236
   Friction constant, 222
   Induced emf, 176
   Insulation types, 178
   Load torque, 222
   Magnetizing inductance, 181
   Measurement of parameters, 193
      core-loss resistance, 194
      leakage inductance, 194
      magnetizing inductance, 194
      rotor resistance, 194
      stator resistance, 193
   Model in flux linkages, 218
   Modified flux linkages, 219
   Moment of inertia, 222
   Multiple cage rotor, 192
   NEMA classification, 192
   Normalization, 220
   Per unit representation, 220
   Phasor diagram, 185
   Pitch factor, 176
   Power factor angle, 184
   Principle of operation, 180
   Random-wound windings, 177
   Reference frames, 209
   Root loci, 244, 248
   Rotating magnetic field, 180
   Rotor current, 182
   Rotor flux linkages phasor, 247
   Rotor leakage inductance, 182
   Rotor reference frames, 214
   Rotor resistance, 181
   Rotor self inductance, 183
   Rotor speed, 181
   Rotor windings, 175
   Shaft, 178
   Shaft power, 186
   Signal flow graph, 244, 247
   Slip, 181
   Slip ring rotor, 175
   Slip speed, 180
   Small signal model, 233
      normalized, 235
   Space phasor model, 246
   Squirrel cage rotor, 176
   Starting torque, 191
   Stator assembly, 175
   Stator current, 184
   Stator flux linkages phasor, 246
   Stator leakage inductance, 182
   Stator reference frames, 213
   Stator referred rotor leakage inductance, 183
   Stator referred rotor resistance, 183
   Stator referred rotor current, 183
   Stator resistance, 181
   Stator self inductance, 183
   Stator windings, 175
   Steady state characteristics, 188
   Steady state stability, 188
   Stray load losses, 186
   Synchronous reference frames, 215
   Synchronous speed, 180
   Transfer functions, 236
   Transformation, 200
      arbitrary reference frames, 211
      constant matrix, 200
      power equivalence, 209
      rotor reference frames, 215
      stator reference frames, 214
      synchronous reference frames, 215
      three to two phase, 203
   Windings, 177
      form-wound, 177
      random-wound, 177
   Winding factor, 176

Inverter, 313, 314, 316, 317, 319
Inverter paralleling, 362
Leakage coefficient, 243
Line commutation, 46
Line harmonics, 98
Linearized controller, 54

Machines
   Dc series, 27
   Dc shunt, 27
   Dc compound, 30

Dc separately excited, 24
Ac induction, 174
Permanent magnet dc, 30
Permanent magnet synchronous, 513
Permanent magnet brushless, 577
Machine tool application, 503
Optimum-torque-per-Ampere control, 537
MOSFET, 5
Mutual flux linkages control, 536

Nonlinear controller gain, 56

Offset voltage, 327
Output vector, 236

PM brushless motor drives, 513, 577
   Air gap torque, 579, 583
   C-dump topology, 601
   Commutation torque ripple, 582
   Design of current and speed controllers, 614
   Drive scheme, 580
   Dynamic model, 580
   Dynamic simulation, 581
   Half-wave drives, 588
   Harmonic torques, 585
   Normalized system equations, 587
   Parameter sensitivity, 614
   Phase advancing control, 586
   PM Brushless machine, 523
      model, 578
   Power density, 524
   Principle of operation, 523
   Sensorless control, 611
   Split-supply-converter topology, 589
   Stator phase current, 582
   Torque-smoothing, 614
   Variable-dc-link converter topology, 607
PM synchronous motor drives, 513
   abc-to-qd transformation, 540
   Air gap line, 514
   Air gap power, 530
   Air gap torque, 527
   Control strategies, 531
      constant torque angle, 531
      constant-mutual-flux-linkages, 536
      optimum-torque-per-ampere, 537
      unity-power-factor, 534
   Flux current, 528
   Flux-weakening, 539
      direct flux-weakening algorithm, 540
      indirect flux-weakening, 548
   Inductance, 519
      direct axis, 519
      quadrature axis, 519
   Inset magnet rotor
   Interior PM rotor, 520
   Interior PM with circumferential rotor, 520
   Line start PM synchronous machines, 522
   Machine configurations, 519
   Machine model, 525
   Magnet flux density, 515
   Maximum speed, 539
   Mutual airgap flux linkages, 529
   Parameter compensation, 569
   Parameter sensitivity, 567
   Permanent magnets, 514
      energy density, 517
      recoil permeability, 514
      volume, 518
   Phasor diagram, 528
   Recoil line, 514
   Remanent flux density, 514
   Rotor flux linkages, 526
   Schematic, 529
   Sensorless control, 562
   Speed controller design, 555
   Stator current phasor, 530
   Surface inset PM rotor, 520
   Surface PM rotor, 519
   Torque angle, 528
   Torque current, 528
   Vector control, 527
Parameter compensation, 439, 475, 569
Parameter sensitivity
   Dc drive, 118
   Induction motor drive, 437, 461
   PM brushless dc drive, 614
   PM synchronous drive, 567
Permanent magnet,
   Brush dc machine, 30
   Brushless dc machine, 577
   Synchronous machine, 519
Phase control, 47, 51
Phase controller, 263
Phase controlled dc motor drive, 36
   Applications, 115
   Application considerations, 114
   Armature control, 38
   Armature losses including harmonics, 109
   Continuous current conduction, 64
   Discontinuous current
      conduction, 67
   Controller design, 76
   Converter gain, 55, 56
   Converter operation, 47, 51
   Converter ratings, 91
      apparent power, 92
   Converter time delay, 55
   Converter transfer characteristic, 49, 52
   Converter transfer function, 55, 75
   Critical triggering angle, 67
   Current controller, 75
      design, 76
   Current feedback gain, 75
   Current loop, 76
      first order approximation, 78

Index **625**

Current source, 56
Current transducer, 75
Device commutation, 46
Dynamic model, 92-95
Flux weakening operation, 37
  two quadrant drive, 88
Four quadrant drive, 59, 89
Four quadrant operation, 43
Four quadrant converter requirement, 45
Harmonic elimination, 104
Harmonic resonance, 98
Harmonic ripple torque, 107
Harmonics and its effects, 98
Line filters, 101
Line commutated converters, 46
Natural frequency of the plant, 101
Overlap conduction, 50
Parameter sensitivity, 118
Self commutating converter, 104
Single phase converter, 47
Single phase converter transfer characteristic, 49
Source inductance effect, 49
Speed controller, 75
  design, 79
Speed feedback gain, 75
Speed loop, 75
Speed transducer, 75
Steady state analysis, 60
  average analysis, 60
  including harmonics, 64
Symmetric optimum, 81
System simulation, 92-95
Three phase converter, 51
  control circuit, 54
  control model, 55
  freewheeling, 58
  linearization of the transfer characteristics, 54
  transfer relationship, 52
  transfer function, 55, 75
Three phase half controlled converter, 57
Transfer function of the machine and load, 73
Twelve pulse converter, 102
Two quadrant drive, 71, 88
Utility impact
Power converters
  Ac to dc, 49, 51
  Buck, *see one quadrant chopper*
  C-dump, 601
  Choppers, 126, 135
  Current source PWM, 316
  Current source six step, 316
  Cycloconverter, 314
  Dc to dc, *see choppers*
  Inverter, 313, 314, 316, 317, 319
  Phase controlled, 49, 51
  Variable voltage variable frequency, 314
  Voltage source PWM, 314
  Voltage source six step, 320

Pulse width modulation control, 152
Pump application, 305

Random wound coils, 177
Reactive power, 323
Reference frames, 209
Regenerative braking, 44
Reluctance, 519
Reverse motoring, 44
Reverse regeneration, 44
Reversible operation, 43
Rotor reference frames, 214

Self commutation, 320
Sensorless control, 562, 611
Silicon controlled rectifier, 4
Slip energy recovery control, 283
  Applications, 305
  Closed loop speed control, 298
  Dc link voltage, 285
  Efficiency, 290
  Electromagnetic torque, 290, 301
  Equivalent circuit, 287
  Filter rating, 297
  Harmonic slip, 300
  Harmonic torques, 303
  Harmonics, 299
  Mechanical power, 291
  Mutual flux linkages, 290
  Performance characteristics, 289
  Power factor, 291
  Power input, 290
  Principle of operation, 283
  Rating of converters, 296
  Recovered power, 288
  Rotor current, 287
  Scheme, 283
  Slip, 287
  Starting, 296
  Static Scherbius scheme, 304
  Stator current, 291
  Steady state analysis, 285
  Torque constant, 290
  Torque pulsation, 299, 301
  Triggering angle, 285
Slip-speed control, 346
Small signal model, 233, 235
Space phasor model, 246
Speed control loop, 71, 79
Stationary reference frames, 213
Stator phase controlled induction machine, 262
  Applications, 282
  Base values, 269
  Closed loop control scheme, 279
  Conduction angle, 267
  Efficiency, 279
  Equivalent circuit model, 267
  Gating control, 263

626   Index

Interaction of the load, 273
Phase voltage, 271
Power circuit, 263
Reversible controller, 263
Rotor current, 271
Stator current, 271
Steady state analysis, 265
Torque-speed characteristics, 273
Symmetric optimum, 81
Synchronous reference frames, 215

Thyristor, 4
Torque angle, 414, 528
Torque command, 71
Torque constant, 22
Torque producing stator current, 413, 528
Transfer functions, 236
Transistor, 3
Tuning of vector controller, 454

Vector control, 411, 527
Volts/hertz control, 325
Vector control of induction machines, 411
  Applications, 503
  Air gap torque, 413
  Constant-power operation, 490
  Direct vector control, 414, 415
    implementation with six step current source, 422
    implementation with voltage source, 425
  Flux and torque processor, 416
  Direct self control, 426
    flux control, 430
    implementation, 434
    instability, 438

    parameter sensitivity, 437
    performance, 434
    stator-resistance compensation, 439
    torque control, 431
Field angle, 412
Flux current, 413
Flux-weakening, 484
  principle, 484
Indirect vector control, 414
  air gap torque, 447
  derivation, 446
  dynamic simulation, 458
  implementation, 450
  parameter sensitivity, 461
    compensation, 475
    reactive power compensation, 476
  scheme, 448
  slip-speed, 447
  speed controller design, 492
  torque constant, 447
  tuning, 454
Phasor diagram, 413
Principle, 412
Rotor flux linkages phasor, 412
Slip angle, 414
Speed controller design
Stator current phasor, 412
Stator phase angle, 412
Switching states of the inverter, 428
Torque angle, 414
Torque current, 413
Tuning, 454